# Injectable Hydrogels for Regenerative Engineering

# Injectable Hydrogels for Regenerative Engineering

Editor

**Lakshmi S Nair**

*University of Connecticut, USA*

ICP

Imperial College Press

*Published by*

World Scientific Publishing Co. Pte. Ltd.
5 Toh Tuck Link, Singapore 596224
*USA office:* 27 Warren Street, Suite 401-402, Hackensack, NJ 07601
*UK office:* 57 Shelton Street, Covent Garden, London WC2H 9HE

**Library of Congress Cataloging-in-Publication Data**
Injectable hydrogels for regenerative engineering / [edited by] Lakshmi S. Nair.
p. ; cm.
Includes bibliographical references and index.
ISBN 978-1-78326-746-0 (hardcover : alk. paper)
I. Nair, Lakshmi S., editor.
[DNLM: 1. Hydrogels--therapeutic use. 2. Tissue Engineering--methods.
3. Hydrogels--chemistry. 4. Injections. 5. Tissue Scaffolds. QT 37.5.P7]
R857.T55
612'.028--dc23
2015016740

**British Library Cataloguing-in-Publication Data**
A catalogue record for this book is available from the British Library.



Typeset by Stallion Press
Email: enquiries@stallionpress.com

Printed in Singapore

# Dedication

To my parents
my husband Prem
my son Bharath

# Foreword

I am delighted to write the foreword for the book *Injectable Hydrogels for Regenerative Engineering*, edited by Dr. Lakshmi S. Nair. The field of regenerative engineering has recently emerged as a distinct platform to integrate the current and future technologies in advanced biomaterials, tissue engineering, regenerative medicine and morphogenesis toward the regeneration of complex tissues. Biomaterials have already impacted the field of tissue regeneration through the development of biodegradable and biomimetic materials. Many of these materials are currently used clinically. Multicomponent complex tissue regeneration, however, presents unique challenges that require further developments in the field of advanced biomaterials particularly as carriers to deliver stem cells and bioactive cues in a spatiotemporal manner. The past few years have seen significant research interest in the design, preparation, and application of injectable hydrogels as advanced biomaterials for cell and factor therapy. The present work is timely and provides a state-of-the-art summary of the status of injectable hydrogels in regenerative engineering. The collection of ten chapters written by experts in their fields presents an excellent overview of the subject for a wide audience. In my opinion, this book will be extremely useful as a reference source for all those working in the area of biomaterials, tissue engineering, regenerative medicine, and drug delivery.

**Cato T. Laurencin, M.D., Ph.D.**
University Professor

**Albert and Wilda Van Dusen**
Distinguished Professor of Orthopaedic Surgery
Professor of Chemical, Materials and Biomedical Engineering

**Director, Raymond and Beverly Sackler**
Center for Biomedical Biological,
Physical and Engineering Sciences

Director, Institute for Regenerative Engineering

Chief Executive Officer
Connecticut Institute for Clinical and Translational Science

The University of Connecticut

# Preface

The aim of *Injectable Hydrogels for Regenerative Engineering* is to provide a state-of-the-art, comprehensive account of research in the rapidly emerging area of injectable hydrogels for regenerative engineering. Regenerative engineering has emerged as a field to take tissue engineering a step further by integrating the elements of tissue engineering, advanced biomaterials, regenerative medicine, and morphogenesis to regenerate complex tissues. The tremendous advances in the field of injectable hydrogels in recent years have significantly impacted the field of regenerative engineering by developing novel translational approaches to enhance tissue regeneration. In addition to its potential for minimally invasive surgical applications, injectable hydrogels are attractive candidates for the emerging field of additive manufacturing which includes tissue or organ printing. This book is intended as a general introduction to the preparation of injectable hydrogels and their various biomedical applications that have the potential to revolutionize tissue/organ engineering in the near future. Each chapter is written with a detailed background as the book covers a highly multidisciplinary area ranging from polymer chemistry, tissue engineering, drug delivery, and cell therapy with audiences from various fields of professions ranging from engineers, clinicians, and scientists to graduate and senior undergraduate students. The chapters also provide an extensive bibliography for the reader who wants to explore the subject in greater depth.

The book is broadly divided into two sections: The first section (Chapters 1–5) is meant to give a general introduction to the field of regenerative engineering and injectable hydrogels, as well as the various synthetic approaches to develop injectable hydrogels; meanwhile, the second section (Chapters 6–10) emphasizes the various biomedical applications of injectable hydrogels toward tissue regeneration. Chapter 1 presents a broad overview of the area of regenerative engineering and the role of scaffolds, cells, and growth factors in supporting tissue regeneration. Chapter 2 reviews the various types of injectable hydrogels and their potential biomedical applications. Chapters 3–5 present an in-depth discussion on physically crosslinked, chemically crosslinked, and enzymatically crosslinked injectable hydrogels, respectively. Chapter 6 vividly presents the use of injectable hydrogels as tissue adhesives. Chapter 7 presents a comprehensive account of intervertebral tissue regeneration and the potential of injectable hydrogels for this application. Chapter 8 summarizes the applications of injectable hydrogels for neural tissue regeneration. Chapter 9 discusses the applications of injectable hydrogels for articular cartilage tissue regeneration, and Chapter 10 presents the potential of injectable hydrogels for promoting cardiac tissue regeneration post-myocardial infarction.

It is my hope that all the chapters, written by eminent experts in the field, will provide a basis to better understand the impact of injectable hydrogels in the field of regenerative engineering.

**Lakshmi S. Nair**
**M.Phil., Ph.D.**

# Editor

**Lakshmi S. Nair, M.Phil., Ph.D.** is an Assistant Professor in the Department of Orthopaedic Surgery, Institute for Regenerative Engineering, and Raymond and Beverly Sackler Center for Biomedical, Biological, Physical and Engineering Sciences at the University of Connecticut Health. She is also a faculty in the Departments of Biomedical Engineering and Materials Science and Engineering at the University of Connecticut. Dr. Nair earned her Ph.D. in Polymer Chemistry from Sree Chitra Tirunal Institute for Medical Sciences and Technology, India. Her research interests include injectable hydrogels and nanostructured materials for tissue engineering and drug delivery. She has edited books and special editions of journals in the area of nanostructured scaffolds for tissue regeneration and has more than 100 publications in the area of biomaterials and tissue engineering.

# Contributors

**Attipettah Jayakrishnan**
Department of Biotechnology
Bhupat and Jyoti Mehta School of Biosciences
Indian Institute of Technology Madras
Chennai, Tamil Nadu

**Biji Balakrishnan**
Department of Biosciences and Bioengineering
Indian Institute of Technology Bombay
Mumbai, Maharashtra

**Bogyu Choi**
Section of Biomaterials
Division of Advanced Prosthodontics
Weintraub Center for Reconstructive Biotechnology
University of California
Los Angeles, California

**Bradley D. Olsen**
Department of Chemical Engineering
Massachusetts Institute of Technology
Boston, Massachusetts

**Casey C. McDermott**
Department of Biomedical Engineering
University of Connecticut
Storrs, Connecticut

**Craig W. Hanna**
Sackler School of Graduate Biomedical Sciences
Tufts University School of Medicine
Boston, Massachusetts

**Debbie Campbell-Rance**
Department of Chemical and Life Science Engineering
Virginia Common Wealth University
Richmond, Virginia

**Farzana Sharmin**
Department of Materials Science and Engineering
University of Connecticut
Storrs, Connecticut

**Gregory N. Grover**
Department of Bioengineering
Sanford Consortium for Regenerative Medicine
University of California
San Diego, California

**Isaac L. Moss**
Department of Orthopaedic Surgery
University of Connecticut Health
Farmington, Connecticut

**Karen L. Christman**
Department of Bioengineering
Sanford Consortium for Regenerative Medicine
University of California
San Diego, California

**Lakshmi S. Nair**
Department of Orthopaedic Surgery
Institute for Regenerative Engineering
UCONN Musculoskeletal Institute
University of Connecticut Health
Farmington, Connecticut

Department of Materials Science and Engineering
Department of Biomedical Engineering
University of Connecticut
Storrs, Connecticut

**Matthew J. Glassman**
Department of Chemical Engineering
Massachusetts Institute of Technology
Boston, Massachusetts

**Min Lee**
Section of Biomaterials
Division of Advanced Prosthodontics
Weintraub Center for Reconstructive Biotechnology
University of California
Los Angeles, California

**Natalie Artzi**
Institute for Medical Engineering and Science
Massachusetts Institute of Technology
Department of Anesthesiology
Brigham and Women's Hospital
Harvard Medical School
Cambridge, Massachusetts

**Ning Zhang**
Department of Biomedical Engineering
Virginia Common Wealth University
Richmond, Virginia

**Nuria Oliva**
Institute for Medical Engineering and Science
Massachusetts Institute of Technology
Cambridge, Massachusetts

**Shalini V. Gohil**
Department of Orthopaedic Surgery
Institute for Regenerative Engineering
University of Connecticut Health
Farmington, Connecticut

**Teja Karukonda**
Department of Orthopaedic Surgery
University of Connecticut Health
Farmington, Connecticut

**Xiaowei Li**
Translational Tissue Engineering Center
Whitaker Biomedical Engineering Institute
Johns Hopkins University
Baltimore, Maryland

**Xiaoyan Liu**
Department of Chemical and Life Science Engineering
Virginia Common Wealth University
Richmond, Virginia

**Xuejun Wen**
Department of Chemical and Life Science Engineering
Virginia Common Wealth University
Richmond, Virginia
Institute for Advanced Materials and Nanomedicine
Tongji University School of Medicine
Shanghai P. R.

**Yusuf M. Khan**
Department of Orthopaedic Surgery, UConn Health
Institute for Regenerative Engineering
UCONN Musculoskeletal Institute
University of Connecticut Health
Farmington, Connecticut

Department of Materials Science and Engineering
Department of Biomedical Engineering
University of Connecticut
Storrs, Connecticut

# Contents

Chapter 1

# Regenerative Engineering: Role of Scaffolds, Cells, and Growth Factors


F. Sharmin

*Department of Materials Science and Engineering*
*University of Connecticut, Storrs, CT*

C. C. McDermott

*Department of Biomedical Engineering*
*University of Connecticut, Storrs, CT*

Y. M. Khan*

*Department of Orthopaedic Surgery*
*University of CT Health, Farmington, CT*
*Institute for Regenerative Engineering*
*University of CT Health, Farmington, CT*
*Department of Materials Science and Engineering*
*University of Connecticut, Storrs, CT*
*UCONN Musculoskeletal Institute*
*University of CT Health, Farmington, CT*
*Department of Biomedical Engineering*
*University of CT Health, Farmington, CT*


## 1.1 Introduction

Tissue engineering (TE) aims to replace or facilitate the regrowth of damaged or diseased tissue by utilizing biomaterials, cells, and bioactive


*Author to whom correspondence should be addressed. Email: ykhan@uchc.edu.

molecules alone or in combination with one another. The need for TE stems from the limited availability of autologous tissue and the potential for immunorejection of allogeneic sources and is bolstered by the numerous and growing availability of advanced material fabrication methods of novel biocompatible biomaterials. The strategies involved in designing and building engineered scaffolds are diverse but often focus on the biomaterial as a synthetic extracellular environment to organize cells into a three-dimensional architecture and to present stimuli that can direct the growth and formation of a desired tissue. The promise of TE would empower scientists to grow tissues and organs in the laboratory and safely implant them when the host body is incapable of self-healing. This possibly offers an alternative to organ transplantation, which, although an important clinical tool, has its limitations. The goals of TE integrate so well with the goals of regenerative medicine that as each field progresses, the distinction between them becomes ever less clear.

Recently, a new field has evolved that seeks to unite TE and regenerative medicine with an additional focus on the integration of stem cells, developmental biology, and advanced materials science. This field, termed 'regenerative engineering', seeks to integrate these fields in pursuit of the current clinical challenge of complex tissue regeneration where more than one tissue type, often in direct proximity to another distinct tissue type, is regenerated [1]. The complexity of regenerating neighboring tissues with distinct structural and chemical requirements cannot be underestimated, and mandates a renewed focus on these specific challenges. Regenerative engineering seeks to build on what has been accomplished thus far in TE and regenerative medicine, and move the field forward towards true clinical implementation.

A key moment in organ transplant history was in 1954 with the first successful kidney transplant. Development in the field continued into the 1960s with successful pancreas–kidney, liver, and later in the 1980s with heart–lung, single lung, double lung, and living-donor liver transplants [2]. Despite the utility of organ transplantation there is a real mismatch in supply and demand in respect of organs and tissues available for transplantation. Data published by United Network for Organ Sharing (UNOS) listed only 11,579 people receiving transplants from January 2013 to May 2013, while 76,223 people were recorded as waiting-list candidates (Figure 1.1) [3]. While the simple hope of TE is that discrepancies like this would be obviated, its potential full impact is far broader than this. In addition to reducing the need for organ replacement TE, advanced biomaterials

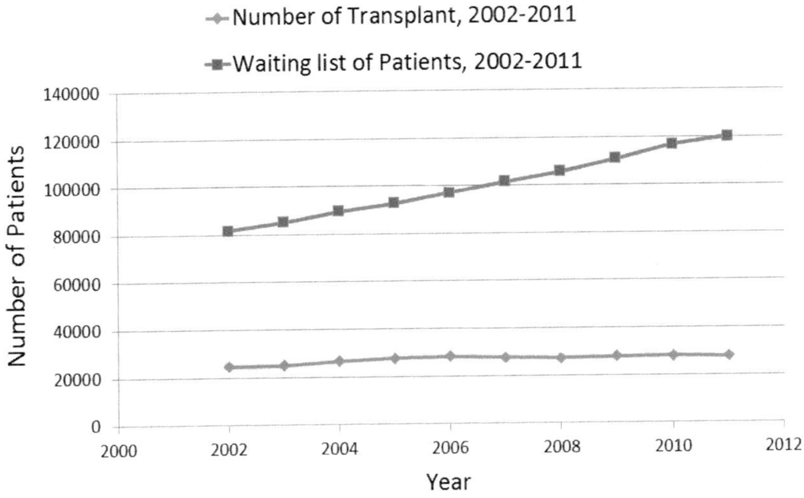


Figure 1.1 UNOS organ transplant statistics for 2002–2011.

science, developmental biology, and a focus on stem cell integration, regenerative engineering could, for instance, greatly accelerate the development of new drug delivery systems that may reduce the need for transplantation through novel drug delivery applications and other novel treatments.

Toward organ replacement, researchers have also used the TE model to form engineered tissue with a structure similar to that of native bladder tissue [4–6] derived from a patient's own cells, cultured *ex vivo*, and successfully transplanted. Similarly, porcine small intestinal submucosa (SIS), a xenogeneic, collagen-based acellular, biocompatible, and biodegradable scaffold has been widely used by physicians to encourage reconstruction of the bladder.

Short of full organ replacement, TE has already become a useful tool for many clinical indications. For instance, in addition to bladder reconstruction, porcine SIS-ECM (small intestinal submucosa ECM) is also currently used, and in some instances approved by the Food and Drug Administration (FDA) for use, in humans for a variety of reconstructions including cardiac patching, carotid artery repair, dural grafting, tendon repair, and lower urinary tract reconstruction [7]. Engineered skin has been utilized for skin replacement, temporary wound cover for burns, and treatment for diseases such as vitiligo, melanoma, psoriasis, and blistering disorders [8]. Growth factor-based tissue engineered scaffolds have also been used in patients to induce long bone regeneration [6] to supplement or in some instances replace clinical options that are currently available for regeneration of degenerated bone tissue such as bone grafts (autografts, allografts, xenografts), soft tissues to reconstruct bony defects, biomaterials, metals, and ceramics like hydroxyapatite (HAp) [9]. For other musculoskeletal tissues like cartilage, repair autologous chondrocyte implantation (ACI) is an FDA-approved, cell-based therapeutic strategy for repairing focal cartilage defects in younger patients [10].

Vascular TE passed several important milestones in the late 1990s and early 2000s toward clinical implementation. One of the biggest achievements in the field was when Shinoka *et al.* successfully implanted a tissue engineered pulmonary artery in a 4-year-old girl born with a single right ventricle and pulmonary atresia in Japan [11]. Following this in August 2011, the FDA approved human clinical trials in the USA to investigate the use of engineered vascular grafts in children with congenital heart defects [12]. Regenerative engineering seeks to build upon this progress by examining complex tissue regeneration, where more than one tissue type is to be regenerated simultaneously. Below is a brief overview of TE state of the art and how regenerative engineering seeks to build upon its success thus far.

## 1.2 TE: An Overview

As indicated above, the TE approach involves three main components: biomaterials to form a substrate onto which cells and biological molecules can reside; proteins and/or peptides that function to influence cell behavior; and, the cells to build the tissue (Figure 1.2). Designing the substrate or scaffold (the first of the three components) onto which new tissue may grow has become a critical part of the development process. Although some scaffolds simply involve the injection of a material into the defect site, those that are designed to provide structural support with a specific architecture require additional preparation prior to implantation. For instance, often the tissue engineer chooses to incorporate biological molecules (the second of the three components) like growth factors or peptides into the scaffold to control proliferation and/or differentiation of progenitor cells to increase the rate of tissue healing. The choice of cells (the third component) that are seeded onto the scaffold to supplement the process of cell migration from the host and therefore accelerate the overall tissue formation can greatly influence the success of the overall repair. These three components — the design of the scaffolds, the use of growth factors and other biological molecules,

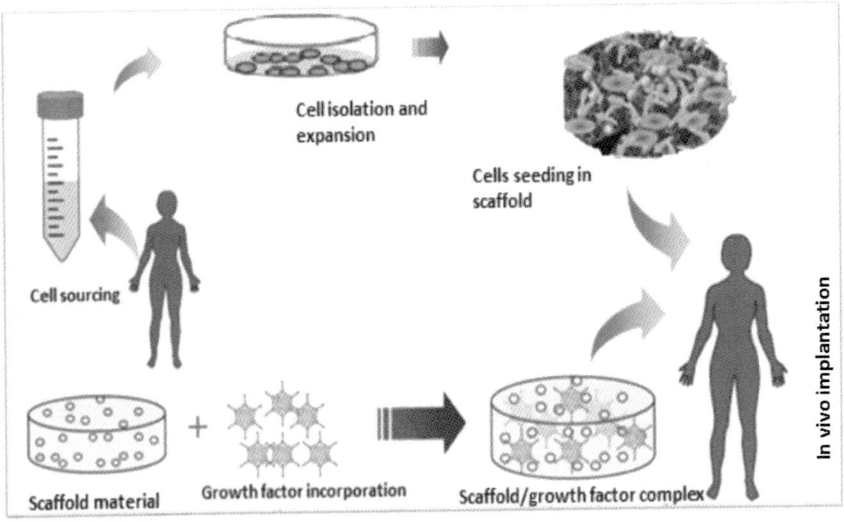


Figure 1.2 TE paradigm.

and the cells chosen to seed onto the scaffold — are reviewed in depth below, with attention paid to those elements that contribute to regenerative engineering. This is not meant as an exhaustive survey of the field as that is beyond the scope of this chapter, but an update on select approaches to tissue regeneration. It will be noted that these three components exist neither alone nor always in combination toward whole tissue regeneration, but rather as a combination of one, two, or three elements.

## 1.3 Scaffolds: Roles and Requirements

Although each tissue retains unique healing processes, there are steps of natural tissue healing that are common in most tissue types. Immediately after injury, the wound site exhibits acute inflammation which signals recruitment of macrophages and other immune cells. This process lasts for between 72 hours and 6 weeks, depending on the injury and tissue type, and is followed by the repair phase. During this, the inflammatory response resolves and the body starts to repair the injury site. The last phase of wound healing, the remodeling phase, is when the repairing tissue starts to remodel and resemble the tissue prior to injury [13]. All these tissue-healing phases involve a complex array of growth factors, cytokines, and cell-adhesion molecules that signal cells under spatial and temporal control, optimally leading to tissue regeneration. Often the design and fabrication of a porous, biodegradable, biocompatible three-dimensional structure as a scaffold capable of exhibiting high porosity, high pore interconnectivity, and uniform pore distribution is considered a necessity to support the intricate healing process, especially when large amounts of tissue have been damaged or removed [14].

When considering the design of a scaffold there are several criteria thought to be critical if not essential: (i) suitable biomaterials should be chosen with both the tissue type to be replaced and the degradation time to match the ingrowth of newly forming tissue; (ii) appropriate initial mechanical strength and stiffness to substitute for the mechanical function of the damaged tissue being repaired/regenerated; (iii) optimization of the scaffold surface to encourage cell attachment, proliferation, differentiation, and migration through the scaffold; (iv) minimal to no immune response, neither from the intact biomaterial nor its degradation by-products; (v) interconnected pore structure such that cell migration and proliferation in initial stages and consequently ECM infiltration of the desired tissue is supported without compromising the mechanical properties; and, (vi) the scaffold should be reproducible at an economic cost and reasonable efficacy.

While these are important guidelines, the final design should match the specific requirements for that tissue. For example, in cardiac tissue both myocardial and endothelial cells range from 10–100 $\mu$m [15] and 8–12 $\mu$m [16], respectively. Hence, scaffolds with pore sizes between 10 and 1,000 $\mu$m are desirable. For skeletal regeneration, bone has a compressive modulus in the range of 0.01–2.0 GPa or 14–18 GPa for trabecular or cortical bone, respectively [17], so when designing a scaffold one would consider these parameters in the final design.

### 1.3.1 *Design variables and fabrication of scaffolds*

A number of fabrication technologies have been applied to process biodegradable and bioresorbable materials into three-dimensional polymeric scaffolds with interconnected pores and appropriate surface area and topography (Table 1.1). These techniques can be further classified into two types: top-down and bottom-up approaches. Traditional TE methods use a top-down approach, in which cells are seeded onto a scaffold with biocompatible and biodegradable properties, and are expected to populate in the scaffold and create their own extracellular matrix (ECM).

Table 1.1 Scaffold fabrication techniques.

| Technique | Porosity (%) | Description of process | Comments |
| --- | --- | --- | --- |
| Gas foaming | <93 | Effervescent salts (e.g., ammonium bicarbonate) are used as porogens and mixed with a polymer or polymer/ceramic solutions. After solvent evaporates, scaffolds are placed in an acidic solution (e.g., citric acid) and ammonium bicarbonate produces gaseous ammonia and carbon dioxide gas upon contact with acid which dissolves and thus removes the porogen. | The porosity and pore structure, which is partially interconnected, is dependent on the amount of gas dissolved in the polymer, the rate and type of gas nucleation, and the diffusion rate of gas molecules through the polymer. The pore diameters are usually in the range of 100–500 $\mu$m. |
| Fiber extrusion and bonding | <97 | Produced by extruding a polymer melt or polymer solution through a spinneret and mechanically drawn onto a winder. Random coiling/heat treatment of fibers generates highly porous interconnected structures. | The fibers have large surface area but lack structural stability. |
| Solvent casting/salt leaching | <82 | Scaffolds prepared by dissolving/suspending polymers/ceramics in organic solvent with added porogens (salt/sugar particles).<br>Porogen is dissolved in water to remove the salt/sugar particles and create pores. | The main advantage of this technique is the use of small amounts of polymer, which minimizes any polymer wastage. The drawback is the use of organic solvents that are not easy to remove after the scaffold fabrication. |
| Electrospinning | 80–95 | Electrospininng process consists of a syringe pump, a high voltage source, a needle tip, and a collector. A polymer solution is held at the needle tip by means of its surface tension. The high voltage source creates an electric field resulting in charge repulsion. When the intensity of the electric field increases, the charge repulsion overcomes the surface tension, forming a jet. Solvents evaporate from the polymer solution forming polymer fibers as the jet stream travels to the collector and deposits on the collector. | Fibers ranging from nanometers to micrometers in diameter can be obtained by regulating parameters such as viscosity of the solvent, the dielectric constant of the solvent conductivity, surface tension of the polymer, molecular weight of the polymer, the distance between the tip and collector, the flow rate etc. Electrospun scaffolds contain high surface area for cell attachment and high mechanical properties. |

(*Continued*)

Table 1.1 (*Continued*)

| Technique | Porosity (%) | Description of process | Comments |
|---|---|---|---|
| Phase separation | 60–98 | Based on the thermodynamic demixing of a homogenous polymer-solvent solution into a polymer-rich phase and polymer-lean phase, under certain temperature conditions. The polymer-rich phase solidifies to form a matrix while the polymer-lean phase turns into pores during solvent removal. | This process holds great potential for fabricating three-dimensional nanofibrous scaffolds with uniform pore structures through multiple phase separation processes. Various growth factors could also be incorporated into porous polymer scaffolds by choosing appropriate solvents. |
| Freeze drying | 30–70 | Scaffolds fabricated by sublimating solution directly to gas phase, which results in pore formation. | This process generates scaffolds with high volume of interconnecting smaller (<200 um) micropores. |
| Self-assembly | 80–90 | Spontaneous self-organization of disorganized molecules into ordered structures. Non-covalent interaction between molecules such as electrostatic, van der Waals, hydrophobic interactions, hydrogen bonds. | A typical example of molecular self-assembly is phospholipids. When anion strength is increased or pH values are raised to neutrality, the peptides spontaneously self-assemble in anti-parallel arrangement, forming a network of interweaving fibers. |

Figure 1.3 Scanning electron micrographs of scaffolds developed using various fabrication techniques. (a) Electrospun polydioxanone: nano-HAp: fibrinogen (PDO: nHA: Fg) blended scaffolds. Scale bar, 10 $\mu$m [18], Reprinted from [18] with permission. (b) SEM images show that gas foaming with nitrogen creates high density of interconnected or open pores to PHEMA–PMMA scaffold. Scale bar 1 mm [19], Reprinted from [19] with permission from Elsevier. (c) Poly(L-lactic acid) biodegradable scaffolds fabricated by thermally induced phase separation (TIPS) technique (image courtesy of Nelson, C.). (d) PLLA microsphere-Collagen fiber-based composite scaffold prepared by freeze-drying technique. Scale bar, 1 mm (image courtesy of Nelson, C.).

Lithography, chemical etching, and electrospinning are popular top-down approaches. In the bottom-up approach, the building blocks (individual atoms or molecules, for instance) interact one by one and organize into novel supramolecular structures and assemblies. Self-assembly is a spontaneous bottom-up process.

Various types of scaffolds can be fabricated using these techniques that are used in different applications. These scaffold architectures can vary in pore size, texture, shape, and mechanical and chemical properties

(Figure 1.3) [18, 19]. For instance, Zoccola *et al.* used a freeze-drying process to produce chitosan/CNT/HA (chitosan/carbon nanotube/hydroxyapatite) composite scaffolds which promoted increased osteoblast proliferation over control groups *in vitro* [20]. In another study, Ba Linh *et al.* demonstrated that different ratios of HAp-loaded electrospun polyvinyl alcohol/gelatin (PVA/GE) scaffolds exhibited varied enhancement of bone formation using a rat *in vivo* model [21]. Using phase separation, Lavik *et al.* fabricated highly porous scaffolds with pores oriented normal to the plane of the scaffold from blends of poly(L-lactic acid) and poly(lactic-*co*-glycolic acid) for retinal regeneration [22]. The study suggested that these polymer scaffolds were capable of delivering retinal progenitor cells to the injury site and may have assisted in the formation of retinal cell phenotypes. Among the different self-assembling peptides, molecular Lego peptides and molecular cargo have major applications in the field of regenerative medicine [23]. These and many other techniques can be employed to design scaffolds that include biomimetic structures native to two or more structures such that multiple tissue types can be regenerated from the same scaffold. This is the challenge of regenerative engineering.

### 1.3.2 *Natural biomaterials as scaffolds for tissue regeneration*

Natural biomaterials are widely available in nature and derived from natural sources, mostly from plants, animals, and microorganisms. Several researchers have examined the host tissue response to natural biomaterials and found in general that these materials evoke a minimal foreign body reaction, with little or no fibrous encapsulation. Formation of normal granulation tissue associated with accelerated angiogenesis appears to be the typical course of the healing response. These materials exhibit chemical versatility and the possibility of generating structures with predictable pore sizes and degradation rates. In fact, the combination of good biocompatibility, intrinsic antibacterial activity, and ability to bind to growth factors make these materials good options for several regenerative engineering applications. Natural polymers can often be degraded by naturally occurring enzymes as well.

Polysaccharides are one of the more widely used natural polymers, constituted by simple sugar monomers [24]. Differences in monosaccharide composition, chain shapes, and molecular weight dictate their physical properties including solubility, gelatination, and surface properties [25].

Recent research has been focused on polysaccharide-derived polymers such as chitin and chitosan, hyaluronic acid (HA), alginate, starch, and cellulose-based polymers for the development of scaffolds for bone, cartilage, and skin regeneration [26]. Chitosan is structurally similar to the glycosaminoglycans (GAGs) found in ECM of cartilage and are very important to modulate the morphology, differentiation, and function of chondrocytes. Therefore, chitosan is an attractive natural-origin polymer for cartilage regeneration. Chitosan also aids in wound healing by enhancing the functionality of inflammatory cells. HemCon dressing (HemCon Medical Technologies) is an FDA-approved chitosan-based wound dressing that is often used in combat emergency medicine and dentistry [27]. Several studies have shown that composites of chitosan and ceramics like HAp and bioglass show enhanced osteoconductivity [28, 29].

Starch is a biodegradable, inexpensive natural polymer which is abundantly available. Due to its intrinsic thermal and mechanical properties, it is difficult to form various scaffolds using starch alone. However, starch can be blended with synthetic materials such as polyethylene, polycaprolactone, and poly(ethylene-co-vinyl alcohol) or natural origin polymers like proteins [30, 31]. Because of the biocompatibility and the non-immunogenicity of these blended materials, a large range of biomedical applications are being developed using starch-based materials, such as partially degradable bone cement [32] and drug delivery systems [33].

Alginates are naturally derived linear unbranched polysaccharides which are produced by several brown algae as well as some bacteria, consisting 1,4 linked $\beta$-D-mannuronic acid (M block) and $\alpha$-L-guluronic acid (G block) [34]. Alginates have high guluronic acid composition; hence they develop stiffer and more porous gels. During cationic crosslinking, they do not undergo excessive swelling and subsequent shrinking, thus they better maintain their form. Alginate-based materials are pH-sensitive: alginate shrinks at low pH and the encapsulated drugs are not released, whereas at higher pH alginate undergoes a rapid dissolution which may result in burst release of proteins/drugs. This pH-dependent behavior of alginate is utilized to tailor release profiles and in the development of 'smart' systems [35].

Hyaluronic acid (HA) or hyaluronan is a major element of the intercellular matrix of most connective tissues such as cartilage, synovial fluid, vitreous of the human eye, and umbilical cord [36]. It is a non-sulfated glycosaminoglycan. Many researchers have demonstrated that HA can stimulate the proliferation and differentiation of bone marrow stromal cells and chondrocytes [37]. Hyaluronan is available for several applications, such as lubrication and mechanical support for the joints in osteoarthritis, viscoelastic gel for surgery and wound healing, and for implantation of artificial intraocular lenses.

Cellulose is abundantly found in nature as it is the primary structural component of plant cell walls. The biodegradability of cellulose is considered to be limited. However, with incorporation of growth factors, this material can further display a desirable enhancement of bone formation. Porous scaffolds fabricated from bacterial cellulose demonstrated the ability to support the proliferation of bovine-derived chondrocytes [38].

The structural and functional properties of peptides and proteins, the biocompatibility of these molecules, the nutritional requirement for protein in animals, and the tissue toxicity of various synthetic organic biopolymers, have contributed to the development of biomaterials based on proteins and polypeptides. Proteins are linear polymers of 20 distinct amino acids linked by amide groups that have adopted a complex three-dimensional structure. In most cases a 'folded' protein will display several levels of structural organization. Proteins are important in cell signaling, immune responses, cell adhesion, and the cell cycle. Protein fibers can be considered building blocks of organismal structure enabling scaffolding, stabilization, protection, elasticity, and motility at length scales ranging from nanometers to meters. Collagen, fibronectin, elastin, and laminin are fibrous proteins that reside outside the cells and make up the bulk of the ECM. These proteins influence adhesion, proliferation, and migration *in vitro* and *ex vivo*, especially in the presence of growth factors and other bioactive molecules, and they are important in embryogenesis, wound healing, and cancer treatment.

Native collagens form the main structural components of the ECM of many tissues, such as bone, skin, teeth, tendons, cartilage, and ligaments [39]. Twenty-seven types of collagens have been identified to date, but collagen type I is the most abundant and widely investigated. A broad range of collagen-based scaffolds have been developed and commercialized in recent times. For example, bilayered collagen gels seeded with human fibroblasts in the upper layer and keratinocytes in the upper layer have been used as the dermal matrix, an artificial skin product commercialized by Organogenesis in the USA under the name of Apligraf in 1998 [40]. Collagen sponges have been used as an osteoconductive carrier of bone morphogenetic protein-2 (BMP-2) for spinal fusion marketed by Medtronic in the USA [41]. Collagraft, commercialized by Angiotech Pharmaceuticals, Inc., Canada, is a mixture of porous HAp and tricalcium phosphate and animal derived collagen I, and has been used clinically for the treatment of long bone fractures for more than a decade. Healos Bone Graft Replacement by DePuy Orthopedics, USA, is an osteoconductive matrix constructed of crosslinked collagen fibers that are fully coated with HAp. It has been approved recently for clinical use as a bone graft substitute in spinal fusions [42].

Elastin, responsible for skin elasticity and recoiling, is a crucial structural component of the ECM in many tissues. It is present in the thin walls of arteries and veins, ligaments, lung parenchyma, and intestines. Tropoelastin, a soluble precursor of elastin, is useful for modulating physical properties of engineered scaffolds and prompting cellular adherence, growth, and responses to mechanical stimuli *in vitro* [43]. A synthetic form of human tropoelastin produced in genetically modified organisms (DermaPlus; DermaPlus, Staten Island, New York) is available for use in wound healing and topical skin care. Much of the work on elastin scaffolds has been done in vascular graft applications.

Silk fibers, obtained from arthropods and lepidopteran insects are composed of two components: fibroin and sericin. Fibroin is a fibrous protein constituting the core of silk, while sericin is a glue-like protein surrounding fibroin. Lu *et al.* investigated the effect of silk fibroin and recombinant human-like collagen in promoting hepatocyte viability. The fibroin/collagen hydrogel showed better hepacytes attachment and mechanical properties [44]. Hino *et al.* prepared a novel basic fibroblast growth factor (bFGF) delivery system using fibroin as a scaffold [45].

Fibrin, a protein matrix produced from fibrinogen that can be harvested from the patient, can be used as a biocompatible carrier for delivery of active biomolecules and cells [46]. Polymerized fibrin is a major component of blood clots and plays a vital role in the subsequent wound healing response. Despite the seemingly abundant benefits of natural origin materials they do have drawbacks: variability from batch to batch, inconsistent mechanical properties, and limitations in processing options limit the applicability of natural materials as biomaterials. However, the biocompatibility, low cost and availability, possession of similar chemical properties as the ECM, and intrinsic cellular interaction make these materials attractive candidates for biomedical applications.

### 1.3.3 *Synthetic biomaterials as scaffolds for tissue regeneration*

The application and use of synthetic polymers in tissue regeneration is relatively recent compared to natural polymers. In the past 50 years with the synthesis of novel synthetic biomaterials, scientists have been able to simulate many key aspects of the extracellular microenvironment [47]. Scaffolds can be produced from a variety of materials, including metals, ceramics, and polymers. Metallic alloys are popular for both dental and bone implants [48] while ceramics with good osteoconductivity have been used for bone TE [49]. However, both metals and ceramics have significant drawbacks. Metals are not biodegradable and do not provide a biomimetic matrix for cell growth and tissue formation. Ceramics, meanwhile, also have limited biodegradability and are difficult to process into highly porous structures due to their brittleness. In contrast, polymers have great processing flexibility and their biodegradability can be imparted through molecular design. Therefore, polymers are the dominant scaffolding materials used in TE. In the process of developing scaffolds, synthetic polymers hold numerous advantages over natural materials. Key among those advantages is the ability to tailor mechanical properties and degradation kinetics for various applications. Synthetic polymers can be engineered into various shapes with optimum pore sizes and surface textures, conducive to tissue formation. Furthermore, polymers can be designed with chemical functional groups that can induce tissue ingrowth.

Aliphatic polyesters, including poly(glycolic acid) (PGA), poly(lactic acid) (PLA), and their copolymer poly[(lactic acid)-co-(glycolic acid)] (PLGA), poly($\varepsilon$-caprolactone) (PCL) and its copolymer are the most widely used synthetic polymeric materials in TE. These polymers are well characterized and have gained FDA approval for certain clinical application (e.g., sutures). PGA is a rigid thermoplastic material with high crystallinity making it less soluble in most organic solvents; a troublesome property if trying to form scaffolds. PGA is widely used in medical application as resorbable sutures (dexon, American Cyanamide Co). The degradation process of the material occurs in two stages: the first stage involves the diffusion of water into the amorphous (non-crystalline solid) regions of the matrix followed by the degradation of the crystalline structure of the polymer. The rapid degradation rate of PGA causes the weak mechanical integrity of the scaffolds [50]. Poly(lactic acid) (PLA) is more hydrophobic and stable against hydrolysis than PGA. PLA degrades to form lactic acid, naturally present

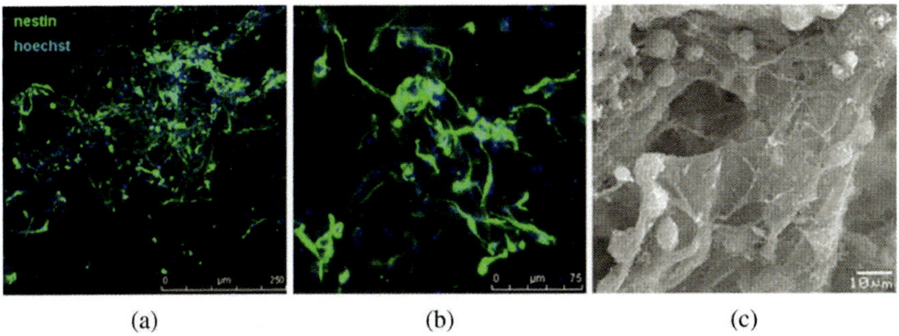


Figure 1.4 The growth of NSCs on the HA–PLGA composite for five days. (a, b) The fluorescent staining of nestin (protein expressed in nerve cells) showed the NSCs (green) adhered on the composite, proliferate and migrate along the periphery of the scaffold (a, overall view), extending long neurites (b, high magnification). (c) The scaffold showed good biocompatibility and support for NSCs by SEM. Reprinted from [58] with permission from Springer Science and Business Media.

in the body. No significant amounts of accumulation of harmful degradation products of PLA have been reported. Both *in vitro* and *in vivo* studies have been carried out to determine the biocompatibility of PLA and PGA. Many studies suggest that these polymers are sufficiently biocompatible [51, 52]. Poly(lactide-co-glycolide) (PLGA), copolymer of PLA and PGA, has been used in many biomedical applications, for example, sutures, drug delivery devices, and TE scaffolds. Due to its good biodegradability, biocompatibility, and process ability, PLGA has been considered for many TE applications, such as the skeletal muscle [53], bone [54], cartilage [55], ligament/tendon [56], and nerve [57]. PLGA can also be combined with other materials to fabricate composites. A composite scaffold of HA hydrogel–PLGA microsphere was developed, loaded with brain-derived neurotrophic factor (BDNF) and vascular endothelial growth factor (VEGF) to study the proliferation of NSCs. The composite scaffold showed promising results in mimicking the microenvironment for new blood vessel formation and neural regeneration, shown in Figure 1.4 [58].

Polycaprolactone (PCL) is a semicrystalline polyester with a melting temperature of 55–60°C [59]. PCL shows very low *in vivo* degradation rate up to 3–4 years. Therefore, it is used in long-term implant delivery devices. PCL films modified with arginine–glycine–aspartate (RGD) peptide (which promotes cell adhesion by acting as a binding site for integrins) served to promote the attachment and spreading of mesenchymal stem cells (MSCs) [60]. Other synthetic biodegradable polymers, including poly(3-hydroxybutyrate) (PHB) [61], polyurethanes (PU) [62], polycarbonate (PC) [63], poly(ortho ester) (POE) [64], poly(propylene fumarate) (PPF) [65], and polyphosphazenes (PP) [66] have also been used as scaffolding biomaterials. Biodegradable bioceramics are designed to aid in the self-repair processes of the living organism, and are subsequently resorbed. Tricalcium phosphates, calcium sulfates, HA, and certain compositions of glasses and glass-ceramics have been recently investigated as biodegradable bioceramics.

### 1.3.4 *Hydrogels for tissue regeneration*

Hydrogels are highly crosslinked, water-insoluble polymer networks capable of storing large amounts of water. Hydrogels can also encapsulate cells during the crosslinking process to function as cell delivery vehicles [67]. Crosslinking is accomplished by either covalent or ionic bonds. Covalent crosslinking tends to allow more precise and accurate construction, but also increases potential toxicity. Formation of hydrogels can also be temperature-induced. The mechanical properties of a hydrogel are mostly determined by the extent of the crosslinking but are generally compatible with conditions necessary for tissue regrowth. Synthetic polymers tend to give greater control of mechanical properties compared to natural polymers, but are also less likely to be biocompatible [68].

The degradation of a hydrogel can occur by either bulk or local dissolution and degrades at a rate partially dependent on the lattice structure [69]. By intentionally introducing network defects, soft hydrogels can be designed to have longer degradation time than the stiffer ones. Local, or surface, dissolution is very useful for drug delivery as it can release a steady amount of the substance over a long time period. Hydrogel degradation can occur in a number of different ways: hydrolysis, breakdown of the chemical bonds by enzymatic reaction, disentanglement of the network, or environmental triggers. The degradation normally involves breaking the bonds of the crosslink, and is therefore determined by the rigidity and extent of the crosslinking. Generally, the more crosslinked a hydrogel is, the stiffer and quicker the degradation time.

Hydrogels can be used as space-filling scaffolds, for cell delivery, and for bioactive molecule delivery [70]. Gilbert *et al.* [71] demonstrated how adult muscle stem cells on engineered hydrogel substrates can be proliferated *in vitro* with higher efficiency, leading to better host tissue integration. These stem cells would normally lose their pluripotency and undergo massive cell death within the first weeks of culture on rigid plastic culture dishes. However, when cultured on chemically crosslinked bioactive poly(ethylene glycol) (PEG) hydrogels, the cells showed signs of self-renewal and were engrafted with substantially better integration in a muscle implant model [71].

Hydrogel-based products have been made available for patient care in the recent years. For example, soft contact lenses are typically made from poly(hydroxyethylmethacrylic) acid [poly(HEMA)] while biological adhesives routinely used in surgical procedures have been made from reconstituted fibrin or albumin. Hydrogels are currently extensively investigated for bone, cartilage, intervertebral, and cardiac regeneration. Moreover, 'smart' hydrogels are raising significant interest due to their ability to change shapes. They can dynamically shrink, swell, or degrade based on exposure to changing stimuli such as pH, temperature, and other activators. They are often used for vascular applications, since their smart capabilities allow them to be maneuvered through small, tight pathways [72].

## 1.4 Cell for Tissue Regeneration

One of the most important aspects of regenerative engineering is to have cells that are able to differentiate and proliferate based not only on the cell type, but also the environment that surrounds them. The scaffold acts similarly to an ECM, creating an environment that is conducive to integration, differentiation, and cell growth. With that in mind, it is extremely important to select a cell source that not only matches the overall goal of the implantation, but that also integrates well with compatible scaffolding. Obtaining cells that are able to differentiate and proliferate to the correct degree is imperative in successful tissue regeneration, as many damaged tissues are not able to regenerate the native tissue to an acceptable extent.

### 1.4.1 *Cell source*

The most standard source of cells comes directly from the patient in what are known as autologous cells. Cells can be extracted from different parts of the body and inserted into the injured area to help repair the damaged tissue. This option is preferred because the cells are guaranteed to be biocompatible and there is no possibility of disease transmission or immune response following the transplantation. There are, however, a few limitations to using autologous cells. Because the cells are being taken from a living person, there is automatically a limitation to the amount of cells that can be safely harvested before causing donor-site morbidity. Also, in order to harvest the cells from the patient, a second surgery is required. This may cause injury to the donor site in the short or long term.

The most common alternative to autologous cells are allogeneic cells where the cells are extracted from a human donor or cadaver. Berner *et al.* compared the use of allogeneic and autologous cells for reconstruction of critical bone defects in sheep, showing similar healing results from the two models [73]. Extracting cells from a cadaver creates an unlimited supply of cells to be used for regenerative purposes and also has the benefit of no risk of injury to the donor site. However, there are still major limitations, for instance the possibility of disease transmission from the donor to the patient and the possibility of immune rejection. Along with this, the differentiating properties of stem cells may also lead them to be tumorigenic. As shown by Ben-David and Benvenisty different stem cell types can have a higher or lower tumorigenicity, making them more or less useful for transplantation [74]. Xenogenic transplant, which incorporates the cells of another species of animal, has also occurred in the past. For example, Niemeyer *et al.* successfully transplanted human MSCs to be used in critical bone defects of sheep [75]. In spite of the potential, such transplantations have not been shown to be as effective as those using autologous or allogenic cells, and translation to the clinical realm would be considerably more challenging.

### 1.4.2 *Cell types*

Along with the choice of where to extract the cells from, the decision of what cell type to use has been researched extensively. The most common choice is still autologous cells from the tissue to be repaired where a biopsy is taken of the tissue and native cells are isolated. They are then expanded in culture in order to create an appropriate cell population to be transplanted back into the body. Because autologous cells from the damaged tissue are not always readily available, researchers are often forced to look elsewhere for cells that can replace or differentiate into the native cells of the tissue.

The cell types that have shown the most promise and potential are the various kinds of stem cells located throughout the body in almost all tissues. Stem cells are highly useful, as they are not only able to proliferate into more stem cells, but are also able to differentiate into many different types of cells as they mature (Figure 1.5) [76].

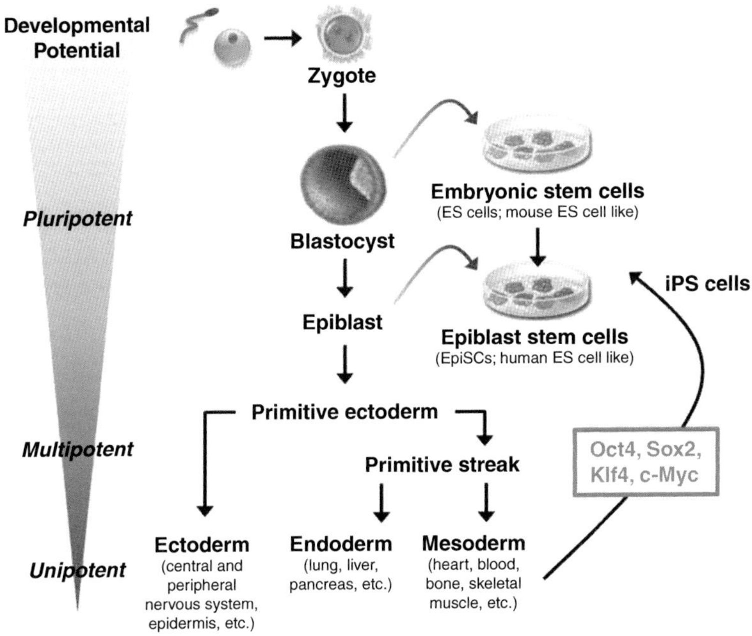


Figure 1.5 Derivation of a human embryonic stem cell line, and differentiation strategies [76]. Reprinted from [76] with permission from John Wiley and Sons.

Multilineage, pluripotent cells can be converted into different phenotypes based on environmental cues by manipulating the culture conditions that the cells are grown in. Different stem cell types have varied multilineage and proliferative capacity [77]. Embryonic stem cells (ESCs) are a major source of interest due to their ability to differentiate into many different lineages [78]. ESCs are not the only choice for regenerative engineering, however: adult stem cells can also be used, but are more limited in their multilineage potential. Another type of stem cell used is amniotic fluid stem cells which, in addition to being able to differentiate into multiple lineages, also offer a lower risk of tumorigenicity.

The most common types of stem cell used are progenitor stem cells. These cells, while having undergone some degree of differentiation, are still able to differentiate into multiple lineages. They can also be isolated, cultured and differentiated into the appropriate lineage before being transplanted into the patient.

Human MSCs are of particular interest due to their multilineage potential, being able to differentiate into many different cell types like osteogenic, adipogenic, and chondrogenic [79]. They are used for vastly different functionalities, treating hundreds of symptoms from a wide spectrum of ailments [80]. NSCs are used for differentiation into almost all types of neural lineages, but a large amount of difficulty has been encountered in creating the desired type and number of cells when expanding the NSCs. Epithelial stem cells have been used for skin grafts, as well as things such as ocular regeneration.

Considerable research efforts have been directed toward human induced pluripotent stem cells (hiPSCs) that are able to differentiate into many lineages, acting very similarly to ESCs [81]. These hiPSCs are capable of different functionalities and are also procured much more easily than most other stem cell types. They are created by inducing a forced expression of specific genes in order to cause a specific expression. Cells are often taken from bone marrow, hair keratinocytes, or skin fibroblasts to create an array of functions, as noted by Streckfuss-Bomeke *et al.* [82]. Somewhat similar to MSCs, it has also been found that it is possible to extract stem cells from the amnion and amniotic fluid that have similar capabilities. Corradetti *et al.* showed that MSCs from the amniotic fluid are capable of many different functions from their original lineages [83].

With so many different types of stem cells to choose from for tissue regeneration procedures, steps must be taken to ensure the correct cell type is selected. Grompe created a tracing system in order to show the differentiation process of stem cell systems [84]. This allowed for better selection of different cells for different functions. In order to harvest the stem cells for implantation, they must first be isolated and expanded in culture. Because it is difficult to harvest a large number of stem cells, expansion is important in order to have a viable level of cells for transplantation. The environment and methods to culture the stem cells is of particular importance, as it can result in changes to the growth rate and extent of the expansion.

### 1.4.3 *Cellular encapsulation in hydrogels*

One of the major ways to incorporate cells into a tissue-engineered scaffold is through the use of hydrogels. As described above, hydrogels combine water-soluble polymer chains by crosslinking them in order to obtain a polymer network that becomes water-insoluble. During the crosslinking process, cells can be encapsulated within the hydrogel, creating a very effective model for cell delivery. This method of delivering cells to an area of injured tissue is quite prevalent given the inherent variability of the hydrogel environment. The hydrogels can be designed in many different ways, allowing them to remain stable or degrade over a designated period of time. Hydrogels are also formed from either natural or synthetic polymers. Naturally derived polymers that are able to form hydrogels include agarose, chitosan, collagen, fibrin, and HA, among others. Natural hydrogels normally have the benefit of being biocompatible, promoting a high amount of tissue development. However, because they are not synthetically made, it is much more difficult to control the physical properties and the crosslinking process.

## 1.5 Growth Factor Mediated Tissue Regeneration

Growth factors are signaling molecules/proteins that modulate cell behavior during tissue growth. Controlled delivery of growth factors can aid in tissue regeneration. Growth factors can be incorporated into TE scaffolds and delivered to cells seeded on the scaffold or localized at the defect site. As proteins, the activity of growth factors depends on their conformation and ability to interact with cellular receptors. They may be denatured if exposed to extremes in temperature or pH and must therefore be loaded under appropriate conditions. Direct injection of growth factors into a defect site is generally a less effective means of delivery since a relatively large quantity of protein is required to stimulate a significant effect. By loading cytokines or chemokines on an implantable scaffold it is possible to reduce the amount of factor necessary to elicit cellular migration or differentiation. In addition, if the loaded scaffold has tunable degradation rates, a controlled release of the factor may be achieved. Additionally, cells can be transfected with a virus containing the gene for a specific growth factor and subsequently implanted into the defect site. The virus forces the seeded cells to produce and secrete the factor, which can then interact with local stem cells to promote cellular differentiation and tissue regeneration. However, associated immunological and other risks like controlling the dose and time period of viral expression have raised concerns regarding the clinical use of gene therapy.

### 1.5.1 *Controlled growth factor delivery methods*

When creating a delivery method for growth factors, it is common to integrate them within the scaffold in several ways (Figure 1.6) [85]. They can be chemically immobilized onto the scaffold or physically encapsulated

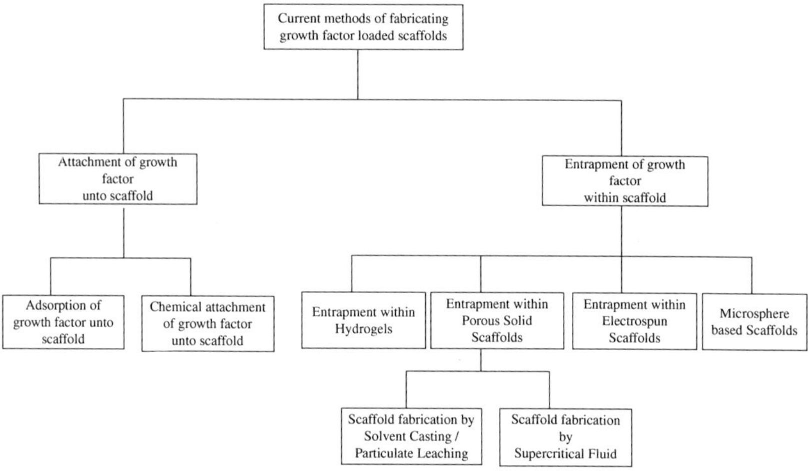


Figure 1.6 A schematic representation of various techniques currently employed in the incorporation of growth factors within polymeric scaffolds [85]. Reprinted from [85] with permission from Elsevier.

within it. In chemical immobilization schemes, strong and localized interactions occur between the growth factor and the cells. Physical encapsulation can be accomplished by surrounding the growth factor with a polymeric material. This will degrade over time, allowing for a slow release of the growth factor depending on the degradation rate of the polymer. This type of growth factor delivery system can be created using techniques such as phase separation, solvent casting and particulate leaching, melt molding, freeze drying, phase emulsion, and gas foaming. Some of the natural and synthetic polymeric materials used with growth factor delivery are discussed in Sections 1.3.2–1.3.4.

While growth factors are most commonly released over time from the scaffolds, they can also be designed to be released in response to local changes in the host tissue or immediate surrounding environment. Local pH and temperature triggers, for instance, can cause a scaffold to grow or shrink in volume and thus mediate payload release. Hydrogels are often used to deliver a growth factor using just such a response as they can have shape memory properties based on pH and temperature. Another trigger could be cleaving a crosslink in the gel. As the crosslink is cleaved, the gel is no longer immobilizing the growth factor and so it becomes activated and is released into the host system. There are drugs, antigens, and antibiotics that can be used to initiate the cleaving process, allowing control of the release in this way.

### 1.5.2 *Growth factors for tissue regeneration*

Growth factors are therapeutic agents for tissue regeneration. Different growth factors are used in order to create differentiation of cells towards a specific phenotype. The following are some of the more commonly used growth factors given their efficacy and the tissue types they have been found to repair and regenerate effectively: nerve growth factor (NGF), epidermal growth factor (EGF), VEGF, and bone morphogenetic proteins (BMPs).

NGF is normally used to help with nerve damage and neurodegenerative disorders. NGF is used to maintain neuronal survival and promote axonal elongation. It has also been used to treat disorders such as Alzheimer's disease or Huntington's disease [86]. By promoting neuronal growth, NGF can be used for brain and spinal cord injuries.

BMPs are used extensively and have been called the gold standard for enhancing bone formation and cell proliferation [87]. Due to their extensive osteoinductive properties and ability to stimulate osteoblast proliferation, BMPs are able to help form new bone, as well as help it continue to grow. They are applicable for bone defects almost anywhere within the body.

VEGF is a growth factor that is used to form blood vessels. Its job is to not only induce endothelialization of existing vessels, but also to promote angiogenesis (the process of forming new blood vessels from pre-existing tissue). VEGF can also recruit endothelial cells to migrate to where they are needed.

EGF is a growth factor used for skin injuries such as incisions and burns. It supports epidermal cell growth, proliferation, and migration. It is also able to catalyze the creation of collagen to help with wound healing. It is used to engineer new skin, as well as heal existing skin [88].

Using these (or similar) growth factors, many different cell types can be proliferated and differentiated. Tissue engineered scaffolds incorporate these growth factors to ensure that the correct type of cells are created to treat the defect. In order to do this, it is important to know how the growth factor will react with the ECM of the scaffold, as well as to the material of the scaffold itself.

### 1.5.3 *Growth factor–biomaterial interactions*

In growth factor-mediated therapy, the specific growth factor used is not the only factor that determines the response elicited by the cells. Depending on the ECM, different messages can be sent by the growth factor. A lot of this is dependent on the receptor molecules within the ECM. These include adhesive molecules, notch signaling molecules, traction-enabling adhesion molecules, and proteoglycan molecules [89]. All of these are able to bind to the growth factor to influence its activity. In controlling the number of active growth factors, it allows for prolonged diffusion of the protein. Along with this, a growth factor can bind to different receptors and send different messages. The interface between the growth factor and ECM is also extremely important for growth factor function as it influences spots for receptors to bind. The relative concentration of the growth factor is also important to control the release profile throughout the healing process.

The degradation of the growth factor is another parameter that can affect its biological activity. The degradation can occur in several ways, such as oxidation, denaturation, and proteolysis. Growth factors are normally encapsulated in some way within the ECM which allows the growth factor to be protected from degradation so that a prolonged release of the protein is possible. In growth factor therapy, everything from the material used to encapsulate the growth factor to the shape and size can affect growth factor degradation rate, release profile and the spatial distributions and these in turn subsequently affect the proliferation and differentiation of the cells [90].

## 1.6 Summary, Conclusions, and Outlook

The biggest problem to date in organ and tissue transplantation continues to be the availability of tissue and the far greater number of people on waiting lists. While the field of TE continues to grow and advance in many different and exciting ways with the hope that it can someday mitigate the discrepancy between required and available transplantation organs and tissues, there are still many complications that hinder advancement. In the clinical realm a successful TE strategy to repair damaged tissues and organs would likely rely, at least in part, on autologous cell harvest. There is, however, a major limitation on the amount of tissue-specific cells that are available to be isolated and expanded for re-implantation. In many patients, a tissue biopsy is not able to yield enough autologous cells for culture and growing certain cell types in the large quantities required for tissue repair is difficult, as shown by Olson *et al.* [91]. Because of this, researchers continue to explore the use of stem cells that can differentiate into those of the diseased tissue. ESCs continue to show vast potential for tissue-engineered constructs partly due to their self-renewal ability while in an undifferentiated state. However, the tumorigenic potential of ESCs and the societal debate over their use raises additional challenges. While other types of stem cells are being used, there are still many issues around cell isolation, expansion, and the nature of differentiation that mandate continued research. Additionally, and somewhat peripheral to the practice of TE is the large volume of new information generated daily in these and related fields. It is vast and can be cumbersome to navigate and incorporate. Regenerative engineering is founded in part on the importance of having highly interdisciplinary expertise from stem cell biology, developmental biology, and advanced materials science, brought together to solve these nagging but critical hurdles. Future strategies for managing new advances in these fields need to be developed as they could greatly advance the way in which engineered stem cells [92] and other elements of the regenerative engineering toolbox are used for the treatment of diseases

Recent developments in polymer chemistry and material science have opened the possibilities to develop a wide range of novel biomaterials. Tissue engineered scaffold designs and the materials used to make them are advancing but challenges exist with incorporating and delivering cells and growth factors that mimic the endogenous tissue repair process. Most growth factor delivery strategies include one or two factors delivered with some degree of kinetic control, but the tissue repair process involves many growth factors with widely varying kinetics. Matching the native tissue healing process is and will continue to be a challenge. Greater understanding of the kinetics of different growth factors will not only allow for improvement of degradation timing, but it may also open up possibilities of different growth factors being utilized in tissue engineered systems, or using them in a complementary fashion with carefully designed scaffolds with surface topographies, mechanical properties, and three-dimensional architecture to direct cellular behavior. Finally, the cost and difficulty of getting a technology approved for commercial use is an ever-present challenge that, although not always at the center of research endeavors, cannot be overlooked. Because TE scaffolds require implantation into the host body, it is subject to regulatory agency review prior to human use, which can increase the cost of bringing accomplishments in the research lab to the bedside. These issues slow the introduction of new technology and products into the market. Many of these challenges, utility of biomaterials, design of scaffolds, use of stem cells, and complementing or replacing the need for growth factors, are at the heart of what defines regenerative engineering. It is the hope that the field will lead to a greater understanding of, tissue regeneration for prolonged and enhanced life.

## References

[1] Amini, A.R., Laurencin, C.T. and Nukavarapu, S.P. (2012). Bone TE: Recent advances and challenges. *Critical Reviews™ in Biomedical Engineering*, **40**(5)

[2] Fact Sheet — Regenerative Medicine, National Institutes of Health, Updated October 2010.

[3] Annual Report of the US Scientific Registry for Transplant Recipients and the Organ Procurement and Transplantation Network: Transplantation Data: 2001–2011. (2011) US Department of Health and Human Services, Health Resources and Services Administration, Office of Special Programs, Division of Transplantation, Rockville, MD; United Network for Organ Sharing, Richmond, VA. Available at: www.unos.org.

[4] Atala, A., Bauer, S.B., Soker, S., Yoo, J.J. and Retik, A.B. (2006). Tissue-engineered autologous bladders for patients needing cystoplasty. *The Lancet*, **367**(9518), pp. 1241–1246.

[5] Oberpenning, F., Meng, J., Yoo, J.J. and Atala, A. (1999). *De novo* reconstitution of a functional mammalian urinary bladder by tissue engineering. *Nat. Biotechnol.*, **17**(2), pp. 149–155.

[6] Moghaddam, A., Elleser, C., Biglari, B., Wentzensen, A. and Zimmermann, G. (2010). Clinical application of BMP 7 in long bone non-unions. *Arch. Orthop. Trauma Surg.*, **130**(1), pp. 71–76.

[7] Lin, H., Godiwalla, S.Y., Palmer, B., Frimberger, D., Yang, Q., Madihally, S.V., *et al.* (2013). Understanding roles of porcine small intestinal submucosa in urinary bladder regeneration: identification of variable regenerative characteristics of small intestinal submucosa. Tissue Engineering Part B: Reviews.

[8] MacNeil, S. (2007). Progress and opportunities for tissue-engineered skin. *Nature*, **445**(7130), pp. 874–880.

[9] Laurencin, C., Khan, Y. and El-Amin, S.F. (2006). Bone graft substitutes. *Expert Rev. Med. Devices*, **3**(1), pp. 49–57.

[10] Ge, Z., Li, C., Heng, B.C., Cao, G. and Yang, Z. (2012). Functional biomaterials for cartilage regeneration. *Journal of Biomedical Materials Research Part A*, **100**(9), pp. 2526–2536.

[11] Shinoka, T., Shum-Tim, D., Ma, P.X., Tanel, R.E., Isogai, N., Langer, R., *et al.* (1998). Creation of viable pulmonary artery autografts through tissue engineering. *J. Thorac. Cardiovasc. Surg.*, **115**(3), pp. 536–546.

[12] Dean, E.W., Udelsman, B. and Breuer, C.K. (2012). Current advances in the translation of vascular tissue engineering to the treatment of pediatric congenital heart disease. *Yale J. Biol. Med.*, June **85**(2), pp. 229–238.

[13] Gurtner, G.C., Werner, S., Barrandon, Y. and Longaker, M.T. (2008). Wound repair and regeneration. *Nature*, **453**(7193), pp. 314–321.

[14] Hutmacher, D.W. (2000). Scaffolds in tissue engineering bone and cartilage. *Biomaterials*, **21**(24), pp. 2529–2543.

[15] Davis, M.E. and Shin, D.M. (2008). Nanoparticle therapeutics: An emerging treatment modality for cancer. *Nature Reviews Drug Discovery*, **7**(9), pp. 771–782.

[16] Christenson, L.K. and Stouffer, R.L. (1996). Proliferation of microvascular endothelial cells in the primate corpus luteum during the menstrual cycle and simulated early pregnancy. *Endocrinology*, January **137**(1), pp. 367–374.

[17] Bayraktar, H.H., Morgan, E.F., Niebur, G.L., Morris, G.E., Wong, E.K. and Keaveny, T.M. (2004). Comparison of the elastic and yield properties of human femoral trabecular and cortical bone tissue. *J. Biomech.*, **37**(1), 27–35.

[18] Rodriguez, I.A., Madurantakam, P.A., McCool, J.M., Sell, S.A., Yang, H., Moon, P.C., *et al.* (2012). Mineralization potential of electrospun PDO-hydroxyapatite-fibrinogen blended scaffolds. *International Journal of Biomaterials* 2012.

[19] Zellander, A., Gemeinhart, R., Djalilian, A., Makhsous, M., Sun, S. and Cho, M. (2013). Designing a gas foamed scaffold for keratoprosthesis. *Materials Science and Engineering: C*, **33**(6), pp. 3396–3403.

[20] Zoccola, M., Aluigi, A., Vineis, C., Tonin, C., Ferrero, F. and Piacentino, M.G. (2008). Study on cast membranes and electrospun nanofibers made from keratin/fibroin blends. *Biomacromolecules*, **9**(10), pp. 2819–2825.

[21] Ba Linh, N.T., Min, Y.K. and Lee, B. (2013). Hybrid hydroxyapatite nanoparticles-loaded PCL/GE blend fibers for bone tissue engineering. *Journal of Biomaterials Science*, Polymer Edition **24**(5), pp. 520–538.

[22] Lavik, E., Klassen, H., Warfvinge, K., Langer, R. and Young, M. (2005). Fabrication of degradable polymer scaffolds to direct the integration and differentiation of retinal progenitors. *Biomaterials*, **26**(16), pp. 3187–3196.

[23] Bolhassani, A. (2011). Potential efficacy of cell-penetrating peptides for nucleic acid and drug delivery in cancer. *Biochimica et Biophysica Acta (BBA)-Reviews on Cancer*, **1816**(2), pp. 232–246.

[24] Nishinari, K., Takahashi, R. (2003). Interaction in polysaccharide solutions and gels. *Current Opinion in Colloid & Interface Science*, **8**(4), pp. 396–400.

[25] Cascone, M., Barbani, N., Cristallini, C., Giusti, P., Ciardelli, G. and Lazzeri, L. (2001). Bioartificial polymeric materials based on polysaccharides. *Journal of Biomaterials Science*, Polymer Edition **12**(3), pp. 267–281.

[26] Khan, F., Ahmad, S.R. (2013). Polysaccharides and Their Derivatives for Versatile Tissue Engineering Application. *Macromolecular Bioscience*, **13**(4), pp. 395–421.

[27] Malafaya, P.B., Silva, G.A. and Reis, R.L. (2007). Natural–origin polymers as carriers and scaffolds for biomolecules and cell delivery in tissue engineering applications. *Adv. Drug Deliv. Rev.*, **59**(4), pp. 207–233.

[28] Kong, L., Gao, Y., Lu, G., Gong, Y., Zhao, N. and Zhang, X. (2006). A study on the bioactivity of chitosan/nano-hydroxyapatite composite scaffolds for bone tissue engineering. *European Polymer Journal*, **42**(12), pp. 3171–3179.

[29] Zhang, N., Li, J., Jiang, W., Ren, C., Li, J., Xin, J., *et al.* (2010). Effective protection and controlled release of insulin by cationic $\beta$-cyclodextrin

polymers from alginate/chitosan nanoparticles. *Int. J. Pharm.*, **393**(1), pp. 213–219.

[30] Gomes, M.E., Ribeiro, A., Malafaya, P., Reis, R. and Cunha, A. (2001). A new approach based on injection moulding to produce biodegradable starch-based polymeric scaffolds: Morphology, mechanical and degradation behaviour. *Biomaterials*, **22**(9), pp. 883–889.

[31] Mendes, S.C., Reis, R., Bovell, Y.P., Cunha, A., van Blitterswijk, C.A., de Bruijn, J.D. (2001). Biocompatibility testing of novel starch-based materials with potential application in orthopaedic surgery: A preliminary study. *Biomaterials*, **22**(14), pp. 2057–2064.

[32] Boesel, L.F., Cachinho, S.C., Fernandes, M.H. and Reis, R.L. (2007). The *in vitro* bioactivity of two novel hydrophilic, partially degradable bone cements. *Acta Biomaterialia*, **3**(2), pp. 175–182.

[33] Silva, G.A. (2007). Nanotechnology approaches for drug and small molecule delivery across the blood *brain* barrier. **Surg. Neurol.**, **67**(2), pp. 113–116.

[34] O'Sullivan, L., Murphy, B., McLoughlin, P., Duggan, P., Lawlor, P.G. and Hughes, H. *et al.* (2010). Prebiotics from marine macroalgae for human and animal health applications. *Marine Drugs*, **8**(7), pp. 2038–2064.

[35] Pongjanyakul, T. and Puttipipatkhachorn, S. (2007). Alginate–magnesium aluminum silicate films: Effect of plasticizers on film properties, drug permeation and drug release from coated tablets. *Int. J. Pharm.*, **333**(1), pp. 34–44.

[36] Liao, Y., Jones, S.A., Forbes, B., Martin, G.P. and Brown, M.B. (2005). Hyaluronan: Pharmaceutical characterization and drug delivery. *Drug Deliv.*, **12**(6), pp. 327–342.

[37] Hegewald, A., Ringe, J., Bartel, J., Krüger, I., Notter, M., Barnewitz, D., *et al.* (2004). Hyaluronic acid and autologous synovial fluid induce chondrogenic differentiation of equine mesenchymal stem cells: A preliminary study. *Tissue and Cell*, **36**(6), pp. 431–438.

[38] Svensson, A., Nicklasson, E., Harrah, T., Panilaitis, B., Kaplan, D., Brittberg M., *et al.* (2005). Bacterial cellulose as a potential scaffold for tissue engineering of cartilage. *Biomaterials*, **26**(4), pp. 419–431.

[39] Awad, H.A., Butler, D.L., Boivin, G.P., Smith, F.N., Malaviya, P., Huibregtse, B., *et al.* (1999). Autologous mesenchymal stem cell-mediated repair of tendon. *Tissue Eng.*, **5**(3), pp. 267–277.

[40] Landsman, A., Taft, D. and Riemer, K. (2009). The role of collagen bioscaffolds, foamed collagen, and living skin equivalents in wound healing. *Clin. Podiatr. Med. Surg.*, **26**(4), pp. 525–533.

[41] Gautschi, O.P., Frey, S.P. and Zellweger, R. (2007). Bone morphogenetic proteins in clinical applications. *ANZ J. Surg.*, **77**(8), pp. 626–631.

[42] Laurencin, C.T. and El-Amin, S.F. (2008). Xenotransplantation in orthopaedic surgery. *J. Am Acad. Orthop. Surg.*, January **16** (1), pp. 4–8.

[43] Debelle, L. and Alix, A.J. (1999). The structures of elastins and their function. *Biochimie*, **81**(10), pp. 981–994.

[44] Lu, Q., Feng, Q., Hu, K. and Cui, F. (2008). Preparation of three-dimensional fibroin/collagen scaffolds in various pH conditions. *J. Mater. Sci. Mater Med.*, **19**(2), pp. 629–634.

[45] Hino, R., Tomita, M. and Yoshizato, K. (2006). The generation of germline transgenic silkworms for the production of biologically active recombinant fusion proteins of fibroin and human basic fibroblast growth factor. *Biomaterials*, **27**(33), pp. 5715–5724.
[46] Aper, T., Schmidt, A., Duchrow, M. and Bruch, H. (2007). Autologous blood vessels engineered from peripheral blood sample. *European Journal of Vascular and Endovascular Surgery*, **33**(1), pp. 33–39.
[47] Ulery, B.D., Nair, L.S. and Laurencin, C.T. (2011). Biomedical applications of biodegradable polymers. *Journal of Polymer Science Part B: Polymer Physics*, **49**(12), pp. 832–864.
[48] Niinomi, M. (2008). Mechanical biocompatibilities of titanium alloys for biomedical applications. *Journal of the Mechanical Behavior of Biomedical Materials*, **1**(1), pp. 30–42.
[49] Ohtsuki, C., Kamitakahara, M. and Miyazaki, T. (2009). Bioactive ceramic-based materials with designed reactivity for bone tissue regeneration. *J. R. Soc. Interface*, June 6; 6 Suppl, 3, pp. S349–S360.
[50] Gunatillake, P., Mayadunne, R. and Adhikari, R. (2006). Recent developments in biodegradable synthetic polymers. *Biotechnol Annu Rev*, **12**, pp. 301–347.
[51] Carletti, E., Motta, A. and Migliaresi, C. (2011). Scaffolds for tissue engineering and 3D cell culture. 3D cell culture: Springer. pp. 17–39.
[52] Pihlajamäki, H.K., Salminen, S.T., Tynninen, O., Böstman, O.M. and Laitinen, O. (2010). Tissue restoration after implantation of polyglycolide, polydioxanone, polylevolactide, and metallic pins in cortical bone: An experimental study in rabbits. *Calcif Tissue Int.*, **87**(1), pp. 90–98.
[53] Wang, P., Wu, T., Tsai, W., Kuo, W. and Wang, M. (2013). Grooved PLGA films incorporated with RGD/YIGSR peptides for potential application on skeletal muscle tissue engineering. *Colloids and Surfaces B: Biointerfaces*, **110**, pp. 88–95.
[54] Barbanti, S.H., Santos, Jr, A.R., Zavaglia, C.A. and Duek, E.A. (2011). Poly ($\varepsilon$-caprolactone) and poly (D, L-lactic acid-co-glycolic acid) scaffolds used in bone tissue engineering prepared by melt compression–particulate leaching method. *J. Mater Sci. Mater Med.*, **22**(10), pp. 2377–2385.
[55] Andreas, K., Zehbe, R., Kazubek, M., Grzeschik, K., Sternberg, N., Bäumler, H., *et al.* (2011). Biodegradable insulin-loaded PLGA microspheres fabricated by three different emulsification techniques: Investigation for cartilage tissue engineering. *Acta biomaterialia*, **7**(4), pp. 1485–1495.
[56] Sahoo, S., Toh, S.L. and Goh, J.C. (2010). A bFGF-releasing silk/PLGA-based biohybrid scaffold for ligament/tendon tissue engineering using mesenchymal progenitor cells. *Biomaterials*, **31**(11), pp. 2990–2998.
[57] Hidayah, N., Fadzli, A., Ng, M., Ruszymah, B., Naicker, A., Shalimar, A., *et al.* (2012). Porous PLGA sheet and acellularized muscle stuffed vein seeded with neural-differentiated MSCs are potential scaffolds for nerve regeneration. *Regen. Res.*, **1**, pp. 1–7.
[58] Wang, Y., Wei, Y.T., Zu, Z.H., Ju, R.K., Guo, M.Y., Wang, X.M., *et al.* (2011). Combination of hyaluronic acid hydrogel scaffold and PLGA

microspheres for supporting survival of neural stem cells. *Pharm. Res.*, **28**(6), pp. 1406–1414.

[59] Patlolla, A., Collins, G. and Livingston, Arinzeh, T. (2010). Solvent-dependent properties of electrospun fibrous composites for bone tissue regeneration. *Acta Biomaterialia*, **6**(1), pp. 90–101.

[60] Hersel, U., Dahmen, C. and Kessler, H. (2003). RGD modified polymers: Biomaterials for stimulated cell adhesion and beyond. *Biomaterials*, **24**(24), pp. 4385–4415.

[61] Chen, G. and Wu, Q. (2005). The application of polyhydroxyalkanoates as tissue engineering materials. *Biomaterials*, **26**(33), pp. 6565–6578.

[62] Santerre, J., Woodhouse, K., Laroche, G. and Labow, R. Understanding the biodegradation of polyurethanes: From classical implants to tissue engineering materials. *Biomaterials*, **26**(35), pp. 7457–7470.

[63] Lee, S.J., Choi, J.S., Park, K.S., Khang, G., Lee, Y.M. and Lee, H.B. (2004). Response of MG63 osteoblast-like cells onto polycarbonate membrane surfaces with different micropore sizes. *Biomaterials*, **25**(19), pp. 4699–4707.

[64] Wang, C., Ge, Q., Ting, D., Nguyen, D., Shen, H., Chen, J., *et al.* (2004). Molecularly engineered poly (ortho ester) microspheres for enhanced delivery of DNA vaccines. *Nature Materials*, **3**(3), pp. 190–196.

[65] Fisher, J.P., Vehof, J.W., Dean, D., van der Waerden, J.P., Holland, T.A., Mikos, A.G., *et al.* (2002). Soft and hard tissue response to photocrosslinked poly (propylene fumarate) scaffolds in a rabbit model. *J. Biomed. Mater Res.*, **59**(3), pp. 547–556.

[66] Carampin, P., Conconi, M.T., Lora, S., Menti, A.M., Baiguera, S., Bellini, S., *et al.* (2007). Electrospun polyphosphazene nanofibers for *in vitro* rat endothelial cells proliferation. *Journal of Biomedical Materials Research Part A*, **80**(3), pp. 661–668.

[67] Elisseeff, J., Ruffner, M., Kim, T. and Williams, C. (2006). Cellular photoencapsulation in hydrogels. *Culture of Cells for Tissue Engineering*, pp. 213–238.

[68] Zustiak, S.P., Wei, Y. and Leach, J.B. (2012). Protein–hydrogel interactions in tissue engineering: Mechanisms and applications. *Tissue Engineering Part B: Reviews*, **19**(2), pp. 160–171.

[69] Lee, K.Y. and Mooney, D.J. (2001). Hydrogels for tissue engineering. *Chem. Rev.*, **101**(7), pp. 1869–1880.

[70] Drury, J.L. and Mooney, D.J. (2003). Hydrogels for tissue engineering: Scaffold design variables and applications. *Biomaterials*, **24**(24), pp. 4337–4351.

[71] Gilbert, P.M., Havenstrite, K.L., Magnusson, K.E., Sacco, A., Leonardi, N.A., Kraft P., *et al.* (2010). Substrate elasticity regulates skeletal muscle stem cell self-renewal in culture. *Science*, August 27, **329**(5995), pp. 1078–1081.

[72] Lutolf, M.P. (2009). Biomaterials: Spotlight on hydrogels. *Nature Materials*, **8**(6), pp. 451–453.

[73] Berner, A., Reichert, J., Woodruff, M.A., Saifzadeh, S., Morris, A., Epari, D., *et al.* (2013). Autologous vs. allogenic mesenchymal progenitor cells for the

reconstruction of critical sized segmental tibial bone defects in aged sheep. *Acta Biomaterialia*, **9**(8), pp. 7874–7884.

[74] Ben-David, U. and Benvenisty, N. (2011). The tumorigenicity of human embryonic and induced pluripotent stem cells. *Nature Reviews Cancer*, **11**(4), 268–277.

[75] Niemeyer, P., Schönberger, T.S., Hahn, J., Kasten, P., Fellenberg, J., Suedkamp N., *et al.* (2009). Xenogenic transplantation of human mesenchymal stem cells in a critical size defect of the sheep tibia for bone regeneration. *Tissue Engineering Part A*, **16**(1), pp. 33–43.

[76] Lyssiotis, C.A., Lairson, L.L., Boitano, A.E., Wurdak, H., Zhu, S. and Schultz, P.G. (2011). Chemical control of stem cell fate and developmental potential. *Angewandte Chemie International Edition*, **50**(1), pp. 200–242.

[77] Metcalf, D. (2007). Concise review: Hematopoietic stem cells and tissue stem cells: Current concepts and unanswered questions. *Stem Cells*, **25**(10), pp. 2390–2395.

[78] Thomson, J.A., Itskovitz-Eldor, J., Shapiro, S.S., Waknitz, M.A., Swiergiel, J.J., Marshall V.S., *et al.* (1998). Embryonic stem cell lines derived from human blastocysts. *Science* November 6, **282**(5391), pp. 1145–1147.

[79] Chen, Y., Shao, J., Xiang, L., Dong, X. and Zhang, G. (2008). Mesenchymal stem cells: A promising candidate in regenerative medicine. *Int. J. Biochem. Cell Biol.*, **40**(5), pp. 815–820.

[80] Caplan, A.I. (2013). Adult mesenchymal stem cells and the NO pathways. *Proc. Natl. Acad. Sci.* USA, February 19, **110**(8), pp. 2695–2696.

[81] Lin, H.T., Otsu, M. and Nakauchi, H. (2013). Stem cell therapy: An exercise in patience and prudence. *Philos. Trans. R. Soc. Lond. B Biol. Sci.*, January 5, **368**(1609):rstb. 2011.0334.

[82] Streckfuss-Bomeke, K., Wolf, F., Azizian, A., Stauske, M., Tiburcy, M., Wagner, S., *et al.* (2013). Comparative study of human-induced pluripotent stem cells derived from bone marrow cells, hair keratinocytes, and skin fibroblasts. *Eur. Heart J.* **34**(33), pp. 2618–2629.

[83] Corradetti, B., Meucci, A., Bizzaro, D., Cremonesi, F. and Lange Consiglio, A. (2013). Mesenchymal stem cells from amnion and amniotic fluid in the bovine. *Reproduction*, April 15, **145**(4), pp. 391–400.

[84] Grompe, M. (2012). Tissue stem cells: New tools and functional diversity. *Cell Stem Cell*, **10**(6), pp. 685–689.

[85] Sokolsky-Papkov, M., Agashi, K., Olaye, A., Shakesheff, K. and Domb, A.J. Polymer carriers for drug delivery in tissue engineering. *Adv. Drug. Deliv. Rev.*, **59**(4), pp. 187–206.

[86] Houeland, G., Romani, A., Marchetti, C., Amato, G., Capsoni, S., Cattaneo, A., *et al.* (2010). Transgenic mice with chronic NGF deprivation and Alzheimer's disease-like pathology display hippocampal region-specific impairments in short- and long-term plasticities. *J. Neurosci.* 29, **30**(39), pp. 13089–13094.

[87] Johnson, E.E. and Urist, M.R. (1998). One-stage lengthening of femoral nonunion augmented with human bone morphogenetic protein. *Clin. Orthop.*, **347**, pp. 105–116.

[88] Shankaran, V., Brooks, M. and Mostow, E. (2013). Advanced therapies for chronic wounds: NPWT, engineered skin, growth factors, extracellular matrices. *Dermatologic Therapy*, **26**(3), pp. 215–221.
[89] Rahmany, M.B. and Van Dyke, M. (2013). Biomimetic approaches to modulate cellular adhesion in biomaterials: A review. *Acta Biomaterialia*, **9**(3), pp. 5431–5437.
[90] Lee, K., Silva, E.A. and Mooney, D.J. (2011). Growth factor delivery-based tissue engineering: General approaches and a review of recent developments. *J. R. Soc. Interface*, 6, **8**(55), pp. 153–170.
[91] Olson, J.L., Atala, A. and Yoo, J.J. (2011). Tissue engineering: Current strategies and future directions. *Chonnam Medical Journal*, **47**(1)pp. 1–13.
[92] Ricotti, L. and Menciassi, A. Engineering stem cells for future medicine. *IEEE Transactions on Biomedical Engineering*, **60**(3), pp. 727–734.

## Chapter 2

# Injectable Hydrogels for Biomedical Applications

B. Balakrishnan

*Department of Biosciences & Bioengineering*
*Indian Institute of Technology Bombay*
*Mumbai, Maharashtra, India-400076*

A. Jayakrishnan*

*Department of Biotechnology,*
*Bhupat and Jyoti Mehta School of Biosciences,*
*Indian Institute of Technology Madras, Chennai,*
*Tamil Nadu, India-600036*

## 2.1 Introduction

Extensive utilization of hydrogel as a biomaterial began with the development of a copolymer of 2-hydroxyethyl methacrylate and ethylene dimethacrylate by Wichterle and Lim in 1954. Soft hydrogel contact lenses developed by Wichterle in 1961 [1] represented the first successful clinical application of hydrogels and still continues to be one of the most important uses of hydrogels today. By making small variations in the structure of the polymer (e.g., changing the type and amount of monomers and crosslinking agents), hydrogels with a wide range of properties can be prepared. Over the years, hydrogels have been successfully developed for use as coatings, soft contact lenses, wound dressings, drug delivery systems, and tissue engineering scaffolds, as well as bulking and augmenting agents. Due to their ease of delivery, injectable hydrogels are becoming increasingly popular and are

*Author to whom correspondence should be addressed. Email: ajk@iitm.ac.in.

potentially useful as surgical tissue adhesives, space-filling injectable materials for hard and soft tissue repair, drug delivery, and tissue engineering. Hydrogels based on natural as well as synthetic polymers and hybrids have been developed as injectable formulations for various biomedical applications. Recent advancements in the field have led to the development of injectable hydrogels such as *in situ* gelling, shear thinning, self-assembled peptides with nanostructured networks, and those derived from decellularized extracellular matrices. This chapter aims to present a general overview of injectable hydrogel formulations with emphasis on their various biomedical applications.

## 2.2 Hydrogels

The gelation process of polymeric chains usually starts with the formation of progressively larger branched yet soluble polymers, depending on the structure and conformation of the precursor material. Such polydisperse soluble branched polymers during the course of gelation process form crosslinked polymers with decreased solubility and increased size. This crosslinked structure called 'gel' is permeated with finite branched polymers. This three-dimensional network can imbibe volume of a liquid medium which then entangles by means of surface tension effects. Gels can thus be considered as a colloidal system in which the dispersion phase is liquid and dispersion medium is solid.

Hydrogels are gels which are formed by the crosslinking of hydrophilic polymers, both natural and synthetic, in which the dispersion phase is water. One of the interesting features of hydrogels is their excellent water-absorbing property; they absorb quantities of water ranging from 10–20% up to thousands of times their dry weight without undergoing dissolution. This water-absorbing ability makes hydrogels flexible and similar to biological soft tissues, and hence attractive biomaterials. Another fascinating property is their porous structure which along with the presence of water allows the transport of low molecular weight solutes in and out of the hydrogel, which is critical for applications related to delivery of cell and drug molecules. Due to the presence of hydrophilic groups, hydrogels exhibit low interfacial free energy in contact with body fluids, leading to poor adsorption of proteins and cells, and good biocompatibility. Furthermore, the soft, rubbery nature of hydrogels contributes to minimal tissue irritation [2, 3].

Hydrogels can be classified into various different categories. For example, hydrogels prepared by crosslinking of polymers using chemical crosslinkers, photocrosslinkers, and enzymes are permanent or chemical gels having strong chemical bonds between polymeric chains [4]. Swelling of these hydrogels is mainly governed by the crosslinking density [5]. Physical or reversible hydrogels are formed due to molecular entanglements or secondary interactions such as ionic, hydrogen bonding or hydrophobic interactions, microcrystal formation, or helix formation between the polymer chains. Physical gels can undergo a sol-gel transition in response to temperature, pH, and solvent. These hydrogels can be reversible and undergo dissolution by changes in physical conditions or application of stress [4, 5]. This terminology is sometimes misleading as there are a small but critical number of examples where covalent bonds are labile and physical interactions lead to dense crosslinks behaving like permanent crosslinks. Nevertheless, this terminology will be followed throughout the chapter for convenience.

Apart from chemical and physical hydrogels, another novel gel type called 'slide-ring gel' or 'topological gel' has been recently developed by endowing the supramolecular architecture with topological characteristics. A good example is a network structure in which cyclic structures are threaded through a single polymer chain and trapped by means of bulky end groups. These types of hydrogels are superior to chemical and physical gels in terms of mechanical properties. In chemically crosslinked gels, tensile stress is concentrated on the shortest chains leading to easy breakage of the gel, whereas in the slide-ring gels, polymer chains are interlocked topologically by figure-of-eight crosslinks which can pass along the polymer chain equalizing the tension of polymer chains like in the case of pulleys [6] (Figure 2.1).

Another classification is based on the charge that hydrogel carries (either ionic (anionic/cationic) or neutral). Classifications are also there based on structure: amorphous, semicrystalline, bound by oxygen molecules, with a supermolecular structure, or hydrocolloidal aggregations [7].

## 2.3 Injectable Hydrogels

Injectability is one of the major requirements for minimally invasive procedures, particularly in tissue engineering and drug delivery applications. A principal advantage of such a system is ease of application using a syringe,

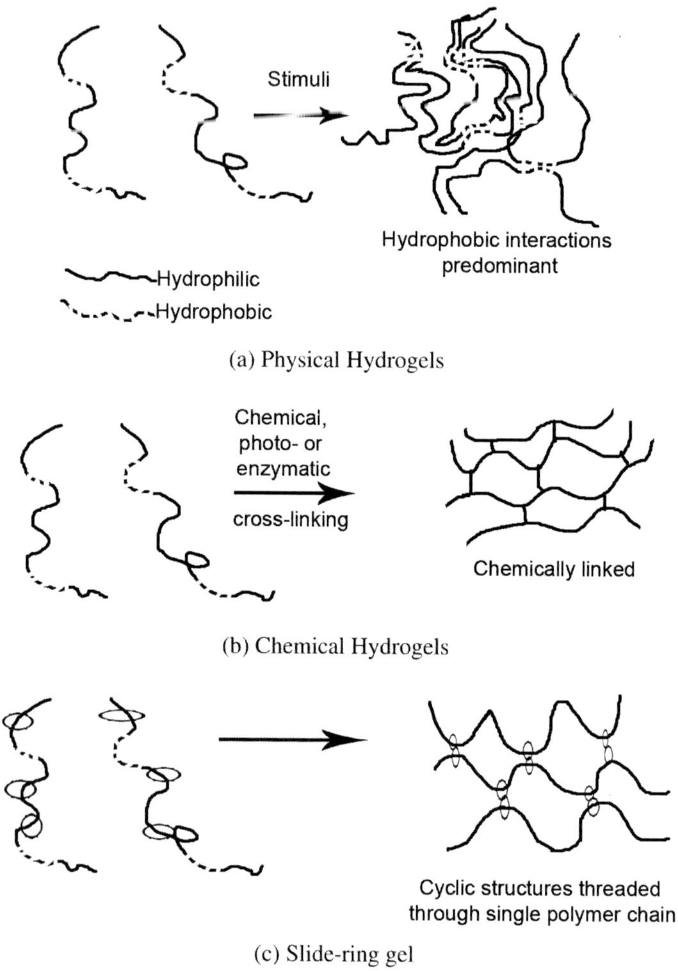


Figure 2.1 Schematic illustration of different types of hydrogels.

where the polymer solution undergoes rapid sol-gel transition at the site of injection, taking the shape of the defect and providing good linkage and interface between gel and host tissue. The incorporation of various therapeutic molecules and cells can be easily achieved by mixing them with the precursor polymer solutions. *In situ* gelling hydrogels and shear-thinning (thixotropic) hydrogels are commonly used as injectables. The success of any injectable approach lies mainly in the gelation process, along with the physical, chemical, and biological properties of the hydrogel. Different strategies currently used to develop injectable hydrogels are schematically represented in Figure 2.2.

To be used for biomedical applications, injectable hydrogels should meet the following criteria:

(a) Viscosity of precursor solution should be sufficiently low to achieve homogenous dispersion of drug molecules/cells;
(b) Precursor solution should flow under pressure;
(c) Optimal gelation time should not be too fast that it clogs the nozzle of the syringe or too slow that it overflows from the defect site to the adjacent tissues;
(d) Gelation reaction should not produce any temperature changes that can cause tissue damage at site of delivery;
(e) Hydrogel should quickly build up mechanical stability and maintain sufficient integrity and strength as long as required;
(f) Precursor solution and degradation products need to be biocompatible;
(g) Precursor solution should be sterilizable; and,
(h) Hydrogel must have controlled degradation.

### 2.3.1 In situ *gelling hydrogels as injectable biomaterials*

Both physical and chemical gels have been employed as injectable hydrogels for various biomedical applications. Injectable hydrogels for repair of bone, cartilage [8, 9], peripheral [10], and central nerve [11], spinal cord [12] and treatment of myocardial infraction [13], have been reviewed elsewhere. Injectable hydrogels based on natural polymers such as alginate, collagen, chitosan, decellularized tissues, fibrin, hyaluronic acid, and a few synthetic materials have shown promise on their own without the addition of therapeutics such as cells or growth factors. While physical gels are formed in response to changes in environment such as pH, temperature, and ion concentration, chemical gels are prepared by crosslinking methods like Schiff base formation, free radical polymerization, Michael-type addition polymerization, and enzyme mediated polymerization. This section will discuss the preparation of those *in situ* gelling hydrogels and their potential as biomaterials.

#### 2.3.1.1 *Crosslinking by physical interactions*

Physical gels also called 'smart hydrogels' are described as having dynamic crosslinks that are continuously formed and broken, bouncing the state

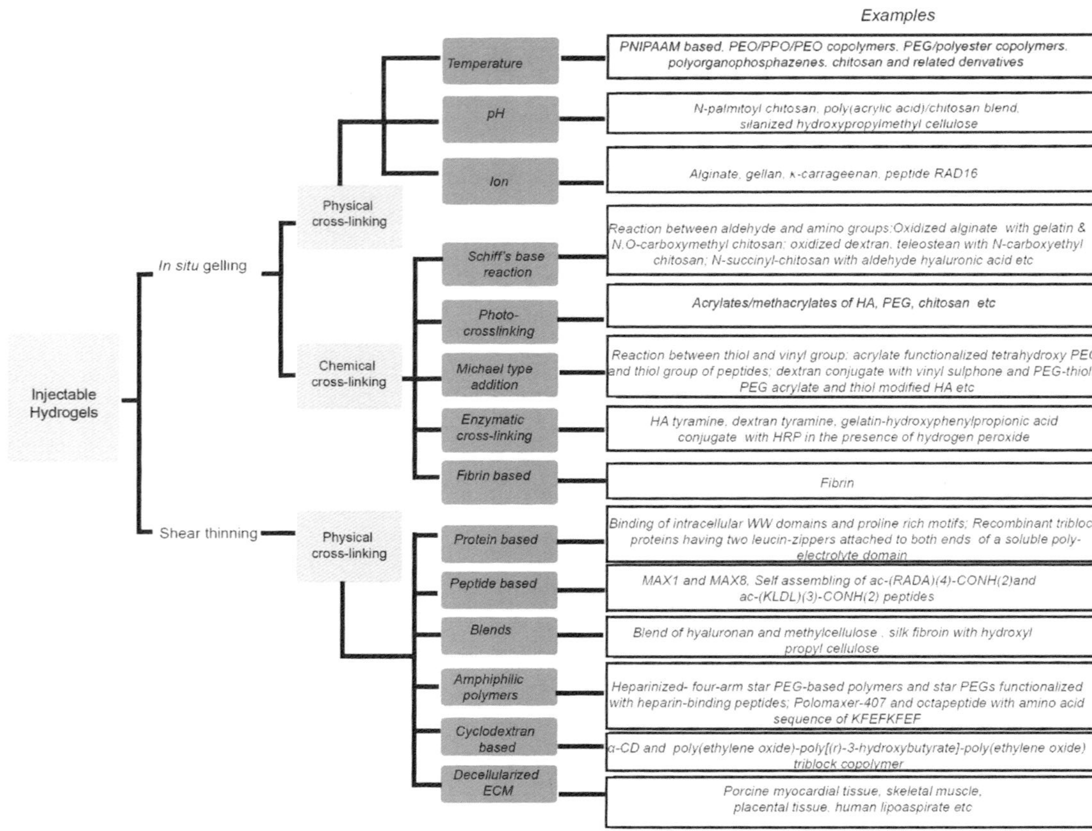


Figure 2.2 Strategies employed for the preparation of injectable hydrogels.

of the gel between solid and liquid under the influence of environmental factors. This unique ability of physical gels makes them an important class of biomaterials in drug delivery and tissue engineering.

2.3.1.1.1 Temperature-induced hydrogel formation

Physical gels, having reversible sol-gel transition at or near the physiological temperature, represent an important class of injectable hydrogels for biomedical applications. Transitions of these polymers having lower critical solution temperature (LCST) in the aqueous medium have been utilized for various biomedical applications. At low temperatures, these polymers will exist in aqueous solution and become gel at temperatures above the LCST value [14]. By proper selection of polymers and appropriate chemistries, LCST values can be designed as a value below body temperature so that gelation occurs upon injection into the body. Polymers with such properties generally exhibit moderate hydrophobic behavior and are called 'gelators'. Gelators, when coupled to hydrophilic polymers either by grafting or copolymerization, lead to the formation of amphiphiles which are water soluble at low temperatures. The temperature dependence of interactions like hydrogen bonding and hydrophobic interactions plays a crucial role in this phenomenon. As the temperature increases, aggregation of hydrophobic segments begins minimizing the hydrogen bonding and amount of structured water surrounding the hydrophobic domains and thus increasing the polymer–polymer and water–water interactions and solvent entropy. When the temperature reaches the LCST value of polymer, entropically favorable abrupt volume phase separation of the polymer solution occurs, dehydrating the solvated polymer to make it more hydrophobic. The LCST values of the polymer depend on the length of the hydrophobic segment and chemical structure of the polymer. The lengthier the hydrophobic domain, the larger the entropic cost of water structuring and the lower the LCST values will be [15]. Some thermoresponsive polymers self-assemble at LCST, leading to micelle packing and gel formation [16]. Biomedical applications of injectable thermosensitive hydrogels has been reviewed elsewhere [17–19]. These hydrogels possess a sol-gel transition temperature around physiologial temperature and upon *in vivo* injection, gel at body temperature. Advantages include ease of drug formulation, shielding of drug from environment, extended and local drug delivery, and ease of application. Examples of thermosensitive hydrogel as drug delivery systems include poly(N-isopropylacrylamide)(PNIPAAm)copolymers,

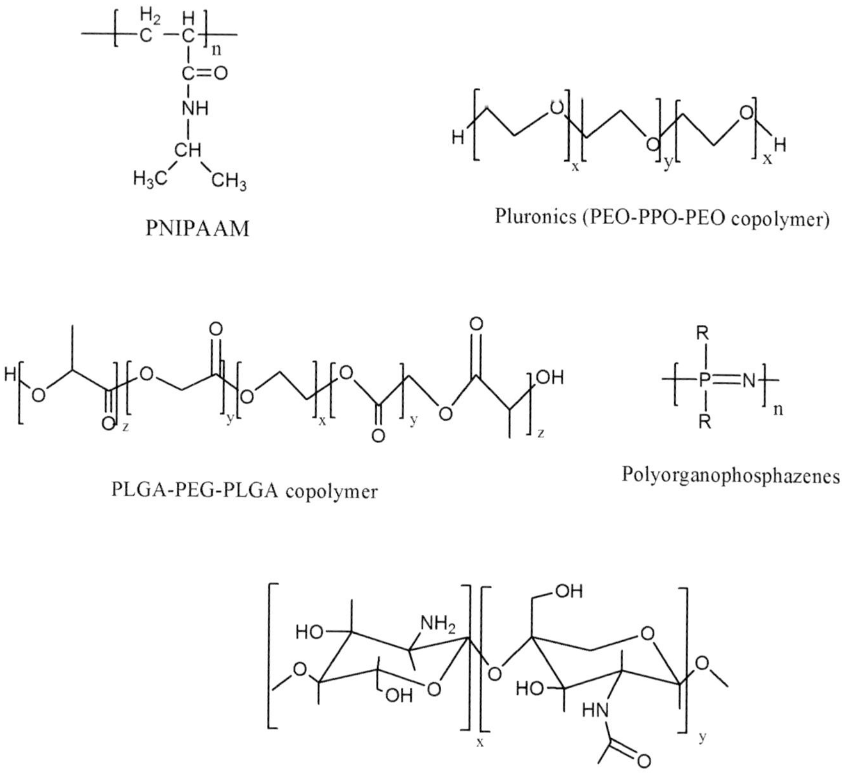


Figure 2.3 Structure of temperature-sensitive polymers.

poly(ethylene oxide)/poly(propylene oxide) (PEO/PPO) copolymers and their derivatives, poly(ethylene glycol)/biodegradable polyester copolymers, poly(organophosphazenes), and polysaccharides and related derivatives. Structures of some commonly investigated temperature responsive polymers are shown in Figure 2.3.

#### 2.3.1.1.1.1 Poly(NIPAAm)-based copolymers

PNIPAAm and its derivatives are the most extensively studied thermosensitive injectable polymers for various biomedical applications [20–25]. The LCST value of PNIPAAm in aqueous solution is 32°C, which can be modulated by copolymerization with hydrophilic monomers and hydrophobic monomers yielding lower and higher LCST values, respectively.

For example, copolymerization with acrylic acid (AAc) increases the LCST value at neutral pH [26, 27], while copolymerization with 2-hydroxyethyl methacrylate-monolactate decreases the LCST values [28, 29].

Thermoresponsive hydrogels based on NIPAAM, 2-hydroxyethyl methacrylate and biodegradable methacrylate-polylactide have been synthesized along with modified copolymers incorporating 1 mol% protein-reactive methacryloxy N-hydroxysuccinimide (MANHS), hydrophilic AAc, or both. MANHS and AAc in the copolymers supported retention of loaded protein during thermal transition (45% vs. 22%) and faster release (2 months), respectively, which may be useful for certain tissue engineering applications [30]. The copolymer of NIPAAM with AAc (poly(NIPAAM-co-AAc)) containing a combination of ascorbate and transforming growth factor $\beta$-3 (TGF$\beta$-3) has been evaluated as an injectable drug carrier and scaffold for cartilage tissue engineering [31]. Poly(NIPAAM-co-AAc) containing hydroxyapapite and bone morphogenic protein-2 has been found to induce osteogenic differentiation of rabbit mesenchymal stem cells leading to homogenous ectopic bone formation [32]. The copolymer of NIPAAM, AAc, dimethyl-$\gamma$-butyrolactone acrylate, and 2-hydroxyethyl methacrylate–poly(trimethylene carbonate) macromer combined with type I collagen were readily injectable at 4°C. The thermal transition temperature of the solution was within the range of 23.6–24.5°C and gelation occurred within 7 s at 37°C to form highly flexible and soft hydrogels with moduli from 39 to 119 KPa and breaking strains >1000%, depending on the copolymer composition and collagen content. Superoxide dismutase loaded within the hydrogel was shown to protect encapsulated human mesenchymal stem cells (MSCs) from attack by superoxide by suppressing its penetration into the hydrogel and subsequently the cell membrane, suggesting the suitability of such systems as delivery vehicles for stem cell therapy and drug delivery [33]. Copolymerization of NIPAAM, AAc and hydroxyethyl methacrylate–poly(trimethylene carbonate) poly(NIPAAM-co-AAc-co-HEMAPTMC) has shown to yield biodegradable and thermoresponsive hydrogel at physiological temperature. The hydrogel has been evaluated as an injectable bulking agent into the ventricular wall to prevent progressive adverse remodeling due to high wall stresses that develop after myocardial infarction. Upon injection into the infarcted left ventricular (LV) wall in a rat chronic infarction model, the LV cavity area and contractility were preserved at eight weeks. Tissue ingrowth with a thicker LV wall and higher capillary density was observed in the hydrogel injected area, as can be seen

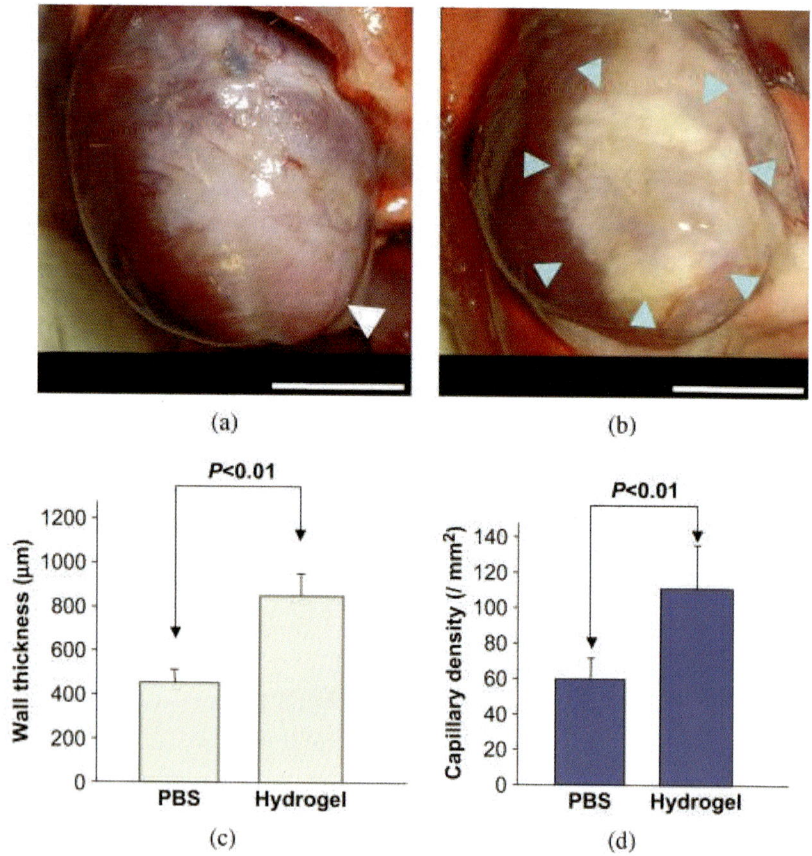


Figure 2.4 Representative images at eight weeks following the injection procedure of the anterior view of (a) PBS injected, and (b) poly(NIPAAm-co-AAc-co-HEMAPTMC) (86/4/10) injected hearts. White arrow shows an aneurysm formation in the apex area (a). Blue arrows indicate the injected hydrogel area (b). The LV myocardial wall thickness (c) and capillary density (d) of the hydrogel and PBS injection groups at eight weeks following injection. Reprinted from [34] with permission from Elsevier.

in Figure 2.4, thus offering an attractive injectable biomaterial-centered treatment option for ischemic cardiomyopathy [34].

Hydrogel composites fabricated from type I collagen, chondroitin sulfate and thermosensitive poly(NIPAAM-co-AAc-co-HEMAPTMC) showed to gel at body temperature, yielding highly flexible hydrogel. MSCs were viable over a period of one week within the gel, which was significantly accelerated by loading insulin growth factor [35]. Thermosensitive PNIPAAM-g-HEMA microgel particles have been used as building blocks for the construction of macroscopic hydrogel scaffold which showed a volume phase transition temperature (VPTT) of about 29°C. The system is colloidally stable above VPTT, but the microgel dispersion can gelate and form macroscopic hydrogel only by the addition of a low concentration of $CaCl_2$, which facilitates ionic interaction [36].

The slightly acidic pH of ischemic myocardium has been shown to induce the gelation of pH- and temperature-responsive random copolymer, poly(*N*-isopro-pylacrylamide-co-propylacrylic acid(PAA)-co-butyl acrylate), which exists as liquid at normal body temperature and pH, and gels at pH 6.8. These hydrogels, loaded with biotinylated basic fibroblast growth factor (bFGF), were found to provide sustained, local delivery of bFGF which improved angiogenesis, and enhanced cardiac function [37]. Injectable hydrogels have also been prepared by employing simultaneous physical and chemical gelling systems using NIPAAM copolymers having acrylate and thiol functional groups. *In situ* gelation was achieved by mixing poly(NIPAAM-co-HEMA-acrylate) and poly(NIPAAM-co-cysteamine) in which physical gelation occurs due to association of PNIPAAM at a temperature near body temperature and chemical gelation through a Michael-type addition reaction. This dually crosslinked gel is envisaged to have potential as an injectable, gelling material for biomedical applications such as endovascular embolization [38]. Poly(ethylene glycol) diacrylate (PEG–DA) crosslinked PNIPAAM hydrogel has been evaluated as an injectable ocular drug delivery platform [39]. These hydrogels have also been used for localized protein delivery [40]. Thermoresponsive hydrogels based on polycaprolactone, *N*-isopropylacrylamide, 2-hydroxyethyl methacrylate, and dimethyl-$\gamma$-butyrolactone acrylate prepared by atom transfer radical polymerization show gelation at body temperature within 5 s. These flexible hydrogels exhibited moduli similar to rat and human myocardium, and supported the proliferation and differentiation of cardiosphere-derived cells into a mature cardiac lineage [41].

Naturally occurring polymers have also been grafted onto PNIPAAM to make thermoresponsive hydrogels. Chitosan-g-PNIPAAM having LCST at around 30°C was shown to form a three-dimensional hydrogel structure with interconnected pores ranging from 10 to 40 $\mu$m for cartilage tissue engineering applications [42]. Efforts have also been made to make thermosensitive copolymer hydrogels based on aminated hyaluronic acid and NIPAAM having an LCST value of 30°C in phosphate buffered saline (PBS). Hydrogels encapsulated with human adipose-derived stem cells showed the preservation of viability of the entrapped cells and demonstrated the potential as an injectable hydrogel for adipose tissue engineering [43]. Injectable, thermoresponsive hyaluronic acid g-chitosan-g-PNIPAAM (HA-CPN) copolymer has been prepared for bone tissue engineering. This biocompatible and bioresorbable HA-CPN hydrogel encapsulated with bone marrow-derived MSCs induced ectopic bone formation upon subcutaneous implantation in nude mice [44]. Mechanical properties of *in situ* forming hydrogels based on PNIPAAM-g-chondroitin sulfate (CS) developed for nucleus pulposus (NP) tissue engineering were found to increase with the increase in crosslinking density and concentration of CS. The unconfined compressive modulus was found to be similar to that reported for the native NP. These hydrogels showed adhesive tensile strength in the range of 0.4–1 kPa [45].

The non-biodegradable nature, and neurotoxicity of acrylamide residues, raise concerns over wide applicabilty of PNIPAAM based hydrogels as a biomaterial. Attempts have been made to make the PNIPAAM bioresorbable by synthesizing copolymers having hydrolysable lactate ester side groups where the byproduct of degradation is lactic acid, which is subjected to renal clearance, but leads to pH reduction and cytotocity [46–48]. Copolymerization with cyclic monomer 2-methylene-1,3-dioxepane [49], poly(ethylene glycol) [50], and dimethyl-$\gamma$-butyrolactone acrylate [51], and crosslinking with poly(amino acid) [52] and methacrylate are other attempts that have been made to make PNIPAAM bioresorbable [53].

##### 2.3.1.1.1.2 Poly(ethylene oxide)/poly(propylene oxide) (PEO/PPO) copolymers and its derivatives

Commercially available macromolecular surfactants based on water soluble triblock copolymers of EPE system where E and P represent poly (ethylene oxide) and poly (propylene oxide), respectively, commonly abbreviated as PEO–PPO–PEO (or (EO)m-(PO)n-(EO)m where m and n are the number of repeating units) have also been exploited as thermoreversible gels for biomedical applications. These copolymers, commercially known as 'Pluronic' or 'Poloxamer' (BASF), are non-toxic in nature with minimal immune response. Nomenclature of commercially available PEO/PPO copolymers include the word 'Pluronic' followed by letter 'L' for liquids, 'P' for paste, 'F' for flakes, and followed by a number in which the first two digits multiplied by three indicates the molecular weight of PPO units and the last digit multipled by ten indicates the percentage of PEO units. For example, Pluronic F127(synonymous to Poloxamer 407) contains 70% PEO units and a molecular weight of PPO around 4,000. Pluronic polymers form micelles which are in equilibrium with their corresponding unimers and once temperature is increased, this equilibrium shifts towards micelle formation and above critical gelation concentration, packing of micelles occurs occupying entire volume leading to gelation [54]. The biodegradable, biocompatible copolymer of poly(ethylene oxide) and poly(propylene oxide) which exists as a liquid below 4°C and polymerizes to a thick gel when it is exposed to physiologic temperature, has been evaluated for the formation of autologous tissue-engineered cartilage in various shapes, which include nipple [55] and human ear helix [56]. This non-ionic polymer surfactant with proven biocompatibility and thermosensitivity has also been used as an injectable drug delivery system. The hydrogel has been found to enhance the bioavailability of short-lived peptide and protein drugs and thus improved the clinical outcome [57]. Poloxamer/Pluronic hydrogel formulation containing dexamethasone showed better retention in the middle ear with no indication of high frequency loss, thereby providing an effective method for prolonged intratymanic delivery [58, 59].

One of the disadvantages of Pluronic F127 is inadequate mechanical stability and various methods have been explored to improve this by: (a) introduction of functional end groups that can undergo chemical crosslinking after thermal gelation which include carbon–carbon double bonds [60], ethoxysilane groups [61, 62], and methacrylate groups [61]; (b) synthesis of PEO–b–PPO–b–PEO copolymers with diacyl chloride as a coupling agent [63] and (c) incorporation of $\varepsilon$-caprolactone [63] or lactide oligoester [64] segments prior to chain extension. The stability of Pluronic F127 can also be improved by incorporating thiol and acrylate groups which can undergo chemical crosslinking [65, 66].

Preparation of Pluronic copolymers linked with D-lactide and L-lactide oligomers with different spacer lengths has been developed in order to exploit the stereocomplexation between corresponding enantiomeric D- or L-lactide oligomer spacer. These hydrogels were found to be suitable for delivery of human growth factors where release was of zero order kinetics and controlled by a combined diffusion/erosion mechanism [67, 68]. Chitosan grated Pluronic polymer having sol gel transition around 25°C exhibited with a storage modulus of $10^4$ Pa suitable has potential as an injectable scaffold for cartilage tissue engineering. Novel thermosensitive linoleic acid (CLA)-coupled poloxamer hydrogel having the benefits of the pro-drug activity and enhanced paclitaxel (PTX) solubility due to the micellar property of the CLA-coupled poloxamer has been evaluated for the local delivery of PTX. Subcutaneous injection of the formulation into tumor-bearing mice showed excellent antitumor activity (Figure 2.5) by inducing stronger cell cycle arrest and apoptosis than the PTX-incorporated poloxamer hydrogel. This enhanced therapeutic effect has been attributed to the synergistic effects of PTX and CLA from the CLA-coupled poloxamer and the enhanced solubility of PTX [69]. Chitosan grafted Pluronic polymer having sol-gel transition around 25°C with a storage modulus of $10^4$ Pa has shown potential as an injectable scaffold for cartilage tissue engineering. Bovine Chondrocytes encapsulated in chitosan grafted pluoronic hydrogel were found to proliferate with an increase in the glycosaminoglycan deposition over a period of 28 days [70]. Another promising injectable hydrogel for controlled delivery of bone morphogenetic protein-2 (BMP-2)-related peptide has been prepared based on poly(trimethylene carbonate)–F127–poly(trimethylene carbonate). The hydrogel-carrying peptide showed elevated alkaline phosphatase activity and expression of osteocalcin mRNA in bone marrow stromal cells (BMSCs) *in vitro* and effective ectopic bone formation *in vivo* [71].

One of the limitations of Pluronic F127 is that the polymer solution must be stored in an icebox and undergoes gelation as soon as it is taken out before reaching body temperature, which can lead to clogging of the nozzle of the syringe. An attempt has been made to overcome this issue by modifying Pluronic F127 by oxidation and incorporation of silver nanoparticles, which has increased the phase transion temperature to 32°C [72]. It has been reported that Pluronics F127 could induce sustained hypercholesterolemia and hypertriglyceridemia in rats by intraperitoneal injection and therefore the amount used in the formulation should be maintained minimal for repeated administration [73].

##### 2.3.1.1.1.3 Poly(ethylene glycol) (PEG)/biodegradable polyester copolymers

Thermoresponsive hydrogels have also been prepared from copolymers of PEG with biodegradable and biocompatible esters by proper adjustment of the hydrophobic polyester block and hydrophilic PEG block length. Aqueous solutions of copolymers of poly(ethylene oxide) and poly(L-lactic acid) were found to exhibit temperature-dependent reversible sol-gel transitions showing upper critical solution temperature (UCST), where they exist

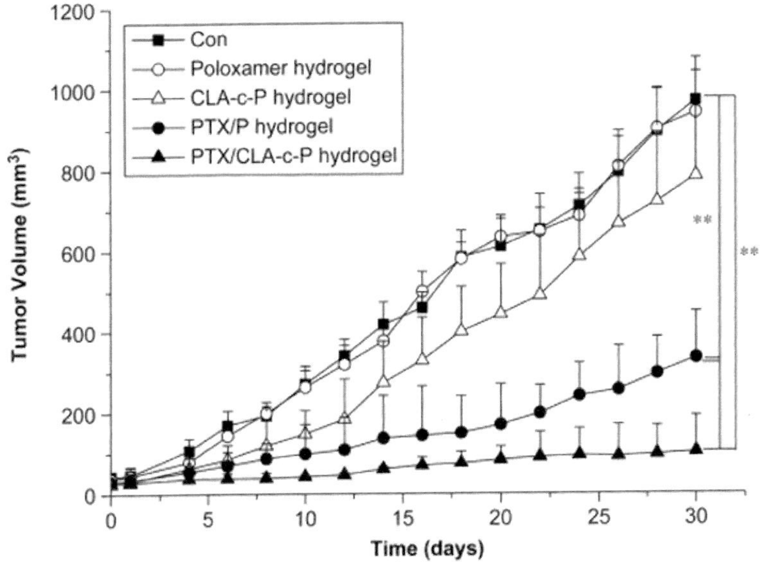


(A)

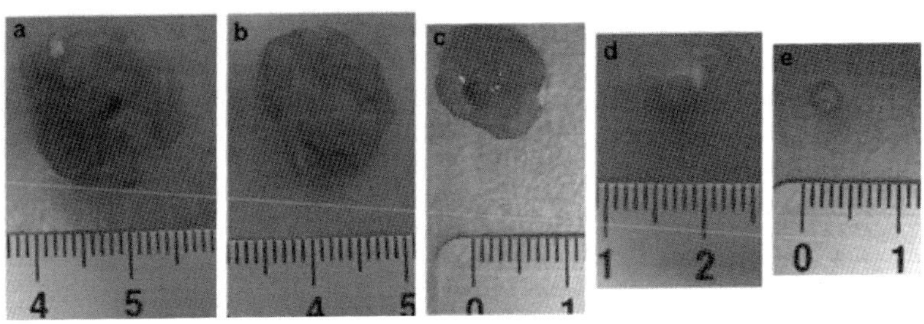

(B)

Figure 2.5 Effect of PTX-incorporated CLA-coupled poloxamer hydrogel on tumor growth. (A) Tumor volume of mice bearing MDA-MB-231 tumors exposured to various treatments ($^*p < 0.05$; $^{**}p < 0.01$, compared with control; $n = 3$). (B) Images of tumors at the time of sacrifice, 30 days post-inoculation. (a) Control; (b) poloxamer hydrogel; (c) CLA-c-P; (d) PTX/P; and (e) PTX/CLA-c-P. Con, control; Poloxamer, poloxamer hydrogel; CLA-c-P, CLA-coupled poloxamer hydrogel; PTX/P, PTX-incorporated poloxamer hydrogel; PTX/CLA-c-P, PTX-incorporated CLA-coupled poloxamer hydrogel. Reprinted from [69] with permission from Elesevier.

in sol form at 45°C and become gel at body temperature. Though having similar phase transition curve of Pluonics, these PEG based triblock copolymers retained their integrity *in vivo* for more than 1 month in contrary to what observed for Pluronics where the hydrogel undergoes erosion upon dilution. These hydrogels have been exploited as injectable sustained release matrices for drugs where loading was done at 45°C, a temperature at which the polymer is injectable and upon subcutaneous injection and subsequent rapid cooling to body temperature, the loaded copolymer forms a gel. These hydrogels were shown to be biodegradable and biocompatible with sufficient mechanical stability. Solubilization of even hydrophobic drugs could be easily achieved because of their surfactant nature [74, 75]. Another biodegradable and biocompatible system based on the triblock of poly(lactide-co-glycolide)-poly(ethylene glycol)-poly(lactide-co-glycolide) (PLGA-PEG-PLGA) triblock copolymers, which exhibit both LCST and UCST behavior and form gel at body temperature, have also been explored as drug delivery vehicles for both hydrophilic and hydrophic drugs [76, 77]. The release was diffusion controlled for hydrophilic drugs, whereas hydrophobic drugs showed a biphasic profile: diffusion controlled followed by matrix erosion [78]. Copolymer hydrogel loaded with interleukin-2 showed antitumor effect *in vivo* demonstrating its biological activity [79]. The potential of these copolymers as drug delivery vehicles has been extensively reviewed elsewhere [80].

Poly(ethyleneglycol)-poly($\varepsilon$-caprolactone)-poly(ethylene glycol) (PEG-PCL-PEG) is another PEG-based polyester copolymer which exists as liquid at room temperature and gels within seconds *in situ* at body temperature. Main advantage of PCL based systems over PLGA systems is that PCL incorporation leads to the formation of polymers in powdery form, instead of sticky paste making it more convenient for practical applications. Loaded with PTX, this hydrogel has been shown to prevent locoregional recurrence of 4TI breast cancer in a mouse model, suggesting this may be a promising approach for local adjuvant therapy in malignant disease [81]. This hydrogel has also shown potential as a delivery vehicle for proteins by preserving their bioactivity [82], for ophthalmic drugs [83], and also as a novel vaccine adjuvant for protein antigen [84]. Injectable and thermosensitive composite hydrogel based on PEG-PCL-PEG copolymer/collagen/nanohydroxy apatite has been evaluated for guided bone regeneration [85]. A dual drug delivery system of self-assembled micelles in a thermosensitive hydrogel composite, where both micelle and hydrogel were based on poly($\varepsilon$-caprolactone)-poly(ethylene glycol)-poly($\varepsilon$-caprolactone) (PCL-PEG-PCL) amphiphilic copolymer, was developed to deliver hydrophilic (5-fluoro uracil (5FU)) and hydrophobic drugs (PTX) simultaneously for colorectal peritoneal carcinomatosis therapy. Micelle/hydrogel composite gels at physiological temperature maintained sustained release of incorporated drugs, effectively inhibited growth and metastasis of CT26 peritoneal carcinomatosis *in vivo* and induced a stronger antitumor effect than that of Taxol® plus 5FU [86]. Attempts have also been made to adjust the degradation and drug release from injectable, thermosensitive amphiphilic PCL-PEG-PCL copolymer hydrogel by incorporating 1,4,8-trioxa[4.6]spiro-9-undecanone [87].

Significant reduction of post-operative peritoneal adhesion was found when a sidewall defect-bowel abrasion in a rabbit model was treated with *in situ* formed poly($\varepsilon$-caprolactone-co-lactide)-b-poly(ethylene glycol)-b-poly($\varepsilon$-caprolactone-co-lactide) (PCLA-PEG-PCLA) hydrogels [88]. Sustained release of peptides (cyclo(-RGDfK-)) from these hydrogels was observed over a period of one week. While the hydrogel offered a physical barrier preventing adhesion between the injured abdominal wall and cecum, the released RGD peptide assisted anti-adhesion [89]. The KRGDKK (Lys-Arg-Gly-Asp-Lys-Lys) peptide was coupled to the ends of a PCLA-PEG-PCLA triblock copolymer to obtain thermoresponsive peptide-PCLA-PEG-PCLA-peptide copolymers, demonstrating promising potential as injectable drug delivery systems with improved mechanical properties and sustained release of the drug for more than a month without initial burst [90].

#### 2.3.1.1.1.4 Poly(organophosphazene) and its derivatives

Poly(organophosphazenes) are inorganic/organic hybrid polymers consisting of alternating phosphorus and nitrogen atoms and organic substituents linked to the phosphorus atoms as side groups. Though poly(dichlorophos phazene), the precursor for the preparation of poly(organophosphazenes), is hydrolytically unstable, by substitution of its chlorine atoms with nucleophilic groups, a wide range of stable poly(organophosphazenes) with an extremely wide range of properties can be synthesized [91]. Depending on the nature and ratio of side substituents, the properties of the resulting material can be varied. These polymers are generally made biodegradable by incorporation of hydrolytically labile substituents like amino esters. The substituted polymers have shown to undergo hydrolysis to give non-toxic products such as ammonia and phosphates, along with the substituted side groups. The synthesis, properties, and applications of poly(organophosphazenes) have been summarized in detail elsewhere [92, 93]. A distinctive property of poly(organophosphazenes) is the ease with which the appropriate hydrophilic, hydrophobic, and other functional pendent groups could be introduced to phosphate atoms, thereby achieving easy modulation of polymer characteristics.

The ability to impart thermoresponsive nature to these polymers by incorporation of hydrophilic and hydrophobic substitutents has led to the preparation of injectable formulations for various biomedical applications. Injectable, biodegradable, and thermosensitive poly(organophosphazene) hydrogel prepared by the stepwise substitution reaction of poly(dichlorophosphazene) with hydrophobic L-isoleucine ethyl ester and hydrophilic, $\alpha$-amino-$\omega$-methyl-poly(ethylene glycol) was found to maintain viability and functionality of spheroidal hepatocytes indicating its potential as a useful three-dimensional cell delivery system for developing bioartificial liver devices [94]. Rat islets entrapped in these gels showed prolonged insulin secretion in response to basal (5.5 mM) glucose concentration and maintained higher cell viability compared to free islets, indicating its potential for tissue engineering applications [95]. This has also been found to be effective as an injectable controlled delivery vehicle for 5FU [96] and doxorubicin [97]. Injectable poly(organophosphazene) hydrogel containing silibinin was found to have the benefits of enhancing the solubility of silibinin up to 2,000 times, diffusion controlled sustained delivery and enhanced antitumor effect [98]. Poly(organophosphazene) used for the delivery of PTX has been shown to overcome the complications related to its poor solubility in water and efflux systems in the gastrointestinal tract. Hydrogel mass was found to be retained at the injection site over six weeks [99]. Poly(organophosphazene)–campothecin conjugates, bearing carboxylic acid, and glycine as linker, were found to show thermoresponsive sol-gel transition behavior favorable for injectable application near room temperature. Campothecin conjugated to the hydrogel was stable over 15 days *in vitro* and inhibited tumor growth upon intratumoral injection in an HT-29 colon cancer cell xenografted mice model [100]. The improved antitumor efficacy of these injectable hydrogels was due to retention of material within tumor tissue with little systemic circulation, sustained release without burst release, and non-toxic degradation products.

Though poly(organophosphazenes) have been suggested as a potential injectable thermosensitive hydrogel for the release of proteins, initial burst release is a limiting factor. This burst release was found to be suppressed when proteins were complexed with chitosan leading to a prolonged release period [101]. A combined system of polyelectrolyte complex (PEC) and poly(organophosphazene) hydrogel has been suggested as an injectable depot for a controlled and sustained delivery of proteins to improve patient compliance [102]. Polyelectrolytes were introduced to minimize the diffusion of proteins from the hydrogels. Polycations like protamine, polyethylenimine (PEI), poly-L-lysine, and poly-L-arginine (PLA) were used to complex with proteins like BSA, gelatin (type B bloom 75), $\alpha$-amylase, and human growth hormone (hGH), out of which the PLA/protein complex displayed the slowest release profile, suggesting the potential of PEC for controlling the protein release profile from poly(organophosphazene) hydrogels [103]. Positively charged PEC prepared with negatively charged hGF and positively charged protamine sulfate, subsequently loaded in anionic poly(organophosphazene) hydrogel has been found to enhance sustained release of hGH by dual ionic interactions [104]. Cationic poly(organophosphazene)–PEI conjugate hydrogels have also been explored as carriers for the sustained delivery of anionic hGH by incorporating different chain lengths of $\alpha$-amino-$\omega$-methoxy-poly(ethylene glycol) (AMPEG550 and AMPEG750) in order to control hydrogel degradation rates. While degradation rates of these hydrogels were in proportion to the molecular weight of AMPEG, the release of hGH was governed by the extent of complexation between PEI and hGH [105] This has also been evaluated for siRNA delivery. Single injection of the polyplex hydrogel with siRNA of cyclin B1 into the tumor xenograft model was shown to maintain slow release of siRNA from the polyplex hydrogel by dissolution and degradation and demonstrated tumor growth inhibition for four weeks [106].

Cell adhesion peptide GRGDS has been conjugated with carboxylic acid-terminated poly(organophosphazene) to improve the potential of the injectable scaffold for bone tissue engineering. Rabbit mesenchymal stem cells encapsulated within the gel upon injection into a nude mouse exhibited expression of osteocalcin and collagen type 1, and high mineralization at the fourth week post-injection [107]. Injectable, self-crosslinkable, and thermosensitive polyphosphazene-based blend systems which form gel with sufficient mechanical strength at body temperature have been prepared by using functional thiolated and acrylated polymer, utilizing both chemical crosslinking and thermosensitive hydrophobic interactions. The system claims to overcome the hurdles associated with the use of monomeric crosslinkers, catalysts, oxidants, pH adjustments, initiators, UV light, heat production, and organic solvent [108].

#### 2.3.1.1.1.5 Chitosan and related derivatives

Chitosan is partially deacetylated chitin consisting of $\beta$ $(1 \rightarrow 4)$-linked D-glucosamine residues with a variable number of randomly located *N*-acetyl-glucosamine groups. Thermosensitive solutions of chitosan (0.5–2%) with $\beta$-glycerol phosphate (GP) (5–20%) undergo sol-gel transition around body temperature and possess pH around the physiological range. Gelation time decreases with an increase in GP and chitosan concentrations. The gel extracts were found to promote cell proliferation when gel was formed using upto 10% solution of GP; beyond that there was a reduction in cell proliferation [109]. *In situ* gel-forming, thermosensitive chitosan/$\beta$-GP has been evaluated for tissue engineering [110] and sustained release of hormones and drugs [77]. The hydrogel utilized for controlled local delivery of (188) Re-Tin colloid showed localization for upto 48 h post-injection indicating an effective treatment mode for regional radiotherapy [111]. These hydrogels have also been reported to have potential as a scaffold for the delivery of adipose-derived MSCs into ischemia/reperfusion-induced acute kidney injuries. The hydrogel improved the retention, survival, and proliferation of MSCs, while reducing apoptosis of host renal cells. A significant improvement of the renal function, microvessel density, and tubular cell proliferation were observed at four weeks [112]. Rapidly gelling, thermosensitive hydrogels based on chitosan/$\beta$-glycerophosphate and 20 vol% contrast agent have been used as a radio-opaque embolizing agent for the treatment of endoleaks after endovascular aneurysm repair (EVAR) and were found to be visible during X-ray-based guided intervention. Short-term radio-opacity was easily obtained with this hydrogel, suggesting the potential of these new radio-opaque thermogels as embolizing agents for EVAR, although their mechanical properties require improvements [113].

Addition of liposomes to the chitosan-GP solution was found to increase the gelation rate and gel strength [114]. Addition of gelatin (1%) to the thermosensitive chitosan-$\beta$-glycerol phosphate solutions has shortened the gelation time and improved the gel strength at body temperature. The non-cytotoxic nature with an improved amount of sulfated glycosaminoglycan deposition and type II collagen and aggrecan expression indicated the potential of hydrogel as a cell carrier for NP regeneration [115]. Grafting of chitosan to PEG (40 wt%) yielded copolymer soluble at low temperature and gels at body temperature. Further *in situ* crosslinking with genipin prolonged the quasi-linear release of bovine serum albumin from the hydrogel for a period of 40 days, demonstrating its potential as a matrix for sustained *in vivo* drug release and tissue engineering [116]. Chitosan-$\beta$-glycerophosphate formulations combined with bioactive glass nanoparticles have been found to promote bone-like apatite formation for orthopedic reconstructive and regenerative medicine applications [117]. Quaternized chitosan also forms a thermogelling formulation with $\beta$-glycerophosphate, undergoing gelation within 3 min at physiological pH and temperature. The formulation exhibited antibacterial activity towards two periodontal pathogens (*Porphyromonas gingivalis* and *Prevotella intermedia*) suggesting potential as a local drug delivery system for periodontal treatment [118]. Further incorporation of 0.1% chlorhexidine into this gel increased the gelling time to 6 mins at body temperature, but exhibited excellent inhibitory acivity against primary periodontal pathogens [119]. Injectable temperature-sensitive zinc-doped chitosan/$\beta$-glycerophosphatehydrogel has been studied for bone tissue engineering [120]. Nanoparticles loaded with insulin have been incorporated with these thermosensitive hydrogels to improve the bioavailability of insulin [121]. By combining physical gelation, Michael-type addition, and disulfide bond formation, injectable hydrogel has also been prepared using thiolated chitosan, $\beta$-glycerophosphate, and PEG-DA without the addition of cytotoxic crosslinkers and catalysts [122]. Another chitosan based thermogelling, injectable nanoscale hydrogel based on amphiphilic carboxymethyl-hexanoyl chitosan (CHC) increased the viability and CD44+ proportion of induced pluripotent stem cells and maintained their stem cell-like gene expression, in the presence of culture media indicating the potential as a cell delivery system [123]. Thermosensitive hydroxybutyl chitosan hydrogels have been found to act as a hydrogel barrier (Figure 2.6) for post-operative adhesions [124].

These hydrogel systems demonstrate thermosensitive property only at low concentrations and show fast release of small hydrophilic drugs due to large water filled matrix pore size, limiting their wide use [22].

#### 2.3.1.1.1.6 Miscellaneous thermogelling systems

Thermally responsive protein-based hydrogels have been prepared by synthesizing recombinant elastin-like polypeptides having periodic cysteine residues that undergoes reversible crosslinking by the formation of

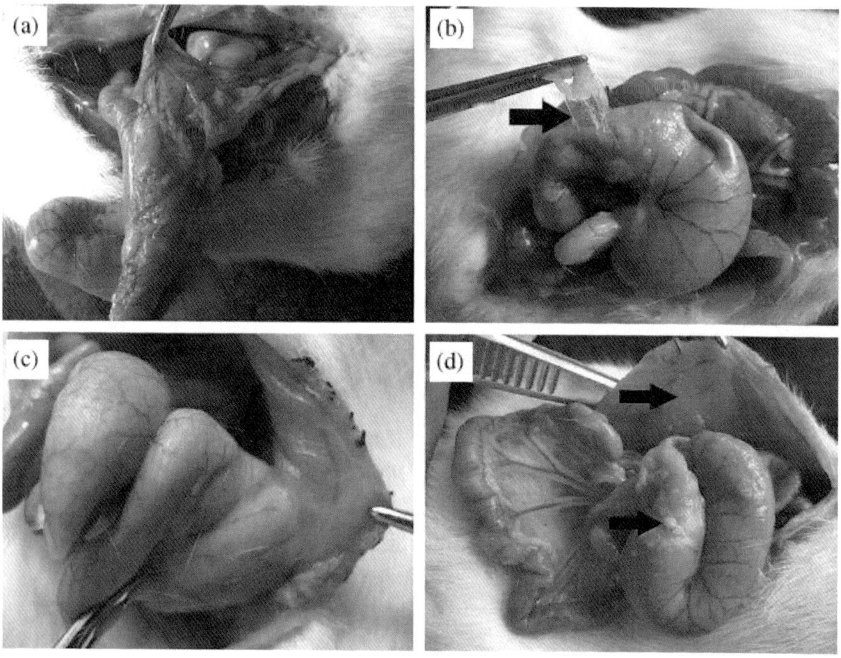


Figure 2.6 Prevention of abdominal adhesions in a rat abrasion model. (a) Adhesions seen between defected cecum and surrounding tissue after two weeks in untreated animals; (b) note the protective membrane over the defected cecum after 2 weeks in untreated animals (arrow); (c) severe adhesions were formed between cecum and abdominal wall after 4 weeks in untreated animals; (d) absence of adhesions after 4 weeks in untreated animals, note the abdominal wall defect (upper arrow) and the residue hydroxybutyl chitosan gels (lower arrow). Reprinted from [124] with permission from Elsevier.

intramolecular disulfide linkages. Rapid gelation occurs under physiologically relevant mild oxidative conditions in the presence of 0.3 wt% hydrogen peroxide. Authors demonstrated the suitability of this hydrogel for delivery of a model protein *in vitro* and intratumoral injectability and retention [125]. Low molecular weight glycosyl-nucleosyl-fluorinated amphiphiles form biodegrable thermosensitive hydrogel within 25 mins upon decreasing the temperature of the solution from 65°C to 37°C by self-assembly of monomers. This has been evaluated for bone tissue engineering [126]. A reverse-thermogelling L/DL-polyalanine block copolymer aqueous solution which undergoes sol-gel transition *in situ* at 37°C has been evaluated for chondrocyte three-dimensional culture [127]. Thermoresponsive MPEG-b-(PCL-ran-PLLA) (MCL) diblock copolymer that turns to gel at body temperature has been used as a biodegradable depot for the delivery of 5-fluorouracil (5FU). Single intratumoral injection of 5FU-loaded MCL was found to induce significant suppression of tumor growth through effective accumulation of 5FU in the tumor compared to repeated injection of the free drug [128]. It has been reported that a low concentration of sodium dodecyl sulfate can induce rapid gelation of silk fibroin (SF) by hydrophobic interactions among the SF chains [129]. Poly(amino acid)s are yet another class of polymers exhibiting thermo sensitive gelation. Solution of poly(*N*-substituted $\alpha/\beta$-asparagine) that forms gel above 35°C has been evaluated as a thermosensitive and an injectable scaffold for cartilage regeneration [130].

A variety of temperature sensitive hydrogels have been prepared and evaluated for various biomedical applications especially as controlled drug delivery and tissue engineering matrices. Out of these, widely exploited hydrogels are based on PNIPAAm, Pluronics, PEG/biodegradable polyesters and chitosan with their own merits and demerits. These temperature sensitive hydrogels have the advantage of ease of preparation and require no external crosslinking agents for gelation. Recent developments in this field could overcome some of the issues related to the use of these thermosensitive hydrogels such as difficulty in handling, nondegradability, toxicity and fast release of incorporated bioactive molecules and drugs.

#### 2.3.1.1.2 pH-induced hydrogel formation

Polymers having ionizable functional moieties can undergo a change in structure from coiled to extended in response to the pH of the surrounding medium. Chitosan, a weak polybase, is reported to have pH sensitivity, dissolve easily at low pH, and remain insoluble at higher pH values. At pH above 6.2, chitosan undergoes precipitation rather than formation of a hydrogel. Neutralization of amine groups present in the polymer at high pH leads to the removal of interchain electrostatic repulsive forces, permitting hydrogen bonding and hydrophobic interactions between chains. Proper balance between the electrostatic repulsive forces and hydrophobic interactions by adjusting pH can lead to rapid gelation. Attempts have been made to improve intermolecular interactions between chitosan molecules by incorporating a hydrophobic palmitoyl group to the free amino group to produce a comb-like polyelctrolyte, retaining their pH sensitivity. The modified system was shown to undergo rapid gelation within a narrow pH range (6.5–7), which is non-cytotoxic to 3T3 fibroblasts and were degradable [131].

pH-triggered *in situ* gelation was observed with a poly(acrylic acid) (Carbopol)/chitosan combination and has been evaluated as an ophthalmic drug delivery vehicle [132].

Injectable scaffolds based on cellulose were fabricated by the selective grafting of silanol groups to hydroxypropyl methyl cellulose (Si-HPMC). Rheological analysis of Si-HPMC clearly indicated that it is initially a viscous liquid, in which cells may be easily mixed and injected at the repair site. The Si-HPMC solution is alkaline in nature and upon injection forms a gel at physiological pH. Studies showed that the hydrogel is biocompatible and can permit the growth of chondrocytes, and support the synthesis of cartilage-like ECM [133]. Even though pH-triggered systems have been evaluated for biomedical applications like drug delivery and tissue engineering, their practical use is debatable as the pH at the time of injection is either lower or higher than the physiological pH, which can create unwanted tissue response.

2.3.1.1.3 Ion-induced hydrogel formation

Alginate hydrogels are the most extensively explored ion-triggered system for biomedical applications. Alginate is an unbranched binary copolymer of (1→4)-linked $\beta$-D-mannuronic acid and $\alpha$-L-guluronic acid, which readily forms bonds with a variety of divalent metal ions, such as calcium, magnesium, and barium. This binding has been used to crosslink alginates for a wide variety of biomedical applications [134, 135]. Alginate hydrogel containing vascular endothelial growth factor was evaluated as an injectable growth factor delivery vehicle for guided bone regeneration and demonstrated enhanced early angiogenesis and bone regeneration [136]. Alginate hydrogel incorporated with retinoids-loaded PLGA microparticles has been employed for the treatment of retinal blinding diseases. Increased electroretinography responses and improved retinal pigmented epithelium (RPE)-rod outer segment (ROS) interface were observed after subcutaneous injection of the drug-loaded delivery combination. Comparable therapeutic efficacy was obtained with just 2% of the systemic dose upon intravitreal injection [137].

Alginate hydrogels have also been used as a carrier for various types of cells for tissue engineering and other therapeutic applications [138, 139]. For example, they have been explored for the encapsulation and differentiation of dental-derived stem cells into adipogenic and osteogenic tissues for bone tissue engineering [140]. RGD-incorporated alginate hydrogel-carrying endothelial cells were used to support cellular proliferation, three-dimensional organization into cellular networks, and cellular migration towards the matrigel kept in contact with the hydrogel [141]. The grafting of MMP-labile peptides like proline-valine-glycine-leucine-isoleucine-glycine to cell-adhesive RGD-alginate hydrogels improved their performance as ECM analogs, giving a more dynamic and physiological three-dimensional cellular microenvironment [142]. Injectable bone regeneration composite (IBRC) comprised of nanohydroxy apatite/collagen (nHAC) particles in an alginate hydrogel carrier was used for the controlled release of recombinant human bone morphogenetic protein-2 (rhBMP-2). It was found that rhBMP-2 released continuously for 21 days promoted bone formation without altering the bioactivity [143]. Collagen/nanohydroxy apatite/alginate hydrogel-carrying nerve growth factor has demonstrated increased regeneration and new bone formation in a rabbit mandibular distraction osteogenesis model [144]. A hybrid hydrogel system composed of alginate particles and a fibrin hydrogel has been studied for soft tissue regeneration by taking advantage of the stability of alginate and cell adhesion, and the migration and tissue integration property of fibrin. Alginate particles could help in retaining 42.4% volume over a period of 12 weeks of otherwise completely absorbed cell-fibrin constructs. Moreover, alginate–fibrin hydrogel showed improved cellular mobility and proliferation *in vitro* and high cellularity, vascular ingrowth, and ECM production *in vivo* [145]. Ionically crosslinked biodegradable oxidized alginate has also been explored for preparing hydrogels for encapsulating adipose-derived stem cells for tissue engineering [146].

Gellan gum is another polysaccharide that can undergo ionotropic gelation. It is an extracellular linear anionic polysaccharide composed of repeating units of glucose, glucuronic acid, and rhamnose and its deacetylated form [147, 148]. As such, it can form thermoreversible gels, but the presence of cations is necessary for the formation of a stable hydrogel structure. It is soluble at high temperature (90°C) and upon temperature decrease the transition of a thermally reversible coil to a double-helix chain occurs, followed by self-assembling leading to the formation of oriented bundles. Cations help to link themselves resulting in the formation of a stable gel. Gellan gels have been explored as injectable drug delivery systems and tissue engineering scaffolds [149, 150]. Gellan gels encapsulated with nasal and articular chondrocytes and adipose-derived stem cells were shown to be useful for cartilage tissue engineering [151]. Gellan has been modified to introduce thiol groups on the back bone to make it suitable for physical and chemical *in situ* gelation. The thiolation-lowered the gelation temperature of gellan under physiological conditions to yield hydrogels with improved stability [152]. An injectable dual growth factor-loaded hydrogel system based on gellan/xanthan prepared by ionic and temperature-induced gelation has been shown to promote osteoblast differentiation [153]. Exploiting the ionotropic gelation of $\kappa$-carrageenan, injectable hydrogels were developed and studied for the delivery of human adipose stem cells (hASCs) for cartilage regeneration. These hydrogels were shown to support the viability, proliferation, and chondrogenic differentiation of hASCs over a period of 21 days [154].

Amphiphilic peptides like RAD16 can form *in situ*, nanoscale hydrogel when exposed to physiological media or salt solution due to the disruption of intramolecular/intermolecular electrostatic interactions. This has been explored as a drug delivery system to stabilize PTX in aqueous solution by decreasing its particle size. This RADA16-PTX hydrogel demonstrated growth inhibition of MDA-MB-435S cancer cells, *in vitro,* and inhibition increased with an increase in peptide concentration [155]. The hydrogel has also been used as three-dimensional cell culture matrix for mesenchymal stem cells [156], neural cells [157], venous endothelial cells [158], chondrocytes [159], and dental pulp stem cells [160].

A limiting factor of ionic crosslinking, particulary that using $Ca^{2+}$ ions, is the exchange of ions with those in the physiological medium, disturbing the stability of hydrogels.

#### 2.3.1.1.4 Multi-stimuli induced hydrogel formation

Multiple triggers have also been used to make *in situ* forming hydrogels. Oligo($\beta$-amino ester urethane) prepared by addition polymerization between the isocyanate groups of 1,6-diisocyanato hexamethylene and the hydroxyl groups of a synthesized monomer piperazine dihydroxyl amino in chloroform in the presence of dibutyltin dilaurate catalyst were shown to undergo gelation as a function of temperature and pH. Gelation occurred at physiological conditions and was found to be non-cytotoxic and biodegradable. Hydrogel remained within the body for two weeks upon subcutaneous injection, and sustained release of loaded doxorubicin was found for more than ten days [161]. Graphene oxide nanosheets self-assemble in the presence of low concentration of Pluronic solution by physical crosslinking near body temeperature. Sol-gel transition occurs in response to various stimuli such as temperature, near infrared light, and pH without imparting any toxic response. Further research must be conducted to understand the utility of this system for biomedical applications [162]. Copolymerization of NIPAAM with PAA with feed ratios of PAA between 0 and 20 mol % by a reversible addition-fragmentation chain transfer polymerization technique yielded a new sharply pH- and temperature-induced hydrogel system designed for delivering drugs to regions of local acidosis, found in wounds, tumor sites, or sites of ischemia. These hydrogels as such did not form gels at physiological pH and temperature, but by incorporation of a hydrophobic monomer raised the pH of gel formation to greater than 6.0 at 37°C. The hydrogel was able to maintain the bioactivity of basic fibroblast growth-factor for 40 h. The hydrogel also maintained a controlled pH-dependent release of growth factors [163].

Another pH- and temperature-sensitive triblock copolymer injectable hydrogel has been synthesized by conjugating poly(amidoamine) with poly(ethylene glycol) [164]. A pH- and temperature-sensitive hydrogel of poly($\beta$-amino ester)-poly($\varepsilon$-caprolactone)-poly(ethylene glycol)-poly ($\varepsilon$-caprolactone)-poly($\beta$-amino ester) pentablock copolymer loaded with insulin, where it forms an ionically linked insulin-PAE complex, has been evaluated as an injectable delivery vehicle. Subcutaneous injection of the insulin loaded polymer solution in normal rats maintained a constant insulin release profile for 15 days. Further studies on a streptozotocin-induced diabetic rat model showed that a single injection of the complex mixture containing 10 mg/mL insulin in a 30 wt% copolymer solution is sufficient for the treatment of diabetic rats for one week [165]. PCLA-PEG-PCLA copolymers with oligomeric sulfamethazine (OSM) end groups (OSM-PCLA-PEG-PCLA-OSM) were shown to have sol-gel transition behavior upon both temperature and pH changes under physiological conditions. Sol-gel transition was found to be tunable by controlling the PEG length, the hydrophobic to hydrophilic block ratio (PCLA/PEG), and the molecular weight of the OSM. The authors envisage that this can be employed as an injectable carrier for protein-based drugs using a long guide catheter into the body [166]. This hydrogel demonstrated easy encapsulation of hMSCs and rhBMP-2 with encapsulation efficiencies of 90 and 80%, respectively. Upon subcutaneous injection of the polymer solution containing hMSCs and rhBMP-2, osteogenic differention with mineralized tissue formation and high alkaline phosphatase activity was observed, demonstrating its suitability as an injectable scaffold for bone tissue engineering [167].

This block copolymer has also been evaluated as a carrier of PTX for the treatment of cancer, and demonstrated good antitumor effect and induced good cellular apoptosis in tumor tissues for two weeks [168].

2.3.1.1.5 Miscellaneous reactions

Injectable biological hydrogel has also been prepared by the self-assembling of 4-arm PEG having thymine and adenine base pairs via Watson–Crick base pairing as tissue engineering scaffolds [169]. Biotin–streptavidin binding was also exploited to make injectable hydrogel using biotin-terminated, star-shaped poly(ethylene glycol) (PEG-Biotin) and streptavidin-functionalized hyaluronic acid (HA-Streptavidin). These nanofibrous hydrogel scaffolds incorporated with nanoparticles and loaded with growth factors have been studied for soft tissue engineering [170].

### 2.3.1.2 *Crosslinking by chemical interactions*

Hydrogels prepared by chemical reaction leading to *in situ* gelation and their biomedical applications are described in this section. Chemical interactions provide precise control over mechanical stability and degradability of hydrogels. Further more, the reaction can be controlled by careful selection of the concentration of the polymer solution and crosslinker to obtain the desired gelling time. The section is organized based on the chemical reactions used for the preparation of hydrogels which include Schiff base reaction, Michael-type addition, photocrosslinking, enzymatic crosslinking, and fibrin-based hydrogels.

2.3.1.2.1 Hydrogel formation by Schiff base reaction

The principle behind this strategy is modification of the polymers of interest in such a way that they can undergo Schiff base reaction between amino groups and aldehyde groups attached to the polymer chains, leading to self-crosslinking to give *in situ* gelling systems. The advantage of such systems is complete avoidance of extraneous toxic crosslinking agents and other triggers which can cause unwanted tissue response. The suitability of self-crosslinking, *in situ* forming biodegradable hydrogels based on oxidized alginate and gelatin as an injectable drug delivery vehicle and as a tissue engineering scaffold has been demonstrated. Aldehyde groups of periodate oxidized alginate and $\varepsilon$-amino groups of gelatin undergo Schiff base reaction in the presence of borax to form *in situ* formed hydrogels [171]. The

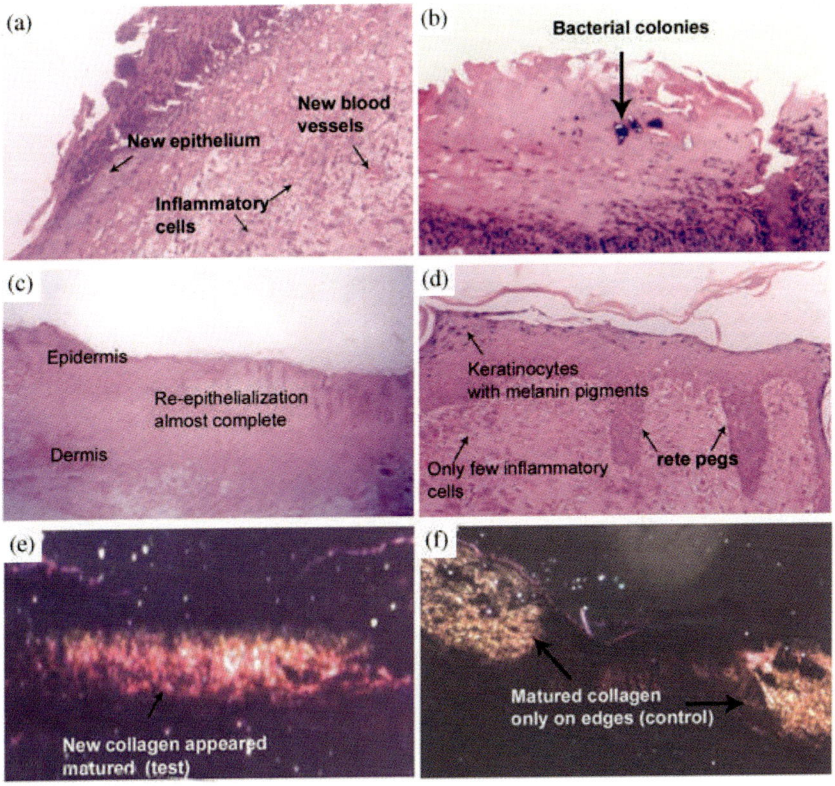


Figure 2.7 Histology of wound sections stained with hematoxylin and eosin. Epithelialization at test wound edges at 5 days (a, 150×); bacterial colonies present in control wounds (b, 300×); neat test wound section at 10 days (c, 60×); test wound with rete pegs at 10 days (d, 150×); test wound (e, 15×) and control wound (f, 15×) under polarized light at 15 days. Reprinted from [172] with permission from Elsevier.

hydrogel was found to maintain a moist environment over a wound surface and enhanced the migration of epithelial cells, resulting in complete filling of the wound within two weeks (Figure 2.7) [172].

Another *in situ*, injectable nanocomposite hydrogel composed of curcumin-loaded nanoparticles, *N*,*O*-carboxymethyl chitosan, and oxidized alginate has been evaluated for use in dermal wound repair [173]. Biocompatible, biodegradable triple interpenetrating network hydrogel was also prepared using partially oxidized dextran (Odex), teleostean, and *N*-carboxyethyl chitosan (CEC) by Schiff's base reaction between aldehyde groups of Odex and amino groups of teleostean and CEC without employing any extraneous crosslinking agents. While triple networking retarded fibroblast-mediated degradation, teleostean enhanced cellular affinity [174]. Biodegradable hydrogel formed by Schiff base reaction between *N*-succinyl-chitosan and aldehyde hyaluronic acid with covalently conjugated glucose oxidase and catalase has been evaluated to deliver insulin *in vitro* for adipose tissue engineering. Enzymes immobilized in the matrix triggered the conversion of glucose, and the resulting pH reduction enhanced hydrolysis and swelling of the network. The release of insulin was found to be dependent on hydrogel swelling [175]. The aldehyde–hydrazide Schiff base reaction has also been employed to make injectable hydrogels based on poly(*N*-isopropylacrylamide) and polysaccharides (hyaluronic acid, carboxymethyl cellulose, dextran, and methylcellulose) by functionalizing the polymers with aldehyde and hydrazide groups. Swelling, degradation, phase transition, drug binding, and mechanical properties of the resultant hydrogels were found to be varied by the number and ratios of polymer prescursors [176]. Injectable hydrogels have also been prepared by crosslinking gelatin-conjugated hyaluronic acid with adipic hydrazide for NP tissue engineering. Advantages include sterilizability, low viscosity for injection, and ease of use. The complex shear modulus of these hydrogels was similar to native tissue and the cultured nucleus pulposus cells in the hydrogels were viable, proliferating, and maintained genetic profile [177]. Poly(vinyl alcohol) modified with aldehyde and hydrazide functional groups undergoes a click reaction under physiological conditions, forming hydrogel within 3 mins. These gels were slow degrading and were cell compatible [178]. Injectable hydrogels were also prepared by using combination of HA/CMC/dextran modified with aldehyde and hydrazide functionalities for the controlled release of proteins like albumin and interleukin-2. The duration of release was found to be extended by the incorporation of gelatin [179]. Dextrin, a glucose polymer with low molecular weight has also been employed to make an injectable carrier of bioactive molecules like interleukin-10 and insulin, by employing periodate oxidation followed by crosslinking with adipic hydrazide [180]. Chondrocytes and mesenchymal cells cultured within *in situ* crosslinkable injectable hyaluronan hydrogels based on aldehyde-modified hyaluronan and hydrazide-modified polyvinyl alcohol were shown to form cartilage-like tissue, rich in glycosaminoglycans, collagen type II, and aggrecan. Hydrogel also demonstrated improved healing upon *in vivo* injection on full-thickness cartilage defect in a rabbit model [181].

One of the limitations of using injectable hydrogels prepared by Schiff's reaction is its pH sensitivity. Schiff's base (imine) linkages are susceptible to hydrolysis under acidic conditions and therefore these hydrogels could not find useful for biomedical applications where stability of hydrogel is crucial even in diseased conditions, where pH is slightly acidic.

#### 2.3.1.2.2 Hydrogel formation by photocrosslinking

The underlying strategy involves introduction of polymerizable vinyl groups onto a polymer using reactive agents (e.g., methacrylol chloride, glycidyl methacrylate, methacrylic anhydride). These vinyl substituted polymers can then undergo polymerization when exposed to UV/visible light in the presence of biocompatible initiators like camphorquinone or Irgacure 2959 (I2959). Copolymerization of a polymer with acrylic monomers in a similar way also results in robust and stable *in situ* gelling systems. The concentration and type of the photoinitiators should be controlled in such a way that the photoirradiation will not produce any unwanted reactions such as tissue necrosis due to heat generation and cell toxicity. *In situ* forming hydrogels based on photocrosslinkable methacrylate derivatives of biopolymers and their copolymers for cartilage tissue engineering has been reviewed elsewhere [9].

Photocrosslinking of thiol-functionalized hyaluronic acid and thiol-functionalized gelatin by PEGDA has been exploited to make *in situ* gelling systems for enhanced remyelination [182]. Stimulation of MSC differentiation into cardiac cells using hydrogels with appropriate matrix modulus has been achieved using a thermosensitive, injectable hydrogel prepared by photocrosslinking of a mixture of *N*-isopropylacrylamide, *N*-acryloxysuccinimide, AAc, and poly(trimethylene carbonate)-hydroxyethyl methacrylate of varying moduli. The hydrogel with 65 kPa modulus maintained the cells with the highest differentiation efficiency that developed calcium channels for electrophysiological property and gap junctions for cell–cell communication. This approach can be envisaged as having higher efficacy than other differentiation approaches such as chemical induction and co-culture, thereby holds great promise for delivering MSCs for the generation of new heart tissue [183]. Mice treated with a photocrosslinked hydrogel of PEGylated denatured fibrinogen fragments with PEG-diacrylates showed enhanced survival and differentiation of encapsulated mesoangioblasts (Figure 2.8), leading to improved cell graftment compared with the same cell population injected using aqueous saline solution. This hydrogel is currently undergoing clinical trials [184]. An

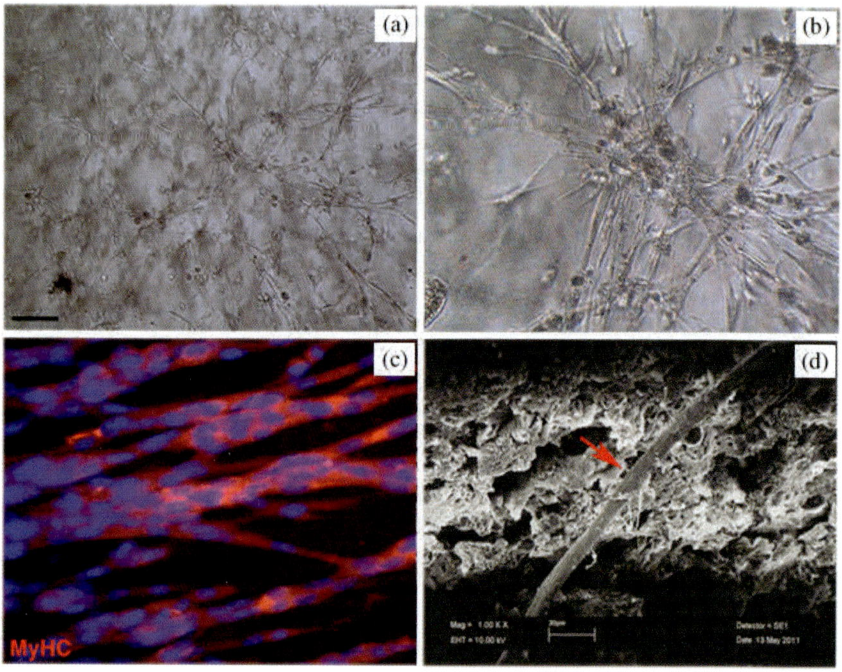


Figure 2.8 Mesoangioblasts cultured in polyethylene glycol-fibrinogen (PF) hydrogels. (a, b) Phase-contrast images of mesoangioblasts in 8 mg/mL PF hydrogel, giving rise to a robust three-dimensional myofiber network. (c) Immunofluorescence showing multinucleated muscle fibers; staining is with an antibody against myosin heavy chain (MyHC; red) and nuclei counterstaining with 4,6-diamidino-2-phenylindole (DAPI; blue). (d) Scanning electron microscopy image revealing the presence of differentiating skeletal-muscle fibers (red arrows) within the PF hydrogel. Scale bar: (a) 200 $\mu$m, (b) 50 $\mu$m and (c) 10 $\mu$m. Reprinted from [184] with permission from Biomed Central.

injectable composite hydrogel prepared by photocrosslinking of a mixture of methacrylated glycol chitosan and hyaluronate has been found to maintain 90% chondrocyte viability over a period of 20 days. Incorporated HA improved the chondrocyte proliferation and glycosaminoglycan deposition [185]. Nanocomposite hydrogels made from poly(ethylene glycol) diacrylate and silicate nanoparticles (Laponite RD) via photopolymerization were transparent with interconnected pores having high elongation and toughness. Addition of silicate enhanced the bioactivity and adhesiveness of the hydrogel and it was found that MC3T3-E1 mouse preosteoblast cells readily adhere and spread on nanocomposite hydrogel surfaces [186].

By means of photo cross linking, hydrogels having different chemistries can be easily prepared by mixing selected derivatized polymers either alone or by copolymerization. Further, it provides better spatial and temporal control over gelation. However, photo cross linking often requires a photoinitiator, which may have toxic effects and prolonged irradiation time, leading to a local temperature increase that can subsequently harm neighboring cells and tissues, limiting their wide use.

#### 2.3.1.2.3 Hydrogel formation by Michael-type addition polymerization

Michael-type addition is yet another reaction used to make *in situ* gelling systems. The strategy involves modification of appropriate polymers by incorporating nucleophilic groups like thiol or amine on one polymer and electrophilic groups like acrylate or maleimide on another so that the polymers can enter into a Michael-type addition reaction leading to crosslinking to give *in situ* gelling formulations. Employing a Michael-type addition reaction between vinyl groups of acrylate functionalized tetrahydroxy PEG and thiol groups of peptides, injectable biodegradable PEG/arginine-glycine-aspartic (RGD) peptide hybrid hydrogels were prepared and used as a biomimetic scaffold for encapsulation of human mesenchymal stem cells for cartilage regeneration where chondrogenesis was found to be RGD dose-dependent [187]. *In situ* forming hydrogel prepared using dextran conjugate with vinyl sulfone and PEG-thiols has been shown to have tunable mechanical and gelling properties by varying the degree of substitution of dextran, its molecular weight, and concentration [188]. Thiol-modified HA and elastin-like polypeptide composite hydrogel crosslinked with poly(ethylene glycol)acrylate has showed promise for NP tissue engineering in early degenerative disc disease. Composite hydrogels showed less stiffness than natural NP, but showed time-dependent loading characteristics, which is a favorable property when injected into intervertebral discs [189]. A hyperbranched, thermoresponsive PEG-acrylate copolymer provided an injectable and *in situ* crosslinkable system via Michael-type thiol-ene reaction with a thiol-modified hyaluronan biopolymer. Rapid gelation occured at body temperature and it was found to maintain cell viability inside the hydrogel [190]. Two component PEG-based systems having complementary reactive end groups of thiol and active vinyl groups, respectively, upon mixing form a chemically crosslinked hydrogel *in situ* and have been explored as a long term vitreous substitute material [191] and for sustained release of methylprednisolone sodium succinate [192]. Injectable hydrogels prepared by *in situ* crosslinking of chitosan acrylate and PEG-thiol have been employed for the delivery of recombinant human bone morphogenetic protein-2 for bone regeneration. Temperature and pH have been found to manipulate the swelling and gelation mechanisms of these hydrogels [193].

#### 2.3.1.2.4 Hydrogel formation by enzyme-mediated crosslinking

Enzymatic crosslinking is another attractive method for the preparation of injectable hydrogels mainly because of its high specificity and mild reaction conditions. Peroxidases (particulary horseradish peroxidase (HRP)) catalyze crosslinking through oxidative polymerization of phenol derivatives in the presence of $H_2O_2$ [194]. Enzyme-catalyzed *in situ* forming hydrogels and their tissue applications have been reviewed elsewhere [195]. Injectable, biodegradable hydrogels prepared by enzymatic crosslinking of HA–tyramine conjugates showed tunable mesenchymal stem cell chondrogenesis by varying the crosslinking density of the hydrogels. Enhanced cellular condensation, chondrogenesis, and biosynthetic rates of glycosaminoglycan and type II collagen was found in softer hydrogels whereas in highly crosslinked matrices, cells exhibited elongated morphology with a reduced degree of cellular condensation [196]. The cell-attracting and self-attaching properties of *in situ* forming dextran–tyramine conjugate hydrogels have been envisaged to favor tissue repair in focal cartilage defects and help in preventing or delaying the onset of osteoarthritis [197]. Platelet lysate supplemention has been found to provide favorable biological signals for these injectable hydrogels for cartilage tissue engineering [198].

Injectable hydrogel prepared by the enzymatic crosslinking of gelatin-hydroxyphenylpropionic acid conjugate in the presence of hydrogen peroxide with tunable stiffness was evaluated for controlling the proliferation rate of human fibroblasts in both two-dimensional and three-dimensional culture environments for potential use as a wound dressing material. In three-dimensional culture the proliferation rate was found to increase with a decrease in stiffness of the hydrogel, unlike in a two-dimensional environment [199]. Hybrid hydrogels prepared by using a gelatin–hydroxyphenyl propionic acid conjugate and 4-arm-poly(propylene oxide)–poly(ethylene oxide) tyramine conjugate in the presence of HRP and hydrogen peroxide also showed cyto-compatibility with human dermal fibroblast. Detailed investigation must be carried out in order to understand the potential of this matrix in tissue engineering [200]. Gelatin-hydroxyphenyl propionic acid conjugated with tyramine also found to undergo enzyme-mediated crosslinking to provide a biodegradable, injectable scaffold for bone tissue engineering [201]. The same hydrogel without tyramine has also been evaluated for its effect on the proliferation, migration, differentiation, and oxidative stress resistance of adult neural stem cells (aNSCs). The matrix showed tunable proliferation and migration of aNSCs in relation to the crosslinking degree [202]. Another hydrogel based on gelatin–poly(ethylene glycol)–tyramine conjugate has been shown to act as an efficient simvastatin delivery vehicle that can trigger osteogenic differentiation [203]. Enzymatically crosslinked hyaluronic acid tyramine conjugates in the presence of $H_2O_2$ and horseradish peroxidase enzyme has been studied for the delivery of interferon-$\alpha$2a (IFN-$\alpha$2a) for liver cancer therapy. Hydrogels with low stiffness were found to maintain the activity of IFN-$\alpha$2a, inhibit the proliferation of liver cancer cells and induced apoptosis by caspase-3/7 pathway. It was found that IFN-$\alpha$2a-incorporated HA-Tyr hydrogels effectively inhibited tumor growth and angiogenesis, while IFN-$\alpha$2a injection did not demonstrate antitumor efficacy [204]. Conjugation of SVVYGLR peptide into injectable gelatin–PEG tyramine hydrogels enhanced the activity of human umbilical vein endothelial cells (HUVEC) *in vitro* and stimulated the formation of new blood vessels in the hydrogel matrices *in vivo* [205]. An enzymatically crosslinked carboxymethyl cellulose-bearing phenolic hydroxyl group showed 92.8% viability, good proliferation, and adipogenic differentiation of adipose derived stem cells *in vitro* and promoted vascularization *in vivo* [206]. Tyramine-conjugated chitosan hydrogels loaded with rutin showed improved wound healing compared PBS treatment, bare hydrogel, and a commercialized wound dressing (Duoderm) [207]. Tyramine-conjugated Pluronic F127 triblock copolymers crosslinked in the presence of tyrosinase showed far lower critical gelation concentration, enhanced gel strength compared to unmodified Pluronic copolymer hydrogels, and tissue adhesive properties suitable for sustained delivery of bioactive agents and cells [208].

Enzymatically crosslinked hydrogels are mainly evaluated as tissue engineering matrix. The concentration of hydrogen peroxide, a reactive oxygen species generator used for enzymatic crosslinking has to be kept minimal to make it useful for various other biomedical applications.

#### 2.3.1.2.5 Fibrin-based hydrogels

Fibrin, the structural component of blood clots, has been utilized for different biomedical applications as it fulfills specific physical and biological requirements. Formation of fibrin involves several steps: initially, the cleavage of small peptides such as fibrinopeptides A and B present in fibrinogen molecule by the serine protease thrombin. The cleaved fibrinogen, which is called 'fibrin monomer', has certain binding sites which then interact with the complementary binding sites of other fibrin monomers, forming oligomers with half-staggered overlapping and further aggregate laterally to form fibers and a three-dimensional network [209, 210]. The main disadvantage of fibrin is its instability and solubility over time *in vitro* and *in vivo*, due to fibrinolysis. However, by varying the fibrinogen concentration, thrombin concentration, and ionic strength, one can generate gels with different appearance, mechanical properties, and stability [211, 212]. The fast gelation of fibrin makes it unsuitable for cellular delivery as it causes mechanical stress on encapsulated cells. The addition of salt like sodium chloride can delay the gelation time and improve physical properties [213, 214]. These gels demonstrated immense potential for tissue regeneration [215–217] and delivery of growth factors [218–220] and drugs [221].

### 2.3.1.3 *Shear-thinning hydrogels as injectable biomaterials*

Hydrogels having shear-thinning behavior enables their delivery *in vivo* by application of shear stress generally by passing through a syringe. The strategy involves the formation of a solid hydrogel with the desired mechanical, morphological, and biological properties *ex vivo* that can flow under stress due to its shear-thinning property like a low viscous material. The interactions involved in shear-thinning hydrogels are mostly physical. Since the crosslinking process leading to the formation of hydrogel is done *ex vivo*, the influence of environment is negligible. One of the important criteria that a shear-thinning hydrogel should meet in order to be used as an injectable hydrogel for biomedical applications is the fast recovery of elastic modulus (self-healing) after injection. Protein- and peptide-based hydrogels, hydrogels from blends, colloidal systems, hydrogels based on CDs and block copolymers, and decellularized extracellular matrix (ECM)-based hydrogels generally show shear-thinning behavior [222]. These gels are mainly evaluated as a vehicle for cell and drug encapsulation and delivery.

#### 2.3.1.3.1 Protein-based hydrogels

In nature there are various examples of amino acid sequences called 'protein domains' that can fold into three-dimensional structures with distinct functions. Formation of physical hydrogels by protein–protein interactions between specific peptide domains is a well-accepted concept. As the protein domain is the controlling factor in determining the material properties of these hydrogels, researchers are trying to engineer various protein or specific peptide sequences, thereby achieving exact control over the identity and sequence of each residue and, consequently, the structural folding patterns of the resultant protein or peptide.

Recombinant proteins having specific peptide sequences are highly flexible building blocks to make shear-thinning hydrogels. Recombinant proteins synthesized by cell expression studies are preferred over peptides synthesized by solid phase chemistry mainly because of their large size and advanced biological activity. Exploiting the protein–protein interactions, a two-component system having separate peptide domains that can associate upon mixing at physiological conditions has been designed (Figure 2.9). Peptide domains like WW (30–50 aminoacid long triple-stranded antiparallel $\beta$-sheet structures having two tryptophan residues denoted as W) domain and proline-rich peptide were selected based on the norms that

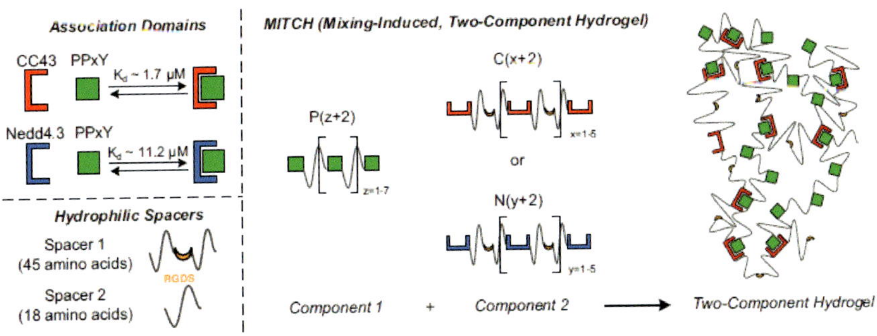


Figure 2.9 Schematic of the mixing-induced, two-component hydrogel. (Top left) Modular association domains assemble via molecular recognition. Two WW domains (CC43 and a Nedd 4.3 variant) bind the same proline peptide (PPxY). (Bottom left) Hydrophilic spacers link multiple repeats of WW domains (spacer 1) or proline peptides (spacer 2). Spacer 1 contains a cell adhesion peptide RGDS. (Right) Three engineered protein families: C(x + 2), N(y + 2), and P(z + 2). Mixing component 1 with component 2 (either C(x + 2) or N(y + 2)) at constant physiological conditions results in hydrogel formation. Reprinted from [223] with permission from the National Academy of Sciences of the United States of America.

their amino acid sequences are specific and tunable, short to allow multiple domains to be repeated in single polymer domains, and found naturally intracellularly so as not to evoke any unwanted responses. Binding of intracellular WW domains and proline-rich motifs by hydrophobic interactions are responsible for the regulation of metabolic and gene functions. These recombinant proteins can self-assemble at constant physiological conditions to form a hydrogel showing shear-thinning properties. A variety of proline rich peptides are available and by their careful selection, the self-healing property of the resulting hydrogels can be modulated. This mixing-induced two-component hydrogel was demonstrated to be effective in successful encapsulation of endothelial and neural stem cells and supported cellular proliferation, differentiation, and three-dimensional cell spreading [223]. Though this method makes use of mild conditions required for encapsulation of cells, preserves the therapeutic efficacy of protein domains and avoids non-specific interactions with ECM components, limiting factors are poor protein yield, extensive purification steps, and risk of endotoxin contamination.

By exploiting a dual-component Dock-and-Lock (DnL) self-assembly mechanism, injectable shear-thinning and self-healing hydrogels have been prepared. Two components involved are: (a) docking domain: RII$\alpha$ subunit of cAMP-dependent kinase-A engineered as a telechelic protein bearing dimerizable end groups; and, (b) locking domain: anchoring domain of A-kinase anchoring protein attached to multi-arm crosslinker polymers. Upon mixing, both components form robust physical hydrogels instaneously at physiological temperature and pH. By changing the peptide sequence, the concentration and ratio of each component and the number of peptides on the crosslinking polymer, mechanical properties and degradation rates could be varied. The self-healing ability of these gels was demonstrated by examining their resistance to yield at strains as high as 400%, with 100% recovery being observed. These molecularly engineered gels maintained the viability of the encapsulated mesenchymal stem cells, envisaging great potential in tissue engineering applications [224].

Hydrogel formation can also be achieved by coiled-coil formation, one of the major structural motifs for proteins which normally involves a seven-residue heptad periodic unit (abcdefg), where positions a and d are hydrophobic residues responsible for interhelical hydrophobic interaction and e and g are charged (Figure 2.9). Recombinant triblock proteins having two leucin-zippers attached to both ends of a soluble polyelectrolyte domain have been demonstrated to form shear-thinning hydrogels [225]. The leucine-zippers were demonstrated to facilitate the formation of trimeric and tetrameric aggregate bundles at physiological temperature and pH, where $\alpha$-helical structure and bundle formation can be easily disrupted by going to very low or high pH, and gelation occurs by changing pH to physiological pH. These gels exhibited cytocompatibility and promoted proliferation of human fibroblasts, HUVEC, and rat neural stem cells with a polarized morphology due to mature actin fibers in RGD modified hydrogels and round morphology in unmodified hydrogels [226, 227]. The advantages of these systems include easy tuning of physical properties, recovery times, and bioactivity by modulating the sequence of leucine-zipper and polyelectrolyte spacer, modifying the protein surface charge and strength of protein associations by changing pH, salt, temperature, and biofunctionalization by incorporating proteins due to specific functions between leucine-zipper blocks [228, 229]. Recombinant silk-elastin-like protein-polymer analogs, SELP-47K and SELP-415K, composed of different lengths and ratios of silk and elastin units that can form injectable hydrogels have been exploited for matrix-mediated controlled adenoviral gene delivery [230, 231]. Human mesenchymal stem cells encapsulated in SELP-47 K have been found to acquire a rounded morphology, embedded in significant amounts of chondrogenic extracellular matrix. The addition of TGF-$\beta$3 to the hydrogels showed upregulation in aggrecan, type II and type X collagen, and SOX9 expression by day 28 [232].

#### 2.3.1.3.2 Peptide-based hydrogels

Peptide-based hydrogels having $\beta$-hairpin (two $\beta$ strands joined by a $\beta$ turn) structures are another class of hydrogels that can be made injectable due to their shear-thinning behavior. These hydrogels have been used for the delivery of cells and therapeutic agents to the site of action via syringe or catheter. It has been shown that the width of the plug flow region has an effect on hydrogel rigidity, but the gel material and cells experienced no shear rate in the central-wide plug flow region while experiencing shear deformation in a narrow shear zone close to the capillary wall [233]. The duration of shear and the strain rate are determining factors of the stiffness of hydrogel immediately recovered after flow and after complete rehealing. The results of structural analysis during flow indicated the presence of large (>200 nm) hydrogel domains [234]. MAX1 and MAX8 are synthetic $\beta$-hairpin peptides that can self-assemble in the presence of a physiologically relevant concentration of salt to form a physically crosslinked hydrogel

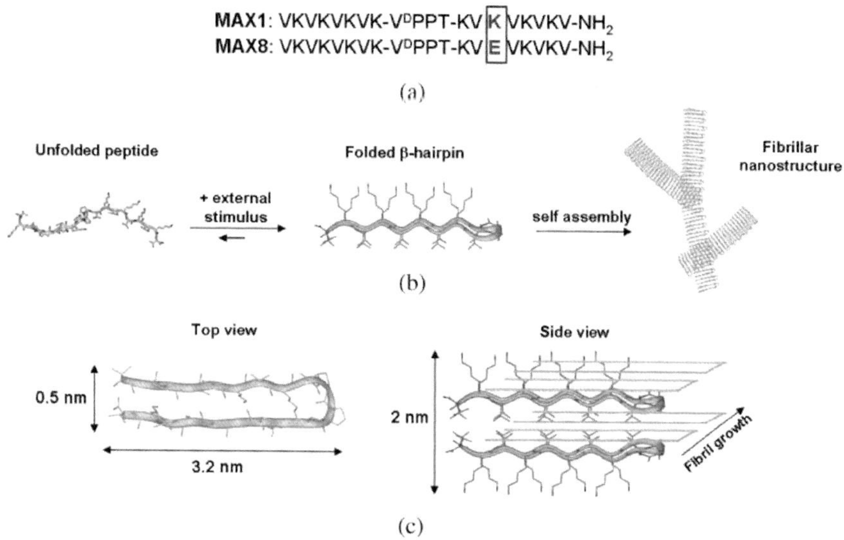


Figure 2.10 (a) Peptide primary sequence for MAX1 and MAX8. (b) Proposed mechanism for the folding and self-assembly of the $\beta$-hairpin peptides. (c) Length scales of the hairpin structure and self-assembled hairpin bilayer. Reprinted from [235] with permission from the American Chemical Society.

network of nanostructured fibrils (Figure 2.10). These peptides contain two blocks of alternating hydrophobic and basic amino acids linked together by a Val$^D$-Pro-Pro-Tyr type II $\beta$-turn forming spacer, while MAX1 peptide contained a (Val-Lys)$_4$-Val$^D$-Pro-Pro-Tyr-(Val-Lys)$_4$ sequence. The MAX8 sequence is similar to that of MAX1 with the exception of Glu replacing Lys at position 15, which reduces net charge on MAX8, resulting in faster folding and self-assembly kinetics and higher modulus than that of MAX1 [235]. Self-assembling MAX8 peptide hydrogel has been employed as an injectable hydrogel for the delivery of proteins having different hydrodynamic diameter and charge. Gelation was found to be triggered by the presence of proteins without affecting the structure of encapsulated proteins. Sheared and non-sheared gels showed similar release profiles to proteins where steric factors governed the release of positive and neutral proteins. Electrostatic interaction with a positively charged peptide fibrillar network restricted the release of negatively charged proteins to 10% of initial payload [236]. Self-assembling $\beta$-hairpin peptide hydrogel has been found to be an effective vehicle for the localized delivery of curcumin over sustained periods of time without altering its bioactivity. The rate of curcumin release and its resultant therapeutic efficacy has also been demonstrated to be easily modulated as a function of the concentration of the MAX8 peptide [237].

The self-assembling of ac-(RADA)(4)-CONH(2) and ac-(KLDL)(3)-CONH(2) peptides that form biocompatible, nanofiber scaffold hydrogels in the presence of salts or biological fluids have been evaluated for long term sustained release of human antibodies without effecting their secondary and tertiary structures and antigen binding efficacy. The hydrogel formed has a two-layered structure with core ac-(RADA)(4)-CONH(2) and shell ac-(KLDL)(3)-CONH(2) [238]. The shear-thinning behavior and rapid rehealing properties of the pre-assembled solutions of BMHP-1 self-assembling peptides make them a preferable choice for injectable scaffolds [239]. Self-assembled multidomain peptides (MDPs) have also been exploited to construct shear-thinning and self-healing hydrogels. These are amphiphilc peptides, consisting of an ABA block motif in which B is a series of amphiphilic (Gln/Ser)-Leu repeats, while terminal A blocks are charged, either positively charged Lys or negatively charged Glu. The hydrophilic (Gln/Ser) residues organize together on one face of the B block and Leu residues stay on the opposite face. The hydrophobic amino acid residues on this amphiphilic block pack together into a $\beta$-sheet structure and fibril-forming peptide sandwiches. Fusing of the charged A blocks to both sides of the amphiphilic B block retards it from sheet formation. However, by reducing the charge either by protonating via pH titration or charge shielding by increasing salt concentration, MDPs fold into $\beta$-sheet fibers of 2 nm height, 6 nm diameter, and 120 nm length. The self-assembled fiber length and viscoelastic properties of the formed hydrogels are controlled by the careful selection of amino acids in the A and B blocks. If A block consists of negatively charged poly-Glu, gelation can be achieved by divalent ions like $Mg^{2+}$, resulting in the formation of hydrogels having storage modulus of 480 Pa and demonstrating 75 and 100% modulus recovery within 13 s and 10 min, respectively, after shear thinning, making it ideal for use as an injectable hydrogel.

Apart from their potential use as a drug cell delivery vehicle [240], MDPs have also been used as sponges to absorb stem-cell secretome when cultured with human H9 embryonic stem cells separated with a protein permeable membrane [241]. Renal microvascular endothelial cells exhibited a low level of apoptosis and hyper permeability in the presence of such preconditioned MDP gels. Furthermore, acute kidney damage induced by lipopolysaccharide treatment was found to be reduced by the intraperitoneal injection of conditioned MDP gels [242]. This strategy has also been employed for cardiac repair. While incorporation of disulfide bridge-forming cysteines in the B block (e.g., $Glu_2(Cys\text{-}Leu\text{-}Ser\text{-}Leu)_3Glu_2$) could increase the storage modulus, incorporation of matrix metalloproteinase-2 cleavable site could make the matrix susceptible to enzymatic degradation [243].

#### 2.3.1.3.3 Hydrogels from blends

A shear-thinning hydrogel blend of hyaluronan and methylcellulose (HAMC) has been employed as an injectable, safe, and localized delivery vehicle for recombinant human epidermal growth factor to the spinal cord [244]. A blend of SF with hydroxyl propyl cellulose (HPC) was also found to have thixotropic properties. The phase separation of the blends prompted the conformational switch of SF from random coil to $\beta$-sheet, leading to gel network formation via $\beta$-sheet crosslinks and the immobilization of the molecular chain in the dispersed phase. This blend hydrogel was found to protect encapsulated cells against high shear force during injection, pointing towards the potential of SF–HPC hydrogel as a promising vehicle for cell delivery [245].

#### 2.3.1.3.4 Amphiphilic polymeric derivatives

Hydrogels with shear-thinning behavior have been prepared with segmented hydrophobic bisurea hard segments and hydrophilic PEG segments. Multiple hydrogen bonds within the hydrophobic segments make them highly stable gels in water at low concentrations. Moduli were found to be tunable via change in the physical crosslink density and the length of the hydrophilic segments so as to make hydrogels that can match the properties of biological host tissues such as muscle tissue and the extracellular matrix [246]. Hydrogels composed of heparinized 4-arm star PEG-based polymers and star PEGs functionalized with heparin-binding peptides are shear thinning and capable of binding and control the release of multiple growth factors without compromising their bioactivity [247]. Injectable hybrid hydrogels prepared using Poloxamer-407 (PO) and octapeptide with an amino acid sequence of KFEFKFEF (PE) utilizing PO's high and tunable mechanical strength and integrity and PE's superior bioactivity, were used to deliver human adipose-derived mesenchymal stem cells (hASCs) for adipose tissue regeneration. It was found that the injected hASCs migrated to the skin and differentiated into epithelial cells *in vivo* [248].

#### 2.3.1.3.5 Cyclodextrin-based hydrogels

Another category of shear-thinning hydrogels is prepared by the self-assembly of inclusion complex formed between cyclodextrin and biodegradable block copolymers. Cyclodextrins (CDs) are natural water-soluble oligosaccharides of D-glucose linked by $\alpha$-1,4 linkages and have been been classified as $\alpha$-, $\beta$-, $\gamma$-CD having a hydrophobic inner cavity and hydrophilic outer surface. Linear polymers having appropriate size can easily form inclusion complexes with CDs wherein the polymer chains pass through CDs and are stabilized by CD/polymer hydrophobic interactions and hydrogen bonding between CDs. Aggregation of these complexes leads to gelation. These hydrogels can also be classified as 'slide-ring gels'. Block copolymers composed of hydrophobic and hydrophilic blocks are kinetically more favorable for making inclusion complexes. Hydrogel prepared by the self-assembly of $\alpha$-CD and a biodegradable poly(ethylene oxide)-poly[(R)-3-hydroxybutyraten-poly(ethylene oxide) triblock copolymer (Figure 2.11) was found to be thixotropic, reversible, and a promising injectable drug delivery system [249]. Injectable, thixotropic cyclodextrin-based polypseudorotaxane hydrogels incorporated with supramolecularly anchored active cationic copolymer/plasmid DNA (pDNA) polyplexes were studied for sustained gene delivery [250]. Hydrogel formed by the interaction of $\beta$-cyclodextrin and a polymeric prodrug PEGylated indomethacin was found to show shear-thinning thixotropic behavior, indicating suitability of the system as an injectable drug delivery vehicle [251].

#### 2.3.1.3.6 Decellularized extracellular matrix-based hydrogels

The decellularized ECM has recently shown much advancement as scaffolds in the field of regenerative medicine as they closely mimic the physical and chemical cues that are experienced by cells *in vivo* [252]. Decellularized ECMs can be processed into injectable formulations by exploiting the self-assembling nature of these materials (which exist in liquid form at low temperature (4°C)) into porous gels upon injection at 37°C forming a complex network of fibers with nanoscale structure. Self-assembling decellularized ECM hydrogel derived from porcine myocardial tissue has been thus developed as a tissue-specific injectable scaffold for cardiac tissue engineering. As collagen is the main component of ECM, various similarities were found between decellularized ECM hydrogel and collagen gels in terms of

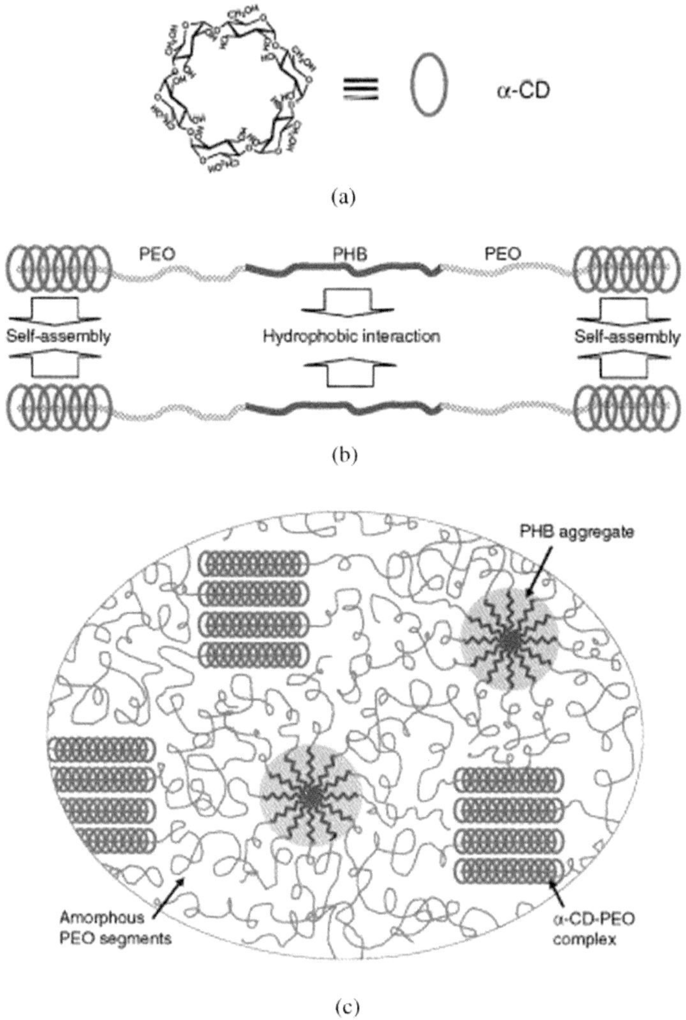


Figure 2.11 The structure of α-CD (a), the schematic illustrations of the proposed structures of α-CD–PEO–PHB–PEO inclusion complex (b), and α-CD–PEO–PHB–PEO supramolecular hydrogel (c). Reprinted from [249] with permission from Elsevier.

the nanofibrous structure and modulation of the properties of the gel by altering the ionic strength [253]. Some of the other tissues such as skeletal muscle and placental tissue have also been decellularized and used for tissue regeneration studies [254, 255]. Human lipoaspirate has also been decellularized and used as an injectable thermally responsive adipose matrix scaffold [256]. Hydrogel based on a decellularized pericardial matrix, having native sulfated GAGs has demonstrated potential for improved delivery of heparin-binding growth factors [257]. A clinically relevant injectable hydrogel derived from a decellularized skeletal muscle extracellular matrix was examined for its potential for treating peripheral artery disease (PAD) in a rat hindlimb ischemia model. Compared to collagen, increased arteriole and capillary density was observed in the skeletal muscle matrix, and more desmin-positive and MyoD-positive cells were recruited, indicating its potential for treating ischemia related to PAD (Figure 2.12), as well as having the potential beneficial effects of restoring muscle mass that is typically

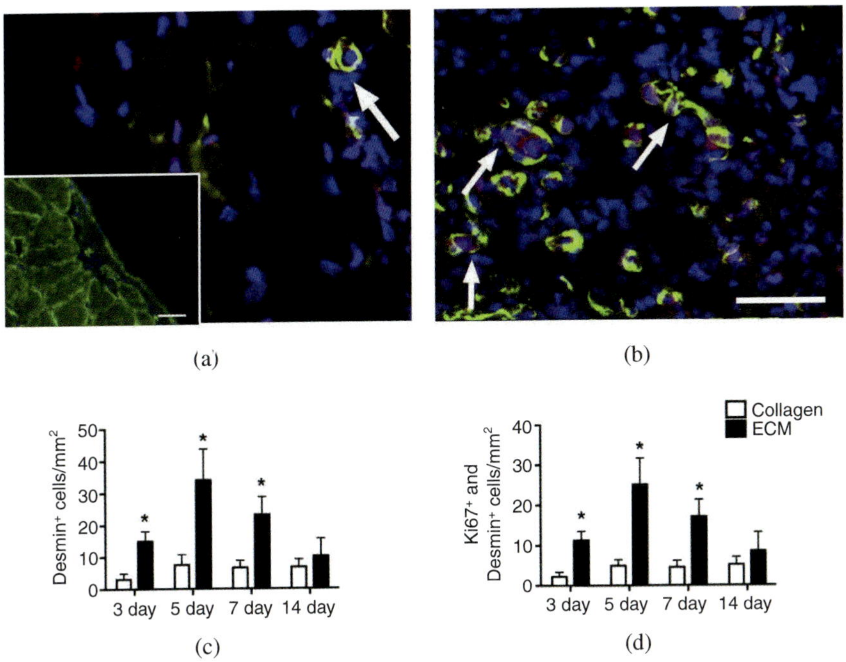


Figure 2.12 Proliferating muscle cell recruitment. Collagen injection region (a) and skeletal muscle matrix injection (b) region at 5 days with desmin-stained cells (green) co-labeled with Ki67 (red). Arrows denote desmin and Ki67 positive cells. Scale bar at 20 $\mu$m. Insert shows positive desmin staining of healthy skeletal muscle, scale bar at 100 $\mu$m. (c) Quantification of desmin-positive cells in the skeletal muscle matrix compared to collagen normalized to area. Note that there are significantly more desmin-positive cells in the skeletal muscle matrix. (d) Of these desmin-positive cells, a majority of the cells are proliferating as seen by Ki67 co-labeling. Reprinted from [258] with permission.

lost in ischemia cases [258]. Hydrogel formed by the self-assembling of a decellularized ventricular extracellular matrix when injected *in vivo* was evaluated for its ability to treat myocardial infarction (MI). It was found that the material injected in the rat MI model increased endogenous cardiomyocytes in the infarcted area and maintained cardiac function without inducing arrhythmias. The feasibility of transendocardial catheter injection of this material in a porcine model was also demonstrated for the first time in the case of injectable materials, a critical step towards the translational approach [259].

## 2.4 Conclusions and Future Prospects

In summary, a wide range of injectable hydrogels have been prepared and evaluated for various biomedical applications. Design approaches for the development of injectable hydrogel are based on *in situ* gelation and the shear-thinning behavior of hydrogels. While *in situ* gelation supports effective circumferential coverage of defects and host integration, use of shear-thinning hydrogels avoids the involvement of any external triggers and related problems. Gelling time and the nature of triggers are critical parameters for *in situ* gelling systems, whereas self-assembling processes, flow under moderate pressure, and self-healing after injection dictate the success of shear-thinning hydrogels. Both systems are highly tunable and friendly towards the incorporation of biological functionalities. Attempts have been made to improve the mechanical properties of injectable systems by employing multitriggers and multi-crosslinking methods. The concept of shear-thinning hydrogel is relatively new in the field of injectable hydrogels and studies are mostly restricted to *in vitro*. Very few systems have been reported for use *in vivo*, having satisfied all the physical, structural, and mechanical properties. So far, the majority of studies conducted on injectable hydrogels have been short term and conducted on animal models. Decellularized extracellular matrix is another exciting strategy which has been used for making shear-thinning hydrogels and has demonstrated very good tissue regeneration potential. Safety concerns and sterility issues related to decellularized ECM must be addressed before wide use of such systems is possible. Tracking the material upon injection and erosion by imaging approaches will prove beneficial. There are only a few products based on injectable hydrogels currently available on the market, apart from those used as soft tissue augmenting agents and drug delivery depots.

The designing of biomimetic hydrogels having tunable mechanical, gelation (for *in situ* gelling systems), self-healing (for shear-thinning systems), and degradation properties is essential for successful clinical translation of injectable hydrogels for various biomedical applications.

## References

[1] Wichterle, O. and Lím, D. (1960). Hydrophilic gels for biological use. *Nature*, **185**(4706), pp. 117–118.
[2] Anderson, J.M. and Langone, J.J. (1999). Issues and perspectives on the biocompatibility and immunotoxicity evaluation of implanted controlled release systems. *J. Control. Release*, **57**(2), pp. 107–113.
[3] Smetana, K. (1993). Cell biology of hydrogels. *Biomaterials*, **14**(14), pp. 1046–1050.
[4] Hennink, W.E. and Nostrum, C.F.V. (2002). Novel cross-linking methods to design hydrogels. *Adv. Drug Deliver. Rev.*, **54**(1), pp. 13–36.
[5] Rosiak, J.M. and Yoshii, F. (1999). Hydrogels and their medical applications. *Nucl. Instrum. Meth. B.*, **151**(40269), pp. 56–64.
[6] Okumura, Y. and Ito, K. (2001). The polyrotaxane gels: a topological gel by figure-of-eight cross-links. *Adv. Mat.*, **13**, pp. 485–487.
[7] Gibas, I. and Janik, H. (2010). Review: Synthetic polymer hydrogels for biomedical applications. *Chemistry Chem. Tech.*, **4**(4), pp. 297–304.
[8] Amini, A.A. and Nair, L.S. (2012). Injectable hydrogels for bone and cartilage repair. *Biomed. Mater.*, **7**(2), 024105.
[9] Balakrishnan, B. and Banerjee, R. (2011). Biopolymer based hydrogels for cartilage tissue engineering. *Chem. Rev.*, **111**(8), pp. 4453–4474.
[10] Lin, Y.C. and Marra, K.G. (2012). Injectable systems and implantable conduits for peripheral nerve repair. *Biomed. Mater.*, **7**(2), 024102
[11] Pakulska, M.M., Ballios, B.G., and Shoichet M.S. (2012). Injectable hydrogels for central nervous system therapy. *Biomed Mater.*, **7**(2), 024101.
[12] Macaya, D. and Spector, M. (2012). Injectable hydrogel materials for spinal cord regeneration: a review. *Biomed. Mater.*, **7**(1), 012001.
[13] Johnson, T.D. and Christman, K.L. (2013). Injectable hydrogel therapies and their delivery strategies for treating myocardial infarction. *Expert Opin. Drug Deliv.*, **10**(1), pp. 59–72.
[14] Ono, T., Sugimoto, T., Shinkai, S., and Sada, K. (2007). Lipophilic polyelectrolyte gels as super-absorbent polymers for nonpolar organic solvents. *Nat. Mater.*, **6**(6), pp. 429–433.
[15] Hoare, T.R. and Kohane, D.S. (2008). Hydrogels in drug delivery: Progress and challenges. *Polymer*, **49**(8), pp. 1993–2007.
[16] Mortensen, K. and Pederson, J.S. (1993). Structural study on the micelle formation of poly(ethylene oxide)-poly(propylene oxide)-poly(ethylene oxide) triblock copolymer in aqueous solution. *Macromolecules*, **26**(4), pp. 805–812.

[17] Gong, C. *et al.* (2013). Thermosensitive polymeric hydrogels as drug delivery systems. *Curr. Med. Chem.*, **20**(1), pp. 79–94.
[18] Nguyen, M.K. and Lee, D.S. (2010). Injectable biodegradable hydrogels. *Macromol. Biosci.*, **10**(6), pp. 563–579.
[19] Van Tomme, S.R., Storm, G., and Hennink, W.E. (2008). *In situ* gelling hydrogels for pharmaceutical and biomedical applications. *Int. J. Pharm.*, **355**(1–2), pp. 1–18.
[20] Doorty, K.B. *et al.* (2003). Poly(*N*-isopropylacrylamide) co-polymer films as potential vehicles for delivery of an antimitotic agent to vascular smooth muscle cells. *Cardiovasc. Pathol.*, **12**(2), pp. 105–110.
[21] Hacker, M.C., Klouda, L., Ma, B.B., Kretlow, J.D., and Mikos, A.G. (2008). Synthesis and characterization of injectable, thermally and chemically gelable, amphiphilic poly(*N*-isopropylacrylamide)-based macromers. *Biomacromolecules*, **9**(3), pp. 1558–1570.
[22] Ruel-Gariépy, E. and Leroux, J.C. (2004). *In situ*-forming hydrogels — review of temperature-sensitive systems. *Eur. J. Pharm. Biopharm.*, **58**(2), pp. 409–426.
[23] Shimizu, K., Fujita, H., and Nagamori, E. (2010). Oxygen plasma-treated thermoresponsive polymer surfaces for cell sheet engineering. *Biotechnol. Bioeng.*, **106**(2), pp. 303–310.
[24] Stile, R.A. and Healy, K.E. (2001). Thermo-responsive peptide-modified hydrogels for tissue regeneration. *Biomacromolecules*, **2**(1), pp. 185–194.
[25] Twaites, B.R. *et al.* (2005). Thermoresponsive polymers as gene delivery vectors: Cell viability, DNA transport and transfection studies. *J. Control. Release*, **108**, pp. 472–483.
[26] Lee, B.H. and Vernon, B. (2005). Copolymers of *N*-isopropylacrylamide, HEMA-lactate and acrylic acid with time-dependent lower critical solution temperature as a bioresorbable carrier. *Polym. Int.*, **54**(2), pp. 418–422.
[27] Stile, R.A., Burghardt W.R., and Healy, K.E. (1999). Synthesis and characterization of injectable poly(*N*-isopropylacrylamide)-based hydrogels that support tissue formation *in vitro*. *Macromolecules*, **32**, pp. 7370–7379.
[28] Neradovic, D., Hinrichs, W.L.J., Kettenes-van den Bosch, J.J., and Hennink, W.E. (1999). Poly(*N*-isopropylacrylamide) with hydrolyzable lactic acid ester side groups: a new type of thermosensitive polymer. *Macromol. Rapid Comm.*, **20**, pp. 577–581.
[29] Neradovic, D. *et al.* (2003). Degradation mechanism and kinetics of thermosensitive polyacrylamides containing lactic acid side chains. *Macromolecules*, **36**(20), pp. 7491–7498.
[30] Nelson, D.M., Ma, Z., Leeson, C.E., and Wagner, W.R. (2012). Extended and sequential delivery of protein from injectable thermoresponsive hydrogels. *J. Biomed. Mater. Res. A.*, **100**(3), pp. 776–785.
[31] Choi, S.J. *et al.* (2007). Combination of ascorbate and growth factor (TGF beta-3) in thermo-reversible hydrogel constructs embedded with rabbit chondrocytes for neocartilage formation. *J. Biomed. Mater. Res. A.*, **83**(4), pp. 897–905.

[32] Na, K. *et al.* (2007). Osteogenic differentiation of rabbit mesenchymal stem cells in thermo-reversible hydrogel constructs containing hydroxyapatite and bone morphogenic protein-2 (BMP-2). *Biomaterials*, **28**(16), pp. 2631–2637.

[33] Li, Z., Wang, F., Roy, S., Sen, C.K., and Guan, J. (2009). Injectable, highly flexible, and thermosensitive hydrogels capable of delivering superoxide dismutase. *Biomacromolecules*, **10**(12), pp. 3306–3316.

[34] Fujimoto, K.L. *et al.* (2009). Synthesis, characterization and therapeutic efficacy of a biodegradable, thermoresponsive hydrogel designed for application in chronic infracted myocardium. *Biomaterials*, **30**(26), pp. 4357–4368.

[35] Wang, F. *et al.* (2010). Injectable, rapid gelling and highly flexible hydrogel composites as growth factor and cell carriers. *Acta Biomater.*, **6**(6), pp. 1978–1991.

[36] Gan, T., Zhang, Y., and Guan, Y. (2009). *In situ* gelation of P(NIPAM-HEMA) microgel dispersion and its applications as injectable 3D cell scaffold. *Biomacromolecules*, **10**(6), pp. 1410–1415.

[37] Garbern, J.C., Hoffman, A.S., Stayton, P.S. (2010). Injectable pH- and temperature-responsive poly(*N*-isopropylacrylamide-co-propylacrylic acid) copolymers for delivery of angiogenic growth factors. *Biomacromolecules*, **11**, pp. 1833–1839.

[38] Bearat, H.H., Lee, B.H., Valdez, J., and Vernon, B.L. (2010). Synthesis, characterization and properties of a physically and chemically gelling polymer system using poly(NIPAAm-co-HEMA-acrylate) and poly(NIPAAm-co-cysteamine). *J. Biomater. Sci. Polym.*, **22**, pp. 1298–1318.

[39] Turturro, S.B. *et al.* (2011). The effects of cross-linked thermo-responsive PNIPAAm-based hydrogel injection on retinal function. *Biomaterials*, **32**(14), pp. 3620–3626.

[40] Drapala, P.W. *et al.* (2011). Role of thermo-responsiveness and poly(ethylene glycol) diacrylate cross-link density on protein release from poly(*N*-isopropylacrylamide) hydrogels. *J. Biomater. Sci. Polym. Ed.*, **22**, pp. 59–75.

[41] Li, Z., Guo, X., Matsushita, S., and Guan, J. (2011). Differentiation of cardiosphere-derived cells into a mature cardiac lineage using biodegradable poly(*N*-isopropylacrylamide) hydrogels. *Biomaterials*, **32**(12), pp. 3220–3232.

[42] Chen, J.P. and Cheng, T.H. (2006). Thermo-responsive chitosan-graft-poly(*N*-isopropylacrylamide) injectable hydrogel for cultivation of chondrocytes and meniscus cells. *Macromol. Biosci.*, **6**(12), pp. 1026–1039.

[43] Tan. H. *et al.* (2009). Thermosensitive injectable hyaluronic acid hydrogel for adipose tissue engineering. *Biomaterials*, **30**(36), pp. 6844–6853.

[44] Liao, H.T., Chen, C.T., and Chen, J.P. (2011). Osteogenic differentiation and ectopic bone formation of canine bone marrow-derived mesenchymal stem cells in injectable thermo-responsive polymer hydrogel. *Tissue Eng. Part C Methods*, **17**(3), pp. 1139–1149.

[45] Wiltsey, C. *et al.* (2013). Characterization of injectable hydrogels based on poly(*N*-isopropylacrylamide)-g-chondroitin sulfate with adhesive properties for nucleus pulposus tissue engineering. *J. Mater. Sci. Mater. Med.*, **24**(4), pp. 837–847.
[46] Bostman, O. *et al.* (1992). Degradation and tissue replacement of an absorbable polyglycolide screw in the fixation of rabbit femoral osteotomies. *J. Bone Joint Surg.*, **74**(7), pp. 1021–1031.
[47] Ignatius, A.A and Claes, L.E. (1996). *In vitro* biocompatibility of bioresorbable polymers: poly(L, DL-lactide) and poly(L-lactide-co-glycolide). *Biomaterials*, **17**(8), pp. 831–839.
[48] Suganuma, J. and Alexander, H. (1993). Biological response of intramedullary bone to poly-L-lactic acid. *J. Appl. Biomat.*, **4**(1), pp. 13–27.
[49] Sun, L.F., Zhuo, R.X., and Liu, Z.L. (2003). Studies on the synthesis and properties of temperature responsive and biodegradable hydrogels. *Macromol. Biosci.*, **3**, pp. 725–728.
[50] Neradovic, D., Soga, O., Van Nostrum, C.F., and Hennink, W.E. (2004). The effect of the processing and formulation parameters on the size of nanoparticles based on block copolymers of poly(ethylene glycol) and poly(*N*-isopropylacrylamide) with and without hydrolytically sensitive groups. *Biomaterials*, **25**(12), pp. 2409–2418.
[51] Cui, Z., Lee, B.H., Pauken, C., and Vernon B.L. (2011). Degradation, cytotoxicity and biocompatibility of NIPAAm-based thermosensitive, injectable and bioresorbable polymer hydrogels. *J. Biomed. Mater. Res. A.*, **98**(2), pp. 159–166.
[52] Yoshida, T., Aoyagi, T., Kokufuta, E., and Okano, T. (2003). Newly designed hydrogel with both sensitive thermoresponse and biodegradability. *J. Polym. Sci. Part A Polym. Chem.*, **41**(6), pp. 779–787.
[53] Ma, Z., Nelson, D.M., Hong, Y., and Wagner, W.R. (2010). Thermally responsive injectable hydrogel incorporating methacrylate-polylactide for hydrolytic lability. *Biomacromolecules*, **11**, pp. 1873–1881.
[54] Alexandridis, P. and Hatton, T.A. (1995). Poly(ethylene oxide)-poly(propylene oxide)-poly(ethylene oxide) block copolymer surfactants in aqueous solutions and interfaces: thermodynamics, structure, dynamics and modeling. *Coll. Surf. A*, **96**, pp. 1–46.
[55] Cao, Y.L. *et al.* (1998). Tissue-engineered nipple reconstruction. *Plast. Reconstr. Surg.*, **102**(7), pp. 2293–2298.
[56] Saim, A.B. *et al.* (2000). Engineering autogenous cartilage in the shape of a helix using an injectable hydrogel scaffold. *Laryngoscope*, **110**(10), pp. 1694–1697.
[57] Liu, Y. *et al.* (2007). Controlled delivery of recombinant hirudin based on thermosensitive Pluronic F127 hydrogel for subcutaneous administration: *In vitro* and *in vivo* characterization. *J. Control. Release*, **117**(3), pp. 387–395.
[58] Piu, F. *et al.* (2011). OTO-104: a sustained-release dexamethasone hydrogel for the treatment of otic disorders. *Otol. Neurotol.*, **32**(1), pp. 171–179.

[59] Salt, A.N., Hartsock, J., Plontke, S., LeBel, C., and Piu, F. (2011). Distribution of dexamethasone and preservation of inner ear function following intratympanic delivery of a gel-based formulation. *Audiol. Neurotol.*, **16**(5), pp. 323–335.
[60] Sosnik, A., Cohn, D., San Román, J., and Abraham, G.A. (2003) Crosslinkable PEO-PPO-PEO-based reverse thermo-responsive gels as potentially injectable materials. *J. Biomater. Sci. Polym. Ed.*, **14**(3), pp. 227–239.
[61] Cohn, D., Sosnik, A., and Garty, S., (2005). Smart hydrogels for *in situ* generated implants. *Biomacromolecules*, **6**(3), pp. 1168–1175.
[62] Sosnik, A. and Cohn, D. (2004). Ethoxysilane-capped PEO–PPO–PEO triblocks: a new family of reverse thermo-responsive polymers. *Biomaterials.*, **25**(14), pp. 2851–2858.
[63] Sosnik, A. and Cohn, D. (2005). Reverse thermo-responsive poly(ethylene oxide) and poly (propylene oxide) multiblock copolymers. *Biomaterials*, **26**(4), pp. 349–357.
[64] Cohn, D., Lando, G., Sosnik, A., Garty, S., and Levi, A. (2006). PEO-PPO-PEO-based poly (ether ester urethane)s as degradable reverse thermo-responsive multiblock copolymers. *Biomaterials*, **27**(9), pp. 1718–1727.
[65] Cellesi, F. and Tirelli, N.A. (2005). A new process for cell microencapsulation and other biomaterial applications: Thermal gelation and chemical cross-linking in "tandem". *J. Mater. Sci. Mater. Med.*, **16**(6), pp. 559–565.
[66] Niu, G. *et al.* (2009). Synthesis and characterization of reactive poloxamer 407s for biomedical applications. *J. Control. Release*, **138**(1), pp. 49–56.
[67] Chung, H.J., Lee, Y., and Park, T.G. (2008). Thermo-sensitive and biodegradable hydrogels based on stereocomplexed Pluronic multi-block copolymers for controlled protein delivery. *J. Control. Release*, **127**(1), pp. 22–30.
[68] Park, S.Y., Chung, H.J., Lee, Y., and Park, T.G. (2008). Injectable and sustained delivery of human growth hormone using chemically modified Pluronic copolymer hydrogels. *Biotechnol. J.*, **3**(5), pp. 669–675.
[69] Guo, D.D. *et al.* (2009). Synergistic anti-tumor activity of paclitaxel-incorporated conjugated linoleic acid-coupled poloxamer thermosensitive hydrogel *in vitro* and *in vivo*. *Biomaterials*, **30**(27), pp. 4777–4785.
[70] Park, K.M. *et al.* (2009). Thermosensitive chitosan-Pluronic hydrogel as an injectable cell delivery carrier for cartilage regeneration. *Acta Biomater.*, **5**(6), pp. 1956–1965.
[71] Tang, S. *et al.* (2012). Bone induction through controlled release of novel BMP-2-related peptide from PTMC(11)-F127-PTMC(11) hydrogels. *Biomed Mater.*, **7**(1), 015008.
[72] Ma, Y. *et al.* (2012). The influence of modified pluronic F127 copolymers with higher phase transition temperature on arsenic trioxide-releasing properties and toxicity in a subcutaneous model of rats *AAPS Pharm. Sci. Tech.*, **13**(2), pp. 441–447.
[73] Wout, Z., Pec, E., Maggiore, J., Williams, R., Palicharla, P., and Johnston, T. (1992). Poloxamer 407-mediated changes in plasma cholesterol and

triglycerides following intraperitoneal injection to rats. PDA *J. Pharm. Sci. Tech.*, **46**(6), pp. 192–200.

[74] Jeong, B.M., Lee, D.S., Bae, Y.H., and Kim, S.W. (1997). Biodegradable block copolymers as injectable drug-delivery systems. *Nature*, **388**, pp. 860–862.

[75] Jeong, B.M., Lee, D.S., Bae, Y.H., and Kim, S.W. (2000). *In situ* gelation of PEG-PLGA-PEG triblock copolymer aqueous solutions and degradation thereof. *J. Biomed. Mater. Res.*, **50**, pp. 171–177.

[76] Chen, S. and Singh, J. (2005). Controlled delivery of testosterone from smart polymer solution based systems: *In vitro* evaluation. *Int. J. Pharm.*, **295**(1–2), pp. 183–190.

[77] Khodaverdi, E., Tafaghodi, M., Ganji, F., Abnoos, K., and Naghizadeh, H. (2012). *In vitro* insulin release from thermosensitive chitosan hydrogel. *AAPS Pharm. Sci. Tech.*, **13**, pp. 460–466.

[78] Qiao, M., Chen, D., Ma, X., and Liu, Y. (2005). Injectable biodegradable temperature-responsive PLGA-PEG-PLGA copolymers: Synthesis and effect of copolymer composition on the drug release from the copolymer-based hydrogels. *Int. J. Pharm.*, **294**(1–2), pp. 103–112.

[79] Qiao, M. *et al.* (2008). Injectable thermosensitive PLGA-PEG-PLGA triblock copolymers-based hydrogels as carriers for interleukin-2. *Pharmazie*, **63**(1), pp. 27–30.

[80] He, C., Kim, S.W., and Lee, D.S. (2008). *In situ* gelling stimuli-sensitive block copolymer hydrogels for drug delivery. *J. Control. Release*, **127**(3), pp. 189–207.

[81] Lei, N. *et al.* (2012). Therapeutic application of injectable thermosensitive hydrogel in preventing local breast cancer recurrence and improving incision wound healing in a mouse model. *Nanoscale*, **4**(18), pp. 5686–5693.

[82] Ma, G., Miao, B., and Song, C. (2010). Thermosensitive PCL-PEG-PCL hydrogels: Synthesis, characterization, and delivery of proteins. *J. Appl. Polym. Sci.*, **16**, pp. 1985–1993.

[83] Yin, H. *et al.* (2010). Toxicity evaluation of biodegradable and thermosensitive PEG-PCL-PEG hydrogel as a potential *in situ* sustained ophthalmic drug delivery system. *J. Biomed. Mater. Res. B. Appl. Biomater.*, **92**(1), pp. 129–137.

[84] Gong, C.Y. *et al.* (2009). Biodegradable thermosensitive injectable PEG-PCL-PEG hydrogel for bFGF antigen delivery to improve humoral immunity. *Growth Factors*, **27**(6), pp. 377–383.

[85] Fu, S. *et al.* (2012). Injectable and thermo-sensitive PEG-PCL-PEG copolymer/collagen/n-HA hydrogel composite for guided bone regeneration. *Biomaterials*, **33**(19), pp. 4801–4809.

[86] Gong, C. *et al.* (2012). Efficient inhibition of colorectal peritoneal carcinomatosis by drug loaded micelles in thermosensitive hydrogel composites. *Nanoscale*, **4**(10), pp. 3095–3104.

[87] Wang, W. *et al.* (2012). Adjustable degradation and drug release of a thermosensitive hydrogel based on a pendant cyclic ether modified poly($\varepsilon$-caprolactone) and poly(ethylene glycol)co-polymer. *Acta Biomater.*, **8**(11), pp. 3963–3973.

[88] Zhang, Z. *et al.* (2011). Biodegradable and thermoreversible PCLA-PEG-PCLA hydrogel as a barrier for prevention of post-operative adhesion. *Biomaterials*, **32**(21), pp. 4725–4736.

[89] Zhang, Z. *et al.* (2012). Encapsulation of cell-adhesive RGD peptides into a polymeric physical hydrogel to prevent postoperative tissue adhesion. *J. Biomed. Mater. Res. B. Appl. Biomater.*, **100**(6), pp. 1599–1609.

[90] Xun, W. *et al.* (2009). Peptide-functionalized thermo-sensitive hydrogels for sustained drug delivery. *Macromol. Biosci.*, **9**(12), pp. 1219–1226.

[91] Allcock, H.R. (2003). *Chemistry and applications of polyphosphazenes*, Wiley, Hoboken, NJ, USA.

[92] Lakshmi, S, Katti, D.S., and Laurencin C.T. (2003). Biodegradable polyphosphazenes for drug delivery applications. *Adv. Drug Deliv. Rev.*, **55**(4), pp. 467–482.

[93] Teasdale, I. and Brüggemann, O. (2013). Polyphosphazenes: Multifunctional, biodegradable vehicles for drug and gene delivery. *Polymers*, **5**(1), pp. 161–187.

[94] Park, K.H. and Song, S.C. (2006). Morphology of spheroidal hepatocytes within injectable, biodegradable, and thermosensitive poly (organophosphazene) hydrogel as cell delivery vehicle. *J. Biosci. Bioeng.*, **101**(3), pp. 238–242.

[95] Park, K.H. and Song, S.C. (2005). A thermo-sensitive poly(organo phosphazene) hydrogel used as an extracellular matrix for artificial pancreas. *J. Biomater. Sci. Polym. Ed.*, **16**(11), pp. 1421–1431.

[96] Lee, S.M., Chun, C.J., Heo, J.Y., and Song, S.C. (2009). Injectable and thermosensitive poly(organophosphazene) hydrogels for a 5-fluorouracil delivery. *J. Appl. Polym. Sci.*, **113**(6), pp. 3831–3839.

[97] Chun, C. *et al.* (2009). Doxorubicin–polyphosphazene conjugate hydrogels for locally controlled delivery of cancer therapeutics. *Biomaterials*, **30**(27), pp. 4752–4762.

[98] Cho, J.K., Park, J.W., and Song, S.C. (2012). Injectable and biodegradable poly(organophosphazene) gel containing silibinin: Its physicochemical properties and anticancer activity. *J. Pharm. Sci.*, **101**(7), pp. 2382–2391.

[99] Kim, J.H. *et al.* (2013). Intratumoral delivery of paclitaxel using a thermosensitive hydrogel in human tumor xenografts. *Arch. Pharm. Res.*, **36**(1), pp. 96–101.

[100] Cho, J.K., Chun, C., Kuh, H.J., and Song, S.C. (2012). Injectable poly(organophosphazene)–camptothecin conjugate hydrogels: Synthesis, characterization, and antitumor activities. *Eur. J. Pharm. Biopharm.*, **81**(3), pp. 582–590.

[101] Kang, G.D. and Song, S.C. (2008). Effect of chitosan on the release of protein from thermosensitive poly(organophosphazene) hydrogels. *Int. J. Pharm.*, **349**(1–2), pp. 188–195.

[102] Park, M.R. *et al.* (2010). Sustained delivery of human growth hormone using a polyelectrolyte complex-loaded thermosensitive polyphosphazene hydrogel. *J. Control. Release*, **147**(3), pp. 359–367.

[103] Park, M.R., Chun, C., Cho, C.S., and Song, S.C. (2010). Enhancement of sustained and controlled protein release using polyelectrolyte complex-loaded injectable and thermosensitive hydrogel. *Eur. J. Pharm. Biopharm.*, **76**(2), pp. 79–88.

[104] Park, M.R., Seo, B.B., and Song, S.C. (2013). Dual ionic interaction system based on polyelectrolyte complex and ionic, injectable, and thermosensitive hydrogel for sustained release of human growth hormone. *Biomaterials*, **34**(4), pp. 1327–1336.

[105] Seo, B.B., Park, M.R., Chun, C., Lee, J.Y., and Song, S.C. (2011). The biological efficiency and bioavailability of human growth hormone delivered using injectable, ionic, thermosensitive poly(organophosphazene)-polyethylenimine conjugate hydrogels. *Biomaterials*, **32**(32), pp. 8271–8280.

[106] Kim, Y.M., Park, M.R., and Song, S.C. (2012). Injectable polyplex hydrogel for localized and long-term delivery of siRNA. *ACS Nano.*, **6**(7), pp. 5757–5766.

[107] Chun, C., Lim, H.J., Hong, K.Y., Park, K.H., and Song, S.C. (2009). The use of injectable, thermosensitive poly(organophosphazene)–RGD conjugates for the enhancement of mesenchymal stem cell osteogenic differentiation. *Biomaterials*, **30**(31), pp. 6295–6308.

[108] Potta, T, Chun, C., and Song, S.C. (2010) Injectable, dual cross-linkable polyphosphazene blend hydrogels. *Biomaterials*, **31**(32), pp. 8107–8120.

[109] Ahmadi, R. and de Bruijn, J.D. (2008). Biocompatibility and gelation of chitosan-glycerol phosphate hydrogels. *J. Biomed. Mater. Res A*, **86**(3), pp. 824–832.

[110] Zheng, L., Ao, Q., Han, H., Zhang, X., and Gong, Y. (2010). Evaluation of the chitosan/glycerol-beta-phosphate disodium salt hydrogel application in peripheral nerve regeneration. *Biomed. Mater.*, **5**(3), pp. 35003–35007.

[111] Huang, F.Y. *et al.* (2009). Development of a thermosensitive hydrogel system for local delivery of (188) Re colloid drugs. *Appl. Radiat. Isot.*, **67**(7–8), pp. 1405–1411.

[112] Gao, J. *et al.* (2012). The use of chitosan based hydrogel for enhancing the therapeutic benefits of adipose-derived MSCs for acute kidney injury. *Biomaterials*, **33**, pp. 3673–3681.

[113] Coutu, J.M., Fatimi, A., Berrahmoune, S., Soulez, G., and Lerouge, S. (2013). A new radiopaque embolizing agent for the treatment of endoleaks after endovascular repair: influence of contrast agent on chitosan thermogel properties. *J. Biomed. Mater. Res. B. Appl. Biomater.*, **101**(1), pp. 153–161.

[114] Ruel-Gariépy, E., Leclair, G., Hildgen, P., Gupta, A., and Leroux, J.C. (2002). Thermosensitive chitosan-based hydrogel containing liposomes for the delivery of hydrophilic molecules. *J. Control. Release*, **82**(2–3), pp. 373–383.

[115] Cheng, Y.H. *et al.* (2010). Thermosensitive chitosan-gelatin-glycerol phosphate hydrogels as a cell carrier for nucleus pulposus regeneration: an in vitro study. *Tissue Eng. Part A*, **16**(2), pp. 695–703.

[116] Bhattarai, N., Ramay, H.R., Gunn, J., Matsen, F.A., and Zhang, M. (2005). PEG-grafted chitosan as an injectable thermosensitive hydrogel for sustained protein release. *J. Control. Release*, **103**(3), pp. 609–624.

[117] Couto, D.S., Hong, Z., and Mano, J.F. (2009). dDevelopment of bioactive and biodegradable chitosan-based injectable systems containing bioactive glass nanoparticles. *Acta Biomater.*, **5**(1), pp. 115–123.

[118] Ji, Q.X. *et al.* (2009). Injectable thermosensitive hydrogel based on chitosan and quaternized chitosan and the biomedical properties. *J. Mater. Sci. Mater. Med.*, **20**(8), pp. 1603–1610.

[119] Ji, Q.X., Zhao, Q.S., Deng, J., and Lu, R. (2010). A novel injectable chlorhexidine thermosensitive hydrogel for periodontal application: preparation, antibacterial activity and toxicity evaluation. *J. Mater. Sci. Mater. Med.*, **21**(8), pp. 2435–2442.

[120] Niranjan, R. *et al.* (2013). A novel injectable temperature-sensitive zinc doped chitosan/$\beta$-glycerophosphate hydrogel for bone tissue engineering. *Int. J. Biol. Macromol.*, **54**, pp. 24–29.

[121] Peng, Q. *et al.* (2013). Injectable and biodegradable thermosensitive hydrogels loaded with PHBHHx nanoparticles for the sustained and controlled release of insulin. *Acta Biomater.*, **9**(2), pp. 5063–5069.

[122] Chen, C., Wang, L., Deng, L., Hu, R., and Dong, A. (2013). Performance optimization of injectable chitosan hydrogel by combining physical and chemical triple crosslinking structure. *J. Biomed. Mater. Res. A.*, **101**(3), pp. 684–693.

[123] Chien, Y. *et al.* (2012). Corneal repair by human corneal keratocyte-reprogrammed iPSCs and amphiphatic carboxymethyl-hexanoyl chitosan hydrogel. *Biomaterials*, **33**(32), pp. 8003–8016.

[124] Wei, C.Z. *et al.* (2009). A thermosensitive chitosan-based hydrogel barrier for post-operative adhesions' prevention. *Biomaterials*, **30**(29), pp. 5534–5540.

[125] Asai, D. *et al.* (2012). Protein polymer hydrogels by *in situ*, rapid and reversible self-gelation. *Biomaterials*, **33**(21), pp. 5451–5458.

[126] Ziane, S. *et al.* (2012). A thermosensitive low molecular weight hydrogel as scaffold for tissue engineering. *Eur. Cell. Mater.*, **23**, pp. 147–160.

[127] Choi, B.G. *et al.* (2010). *In situ* thermal gelling polypeptide for chondrocytes 3D culture. *Biomaterials*, **31**(35), pp. 9266–9272.

[128] Seo, H.W. *et al.* (2013). Injectable intratumoral hydrogel as 5-fluorouracil drug depot. *Biomaterials*, **34**(11), pp. 2748–2757.

[129] Wu, X. *et al.* (2012). Sodium dodecyl sulfate-induced rapid gelation of silk fibroin. *Acta Biomater.*, **8**(6), pp. 2185–2192.

[130] Chueh, S.K., Tachibana, Y., Uyama, H., Kobayashi, S., and Tomita, N. (2007). Evaluation of chondrocytes expression embedded in thermoresponsive poly(amino acid)s with sol-gel transition. *Biomed. Mater. Eng.*, **17**(3), pp. 137–146.

[131] Chiu, L. *et al.* (2009). pH-triggered injectable hydrogels prepared from aqueous *N*-palmitoyl chitosan: *In vitro* characteristics and *in vivo* biocompatibility. *Biomaterials*, **30**(28), pp. 4877–4888.

[132] Gupta, S. and Vyas, S.P. (2010). Carbopol/chitosan based pH triggered *in situ* gelling system for ocular delivery of timolol maleate. *Sci. Pharm.*, **78**(4), pp. 959–976.
[133] Vinatier, C. *et al.* (2005). A silanized hydroxypropyl methylcellulose hydrogel for the three-dimensional culture of chondrocytes. *Biomaterials*, **26**(33), pp. 6643–6651.
[134] Augst, A.D., Kong, H.J., and Mooney, D.J. (2006). Alginate hydrogels as biomaterials. *Macromol. Biosci.*, **6**(8), pp. 623–633.
[135] Cao, L., Arany, P.R., Wang, Y.S., and Mooney, D.J. (2009). Promoting angiogenesis via manipulation of VEGF responsiveness with notch signaling. *Biomaterials*, **30**(25), pp. 4085–4093.
[136] Kaigler, D., Silva, E.A., and Mooney, D.J. (2013). Guided bone regeneration using injectable vascular endothelial growth factor delivery gel. *J. Periodontol.*, **84**(2), pp. 230–238.
[137] Gao, S.Q., Maeda, T., Okano, K., and Palczewski, K. (2012). A microparticle/hydrogel combination drug-delivery system for sustained release of retinoids. *Invest. Ophthalmol. Vis. Sci.*, **53**(10), pp. 6314–6323.
[138] Bron, J.L., Vonk, L.A., Smit, T.H., and Koenderink, G.H. (2011). Engineering alginate for intervertebral disc repair. *J. Mech. Behav. Biomed. Mater.*, **4**(7), pp. 1196–1205.
[139] Zhao, L., Weir, M.D., and Xu, H.H. (2010). An injectable calcium phosphate-alginate hydrogel-umbilical cord mesenchymal stem cell paste for bone tissue engineering. *Biomaterials*, **31**(25), pp. 6502–6510.
[140] Moshaverinia, A. *et al.* (2012). Alginate hydrogel as a promising scaffold for dental-derived stem cells: an *in vitro* study. *J. Mater. Sci. Mater. Med.*, **23**(12), pp. 3041–3051.
[141] Bidarra, S.J. *et al.* (2011). Injectable in situ crosslinkable RGD-modified alginate matrix for endothelial cells delivery. *Biomaterials*, **32**(31), pp. 7897–7904.
[142] Fonseca, K.B., Bidarra, S.J., Oliveira, M.J., Granja, P.L., and Barrias, C.C. (2011). Molecularly designed alginate hydrogels susceptible to local proteolysis as three-dimensional cellular microenvironments. *Acta Biomater.*, **7**(4), pp. 1674–1682.
[143] Tan, R. *et al.* (2012). Repair of rat calvarial bone defects by controlled release of rhBMP-2 from an injectable bone regeneration composite. *J. Tissue Eng. Regen. Med.*, **6**(8), pp. 614–621.
[144] Cao, J. *et al.* (2012). Local injection of nerve growth factor via a hydrogel enhances bone formation during mandibular distraction osteogenesis. *Oral Surg. Oral Med. Oral Pathol. Oral Radiol.*, **113**(1), pp. 48–53.
[145] Hwang, C.M. *et al.* (2013). Assessments of injectable alginate particle-embedded fibrin hydrogels for soft tissue reconstruction. *Biomed. Mater.*, **8**(1), 014105.
[146] Kim, W.S. *et al.* (2012). Adipose tissue engineering using injectable, oxidized alginate hydrogels. *Tissue Eng. Part A.*, **18**(7–8), pp. 737–743.
[147] Jansson, P.E., Lindberg, B., and Sandford, P.A. (1983). Structural studies of gellan gum, an extracellular polysaccharide elaborated by *Pseudomonas elodea*. *Carbohydr. Res.*, **124**, pp. 135–137.

[148] Moorhouse, R., Colegrove, G.T., Sandford, P.A., Baird, J.K., and Kang, K.S. (1981). PS-60: A new gel-forming polysaccharide. *ACS Symposium Series*, **150**, pp. 111–124.
[149] Oliveira, J.T. *et al.* (2010). Gellan gum injectable hydrogels for cartilage tissue engineering applications: *in vitro* studies and preliminary in vivo evaluation. *Tissue Eng. Part A*, **16**(1), pp. 343–353.
[150] Sanzgiri, Y.D. *et al.* (1993). Gellan-based systems for ophthalmic sustained delivery of methylpredni-solone. *J. Control. Release*, **26**, pp. 195–201.
[151] Oliveira, J.T. *et al.* (2010). Injectable gellan gum hydrogels with autologous cells for the treatment of rabbit articular cartilage defects. *J. Orthop. Res.*, **28**(9), pp. 1193–1199.
[152] Du, H., Hamilton, P., Reilly, M., and Ravi, N. (2012). Injectable *in situ* physically and chemically crosslinkable gellan hydrogel. *Macromol. Biosci.*, **12**(7), pp. 952–961.
[153] Dyondi, D., Webster, T.J., and Banerjee, R. (2012). A nanoparticulate injectable hydrogel as a tissue engineering scaffold for multiple growth factor delivery for bone regeneration. *Int. J. Nanomedicine*, **8**(1), pp. 47–59.
[154] Popa, E.G., Caridade, S.G., Mano, J.F., Reis, R.L., and Gomes, M.E. (2015). Chondrogenic potential of injectable? $\kappa$-carrageenan hydrogel with encapsulated adipose stem cells for cartilage tissue-engineering applications. *J. Tissue Eng. Regen. Med.*, **9**(5), pp. 550–563. (in press).
[155] Liu, J., Zhang, L., Yang, Z., and Zhao, X. (2011). Controlled release of paclitaxel from a self-assembling peptide hydrogel formed *in situ* and antitumor study *in vitro*. *Int. J. Nanomedicine*, **6**, pp. 2143–2153.
[156] Hamada, K., Hirose, M., Yamashita, T., and Ohgushi, H. (2008). Spatial distribution of mineralized bone matrix produced by marrow mesenchymal stem cells in self-assembling peptide hydrogel scaffold. *J. Biomed. Mater. Res. A.*, **84**(1), pp. 128–136.
[157] Semino, C.E., Kasahara, J., Hayashi, Y., and Zhang, S. (2004). Entrapment of migrating hippocampal neural cells in three-dimensional peptide nanofiber scaffold. *Tissue Eng.*, **10**(3–4), pp. 643–655.
[158] Sieminski, A.L., Semino, C.E., Gong, H., and Kamm, R.D. (2008). Primary sequence of ionic self-assembling peptide gels affects endothelial cell adhesion and capillary morphogenesis. *J Biomed. Mater. Res. A*, **87**(2), pp. 494–504.
[159] Kisiday, J. *et al.* (2002). Self-assembling peptide hydrogel fosters chondrocyte extracellular matrix production and cell division: implications for cartilage tissue repair. *Proc. Natl. Acad. Sci. USA*, **99**(15), pp. 9996–10001.
[160] Cavalcanti, B.N., Zeitlin, B.D., and Nör, J.E. (2013). A hydrogel scaffold that maintains viability and supports differentiation of dental pulp stem cells. *Dent Mater.*, **29**(1), pp. 97–102.
[161] Huynh, C.T., Nguyen, M.K., and Lee, D.S. (2011). Biodegradable pH/temperature-sensitive oligo($\beta$-amino ester urethane) hydrogels for controlled release of doxorubicin. *Acta Biomater.*, **7**(8), pp. 3123–3130.
[162] Sahu, A., Choi, W.I., and Tae, G. (2012). A stimuli-sensitive injectable graphene oxide composite hydrogel. *Chem. Commun. (Camb.)*, **48**(2), pp. 5820–5822.

[163] Garbern, J.C., Minami, E., Stayton, P.S., and Murry, C.E. (2011). Delivery of basic fibroblast growth factor with a pH-responsive, injectable hydrogel to improve angiogenesis in infarcted myocardium. *Biomaterials*, **32**(9), pp. 2407–2416.

[164] Nguyen, M.K., Park, D.K., and Lee, D.S. (2009). Injectable poly(amidoamine)-poly(ethylene glycol)–poly(amidoamine) triblock copolymer hydrogel with dual sensitivities: pH and temperature. *Biomacromolecules*, **10**(4), pp. 728–731.

[165] Huynh, D.P., Im, G.J., Chae, S.Y., Lee, K.C., and Lee, D.S. (2009). Controlled release of insulin from pH/temperature-sensitive injectable pentablock copolymer hydrogel. *J. Control. Release*, **137**(1), pp. 20–24.

[166] Shim, W.S., Kim, S.W., and Lee, D.S. (2006). Sulfonamide-based pH- and temperature-sensitive biodegradable block copolymer hydrogels. *Biomacromolecules*, **7**(6), pp. 1935–1941.

[167] Kim, H.K. *et al.* (2009). Injectable *in situ*-forming pH/thermo-sensitive hydrogel for bone tissue engineering. *Tissue Eng. Part A*, **15**(4), pp. 923–933.

[168] Shim, W.S. *et al.* (2007). pH- and temperature-sensitive, injectable, biodegradable block copolymer hydrogels as carriers for paclitaxel. *Int. J. Pharm.*, **331**(6), pp. 11–18.

[169] Tan, H., Xiao, C., Sun, J., Xiong, D., and Hu, X. (2012). Biological self-assembly of injectable hydrogel as cell scaffold via specific nucleobase pairing. *Chem. Commun. (Camb.)*, **48**(33), pp. 10289–10291.

[170] Tan, H., Shen, Q., Jia, X., Yuan, Z., and Xiong, D. (2012). Injectable nanohybrid scaffold for biopharmaceuticals delivery and soft tissue engineering. *Macromol. Rapid Commun.*, **33**(23), pp. 2015–2022.

[171] Balakrishnan, B. and Jayakrishnan, A. (2005). Self-cross-linking biopolymers as injectable *in situ* forming biodegradable scaffolds. *Biomaterials*, **26**(18), pp. 3941–3951.

[172] Balakrishnan, B., Mohanty, M., Umashankar, P.R., and Jayakrishnan, A. (2005). Evaluation of an *in situ* forming hydrogel wound dressing based on oxidized alginate and gelatin. *Biomaterials*, **26**(32), pp. 6335–6342.

[173] Li, X. *et al.* (2012). *In situ* injectable nano-composite hydrogel composed of curcumin, *N*,*O*-carboxymethyl chitosan and oxidized alginate for wound healing application. *Int. J. Pharm.*, **437**(1–2), pp. 110–119.

[174] Zhang, H., Qadeer, A., Mynarcik, D., and Chen, W. (2011). Delivery of rosiglitazone from an injectable triple interpenetrating network hydrogel composed of naturally derived materials. *Biomaterials*, **32**(3), pp. 890–898.

[175] Tan, H., Rubin, J.P., and Marra, K.G. (2010). Injectable *in situ* forming biodegradable chitosan-hyaluronic acid based hydrogels for adipose tissue regeneration. *Organogenesis*, **6**(3), pp. 173–180.

[176] Patenaude, M. and Hoare, T. (2012). Injectable, mixed natural-synthetic polymer hydrogels with modular properties. *Biomacromolecules*, **13**(2), pp. 369–378.

[177] Chen, Y.C., Su, W.Y., Yang, S.H., Gefen, A., and Lin, F.H. (2013). *In situ* forming hydrogels composed of oxidized high molecular weight hyaluronic acid and gelatin for nucleus pulposus regeneration. *Acta Biomater.*, **9**(2), pp. 5181–5193.

[178] Alves, M.H., Young, C.J., Bozzetto, K., Poole-Warren, L.A., and Martens, P.J. (2012). Degradable, click poly(vinyl alcohol) hydrogels: characterization of degradation and cellular compatibility. *Biomed. Mater.*, **7**(2), 024106.

[179] Epstein-Barash, H., Stefanescu, C.F., and Kohane, D.S. (2012). An *in situ* cross-linking hybrid hydrogel for controlled release of proteins. *Acta Biomater.*, **8**(5), pp. 1703–1709.

[180] Molinos, M., Carvalho, V., Silva, D.M., and Gama, F.M. (2012). Development of a hybrid dextrin hydrogel encapsulating dextrin nanogel as protein delivery system. *Biomacromolecules*, **13**(2), pp. 517–527.

[181] Aulin, C. *et al.* (2011). *In situ* cross-linkable hyaluronan hydrogel enhances chondrogenesis. *J. Tissue Eng. Regen. Med.*, **5**(8), pp. 188–196.

[182] Li, X. *et al.* (2012). Engineering an *in situ* crosslinkable hydrogel for enhanced remyelination. *FASEB J.*, **27**(3), pp. 1127–1136.

[183] Li, Z., Guo, X., Palmer, A.F., Das, H., and Guan, J. (2012). High-efficiency matrix modulus-induced cardiac differentiation of human mesenchymal stem cells inside a thermosensitive hydrogel. *Acta Biomater.*, **8**(10), pp. 3586–3595.

[184] Fuoco, C. *et al.* (2012). Injectable polyethylene glycol-fibrinogen hydrogel adjuvant improves survival and differentiation of transplanted mesoangioblasts in acute and chronic skeletal-muscle degeneration. *Skelet. Muscle*, **2**(1), pp. 24–32.

[185] Park, H., Choi, B., Hu, J., and Lee, M. (2013). Injectable chitosan hyaluronic acid hydrogels for cartilage tissue engineering. *Acta Biomater.*, **9**(1), pp. 4779–4786.

[186] Gaharwar, A.K., Rivera, C.P., Wu, C.J., and Schmidt, G. (2011). Transparent, elastomeric and tough hydrogels from poly(ethylene glycol) and silicate nanoparticles. *Acta Biomater.*, **7**(12), pp. 4139–4148.

[187] Liu, S.Q. *et al.* (2010). Injectable biodegradable poly(ethylene glycol)/RGD peptide hybrid hydrogels for *in vitro* chondrogenesis of human mesenchymal stem cells. *Macromol. Rapid Commun.*, **31**(13), pp. 1148–1154.

[188] Hiemstra, C., van der Aa, L.J., Zhong, Z., Dijkstra, P.J., and Feijen, J. (2007). Novel *in situ* forming, degradable dextran hydrogels by Michael addition chemistry: synthesis, rheology, and degradation. *Macromolecules*, **40**(4), pp. 1165–1173.

[189] Moss, I.L., Gordon, L., Woodhouse, K.A., Whyne, C.M., and Yee, A.J. (2011). A novel thiol-modified hyaluronan and elastin-like polypeptide composite material for tissue engineering of the nucleus pulposus of the intervertebral disc. *Spine (Phila Pa 1976)*, **36**(13), pp. 1022–1029.

[190] Dong, Y. *et al.* (2011). "One-step" Preparation of thiol-ene clickable PEG-based thermoresponsive hyperbranched copolymer for *in situ* crosslinking hybrid hydrogel. *Macromol. Rapid Commun.*, **33**, pp. 120–126.

[191] Tao, Y. *et al.* (2013). Evaluation of an *in situ* chemically crosslinked hydrogel as a long-term vitreous substitute material. *Acta Biomater.*, **9**(2), pp. 5022–5030.
[192] Pritchard, C.D. *et al.* (2011). An injectable thiol-acrylate poly(ethylene glycol) hydrogel for sustained release of methylprednisolone sodium succinate. *Biomaterials*, **32**(2), pp. 587–597.
[193] Jo, S. *et al.* (2013). Effects of recombinant human bone morphogenic protein-2 and human bone marrow-derived stromal cells on *in vivo* bone regeneration of chitosan poly(ethylene oxide) hydrogel. *J. Biomed. Mater. Res. A*, **101**(3), pp. 892–901.
[194] Kobayashi, S., Uyama, H., and Kimura, S. (2001). Enzymatic polymerization. *Chem. Rev.*, **101**(12), pp. 3793–3818.
[195] Moreira Teixeira, L.S., Feijen, J., van Blitterswijk, C.A., Dijkstra, P.J., and Karperien, M. (2012a). Enzyme-catalyzed crosslinkable hydrogels: emerging strategies for tissue engineering. *Biomaterials*, **33**(5), pp. 1281–1290.
[196] Toh, W.S., Lim, T.C., Kurisawa, M., and Spector, M. (2012). Modulation of mesenchymal stem cell chondrogenesis in a tunable hyaluronic acid hydrogel microenvironment. *Biomaterials*, **33**(15), pp. 3835–3845.
[197] Moreira Teixeira, L.S. *et al.* (2012). Self-attaching and cell-attracting *in situ* forming dextran–tyramine conjugates hydrogels for arthroscopic cartilage repair. *Biomaterials*, **33**(11), pp. 3164–3174.
[198] Moreira Teixeira, L.S. *et al.* (2012). The effect of platelet lysate supplementation of a dextran-based hydrogel on cartilage formation. *Biomaterials*, **33**(14), pp. 3651–3661.
[199] Wang, L.S., Chung, J.E., and Kurisawa, M. (2011). Controlling fibroblast proliferation with dimensionality-specific response by stiffness of injectable gelatin hydrogels. *J. Biomater. Sci. Polym. Ed.*, **12**, pp. 1793–1806.
[200] Park, K.M. *et al.* (2012). Synthesis and characterizations of *in situ* cross-linkable gelatin and 4-arm-PPO-PEO hybrid hydrogels via enzymatic reaction for tissue regenerative medicine. *Biomacromolecules*, **13**(3), pp. 604–611.
[201] Wang, L.S., Du, C., Chung, J.E., and Kurisawa, M. (2012). Enzymatically cross-linked gelatin-phenol hydrogels with a broader stiffness range for osteogenic differentiation of human mesenchymal stem cells. *Acta Biomater.*, **8**(5), pp. 1826–1837.
[202] Lim, T.C., Toh, W.S., Wang, L.S., Kurisawa, M., and Spector M. (2012). The effect of injectable gelatin-hydroxyphenylpropionic acid hydrogel matrices on the proliferation, migration, differentiation and oxidative stress resistance of adult neural stem cells. *Biomaterials*, **33**(12), pp. 3446–3455.
[203] Park, Y.S. *et al.* (2012). Controlled release of simvastatin from *in situ* forming hydrogel triggers bone formation in MC3T3-E1 cells. *AAPS J.*, **15**(2), pp. 367–376.
[204] Xu, K. *et al.* (2013). Injectable hyaluronic acid-tyramine hydrogels incorporating interferon-$\alpha$2a for liver cancer therapy. *J. Control. Release*, **163**(10), pp. 203–210. *Bioconjug. Chem.*, **23**(10), pp. 2042–2050.
[205] Park, K.M., Lee, Y., Son, J.Y., Bae, J.W., and Park, K.D. (2012). *In situ* SVVYGLR peptide conjugation into injectable gelatin-poly(ethylene

glycol)-tyramine hydrogel via enzyme-mediated reaction for enhancement of endothelial cell activity and neo-vascularization.

[206] Ogushi, Y., Sakai, S., and Kawakami, K. (2013). Adipose tissue engineering using adipose-derived stem cells enclosed within an injectable carboxymethyl cellulose-based hydrogel. *J. Tissue Eng. Regen. Med.*, **7**(11), pp. 884–892.

[207] Tran, N.Q., Joung, Y.K., Lih, E., and Park, K.D. (2011). *In situ* forming and rutin-releasing chitosan hydrogels as injectable dressings for dermal wound healing. *Biomacromolecules*, **12**(8), pp. 2872–2880.

[208] Lee, S.H. *et al.* (2011). Enzyme-mediated cross-linking of Pluronic copolymer micelles for injectable and *in situ* forming hydrogels. *Acta Biomater.*, **7**(4), pp. 1468–1476.

[209] Bereczky, Z., Katona, E., and Muszbek, L. (2003). Fibrin stabilization (factor XIII), fibrin structure and thrombosis. *Pathophysiol. Haemost. Thromb.*, **33**(5–6), pp. 430–437.

[210] Janmey, P.A., Winer, J.P., and Weisel, J.W. (2009). Fibrin gels and their clinical and bioengineering applications. *J. R. Soc. Interface*, **6**(10), pp. 1–10.

[211] Kjaergard, H.K. and Weis-Fogh, U.S., (1994). Important factors influencing the strength of autologous fibrin glue; the fibrin concentration and reaction time-comparison of strength with commercial fibrin glue. *Eur. Surg. Res.*, **26**(5), pp. 273–276.

[212] Sidelmann, J.J., Gram, J., Jespersen, J., and Kluft, C. (2000). Fibrin clot formation and lysis: Basic mechanisms. *Semin. Thromb. Hemost.*, **26**(6), pp. 605–608.

[213] Adams, T.E. and Huntington, J.A. (2006). Thrombin-cofactor interactions: structural insights into regulatory mechanisms. *Arterioscler. Thromb. Vasc. Biol.*, **26**(8), pp. 1738–1745.

[214] Lechtenberg, B.C., Johnson, D.J.D., Freund, S.M.V., and Huntington, J.A. (2010). NMR resonance assignments of thrombin reveal the conformational and dynamic effects of ligation. *Proc. Natl. Acad. Sci. USA*, **107**(32), pp. 14087–14092.

[215] King, V.R., Alovskaya, A., Wei, D.Y.T., Brown, R.A., and Priestley, J.V. (2010). The use of injectable forms of fibrin and fibronectin to support axonal ingrowth after spinal cord injury. *Biomaterials*, **31**(15), pp. 4447–4456.

[216] Lee, Y.B. *et al.* (2010). Bio-printing of collagen and VEGF-releasing fibrin gel scaffolds for neural stem cell culture. *Exp. Neurol.*, **223**, pp. 645–652.

[217] Li, Y., Rodrigues, J., and Tamas, H. (2012). Injectable and biodegradable hydrogels: gelation, biodegradation and biomedical applications. *Chem. Soc. Rev.*, **41**, pp. 2193–2221.

[218] Willerth, S.M., Rader, A., and Sakiyama-Elbert, S.E. (2008). The effect of controlled growth factor delivery on embryonic stem cell differentiation inside fibrin scaffolds. *Stem Cell Res.*, **1**(3), pp. 205–218.

[219] Wong, C., Inman, E., Spaethe, R., and Helgerson, S. (2003). Fibrin-based biomaterials to deliver human growth factors. *Thromb. Haemost.*, **89**(3), pp. 573–582.

[220] Yang, H.S., Bhang, S.H., Hwang, J.W., Kim, D.I., and Kim, B.S. (2010). Delivery of basic fibroblast growth factor using heparin-conjugated fibrin for therapeutic angiogenesis. *Tissue Eng. Part A*, **16**(6), pp. 2113–2119.

[221] Spicer, P.P. and Mikos, A.G. (2010). Fibrin glue as a drug delivery system. *J. Control. Release*, **148**(1), pp. 49–55.

[222] Guvendiren, M., Lu, H.D., and Burdick, J.A. (2012). Shear-thinning hydrogels for biomedical applications. *Soft Matter*, **8**, pp. 260–272.

[223] Foo, C., Lee, J.S., Mulyasasmita, W., Parisi-Amon, A., and Heilshorn, S.C. (2009). Two-component protein-engineered physical hydrogels for cell encapsulation. *Proc. Natl. Acad. Sci. USA*, **106**(52), pp. 22067–22072.

[224] Lu, H.D., Charati, M.B., Kim, I.L., and Burdick, J.A. (2012). Injectable shear-thinning hydrogels engineered with a self-assembling Dock-and-Lock mechanism. *Biomaterials*, **33**(7), pp. 2145–2153.

[225] Petka, W.A., Harden, J.L., McGrath, K.P., Wirtz, D., and Tirrell, D.A. (1998). Reversible hydrogels from self-assembling artificial proteins. *Science*, **281**(5375), pp. 389–392.

[226] Fischer, S.E., Liu, X., Mao, H.Q., and Harden, J.L. (2007). Controlling cell adhesion to surfaces via associating bioactive triblock proteins. *Biomaterials*, **28**(22), pp. 3325–3337.

[227] Mi, L., Fischer, S., Chung, B., Sundelacruz, S., and Harden, J.L. (2006). Self-assembling protein hydrogels with modular integrin binding domains. *Biomacromolecules*, **7**(1), pp. 38–47.

[228] Shen, W., Kornfield, J.A., and Tirrell, D.A. (2007). Structure and mechanical properties of artificial protein hydrogels assembled through aggregation of leucine zipper peptide domains. *Soft Matter*, **3**, pp. 99–107.

[229] Wheeldon, I.R., Barton, S.C., and Banta, S. (2007). Bioactive proteinaceous hydrogels from designed bifunctional building blocks. *Biomacromolecules*, **8**(10), pp. 2990–2994.

[230] Cresce, A., Dandu, R., Burger, A., Cappello, J., and Ghandehari, H., (2008). Characterization and real-time imaging of gene expression of adenovirus embedded silk-elastinlike protein polymer hydrogels. *Mol. Pharm.*, **5**(5), pp. 891–897.

[231] Megeed, Z., Cappello, J., and Ghandehari, H. (2002). Controlled release of plasmid DNA from a genetically engineered silk-elastinlike hydrogel. *Pharm. Res.*, **19**(7), pp. 954–959.

[232] Haider, M., Cappello, J., Ghandehari, H., and Leong, K.W. (2008). *In vitro* chondrogenesis of mesenchymal stem cells in recombinant silk-elastin-like hydrogels. *Pharm. Res.*, **25**(3), pp. 692–699.

[233] Yan, C. *et al.* (2012). Injectable solid peptide hydrogel as a cell carrier: effects of shear flow on hydrogels and cell payload. *Langmuir*, **28**(14), pp. 6076–6087.

[234] Yan, C. *et al.* (2010). Injectable solid hydrogel: mechanism of shear-thinning and immediate recovery of injectable $\beta$-hairpin peptide hydrogels. *Soft Matter*, **6**(20), pp. 5143–5156.

[235] Branco, M.C., Nettesheim, F., Pochan, D.J., Schneider, J.P., and Wagner, N.J. (2009). Fast dynamics of semiflexible chain networks of self-assembled peptides. *Biomacromolecules*, **10**(6), pp. 1374–1380.

[236] Branco, M.C., Pochan, D.J., Wagner, N.J., and Schneider, J.P. (2010). The effect of protein structure on their controlled release from an injectable peptide hydrogel. *Biomaterials*, **31**(36), pp. 9527–9534.
[237] Altunbas, A., Lee, S.J., Rajasekaran, S.A., Schneider, J.P., and Pochan, D.J. (2011). Encapsulation of curcumin in self-assembling peptide hydrogels as injectable drug delivery vehicles. *Biomaterials*, **32**(25), pp. 5906–5914.
[238] Koutsopoulos, S. and Zhang, S. (2012). Two-layered injectable self-assembling peptide scaffold hydrogels for long-term sustained release of human antibodies. *J. Control. Release*, **160**(3), pp. 451–458.
[239] Silva, D. *et al.* (2013). Synthesis and characterization of designed BMHP1-derived self-assembling peptides for tissue engineering applications. *Nanoscale*, **5**(2), pp. 704–718.
[240] Miura, M. *et al.* (2003). SHED: Stem cells from human exfoliated deciduous teeth. *Proc. Natl. Acad. Sci. USA*, **100**(10), pp. 5807–5812.
[241] Bakota, E.L., Wang, Y., Danesh, F.R., and Hartgerink, J.D. (2011). Injectable multidomain peptide nanofiber hydrogel as a delivery agent for stem cell secretome. *Biomacromolecules*, **12**, pp. 1651–1657.
[242] Wang, Y. *et al.* (2011). Peptide nanofibers preconditioned with stem cell secretome are renoprotective. *J. Am. Soc. Nephrol.*, **22**(4), pp. 704–717.
[243] Galler, K.M., Aulisa, L., Regan, K.R., D'Souza, R.N., and Hartgerink, J.D. (2010). Self-assembling multidomain peptide hydrogels: designed susceptibility to enzymatic cleavage allows enhanced cell migration and spreading. *J. Am. Chem. Soc.*, **132**(9), pp. 3217–3223.
[244] Shoichet, M.S., Tator, C.H., Poon, P., Kang, C., and Baumann, M.D. (2007). Intrathecal drug delivery strategy is safe and efficacious for localized delivery to the spinal cord. *Prog. Brain Res.*, **161**, pp. 385–392.
[245] Gong, Z., Yang, Y., Ren, Q., Chen, X., and Shao, Z. (2012). Injectable thixotropic hydrogel comprising regenerated silk fibroin and hydroxypropylcellulose. *Soft Matter*, **8**(10), p. 2875.
[246] Pawar, G.M. *et al.* (2012). Injectable hydrogels from segmented PEG–bisurea copolymers. *Biomacromolecules*, **13**(12), pp. 3966–3976.
[247] Kiick, K.L. (2008). Peptide- and protein-mediated assembly of heparinized hydrogels. *Soft Matter*, **4**, pp. 29–37.
[248] Huang, J. *et al.* (2011). *In vivo* differentiation of adipose-derived stem cells in an injectable poloxamer-octapeptide hybrid hydrogel. *Tissue Cell.*, **43**(6), pp. 344–349.
[249] Li, J. *et al.* (2006). Self-assembled supramolecular hydrogels formed by biodegradable PEO-PHB-PEO triblock copolymers and alpha-cyclodextrin for controlled drug delivery. *Biomaterials*, **27**(2), pp. 4132–4140.
[250] Li, Z., Yin, H., Zhang, Z., Liu, K.L., and Li, J. (2012). Supramolecular anchoring of DNA polyplexes in cyclodextrin-based polypseudorotaxane hydrogels for sustained gene delivery. *Biomacromolecules*, **13**(10), pp. 3162–3172.
[251] Ma, D. and Zhang, L.M. (2011). Supramolecular gelation of a polymeric prodrug for its encapsulation and sustained release. *Biomacromolecules*, **12**(9), pp. 3124–3130.

[252] Gilbert, T.W., Sellaro, T.L., and Badylak, S.F. (2006). Decellularization of tissues and organs. *Biomaterials*, **27**(19), pp. 3675–3683.
[253] Johnson, T.D., Lin, S.Y., and Christman, K.L. (2011). Tailoring material properties of a nanofibrous extracellular matrix derived hydrogel. *Nanotechnology*, **22**(49), 494015.
[254] Abberton, K.M. *et al.* (2008). Myogel, a novel, basement membrane-rich, extracellular matrix derived from skeletal muscle, is highly adipogenic *in vivo* and *in vitro*. *Cells Tissues Organs,* **188**, pp. 347–358.
[255] Flynn, L., Semple, J.L., and Woodhouse, K.A. (2006). Decellularized placental matrices for adipose tissue engineering. *J. Biomed. Mater. Res.*, **79**(2), pp. 359–369.
[256] Young, D.A., Ibrahim, D.O., Hu, D., and Christman, K.L. (2011). Injectable hydrogel scaffold from decellularized human lipoaspirate. *Acta Biomater.*, **7**(3), pp. 1040–1049.
[257] Seif-Naraghi, S.B., Horn, D., Schup-Magoffin, P.J., and Christman, K.L. (2012). Injectable extracellular matrix derived hydrogel provides a platform for enhanced retention and delivery of a heparin-binding growth factor. *Acta Biomater.*, **8**(10), pp. 3695–3703.
[258] DeQuach, J.A. *et al.* (2012). Injectable skeletal muscle matrix hydrogel promotes neovascularization and muscle cell infiltration in a hindlimb ischemia model. *Eur. Cell Mater.*, **23**, pp. 400–412.
[259] Singelyn, J.M. *et al.* (2012). Catheter-deliverable hydrogel derived from decellularized ventricular extracellular matrix increases endogenous cardiomyocytes and preserves cardiac function post-myocardial infarction. *J. Am. Coll. Cardiol.*, **59**(8), pp. 751–763.

## Chapter 3

# Injectable Hydrogels by Physical Crosslinking

M. J. Glassman
*Department of Chemical Engineering*
*Massachusetts Institute of Technology, Cambridge, MA, USA*

B. D. Olsen*
*Department of Chemical Engineering*
*Massachusetts Institute of Technology, Cambridge, MA, USA*

### 3.1 Introduction

Hydrogels consist of water-soluble molecules that are physically or chemically crosslinked to form an infinite, volume-spanning network. The nature of the crosslinks that are formed is a critical parameter that controls the material's bulk mechanical response and consequently its biomedical applications. When hydrophilic substituents are crosslinked by irreversible chemical reactions, the resulting network is covalently bonded; when the crosslinking occurs by reversible physical interactions, such as hydrogen bonding, ion complexation, or hydrophobic interactions, a transient network is formed with crosslinks that are not fixed in either space or time. These physically crosslinked hydrogels are viscoelastic materials whose network junctions can restructure in response to changes in environmental conditions that affect bond dynamics and equilibrium, including changes in pH, temperature, or applied stress (Figure 3.1). An important and biomedically

*Author to whom correspondence should be addressed. Email: bdolsen@mit.edu.

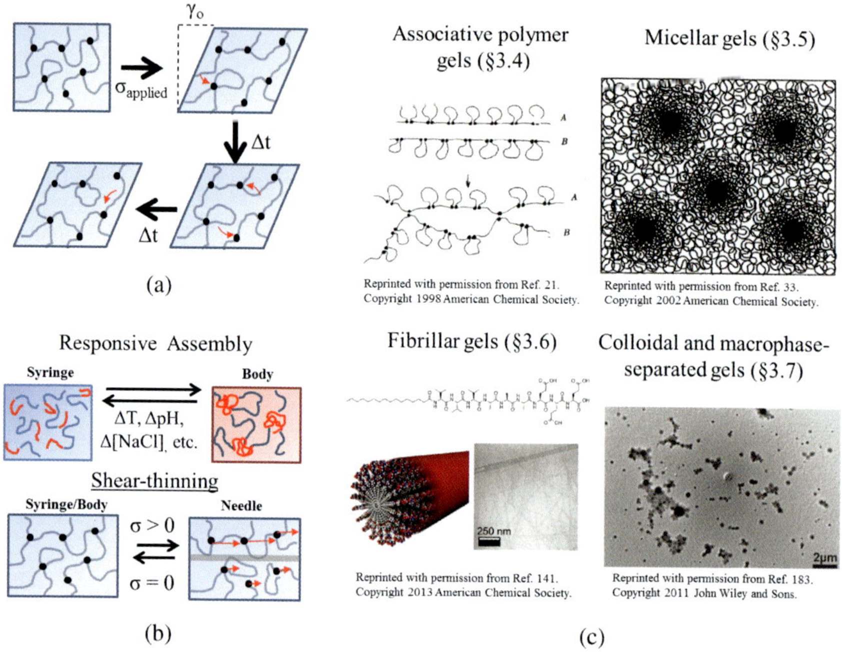


Figure 3.1 Overview of physical gelation. (a) Molecular rearrangements that lead to time-dependent mechanical response (viscoelasticity) in physically crosslinked systems. (b) Injectable flow mechanisms in physical gels. (c) Major classes of injectable physical hydrogels used in biomedical applications.

relevant subset of physically crosslinked gels is injectable physical hydrogels, whose network formation, structure, and mechanical properties have been specifically designed to enable facile delivery of the hydrogel through narrow gauge needles for minimally invasive surgical procedures.

While physical hydrogels are not the only injectable biomaterials, they offer certain advantages over the delivery of liquid-like solutions that are chemically crosslinked post-injection. Physical crosslinking can be triggered by relatively mild stimuli and is typically cytocompatible, although recent advances in chemical crosslinking strategies have been developed to achieve similar properties [1–3]. Physical crosslinking of liquid-like precursor solutions can also occur rapidly and reversibly, and the gels can exhibit self-healing, enabling the delivery of well-defined implants that can be reshaped post-implantation if desired. Certain physical hydrogels can be formulated as solid-like materials prior to injection, will shear-thin during flow through narrow needles, and self-heal into solid materials once implanted [4].

These shear-thinning gels are of particular interest as cell delivery platforms because they are typically cytoprotective during delivery due to the flow profile within the syringe during injection: low shear occurs across the majority of the gel to preserve cells, and high shear is isolated near the walls of the needle [5, 6]. Despite these advantages, physical hydrogels are typically less robust under physiological conditions than comparable chemically crosslinked systems. They often experience high creep and erosion in open environments and typically have a low yield stress, particularly in the case of shear-thinning gels. A number of approaches have been trialed to mitigate these limitations [7–9], but the development of robust, injectable physical gels is an ongoing challenge.

The design of injectable physically gelling systems is characterized by the broad structural and dynamical complexity of the materials, leading to a comparably broad set of applications. Synthetically, physical associations can be attached to hydrophilic polymers using chemical groups, such as the introduction of hydrophobes [10], or by utilizing biomolecules, such as nucleic acids or proteins that can undergo sequence-specific associations [11]. Various responsive groups can be incorporated, such as thermoresponsive synthetic polymers [12] or specific protein domains that associate in the presence of certain ligands [13]. In addition, the binding energy of the associative domains or the gel architecture can be controlled [14]. This diversity of designs has been shown to span a broad range of moduli, toughness, and creep [9, 15], and they have found application in the delivery of encapsulated cells [16], the implantation of drug-depots [17], as well as three-dimensional printing [18]. The complex dimensions of material design space can be navigated systematically with an understanding of the origins of viscoelasticity and injectability and how these are tied to the macromolecular structure and dynamics.

This chapter begins with an overview of the general concepts of the structure and mechanics of injectable physical hydrogels, specifically focusing on practical issues regarding gel design (Section 3.2). A discussion of common experimental techniques employed in the study of injectable gels follows in Section 3.3. The remainder of the chapter is dedicated to a survey and discussion of important examples of the synthesis, characterization, and application of various classes of injectable physical hydrogels: associating polymers (Section 3.4), micellar gels (Section 3.5), fibrillar gels (Section 4.6), and colloidal and macrophase-separated gels (Section 3.7) (Figure 3.1c). The chapter concludes with perspectives on future research into these materials (Section 3.8). Some details of the fundamental soft matter physics that forms the foundation of our understanding of these useful materials is beyond the scope of this review, but the reader is directed to several excellent textbooks, reviews, and original research that cover in more detail the thermodynamics of associative polymers [10, 19–22], transient network theories [23, 24] and the rheology of viscoelastic materials [25–28].

## 3.2 Design of Injectable Physical Hydrogels

The thermodynamics, structure, and mechanics of physical hydrogels are rich areas under active research. This section provides an overview of the different categories of physical hydrogels that have been studied to date and discusses the two types of mechanisms by which these gels can be delivered via injection: responsive gelation, or network disruption under shear (Figure 3.1b). In later sections, greater detail will be provided about the relevant physics and synthetic approaches for each class of gel through a selection of salient examples from the recent literature.

### 3.2.1 *Mechanisms of physical gelation*

In the design of physical hydrogels, and in any crosslinked system in general, an understanding of the relationship between molecular design and hierarchical structure is crucial to optimizing the mechanical response. With the correct design rules, injectable formulations can be developed to meet various constraints for clinical relevance. While the principle feature in the design of injectable physical gels is non-covalent bonding among the molecular or macromolecular constituents, the precise manner in which these non-covalent associations are incorporated into the gel structure plays an important role in determining their gelation behavior (in particular, the gelation concentration) as well as their response to linear and nonlinear deformations.

Physical gels can be categorized into four common classes: associating polymers, micellar gels, fibrillar gels, and colloidal gels. Each of these gels will generally form a complex structure from the molecular to the micron-scale. Associating polymers used in hydrogels are made from hydrophilic polymers containing chemical moieties that form transient, attractive interactions in solution with a broad range of average lifetimes [29, 30]. Depending on the strength, lifetime, and concentration of the associating moieties, as well as the polymer architecture, associating polymers in solution can behave as a broad array of viscoelastic materials, from viscous mixtures like paint thickeners [22] to stiff solids like alginate gels [31]. While associating polymer gels have interacting domains with average sizes on the order of angstroms, micellar gels form networks with nanoscale-sized aggregates. Commonly, these aggregates are formed from the association of hydrophobic polymers, and gelation results from either excluded-volume repulsion of close-packed micelles or micelle bridging [32, 33]. The micellar aggregates can take on various shapes (spherical, cylindrical, etc.) with important consequences for the mechanical response of the hydrogel [34]. Fibrillar gels are formed from a broad range of synthetic systems, such as assemblies of low molecular weight peptides [35], helix-forming proteins [36], or nanotubules, but the gelation mechanism always involves the complex interactions (branching, bundling, entanglement, etc.) of these extended, supramolecular structures. Similarly, colloidal gels can be made from many different types of atomic or macromolecular aggregates, and form fractal-like percolating networks from non-covalent interactions [37].

Hybrid gels can be formed by different combinations of each of these gels or a combination of physical and chemical crosslinks [38] sometimes resulting in high performance soft materials with impressive mechanical properties, such as superior toughness [39]. Among the most impressive hierarchically structured composite gels are biological connective tissues, such as articular cartilage, which exhibits compressive resistance, wearability, and toughness that is only beginning to be mimicked by synthetic systems [40]. While the synthetic and structural diversity in physical hydrogels can be arbitrarily complex, approaching the engineering of gels based on their physical gelation mechanisms provides a simplifying design framework for understanding the common underlying principles.

### 3.2.2 *Mechanisms of injectability*

Injectability of physical gels is typically accomplished in two ways: responsively and rapidly forming a physical network under physiological conditions after injection, or temporarily disrupting the network structure during injection. Responsive gelation post-injection relies on a trigger that modulates the strength of attractive interactions in the physical network once implantation has occurred. In this case, the gel is a liquid in the syringe and forms a solid gel in the implant site after some time depending on the kinetics of association. Common examples of such triggers include changes in temperature [41], pH [42], or ionic strength [43], a specific molecular recognition event [44], or a change in solvent composition [45]. In some examples, a trigger that begins a slow physical crosslinking process, such as the input of mechanical energy, can be applied immediately prior to injection [46, 47]. The alternate mechanism, network disruption during injection, generally involves the use of materials with a sufficiently low yield stress such that they will flow through a needle under only modest stress. In this case, the gel exists as a solid material even prior to injection. Such materials are described as shear-thinning as the viscous and elastic stresses borne by the gel plateau or decrease with increasing strain, resulting in yielding and flow. While in principle most physically associating systems will shear-thin at high strain, in practice a given formulation may not have a sufficiently low yield stress such that it can be handled conveniently in the clinic.

The ability to perform either injection process under clinically relevant conditions is essential to developing new tools for addressing biomedical challenges in a minimally invasive manner. There are a number of important differences between the two injectable approaches, and the benefits and limitations must be considered carefully when designing a gel for a particular application. First, the time required to set post-injection can vary significantly by injectable flow mechanism and gel class. Measuring the kinetics of gelation for materials that require the diffusion of molecules may be difficult, but control over all of the appropriate timescales (for instance, the dilution of pH-buffering salts and the assembly of the associative network) is important to ensure reproducibility of the biomaterial composition. In certain applications, such as in the delivery of encapsulated drugs or when serving as a bioactive scaffold for tissue regeneration, localization of the gel at the implant site must be carefully controlled. Shear-thinning gels have been shown to exhibit extremely fast recovery times post-injection, on the order of seconds [6, 48], but their behavior during and immediately after injection (specifically, their yield stress and gelation kinetics) are intimately tied to their overall mechanical properties. In general, it can be difficult to independently design all aspects of the mechanical response of these materials.

A second important difference between these two injectable mechanisms is the flow profile in the syringe during injection. Responsive physical gels will generally behave like viscous liquids with parabolic cross-sectional flow profiles. In such flows, the shear gradient is zero at the center of the needle and gradually increases to its maximum at the needle walls. On the other hand, some shear-thinning gels have been observed to exhibit plug-type flows during injection, where the shear gradient remains near zero across a bulk of the needle cross-section and increases sharply very close to the walls (Figure 3.2) [49]. The ability to minimize shear throughout the injected volume of material can be beneficial for protecting delicate encapsulated materials that are sensitive to shear stresses. Indeed, flow through constricted geometries is a mechanism for achieving membrane perforation [50] and cell lysis [51] and the cytoprotective nature of certain shear-thinning hydrogels can be particularly advantageous when the biomaterial is intended as a cell delivery platform.

A third set of factors by which the choice of injectable flow mechanism should be judged is the demand that is placed on the clinic, clinician, and patient. Simple gel designs that can be packaged as a ready-to-use formulation with a long shelf life minimize the burden on the healthcare delivery infrastructure, but for a number of applications, including the delivery of cell-laden materials, it is advisable to plan for more complex handling procedures during the material selection and design stage. If multiple components need to be combined immediately prior to implantation, the timescales of hydration or mixing should be sufficiently fast to minimize preparation time. In addition, a plan to maintain sterility during handling procedures should be considered at an early stage in the design process. In a number of cases, certain injectable flow mechanisms may be prohibited by the targeted tissue site. For instance, a viscous liquid that gels thermoresponsively upon heating may not be suitable for injectable implantation into a tissue that cannot be kept sufficiently cold during the procedure, or may be inappropriate for a cosmetic application such as a craniofacial implant where thermal variation cannot be controlled well over the implant lifetime. While these considerations will not often be necessary in the laboratory setting, they can be crucial and costly decisions as a biomaterial progresses through the development phases. Because the chemical composition, synthesis, and mechanical performance of the gel are closely related to the choice of injectable flow mechanism, it is often prohibitively difficult to change this aspect of gel design at the later stages of a project.

### 3.2.3 *Summary of general design considerations*

The choice of injectable flow mechanism and type of physical hydrogel are ultimately dictated by the target surgical application, whether it be for the

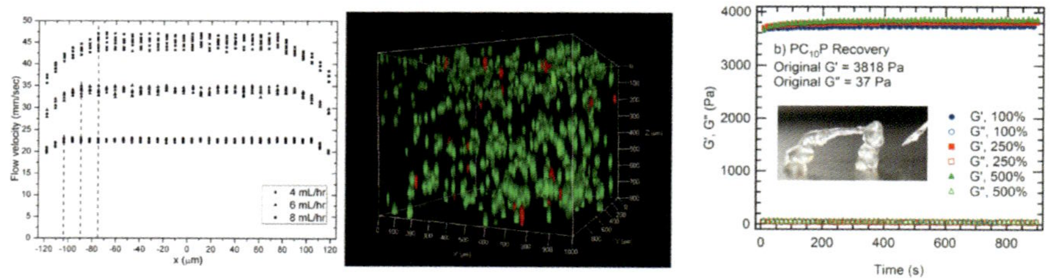


(a) (b)

Figure 3.2 Studies of the flow during injection of shear-thinning physical hydrogels and the implications for biomedical materials. (a) Direct measurement of the capillary flow velocity of a $\beta$-hairpin peptide fibrillar gel. Three-dimensional confocal laser scanning microscopy of live-dead assay on encapsulated HUVECs following injection, indicating homogeneous cell distribution and high survival rates. Adapted from [49] with permission from the American Chemical Society. (b) Linear viscoelastic properties of artificially engineered associative protein hydrogels following large amplitude oscillatory shear, indicating near-immediate recovery that can be used to form self-supporting structures directly after injection. Adapted from [6] with permission from the American Chemical Society.

implantation of a tissue filler, a bioactive matrix for tissue regeneration, a cell-laden scaffold, or a depot for controlled drug release. There are two primary mechanisms by which non-covalently crosslinked systems can be injected: responsive gelation or shear-induced network disruption. Furthermore, there are roughly four classes of physical hydrogels that, in principle, can be delivered using either mechanism: associating polymers, micellar gels, fibrillar gels, and colloidal gels. Additional factors, such as the chemical composition and synthetic approach, also have an important role in controlling gel structure and mechanical behavior by influencing important parameters such as the yield stress, the elastic modulus, creep compliance, erosion, and biodegradability of the gel in the implanted state [9, 52]. Of course, there are numerous issues beyond the structure and mechanics of injectable hydrogels that will play a role in proper design. For rapid adoption of a new implantable biomaterial, the chemical composition and sourcing of the macromolecular constituents must take into consideration safety and regulatory concerns. Engineering the hydrogel for the desired bioactivity involves the appropriate choice of cytocompatible components, cell adhesion ligands, cell signaling factors, and enzymatic degradation sites. In a broad sense, these issues affect most implantable biomaterials, and there are several examples and reviews in the literature concerning engineering bioactivity into hydrogels [52–57].

## 3.3 Experimental Techniques to Understand the Structure and Mechanics of Physical Hydrogels

Physical hydrogels adopt diverse structures over a broad range of length scales, from the atomic to the micron scale, and these structures will influence the gel's mechanical behavior. While a growing number of advanced techniques are available for investigating the structure and mechanics of crosslinked systems, this section highlights some of the most common tools employed in the study of injectable physical hydrogels. The behavior of gels can be particularly sensitive to the hydration level and processing history, so sample handling and preparation requires careful attention to avoid the introduction of artifacts due to effects such as drying, pre-shear, or phase separation. When applied appropriately, these techniques reveal a great deal of information about the relationship between molecular level design, hierarchical structure, and the mechanics of high performance gels.

### 3.3.1 *Mechanical testing*

The reversible nature of physical crosslinking, whether its origin is due to intermolecular associating groups or supramolecular interactions, can add substantial complexity to the mechanical response of physical hydrogels (Figure 3.3). Chemically crosslinked hydrogels behave like elastic solids: in response to an applied stress, the gel will typically deform by a specific amount determined by its modulus or stiffness. However, compared with

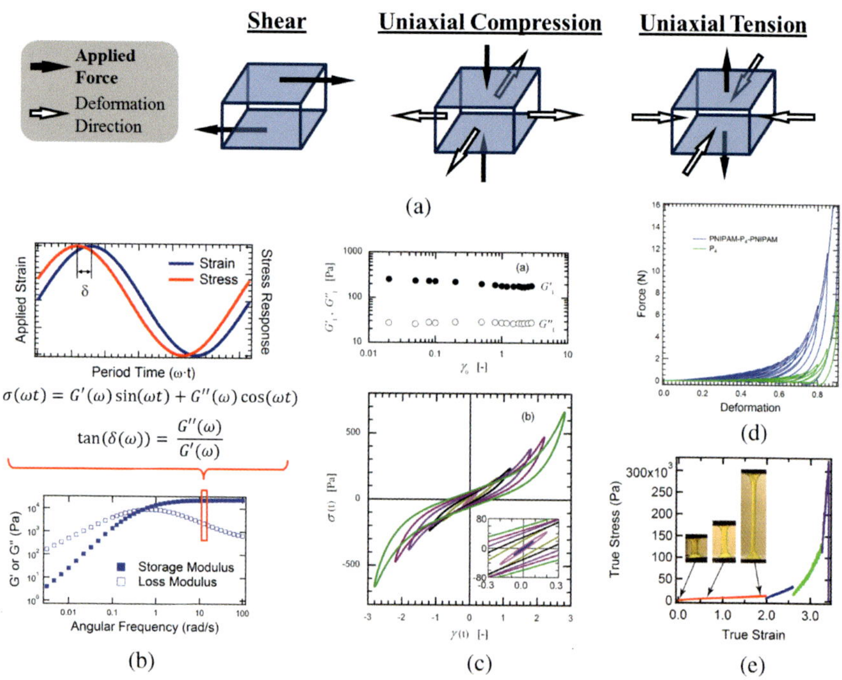


Figure 3.3 Mechanical testing of viscoelastic materials used for studying physical hydrogels. (a) Schematic of applied forces during common mechanical tests. (b) Small amplitude oscillatory shear rheology to study the time-dependent mechanical response in the linear regime. G′(ω) and G″(ω) are extracted from the relationship between the stress and strain waveforms at each experimental frequency (ω). Frequency-sweep data at bottom show typical response for a viscoelastic solid with a Maxwell relaxation mode. (c) Large amplitude oscillatory shear to study the nonlinear mechanical response. At bottom, Lissajous plots constructed from the raw waveforms. Reprinted from [66] with permission from the Society of Rheology. (d) Cyclic uniaxial compression testing of cylindrical hydrogel samples to successively increasing deformation. Adapted from [9] with permission from John Wiley and Sons. (e) Tensile testing of a dogbone-shaped hydrogel sample to failure. Reprinted from [8] with permission from the American Chemical Society.

chemical crosslinking, physical bonding gives the hydrogel time-dependent relaxation behavior because with slow enough perturbations, the material will deform completely to alleviate any internal stresses as the physical crosslinks break and reform; in other words, the material will flow. On the other hand, with perturbations fast enough that the physical crosslinks do not have time to relax, the gel will still exhibit solid-like behavior, deforming elastically and returning to its original dimensions upon the removal of any applied stresses. For intermediate perturbation timescales, a spectrum of responses is manifest. Materials with such behavior are said to be viscoelastic, and an accurate description of their mechanical response requires an analysis of the time-dependent response [58, 59]. This concept is critical for injectable gel design because in biomedical applications, hydrogels are perturbed at various timescales corresponding to flow processes and cell–material interactions, meaning the gel design sets ultimate limitations on the gel's processability as well as the long-term performance of the implanted material in response to physiological loads applied at the cellular, tissue, and organ level. The time-dependent behavior of viscoelastic materials is commonly studied using an oscillatory measurement in the linear regime, where the material is perturbed with a small sinusoidal stress or strain, and the sinusoidal stress or strain response is recorded. In this way, a large amount of information can be extracted efficiently to learn about the structure and dynamics of the material over several orders of magnitude of time simply by changing the frequency of the oscillation (typically $10^{-3}$ to $10^{2}$ $s^{-1}$). Shear is a common, well-controlled geometry to perform such small-angle oscillatory shear (SAOS) experiments, but tension/compression is also extensively employed using dynamic mechanical analysis (DMA). Oscillatory rheology is a convenient experimental technique because the liquid-like or solid-like response of the material can be easily decomposed in the linear regime (Figure 3.3b). For solids, the applied stress is directly proportional to the measured strain, while for liquids, the applied stress is directly proportional to the measured strain rate. For sinusoidal perturbations, the strain and strain rate waveforms are perfectly out of phase with each other. Thus, if the measured strain response is completely in phase with the applied stress, then the material is perfectly elastic with a stiffness directly proportional to the amplitude of the response waveform ($G' \gg G''$). Alternatively, if the strain response is completely out of phase with the applied stress ($G' \gg G'$), then the material is purely viscous with a viscosity directly proportional to the amplitude of the output waveform. For viscoelastic materials, the phase difference between the input and output waveforms directly indicates the liquid-like or solid-like nature of the material being studied. Mathematically, the waveform amplitudes and phase difference are used to compute the elastic or storage modulus (G′), the viscous or dissipative modulus (G″), and the complex viscosity ($\eta^*$). In general, these parameters are a function of the frequency of the input oscillation ($\omega$): a reminder that materials can have complex time-dependent properties.

By changing the frequency of the input oscillation, the rate dependence of the liquid-like or solid-like nature of the material can be studied. Considering for now only the linear response of materials (strain amplitudes low enough that the mechanical response is independent of the strain amplitude), data from a typical SAOS experiment are shown for a physical hydrogel with a single relaxation mechanism, i.e., a gel that can be conceptualized as a simple Maxwell fluid [60] containing one elastic mode and one viscous mode in series (Figure 3.3c). At the high end of the frequency spectrum, it is clear that the material behaves elastically, with G' over an order of magnitude larger than G″. At the low end of the frequency spectrum, the material exhibits viscous behavior, with G″ > G′. The crossover point, $\omega_c$ where $G'(\omega_c) = G''(\omega_c)$, is related to the characteristic relaxation time of the network.

Often, it is important to understand the thermodynamic conditions under which an associative system begins to exhibit gel-like properties, and this phase boundary can be determined using rheological methods. This process is typically referred to as 'gel point determination', and is regularly used to measure conditions such as the gelation concentration or temperature ($c_g$ or $T_g$, respectively) [61, 62]. Oscillatory shear rheology experiments are particularly convenient for this purpose as the system will exhibit a very specific signature in a finite-time frequency-sweep experiment. Specifically, the ratio of the elastic and viscous moduli will become independent of frequency (i.e., they will become congruent), with a characteristic response: $G'(\omega) \sim G''(\omega) \sim \omega^{1/2}$ [62]. For certain types of crosslinked systems, $G'(\omega) = G''(\omega)$ at the gel point [63], but this is not universal and it is prudent to examine the frequency-dependent moduli for this purpose.

From an engineering perspective, we are typically interested in making changes to the gel design to change the magnitude of G′ or G″ as well as the frequency of the crossover point. The origins of these values are intricately linked to the specific class of hydrogel and the network that is formed. As a set of general guidelines, G′ is related to the number of modes in which the network can store energy, and G″ is related to the number of modes for energy dissipation via non-conservative interactions. In a viscoelastic solid with a single relaxation time, the crossover point is related to the timescale at which any physical interactions transition from energy storing to energy dissipating. For physical networks with long relaxation times ($>10^4$ s), an oscillatory measurement may be impractical for studying the low frequency behavior. One approach is to study the network behavior at higher temperature where the dynamics are faster and apply the time-temperature superposition principle [26]. This approach requires the various modes in the relaxation spectrum not to shift with respect to each other, as in the case of networks containing multiple Arrhenius-activated processes or when temperature-dependent structural changes occur in the system [15, 64]. In these cases, a convenient alternative is to perform creep shear experiments in the linear regime to capture the long time behavior. Numerical techniques exist for converting creep compliance data (J(t)) into the frequency-dependent viscoelastic moduli (G'($\omega$) and G"($\omega$)) [65]. The combination of these experiments can allow the direct study of the viscoelastic moduli over a broad frequency spectrum even when time-temperature superposition cannot be applied.

While oscillatory perturbations in the linear regime provide indispensable information for understanding the structure of physical hydrogels and their response to small deformations, injection from a syringe is a highly nonlinear process involving a combination of shear and extensional flow. Thus, understanding the non-linear viscoelastic response of physically associating gels is crucial to developing optimized formulations that are useable in the clinic. This is particularly true for shear-thinning hydrogels, which are typically solid materials prior to and immediately after injection, but must flow through a narrow-gauge needle under gentle hand pressure. The complex fluid phenomena that can occur during start-up, after yielding, and under steady-state flow through syringes have been examined for several geometries and fluid properties, but a full description of these highly nonlinear processes remains the subject of vigorous experimental studies and theoretical development [66–68]. Common nonlinear tests performed on a shear rheometer are: (1) viscosity as a function of shear rate; (2) stress versus time at constant shear rate; and, (3) large amplitude oscillatory shear (Figure 3.3c) [69]. Lessons on how to engineer the linear and nonlinear mechanical behavior will be discussed briefly for each particular class of hydrogels by relating the mechanical response to their molecular, nanoscale, and microscale structure in later sections.

The behavior of solid materials under compressive or tensile loads (Figure 3.3d,e) provides additional important information about how energy is stored or dissipated in the sample, particularly in the case of nonlinear deformations that result in permanent deformation of the material. In addition, compressive and tensile testing reveal more direct information about how samples may perform in a desired application. For instance, several important load-bearing tissues that are targets for surgical repair and regeneration are connective tissues such as ligaments, cartilage, and bone, which are frequently stressed in tension and compression. As in shear, the strain-rate dependence of the tensile or compressive mechanical response is important. Real biological tissues exhibit complex strain-rate-dependent properties, and these responses are often difficult to mimic with simple synthetic systems [70].

Unconfined uniaxial compression is the simplest and most common compressive test to perform on hydrogels [71]. In unconfined compression, a cylindrical sample is allowed to expand freely in the x–y plane as it is compressed along the z-axis. In order to ensure that the measured response is related to pure compression, the surface of the sample needs to be sufficiently lubricated to ensure that sliding friction and bowing of the free surface do not contribute to the applied force. One simple measurement involves a single loading/unloading cycle to a predefined deformation (defined as a percentage of the initial sample height) at a specified rate. For relatively small deformations, typically 10–20%, the load-deformation data can be fit to models such as the neo-Hookean model in order to extract the compressive modulus [72]. Analysis of the unloading curve is important for understanding how energy is stored or dissipated in the sample during the compression cycle: the hysteresis between the loading and unloading curves is related to energy that is lost during the compression cycle [73]. Differences in sample height post-compression indicate that plastic deformation has occurred during the cycle. For large enough deformations, deviations from simple models may be large, particularly if significant yielding has occurred in the sample. Unconfined compression is rarely used to quantitatively analyze hydrogels that can withstand greater than 80–90% deformation without apparent failure, particularly because it is difficult to ensure that the geometric assumptions remain satisfied under high strain conditions. Thus, studies of the mechanical response of gels at large deformations are typically performed in tension, particularly when studying toughened hydrogels. A single sample can be investigated from the linear through nonlinear regimes, and the macroscopic details of fracture and crack propagation can be studied visually using high-resolution video extensometry combined with the appropriate labeling techniques and analysis [71, 74]. In addition to extension-to-failure, strain-rate-dependent tension and cyclic loading/unloading are common experiments for studying sample performance.

### 3.3.2 *Structural techniques*

Physical hydrogels exhibit structure over a very broad range of length scales depending on the gel class (Figure 3.4). Polymers crosslinked via small molecule associations will have aggregates with angstrom-scale sizes, while colloidal gels may form micron-scale aggregates. Many sources of radiation, including visible light, X-rays, electrons, and neutrons have been utilized to study the structure of gels directly in real space (via microscopy), or to interrogate structural correlations in reciprocal space (via scattering) [75]. Often, gel structure is strongly dependent on hydration, so studying gels under physiologically relevant conditions is critical to obtaining meaningful results. When possible, a combination of microscopy and scattering is a powerful approach for a detailed understanding of the structure of physical hydrogels. Numerous spectrophotometric techniques can also provide crucial evidence about atomic-level bonding interactions and protein secondary structure involved in network formation, although the correlation to larger length scale structures is often indirect. For more information about spectroscopic analyses commonly applied to physical gels, the reader is directed to the following references for circular dichroism (CD), Raman, and Fourier transform infrared spectroscopy (FTIR) [76–78].

When successful, imaging of the hydrogel structure provides the most direct information about how the gel network is formed. Since many of the important structural features in physical hydrogels form below the diffraction limit of visible light, traditional optical microscopy is largely uninformative. However, super-resolution techniques are developing rapidly, making these of potential interest in the future [79, 80]. Instead of optical microscopy, electron microscopy is commonly used to study gel structure, largely because of its ability to access a broad range of length scales and its widespread availability. Scanning electron microscopy (SEM) is a surface imaging technique that is frequently utilized to report gel properties such as porosity and interconnectedness. However, a serious but underappreciated difficulty with interpreting SEM is that it is a high vacuum technique and the effect of dehydration on gel structure is often not accounted for.

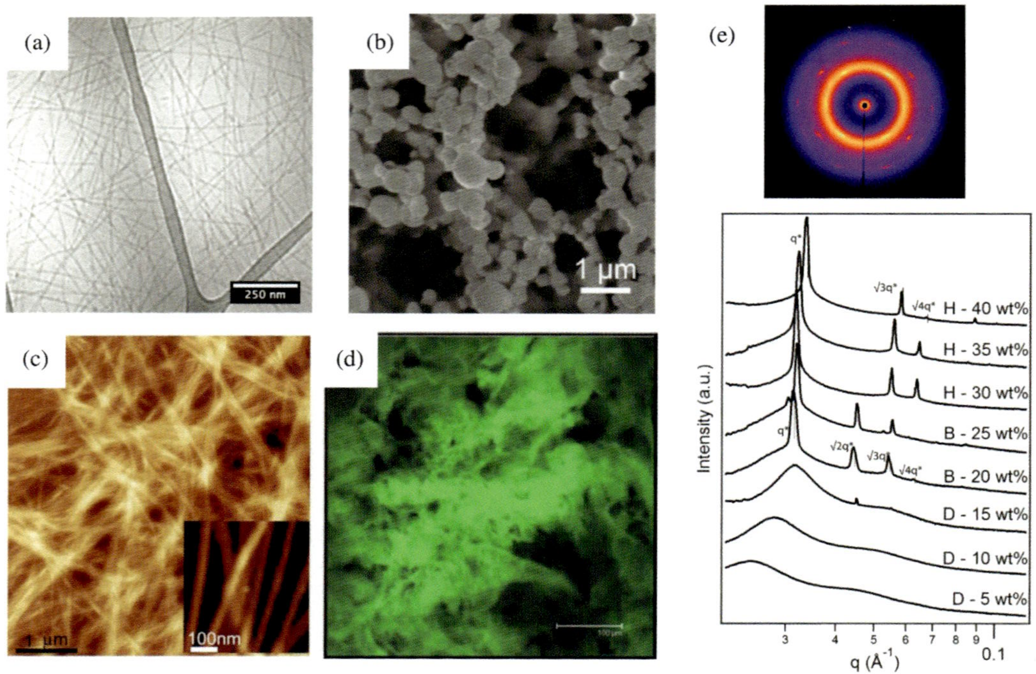


Figure 3.4 Common techniques to study the broad range of length scales involved in physical hydrogels. Microscopy techniques span tens of nanometers to hundreds of $\mu$m: (a) Cryo-TEM of peptide-amphiphile fibrillar gels. Reprinted from [157] with permission from the American Chemical Society, and (b) SEM of gelatin nanoparticle gels. Reprinted from [184] with permission from the American Chemical Society and (c) AFM of fibrillar gels from Fmoc-functionlized short peptides. Reprinted from [85] with permission from Elsevier (d) and CLSM of block copolypeptide gels. Reprinted from [87] with permission from the Royal Society of Chemistry, and (e) Small angle scattering techniques provide information in reciprocal space. SAXS patterns for well-ordered triblock copolymer coacervate hydrogels are shown, exhibiting spots and rings in the two-dimensional scattering pattern (top) and sharp peaks in the one-dimensional reduced data. Adapted from [123] with permission from the American Chemical Society.

Environmental SEM (ESEM) [81] has been developed to allow for imaging in humidified environments, and cryogenic SEM (cryo-SEM) [82] allows for the imaging of samples in the presence of frozen or vitrified water, depending on sample preparation. The appropriate treatment of artifacts from either sample processing, dehydration, or ice crystal formation remains a challenge in using EM techniques to study gel structure. Traditional or cryogenic transmission electron microscopy (TEM, or cryo-TEM) [83, 84] are effective on dilute and particulate gels to examine the structure of individual gel constituents rather than their complete assembly into a bulk gel. Because this technique inherently requires thin samples, there are limits to the mechanical properties that can be understood from this structural perspective. Thus, TEM is a more powerful approach when used in combination with other techniques to understand how molecular components assemble to form the bulk hydrogel.

Atomic force microscopy (AFM) is an alternative surface imaging technique that can be applied to understand the structure of gel substituents [85]. Typical studies involve coating a surface from a dilute solution and using height or phase contrast to identify individual molecules or small aggregates. Because AFM is a surface technique, caution should be exercised in interpreting the three-dimensional assembly of gel aggregates in the absence of the imaging substrate. However, AFM can also provide information about the mechanical behavior of the gel surface, which can be particularly useful for understanding cell–matrix interactions [86]. Finally, laser scanning confocal microscopy (LSCM) can be used *in situ* to investigate gel heterogeneity at micron-length scales [87], although in many systems the relevant structures are difficult to image with traditional resolution instruments.

Scattering techniques, particularly light, X-ray, or neutron scattering, are complementary to imaging/microscopy techniques, and they can easily provide structural data *in situ* and non-destructively for bulk samples. Additionally, scattering samples a macroscopically large volume (typically 1 $mm^3$), avoiding sampling bias potentially involved in the presentation of nanoscale imaging data. However, the drawback is that scattering experiments provide information in reciprocal space that cannot be directly transformed into real space structures. Structural models must often be used to fit the pattern of scattered intensity for any given experiment, extracting structural parameters for a given assumed model. Thus, scattering is best employed when there is strong evidence favoring a given class of gel structure, either from similarities in molecular architecture to known systems or from direct imaging, which can inform the selection of an appropriate structural model. The choice of scattering technique depends primarily on the desired length scale, which is generally within several orders of magnitude of the wavelength of the radiation used in the experiment. X-ray scattering can be used to access atomic and nanometer length scales, as is done in X-ray diffraction (XRD) and small-angle X-ray scattering (SAXS), while visible light scattering can probe micron-scale objects. In addition, the choice of scattering technique is informed by the origin of contrast in the sample. For instance, the scattering of X-rays is due to electron density contrast within the sample and is useful when high atomic weight species are present, whereas the scattering of neutrons is due to nuclear interactions and can discriminate even between low atomic weight species and their isotopes. Neutron scattering is particularly useful in hydrogels where deuterium/hydrogen contrast can easily be generated between the gel and solvent through blending of heavy water. Scattering can also provide unambiguous information regarding orientational correlations among structural elements in hydrogels, specifically from the angular dependence of scattered intensity in the two-dimensional pattern (i.e., the presence of spots rather than rings) [48]. Thus, scattering is an extremely powerful tool for understanding many aspects of hydrogel structure relevant to its bulk mechanical properties and ability to interact with cells or tissues in biomedical applications.

### 3.3.3 *Mechanics during injection*

While the structure and mechanics of physical hydrogels has been studied experimentally and theoretically in detail, the physics of injectability has drawn more attention recently because of the increased demand for novel biomedical solutions. For responsively associating physical hydrogels that are prepared as viscous liquids and only gel under physiological conditions (such as temperature or pH), injectability is related to the solution viscosity. Viscometry can be performed on various instruments and is commonly performed on an oscillatory shear rheometer in steady-shear mode with a cone-and-plate or Couette geometry. However, for physical gels that are made injectable via a yielding mechanism, such as shear-thinning gels, injectability is directly controlled by the yield stress under flow in a syringe–needle geometry, which is a combination of shear and extensional flow. The physical details of the yielding of a shear-thinning gel can be quite complex, related to strain localization [70, 88, 89], shear banding [90], and fracture [91–93], and is beyond the scope of this review. Interested readers are directed to a recent review on the subject [94].

While studies of the detailed mechanics of the injectability of viscoelastic solids are ongoing, in practice there are a number of specific concerns that must be addressed in the design of shear-thinning hydrogels for injectable applications. First, it is necessary to be sure that the gel can be injected using a manual force suitable for clinical applications. In addition to performing test injections using syringes, measurements of shear and extensional stresses during start-up or steady flow can be made using the appropriate rheometers and correlated with injectability. Second, the responsive transition or recovery and self-healing kinetics of the gel must be rapid enough to ensure the formation of a solid implant post-injection in the tissue environment. Typically, the timescale for recovery of shear-thinning gels after network disruption is studied following deformation to high strain in steady or oscillatory shear [6]. In some cases, multiple recovery timescales may be evident when there are multiple mechanisms for healing [48, 95]. These measurements can be used to quantitatively identify promising formulations suitable for the complex process of implant formation. Finally, it can be important to understand the flow profile in the syringe during injection. This issue was relatively underappreciated until it was discovered that certain shear-thinning gels could be cytoprotective for injectable delivery [4–6, 49]. While the historical polymer literature on injection molding is a relevant reference for describing flow under such conditions, only recently has the flow profile during injection been computationally investigated in order to understand how shear-thinning and self-healing physical gels behave during injection [4]. A qualitative understanding about the correlation between properties such as the gel plateau modulus and cell viability can be developed for one particular example of a physically crosslinked gel, but further development is needed to generalize this understanding to the diverse injectable systems seen in the literature.

## 3.4 Associative Polymer Gels

In the previous sections, general concepts in the design and study of injectable physical hydrogels were introduced. The next four sections will provide an overview of each of the primary classes of physical hydrogels. These sections will include details on the fundamental physics governing gel structure and mechanics, as well as present several examples of promising systems from the literature organized by their injectable flow mechanism: responsive self-assembly, or shear-thinning. The discussion will begin with hydrogels with physical interactions on the smallest length scales: associative polymer hydrogels.

Associative polymers are made of a relatively large hydrophilic backbone with many possible chain architectures (linear, branched, star, etc.) decorated with functional groups that tend to aggregate in solution. The aggregating moieties can be incorporated along the backbone, be attached as side chains [30], or exist on the ends of the polymers (telechelic polymers) [79, 80]. Associating moieties come in a variety of forms: short hydrocarbon chains [10], aromatic functionalities that undergo $\pi$–$\pi$ stacking [96], ionically crosslinkable acidic or basic functionalities [97], metal-coordinating moieties [98], or hydrogen bond-forming groups [99]. Associating domains with higher complexity can be formed by combinations of the smaller moieties, such as is the case for DNA or RNA complementation, polypeptide folding and association, and protein–protein binding. In the case of these biological macromolecules, the ability to specify the sequence of associating moieties along the linear polymer chain can lead to specific interactions that provide an improved level of control over the topology of the physical networks that form. In all cases, the association energetics affect the amount of work it takes to perturb or deform physical networks formed by these polymers, but gel behavior is also governed by the association valency and geometry: the number and orientation of associating moieties that can or need to come together to form a network junction. For example, network junctions formed from the bundling of $\alpha$-helices tend to contain a well-defined number of associating units and introduce a strong bias in the chain configurations surrounding the bundle, while junctions formed from the aggregation of hydrophobes contain a broader number of moieties per junction and impose fewer constraints on the linker chains [7]. In all cases, the associating groups can be incorporated along the backbone [11] of the hydrophilic polymer or attached as side chains, [100] and they can be arranged randomly [4] or regularly [8] spaced along the chain. They can be synthesized by incorporating the associating moiety as a monomer prior to polymerization [101], or a hydrophilic polymer can be post-functionalized with the desired group [10, 102]. In the remainder of this section, some core concepts on the physics of associative polymer gels will be discussed, followed by specific examples from this structural class of physical hydrogels. These examples will be grouped according to the mechanism of injectability: responsively triggered associations, or shear-thinning and self-healing gels.

### 3.4.1 *Overview of the physics of associative polymer hydrogels*

Despite the variety of architectures, associating groups, and responsive behaviors seen in associating polymer hydrogels, these materials tend to have phenomenologically similar mechanical behavior. Solutions of these polymers typically show viscous behavior on long timescales due to the transient interactions that occur between the associating groups, but when the average lifetime of the association is sufficiently long (longer than the observation time), a transient network is formed and the solution exhibits elastic responses. A distinguishing aspect of associative polymer gels is the low valency or aggregation number of each node in the transient network, but the physical dimensions of the node can range from the atomic to the nanoscale while retaining similar phenomenological behavior (Figure 3.5). For instance, cyclodextran (~1 kDa) can be used to form shear-thinning physical hydrogels through interactions with small guest molecules [103], and even bundle-forming protein helices [11] (~10 kDa) and antibody–ligand associations (~150 kDa) will form hydrogels with similar viscoelastic responses.

The choice of the associating group determines the solution conditions (temperature, pH, ionic strength, etc.) under which elastic and viscous responses will be observed. For relatively weakly binding groups, warming to physiological temperatures can soften the gel by increasing the exchange rate among associating groups. In these cases, changing the local density of associating groups to increase cooperativity can have significant effects on the relaxation behavior of the gel [97]. When weakly acidic or basic groups are involved as associating functionalities, relatively small changes in pH can affect (or be used to influence) the transient network. Charged moieties will also be affected by ionic strength due to electrostatic screening that can occur at high salt concentration, which may be chosen to be different in the injectable formulation relative to the tissue environment. In addition, certain associating groups require the presence of soluble small molecules (such as divalent cations) to form the physical crosslink, potentially imposing some constraints on the salt content in the targeted tissue environment.

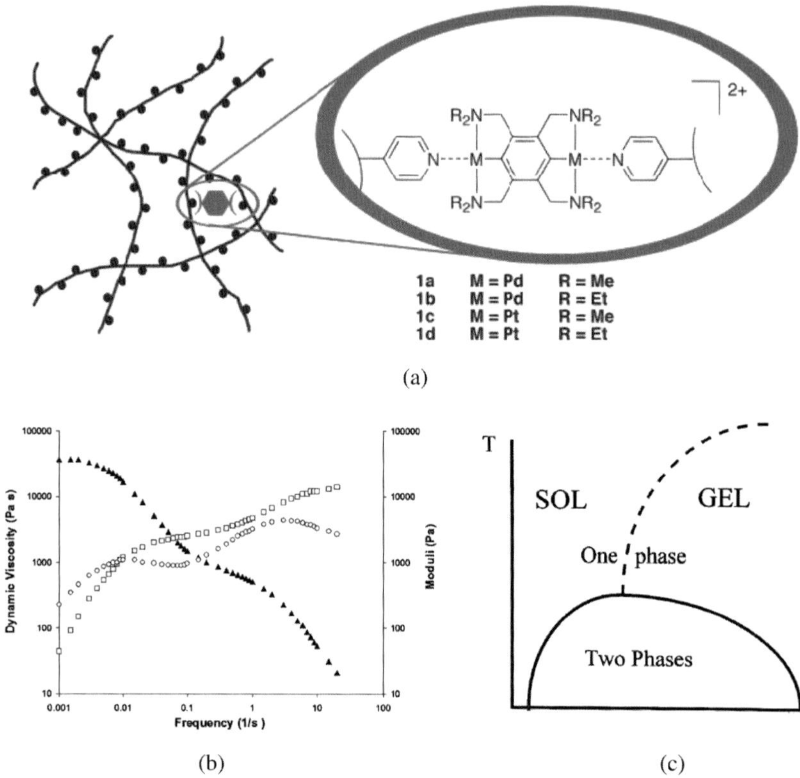


Figure 3.5 Associative polymer gel physics. (a) Schematic of multicomponent associative polymer crosslinked using metalion coordination chemistry with various combinations of ligands and metal cations. Reprinted from [14] with permission from the American Chemical Society. (b) Linear oscillatory shear rheology of multicomponent crosslinked gels with two primary relaxation modes spanning a broad frequency range. (c) A general description of the phase behavior of associative polymer gels. Reprinted from [20] with permission from Elsevier.

Both the chemistry of association and the molecular structure of the associating polymer are critical in determining the thermodynamic conditions under which gelation will occur. An important problem is understanding the macroscopic phase behavior of the physical gel based on its association dynamics and molecular architecture. In particular, it is important to know what conditions will result in a heterogeneous or two-phase region, which may be undesirable as the result could be the formation of a dense gel phase and a dilute sol phase below the gelation concentration ($c_g$). Theoretical treatments of the thermodynamics of physical gels can be used to predict the $c_g$ as well as the concentration dependence of the bulk properties such as the elastic modulus [19, 21, 104]. Several trends are important to highlight. First, increasing either the strength or density of associations with all other parameters held constant tends to promote macrophase separation, effectively increasing $c_g$. Additionally, the water solubility of the backbone polymer can influence the $c_g$: chains that lead to high swelling between associating groups can counteract the tendency to form macroscopic aggregates or precipitates that are driven by phase separation. Furthermore, the potential for intramolecular associations can lead to chain looping that does not contribute productively to the gel network, promoting macrophase separation and raising the $c_g$. Slow macrophase separation upon triggering gelation typically results in turbid gels that will exhibit syneresis, slowly expelling water as the gel phase compacts.

The origin of elasticity in associative polymer hydrogels is typically described quantitatively using classical transient network theory, which relates molecular structure and association energetics to the bulk viscoelastic properties of the gel [23, 24]. It is well established that the proportions of various network topological structures, such as bridges, loops, superloops, and dangling chains will change how the polymeric chains contribute to the viscoelastic behavior of the gels. In short, the greater the number of chains that behave as true bridges linking the nodes of the network at a particular experimental frequency, the stiffer the gel. A typical approach is to estimate the fraction of elastically effective chains in the network, $\nu_{\text{eff}} = \text{G}'/\text{nkT}$, where n is the number density of chains between associating groups in the system and G′ is the plateau modulus. This equation applies in the affine network limit, where crosslinks are assumed to be permanently connected to the bulk network, and where microscopic deformations match bulk deformations [105]. $\nu_{\text{eff}} < 1$ indicates that not all of the polymer chains between associating groups are contributing to the gel network.

A final consideration is that of gel degradation and erodability, both of which are critical determinants of gel lifetime following implantation. For chemically crosslinked gels, network degradation is due to the chemical degradation of the polymer chains or the crosslinks themselves, typically due to hydrolysis. For physical hydrogels, the transient nature of the association means that polymers in the gel are in a state of dynamic exchange between free and bound states. Associating group energetics can be modified to manipulate the equilibrium rates of exchange, but network topology also plays a very important role in controlling gel erosion [7]. In this study, it was shown that the erosion of associative polymer gels typically involves the release of clusters of macromolecules, and introducing topological constraints can increase the minimum cluster size that is likely to disengage from the gel surface and result in significantly lower effective erosion rates.

### 3.4.2 *Responsively self-assembling injectable associative polymer gels*

In general, a responsively associating polymer gel can be engineered to respond to numerous complex stimuli. In practice, biomedically relevant materials in this class are primarily designed to utilize electrostatic interactions (Figure 3.6a). Because thermoresponsive gelation usually involves larger polymeric associating moieties [12, 106], they tend to form gels with

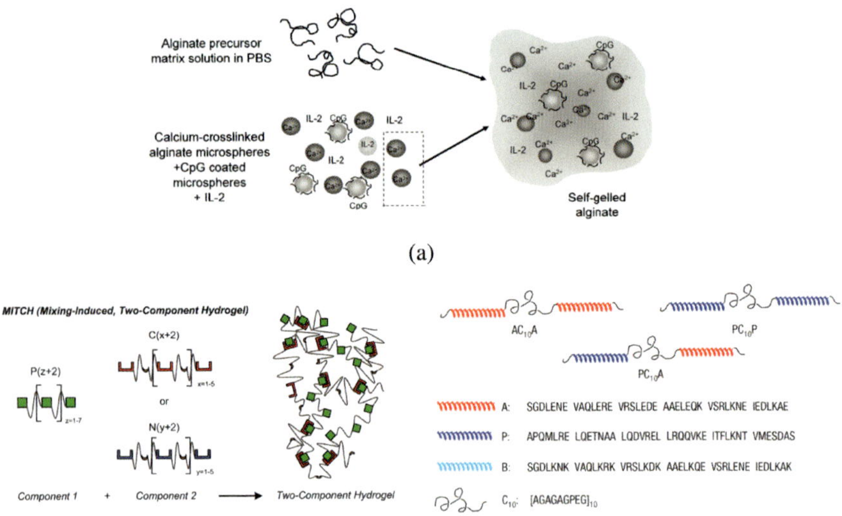


Figure 3.6 Examples of associative polymer hydrogels used in injectable biomedical applications. (a) Responsive crosslinking of soluble alginate due to calcium ion redistribution from microspheres. Reprinted from [111] with permission from Elsevier. Shear-thinning associative protein hydrogels prepared using (b) molecular recognition domains in a two-component formulation. Reprinted from [16] with permission from the National Academy of Sciences or (c) $\alpha$-helical self-associating domains, joined by soluble linker sequences. Reprinted from [7] with permission from Macmillan Publishers, Ltd.

nanoscale associating domains that are better described by the physics of micellar gels (these examples will be discussed with that class of materials). Charge interactions can be tuned by pH, ionic strength, and salt type, offering a number of options for achieving a sharp transition in rheological response based on solution conditions [107]. In addition, electrostatic associative domains with relatively strong individual interactions can be constructed using a handful of atoms (i.e., carboxylic acid or primary amine groups), making them well suited as sticker-type modifications to hydrophilic polymers [97].

There are a number of important examples from living systems where transient network formation due to electrostatic complexation and metal-ion coordination play an important role in tough, self-healing, and adhesive biomaterials [98]. Synthetic hydrogels that incorporate these moieties, such as catechol functional groups, demonstrate important control over gel mechanics with pH and metal cation identity [102, 108]. This level of topological and energetic control of transient networks enables precise design of the mechanical response of physical hydrogels [14, 29] potentially an important tool in developing biomaterials given the complex viscoelastic response of living tissues. A common choice of cation for the responsive crosslinking of multicomponent physical hydrogel formulations is $Ca^{2+}$ [31, 107], another bioinspired strategy whereby the cations will crosslink natural anionic polysaccharides such as alginate by forming aggregates randomly along the linear polymer backbone. Calcium-alginate gels have attracted particular interest for biomedical applications because of their synthetic simplicity, biocompatibility, and more recently their incorporation into extremely tough physical hydrogels [109]. Responsive crosslinking can be achieved by the triggered release [110] or redistribution [111] of calcium from depots stored within the pre-gel formulation. To date, injectable formulations of these toughened materials have not been developed. However, an important concern when using non-physiological concentrations of small-molecule crosslinkers is their ability to diffuse out of the gel into the surrounding tissue environment. This can be addressed by the preparation of single component systems containing both charged species distributed along a polymer backbone [97].

### 3.4.3 *Shear-thinning injectable associative polymer gels*

Because the transient interactions in associative polymer gels can often be disrupted by shear under clinically relevant conditions, many materials in this class do not need to be responsive to be injectable. The shear-thinning behavior of associative polymer systems has been investigated in detail in many fundamental studies [10, 20, 112], but recent work targeting biomedical applications has involved many examples of protein-based associative polymers utilizing specific molecular recognition [13]. The perfect synthesis of sequence-controlled polymers via biosynthesis enables the incorporation of network junctions with very specific association rules (i.e., controlled valency and relative spatial orientation) that can allow for excellent control over network topology [6, 7, 113, 114]. Two important groups of protein gels in this category contain networks made from: (1) bundle-forming $\alpha$-helices or (2) specific protein–protein binding domains (Figure 3.6b,c).

Hydrogels made from polymers containing $\alpha$-helical associating domains can be synthesized using bacterial expression or by a combination of biological and chemical synthesis techniques [115]. The helical sequences can be derived from natural sources, such as proteins that form coiled-coil domains like collagen [7] or the leucine zipper motif from DNA-binding proteins [11, 116] and hydrophilic linking chains can be made using neutral synthetic polymers such as PEG or negatively charged artificial polypeptides [114]. Gels with relatively large fractions of elastically effective chains can be prepared by controlling the geometry of the physically associating junction [7]. The shear-thinning and injectable nature of these hydrogels has been studied in detail, specifically for the purpose of cell encapsulation and delivery. A large fraction of encapsulated cells survived the injection process, which was found to be a result of shear banding during injection that minimized the velocity gradient across the diameter of the needle [6]. Despite the fact that they are typically brittle materials, these $\alpha$-helical associative proteins can also be engineered to exhibit high extensibility by oxidatively polymerizing the chains to prepare a broad distribution of high molecular weight polymers that become physically entangled [8]. Hydrogels prepared from these proteins could be extended up to 3,000% engineering strain.

Associative polymer hydrogels utilizing protein–protein recognition and binding events can also be prepared biosynthetically, forming shear-thinning gels that can be used for the convenient encapsulation and delivery of delicate cargo. A common synthetic strategy employed with these gels is to prepare multicomponent mixtures of hetero-associating polymers. This approach allows for gelation only after complete mixing of separate solutions containing the polymers, enabling facile cell encapsulation under mild conditions without requiring shifts in temperature, pH, salt, or other triggers that might be problematic for handling sensitive cell types such as pluripotent stem cells. The typical products of this approach, such as mixing-induced two-component hydrogels (MITCHs), are soft gels that can maintain cell viability during injection and have been used for the delivery of many delicate cell types [16, 117]. Similarly, Dock-and-Lock (DnL) protein hydrogels have achieved success as encapsulation and delivery scaffolds containing stem cells [118] or immunomodulatory factors such as interleukin-10 [119].

### 3.4.4 *Summary of design considerations*

Associating polymers can be made from a chemically diverse set of functional groups using at least one of many physical interaction mechanisms, and a number of these associative systems can be formulated into injectable hydrogels. These association rules, molecular architecture, and interaction strength are extremely important in controlling the structure and dynamics of the transient network. For network structure, it is crucial to control the balance of intra- and intermolecular associations, which in practice will determine important properties such as gel stiffness and swelling behavior. For network dynamics, it is critical to recognize that relaxation of the transient network will cause time- and rate-dependent mechanical responses, which in practice will influence gel properties such as creep response. Ultimately, both network structure and dynamics will control the gel's yielding and erosion behavior. Advanced network engineering strategies, such as mixing types of associating groups, can produce extensible and tough gels, but it can be challenging to find an approach that simultaneously maintains injectability over a broad range of conditions. While the synthetic diversity within even this one class of gels can lead to increasingly complex designs, engineering the mechanics of associative polymer gels will involve significant uncertainties. In particular, for the design of biomedically relevant injectable gels, complexity can be a significant source of risk for both safety and efficacy.

## 3.5 Micellar Gels

When the network junction is relatively small compared to the effective chain length between active junctions, as for the associative polymer gels discussed in the previous section, node size effects are typically neglected and the behavior of the gel is assumed to be dominated by the deformation of the linker chains. However, as the size of the associating domains grows from a point junction to a micellar aggregate of associating groups, new contributions to the viscoelastic response of the gel arise. Translocation of entire micelles, corona swelling, core chain pullout, and entanglement of corona chains will all make time-dependent contributions to the stress response of the bulk gel [33]. Each of these effects will have strong concentration-dependent behavior: for instance, the large increase in stiffness associated with the close-packing of hydrophobic domains occurs above a critical micellar volume fraction that is typically much larger than the $c_g$ at which elastic effects are first manifest [32].

### 3.5.1 *Overview of the physics of micellar gels*

Micellar gels are hierarchically structured materials, made from interacting nanoscale aggregates of non-covalently bonded molecules (Figure 3.7). They can be formed from amphiphilic block copolymers [120], peptide–lipid conjugates, [46] or surfactants [121, 122] at sufficiently high concentration. The molecules must be designed such that the supramolecular aggregates are repulsive on some length scale, for example, by having a highly hydrophilic or charged surfaces or corona, otherwise a precipitate will tend to form rather than a swollen hydrogel. Often, micellar repulsion is achieved by using amphiphilic molecules and relying on the hydrophobic component to drive assembly, although ionic [123] or hydrogen bonding [124] interactions can also be used to stabilize the nanoscale aggregate.

The thermodynamic behavior of micellar gels typically has a strong dependence on the aggregate structure. Unlike fibrillar gels that can often have extremely low gelation concentrations ($c_g$) around 1%, micellar gels typically require higher concentrations around approximately 10% because the micelle aspect ratio is not as large as is achieved in fibrillar gels, and close packing is required for the excluded-volume interactions to introduce a viscoelastic network. However, micelle bridging via the use of ABA or ABC triblock copolymers with hydrophobic endblocks can lower the $c_g$ [42]. In general, the phase behavior of micellar gels is complex, and solgel coexistence can occur depending on the attractions between the micelles [125, 126]. The onset of macrophase separation upon entering a coexistence regime will yield turbid gels that tend to separate into a dense gel phase and a dilute sol phase. In addition, micelle morphology (spherical, cylindrical, etc.) can be controlled through molecular architecture

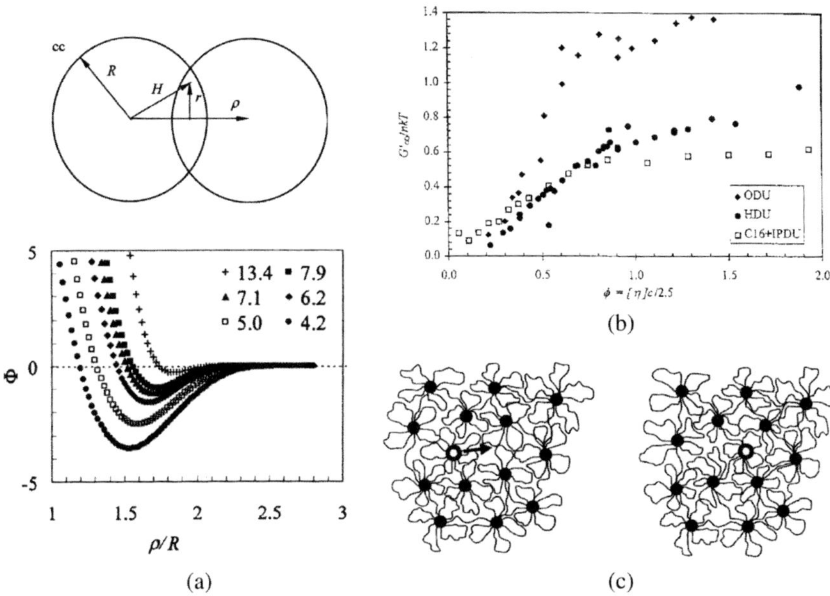


Figure 3.7 Micellar gel physics. (a) Calculation of the intermicellar potential for solutions of telechelic polymers for different degrees of chain stretching (R/N1/2), exhibiting minima due to chain bridging. Reprinted from [126] with permission from the American Chemical Society. (b) Fraction of elastically effective chains ($\nu_{\text{eff}}$) in end-capped polymer gels as a function of micelle volume fraction. Reprinted from [32] with permission from the American Chemical Society. At sufficiently high concentrations $\nu_{\text{eff}} > 1$, indicating that excluded volume interactions among micelles are beginning to contribute to the elastic modulus of the network. (c) Schematic example of network relaxation by micelle hopping, requiring micelle deformation, rearrangement of bridging chains, and relaxation of any corona chain entanglements. Reprinted from [33] with permission from the American Chemical Society.

and solution conditions generally by manipulating the interfacial curvature of the micelle [127], having important consequences for the linear and nonlinear mechanics of the resulting gel [15, 34].

Unlike gels from associative polymers, for which elasticity originates from the entropic elasticity of the polymer chain segments between crosslinks, micellar gels behave as viscoelastic solids because of the combination of repulsive and attractive forces exerted by neighboring micelles. Specifically for block copolymer micelles, theory predicts contributions to the modulus that arise from multiple mechanisms, including the osmotic pressure, stretching, and entanglement of chains in the micelle corona

[32, 33]. Terminal stress relaxation occurs as micelles deform and migrate in the network. In addition, the bridging of micelle cores will influence the gel's viscoelastic behavior, as is possible in the case of micellar gels from ABA triblocks with hydrophobic A blocks. For bridged micelles, endblock dynamics can play a significant role in controlling the longest stress relaxation time of micellar gels because micelle migration cannot occur on timescales faster than endblock pull-out. In certain cases, endblock pull-out will be the slowest relaxation timescale in the gel, as is the case for ABA triblocks with glassy endblocks in a midblock selective solvent [128]. For non-spherical micelles, additional mechanisms such as micelle breakup and reformation can be involved in stress relaxation, particularly under flow conditions similar to injection through a syringe [129–131].

### 3.5.2 *Responsively self-assembling injectable micellar gels*

Injectable micellar gels are most commonly prepared as liquid formulations that responsively self-assemble under physiological conditions (Figure 3.8). Thermoresponsive polymers such as poly(N-isopropylacrylamide) (PNIPAM), poly(oligoethyleneglycol methacrylate) (POEGMA), or elastin-like polypeptide (ELP) [132] are commonly used to trigger nanoscale assembly upon heating to body temperature. These polymers exhibit lower critical solution behavior, with a transition temperature (typically below 37°C) above which they become predominately hydrophobic. The choice of thermoresponsive polymer can be a matter of synthetic convenience. PNIPAM and POEGMA are biocompatible synthetic polymers that can be prepared using free radical polymerization techniques. Well-defined triblock copolymers with thermoresponsive endblocks can be prepared using living techniques such as reversible addition fragmentation chain-transfer (RAFT) polymerization and will self-assemble into hydrogels at physiological temperatures [12, 133]. ELPs are biosynthetic proteins that can be genetically fused in various architectures to synthesize micelle-forming block copolymer hydrogels [134]. Polycaprolactone-poly(ethylene glycol) (PCL-PEG) block copolymers can also form biocompatible thermoresponsive hydrogels for various applications, including serving as a moldable gel to prevent undesired tissue adhesion following surgical procedures [135]. Other PEG-based combinations with thermoresponsive polymers, such as the family of Pluronics® made with poly(propylene oxide) (PPO), have been extensively studied for drug delivery and tissue engineering applications. In addition, pH-sensitive monomers can be

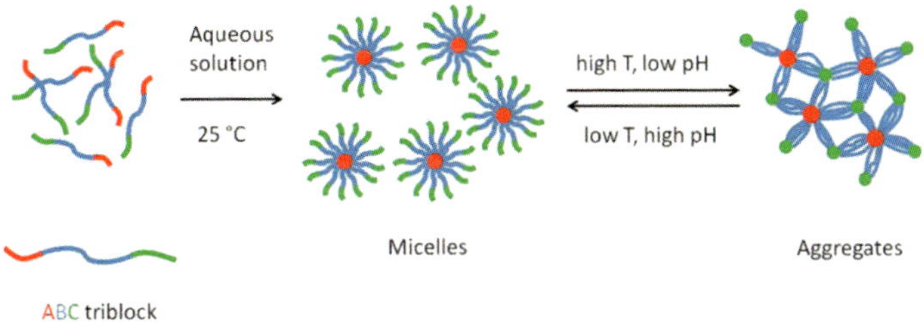


(a)

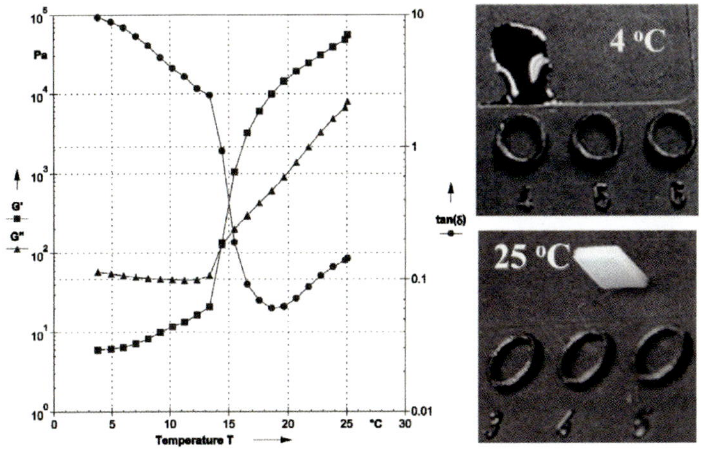


(b)

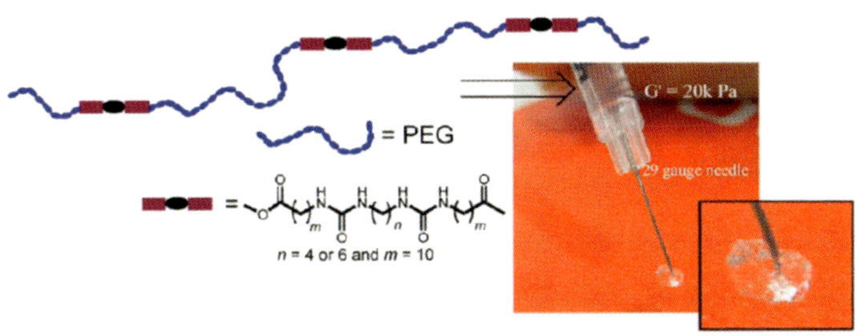


(c)

Figure 3.8 Examples of injectable micellar gels. (a) Thermal and pH-responsive hydrogels prepared from ABC triblock copolymers. Reprinted from [42] with permission from the American Chemical Society. (b) Thermoresponsive hydrogels from ABA-type ELP triblock copolymers. Reprinted from [134] with permission from Elsevier. (c) Shear-thinning injectable hydrogels prepared from the polycondensation of PEG with hydrophobic and hydrogen-bonded bisurea linkages. Reprinted from [124] with permission from the American Chemical Society.

incorporated into block copolymers to provide an additional level of control over micelle assembly. For example, acrylic acid copolymerized with N-isopropylacrylamide will yield temperature- and pH-sensitive polymers that impart a dual-responsive nature to gel-forming triblock copolymers [42]. Furthermore, block copolymer self-assembly induced by a change in solvent composition can be used to produce biomedical hydrogels. While a less frequently exploited approach, triblock copolymers with fluorinated endblocks can be prepared with a biocompatible organic solvent such as N-methyl pyrrolidone: they will gel upon injection into the body due to the exchange of the solvent with water from the tissue environment [45]. This strategy was used to implant physical hydrogels for the delivery of human growth hormone (hGH) with a favorable release profile (release over two weeks with no initial burst), suggesting that the use of cytocompatible organic solvents is still useful for preparing bioactive, implantable hydrogels.

### 3.5.3 *Shear-thinning injectable micellar gels*

Shear-thinning injectable micellar gels are relatively uncommon, likely because of the difficulty of shear-thinning the micelle network during injection under clinically relevant conditions due to the typically high yield stress of the materials. In one case, a PEG step-growth polymer was prepared using bisurea crosslinkers with various alkyl chain lengths [124]. The micellar self-assembly of the hydrophobic segments, stabilized by hydrogen bonding among the urea linkages, led to the formation of a gel that would shear-thin, self-heal on the order of minutes, and was cytocompatible. However, the yield stress increases rapidly with increasing concentration, so there is a practical limit to formulation of the material. Thus, while micellar gels will shear-thin at low enough concentrations, the predominant approach to forming injectable biomedical gels is to use responsive self-assembly to form the micellar network under physiological conditions.

### 3.5.4 *Summary of design considerations*

Conceptually, a micellar gel is closely related to an associative polymer gel, except that repulsive or attractive interactions between the aggregates of network associating domains contribute to the gel's mechanical behavior. This particular network structure is usually formed when responsively hydrophobic block polymers are utilized to drive physical gelation. In particular, thermoresponsive moieties with transitions in a clinically relevant temperature range tend to have molecular weights above 5 kDa that will lead to the formation of nanoscale aggregates when assembled. The properties of these gels have a very strong concentration dependence, undergoing transitions between a weakly interacting liquid of micelles to a strongly interacting viscoelastic network. These transitions are accompanied by changes in the elastic modulus of the gel by several orders of magnitude, and will result in structural changes in micelle morphology and packing as well. It is crucial to recognize that the phase behavior of micellar gels can be quite complex and that triggering gelation can result in phase separation and syneresis depending on macromolecular design and solution conditions, having a significant effect on the gel porosity and mechanical response.

## 3.6 Fibrillar Gels

Fibrillar gels are supramolecular, hierarchically structured materials that exhibit viscoelastic behavior due to the transient effects of complex fiber interactions, including bundling, branching, and entanglement. These nanostructured materials can manifest elastic-like behavior at significantly lower concentrations (often less than 1% (w/v)) than either micellar gels or associative polymer gels due to efficient fiber formation and the high aspect ratio of the stiff nanostructures that produces relatively strong interactions at low densities. A common synthetic motif for robust fiber formation is the use of peptide-based materials, using short peptides [136], peptide amphiphiles [137, 138], or block copolypeptides [139]. Other synthetic moieties can be utilized to form nanofibers from physical interactions such as $\pi$–$\pi$ stacking in coronenes and hydrogen-bonded pairs [140, 141]. A number of examples in the recent literature have demonstrated the particularly promising application of these hydrogels as implantable biomaterials in a diverse array of medical applications, serving as drug delivery depots [142, 143], biofunctionalized scaffolds for tissue repair, [85, 144] and even vaccine adjuvants [145]. The combination of rich chemistry, complex physical behavior, and promising biomedical applications has supported an explosion of research into this particular class of injectable physical hydrogels.

### 3.6.1 *Overview of the physics of fibrillar gels*

Fibrillar gels are commonly assembled from well-defined oligomers and macromolecules such as peptides, which exhibit anisotropic interactions that lead to efficient fiber formation (Figure 3.9) [146]. Examples of these interactions include hydrogen bonding along a peptide backbone that can lead to secondary structure formation, $\pi$–$\pi$ stacking, electrostatic

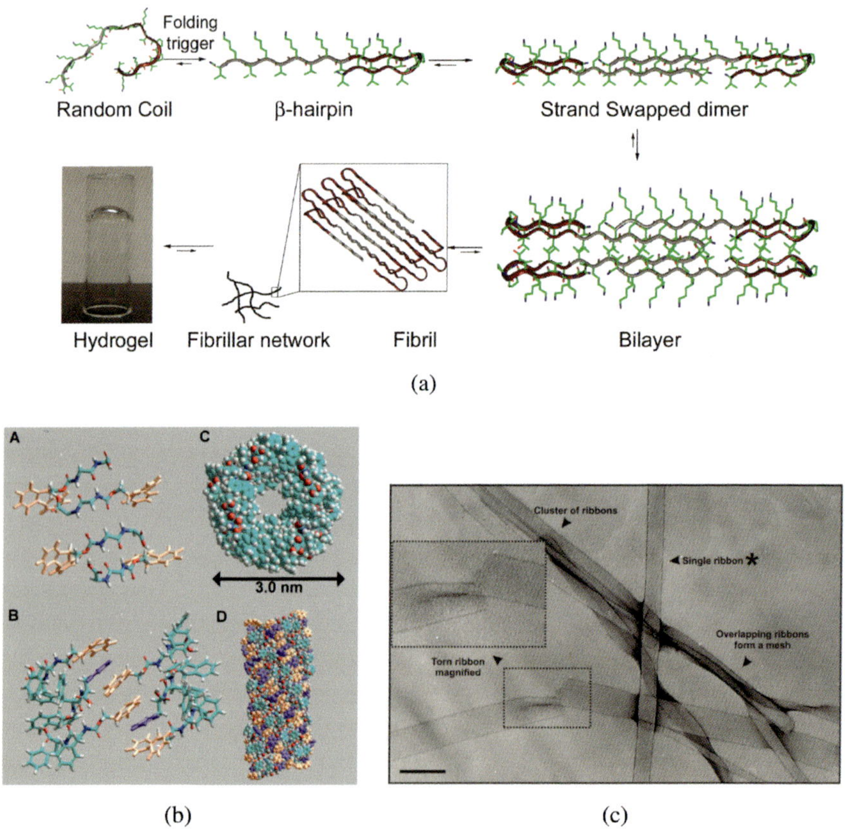


Figure 3.9 Examples of inter- and intramolecular interactions leading to nanofiber assembly. (a) Folding of β–hairpin peptide into fibrillar network due to hydrogen bonding along the peptide backbone and arrangement of hydrophobic/hydrophillic side chains. Reprinted from [146] with permission from the American Chemical Society. (b) Assembly of Fmoc-dipeptide moleucles into individual fibers due to β–sheet interactions among the peptides and $\pi$–$\pi$ stacking of the fluorenyl groups. (c) TEM of Fmoc-dipeptide fibrils indicating the formation ribbon-type structures. Figures in (b) and (c) are reprinted from [82] with permission from John Wiley and Sons.

attraction, and the packing of hydrophobic moieties [141]. Such interactions can lead to fibrillar gels even when very short peptides or other small molecules are employed [85]. Supramolecular self-assembly for fiber formation is an extremely rich and diverse field that involves an array of synthetic, bioinspired, and purely biological molecules used in a broad array of applications [147]. While the design space for fiber-forming molecules is astronomically large, even when considering peptide-based systems alone, several important lessons for efficient fiber formation have been identified through systematic studies. For instance, using alternating hydrophobic/hydrophilic residues and adding $\beta$-turns are common approaches to control peptide folding and fiber assembly in amphiphilic peptides [148]. Maintaining the appropriate balance of charges is key to controlling the assembly when charged amino acids such as lysine are used as the hydrophilic components. In this case, high salt and relatively high pH are needed to prevent electrostatic repulsion and disruption of the supramolecular assembly. The influence of charge balance and screening on assembly can be even more complex, with the formation of twisted fibers, nanotapes, and nanotubes reported for peptides based on sequences implicated in fiber formation in amyloid diseases [149] or derived from silk fibroin [150].

Synthetic polypeptides with blocky structures can also assemble into nanofibers, and control over block ratios is an important route to tuning gel morphology and mechanics. Studies of ABA-type peptides with positively charged A blocks (containing lysine) and alternating hydrophobic/hydrophilic B blocks (containing glutamine and leucine) demonstrated that fiber length could be controlled by balancing the forces along the supramolecular assembly [151]. While these peptides were fairly short, N-carboxyanhydride polymerization can be employed to prepare much larger block copolypeptides. Using multiblock copolypeptides of lysine and leucine, the role of the central hydrophilic block enabled control over fiber morphology; consequentely, the gel mechanical properties could be manipulated by over an order of magnitude up to 1 kPa [87].

Supramolecular structure can also be manipulated by the conjugation of aliphatic moieties to peptides. With this synthetic approach, one level of molecular association is driven by hydrophobic packing of the hydrocarbon tail(s), and peptide interactions play a critical role in guiding assembly. For instance, charged amino acids and solution conditions (pH and ionic strength) will control the transition between dispersed micelles and high aspect ratio fibers, and hydrogen bonding among the peptide chains can influence the fiber nucleation and growth as well as gelation kinetics [152]. Additionally, peptide conjugates containing single and double aliphatic tails have been synthesized to form structures other than fibers, such as micelles and vesicles [148].

In theoretical treatments of semiflexible polymer networks such as fibrillar gels, the elastic modulus is related to number density of physical crosslinks as well as the stiffness and solution behavior of the fibers [153]. Additional fiber properties, such as the longitudinal stretch modulus and length distribution, will influence the nonlinear mechanics and can be responsible for the strain stiffening that is seen in biopolymer networks [154]. The complexity of hierarchical structures, such as fiber branching, twisting, and bundling, has demonstrated consequences for the linear and non-linear mechanical response of the resulting gel. Deconvoluting the contributions from the fiber network (entanglements distribution) and individual fiber mechanics (bending modulus) can be difficult since changes to many parameters in the gel formulation will influence network assembly and can be difficult to generalize [155]. In one study on entangled peptide-amphiphile gels, inter- and intrafiber physical crosslinking performed using divalent cations was reported to influence the linear elastic modulus, yield strain, and recovery behavior of the gels with minimal changes to fiber morphology observed in SEM [156]. The ability to influence fiber rigidity independent of morphology was also reported by controlling secondary structure formation in the hydrophobic sequence of peptide-amphiphile gels [157]. This study demonstrated that peptide-amphiphiles with stronger $\beta$-sheet signatures resulted in stiffer gels, and a model was proposed in which the predominantly linear arrangement of molecules along the fiber axis resulted in stiffer nanofibers than when molecules were organized in a twisted arrangement. The mechanical properties of $\beta$-hairpin peptide fiber hydrogels can also be tuned through chirality, with racemic mixtures of peptides demonstrating up to a four-fold increase in stiffness for soft networks compared to gels made from either of the pure enantiomers [158].

Examples of injectable fibrillar gels from peptidic molecules are abundant in the literature, prepared from peptide-amphiphiles, $\beta$-hairpin peptides, and block copolypeptides via responsive self-assembly or as shear-thinning formulations (Figure 3.10). Responsive nanofiber assembly can be controlled by either turning on aggregation or turning off disaggregation. For example, highly charged gelators may exist as monomers or low valency micelles in solution due to electrostatic repulsion at low ionic strength;

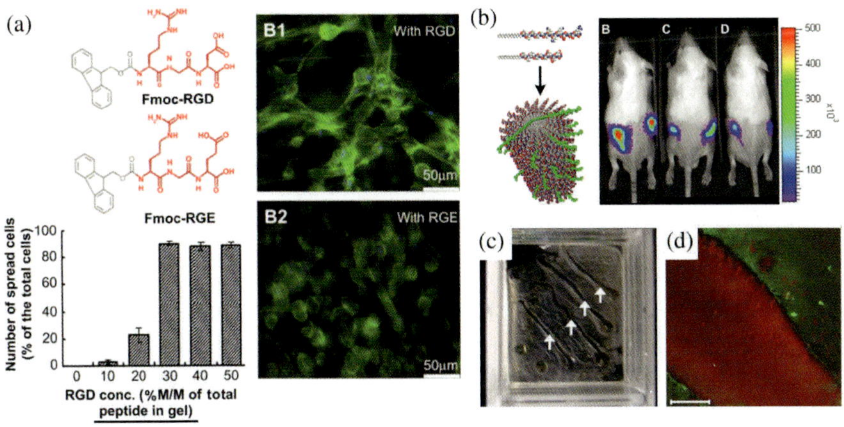


Figure 3.10 Examples of fibrillar hydrogels used in injectable biomedical applications. (a) RGD-functionalized fibrils made from the assembly of short peptides promoted adhesion and spreading of human adult dermal fibroblasts (HDFa). Reprinted from [85] with permission from Elsevier. (b) RGDS-decorated peptide-amphiphile hydrogels supported encapsulation, delivery, and viability of bioluminescent bone marrow mononuclear cells (BMNCs) *in vivo*. Adapted from [174] with permission from Elsevier. (c) Antimicrobial activity of an injectable β-hairpin hydrogel (white arrows) against an MRSA culture. (d) Only bacteria that have come into contact with the hydrogel are killed (stained red). Figures in (c) and (d) are reprinted from [166] with permission from John Wiley and Sons.

after shifting to physiological conditions following injection, molecular aggregation and elongation of fibers can occur. Important structural events during responsive gelation, such as fiber growth, branching, and entanglement, have been observed directly in β-hairpin peptide gels at high ionic strength via cryo-TEM, revealing important relationships between peptide folding, cluster growth, and gel rheology [159]. The rate of gelation in this class of materials could also be increased by sequence mutations that decrease the overall charge of the assembling molecule, or solution conditions that help screen the charges [160]. The dynamics of fiber entanglement under shear has also been studied for the development of shear-thinning hydrogels. In β-hairpin peptide gels, the absence of fiber alignment under shear suggested that the mechanism of shear-thinning could be attributed to the breakup of fibers bridging nanoscale clusters [48]. Recovery following the cessation of shear was characterized by two timescales: an initial rapid bridging of clusters, followed by a slower aging process that reformed the original network structure.

### 3.6.2 *Responsively self-assembling injectable fibrillar gels*

The self-assembly of fibrillar gels in response to diverse triggers has been demonstrated for many peptidic molecules. Peptides that are dispersed under acidic conditions have been shown to fold into shear-thinning $\beta$-hairpin gels when shifted to pH = 9 [161]. The principle of electrostatic control over self-assembly is quite general, and by simultaneously controlling pH and ionic strength under conditions relevant to cell culture (DMEM media, pH = 7.4, $\sim$ 165 mM NaCl), similar peptides were shown to assemble into rigid scaffolds that supported cell proliferation [43, 162].

Biofunctionalization of $\beta$-hairpin peptide gels can broaden the utility of these gels in medical applications. The incorporation of matrix metalloprotease (MMP) recognition sequences into peptides assembled under physiological conditions enabled controlled enzyme-responsive fiber disassembly due to peptide degradation [163]. Furthermore, while certain sequences were found to exhibit favorable immunogenic profiles due to the lack of macrophage activation, [164] they can exhibit inherent antimicrobial activity against a number of different bacteria due to membrane disruption [165]. However, in co-culture experiments, this cytotoxicity was found to be selective and did not appear to affect the viability of mammalian cells. A particularly promising result was that an injectable hydrogel was designed to kill certain drug-resistant bacteria, such as methicillin-resistant *Staphylococcus aureus* (MRSA) [166]. An interesting mechanism for controlling the electrostatic interactions driving peptide folding and assembly involved the use of a photocaged cysteine that was positioned in what would be the hydrophobic face of a self-assembled peptide fiber. The negative charge on the photocaged cysteine inhibited fibrillar assembly, but the peptide readily gelled following photolysis [167]. $\beta$-hairpin peptide gels can also be designed to be thermoresponsive by the substitution of threonine for valine in select positions in the peptide sequence [35].

A variety of other engineered peptides have also been designed to responsively self-assemble. Even in the absence of the type II' turn sequence, the alternation of hydrophilic/hydrophobic residues is sufficient to drive self-assembly in the presence of monovalent and divalent salt. In particular, the release of calcium from the responsive destruction of liposomes could be used to trigger gelation due to temperature, near-infrared (NIR) light, or osmotic shock stimuli [168]. Peptide amphiphiles that contained histidine, serine, and glycine residues, and conjugated to a fatty acid moiety, formed pH-responsive fibrillar gels that assembled above pH 6.5, near the p$K_a$ of the histidine [169]. In addition to supporting two- and three-dimensional cell culture, pH responsive peptide amphiphiles have been demonstrated to support the biomineralization of hydroxyapatite, analogous to the templating function of collagen in bone development [170]. Finally, responsive self-assembly of short ionic dipeptide and tetrapeptide sequences can lead to gelation in response to the proteolysis [171] or even reverse-hydrolysis behavior of proteases [172].

### 3.6.3 *Shear-thinning fibrillar gels*

The distinction between peptide fibrillar gels that can be responsively self-assembled post-injection or injected by shear thinning is largely irrelevant because many of these systems can be implanted in either manner. In fact, there are a number of cases where it is useful to leverage responsive gelation of a shear-thinning network to design a biomedically relevant treatment approach that uses responsive gelation to encapsulate cells and shear thinning to facilitate injection. Salt-responsive, shear-thinning gels for delivering homogeneously encapsulated cells have been prepared from alternating leucine-lysine peptides [173], as well as biofunctionalized peptide-amphiphilies [174]. Additionally, responsive self-assembly of a $\beta$-hairpin peptide gel has been used to achieve similar effects under mild conditions, simply by adding the solution of peptide to a cell suspension in media, and the resulting shear-thinning gel can be injected [5]. Proper control over the gelation kinetics has been shown to be important to maintain the homogeneous cell distribution, and cell viability and retention at the implant site was demonstrated. Particle tracking experiments enabled flow visualization in capillary flow mimicking the injection process, revealing that the yielding mechanism of the gel minimizes shear across the diameter of the capillary [49]. This plug flow-like profile is reported to be responsible for the cytoprotective nature of these gels, similar to earlier studies on associative protein hydrogels [6]. However, certain cell types can be sensitive to the positive charge on the nanofibers during the injection process [175].

Block copolypeptides, prepared using NCA polymerization to access higher molecular weights, will also form shear-thinning hydrogels when assembled into fibers. Amphiphilic diblock copolypeptides with either positively charged lysine blocks or negatively charged glutamic acid blocks readily formed shear-thinning hydrogels at low concentrations, and are unusually stable at high salt and near boiling temperatures [176]. Similar gels did not support cell growth, but gel formation appeared to completely suppress the moderate cytotoxicity seen in dilute solutions of lysine-containing block copolypeptide assemblies [177]. However, poly(lysine-b-leucine) diblock gels demonstrated promise when tested as injectable scaffolds [178] or depots [179] for applications in central nervous system (CNS) tissues. Changing block ratios enabled gel porosity to be manipulated while keeping the gel's modulus constant, and the materials demonstrated low toxicity, good tissue integration, and long-term release of encapsulated growth factors.

### 3.6.4 *Summary of design considerations*

Despite their low molar mass, the sequence space for peptide fibrillar hydrogels can be quite large. Several robust design strategies have been discovered to support the development of many promising formulations that can serve as implantable regeneration scaffolds and templates, cell encapsulation and delivery matrices, and drug delivery depots. Assembly of these amphiphilic molecules is typically driven by the formation of intra- and/or intermolecular $\beta$-sheet interactions, governed by the packing of hydrophobic and hydrophilic side chains. An alternating pattern of hydrophilic/hydrophobic amino acids is a common motif among fiber-forming peptides, and other elements can be added to control fiber morphology, rigidity, and gelation behavior. For instance, $\beta$-hairpin peptide fibers containing type II' turn-promoting motifs tend to organize into fibrils consisting of stacks of planar folded peptides, while peptide-amphiphiles containing aliphatic tails tend to form cylindrical fibril assemblies with radially oriented molecules. Finally, high molecular weight block copolypeptides have also been shown to assemble into extended nanofibers that form rigid hydrogels.

Secondary structure formation causes the high aspect ratio of the supramolecular aggregates, leading to the formation of fibers rather than spherical micelles typically seen in amphiphilic block copolymers at low concentrations. However, without proper balancing of the electrostatic repulsion of the hydrophilic residues, fiber assembly is inhibited, and gelation does not occur. The consequence of this assembly behavior is that many peptide fibrillar gels are pH and salt sensitive, endowing them with responsive assembly that is well suited for homogeneous cellular and molecular encapsulation and injectable delivery. Furthermore, many of these fibrillar gels are shear-thinning, and this effect has been employed throughout the literature for a variety of biomedical applications. Because peptide nanofibers are often positively charged, electrostatic interactions will play a role on the release of negatively charged macromolecular drugs from peptide-gel depots [180], and this charge can be utilized to make antimicrobial gels [181].

## 3.7 Colloidal and macrophase-separated gels

Colloidal gels can be prepared from a broad array of synthetic materials, including tissue-extracted extracellular matrix components, synthetic polymers, metallic nanoparticles, and inorganic clays, yet they exhibit rich physical behavior with many universal concepts (Figure 3.11). Gelation requires the flocculation of particles, which can be nano- or microscale in size, due to a balance of repulsive and attractive interactions [182], yielding a thermodynamically stable or kinetically arrested network of aggregates [37]. A description of kinetically arrested phase separation is also appropriate for accounting for the handful of examples from the literature that describe responsive macrophase-separated formulations that can be used as injectable and reversible viscoelastic solids. *In lieu* of a detailed discussion of the thermodynamics, structure, and mechanics of these materials, a brief overview of select examples will be used to discuss the biomedical applications of these injectable materials.

Injectable colloidal gels can be prepared from extremely simple starting materials, such as mixtures of oppositely charged gelatin microparticles [183]. These gels were shown to be shear-thinning and rapidly self-healing, although injection through narrow-gauge needles was not demonstrated. Simple formulations of electrostatically interacting, degradable dextran nanoparticles can also be prepared as smart materials that can responsively release a cargo of insulin under hyperglycemic conditions [184]. While gelatin can be harvested in either positively or negatively charged forms, glucose oxidase or insulin-loaded dextran nanoparticles could be coated with chitosan or alginate, respectively. Mixing of oppositely charged nanoparticles resulted in hierarchically structured, microporous hydrogels, which are advantageous for delivery applications. Furthermore, these networks could shear-thin and were readily injected through needles.

The behavior of macrophase-separated physical gels is difficult to generalize, but these materials tend to be prepared from simple blends of biologically derived or biocompatible polymers, occasionally modified via simple chemical processing. For example, regenerated silk fibroin can be blended with hydroxypropylcellulose, leading to physical gelation due to

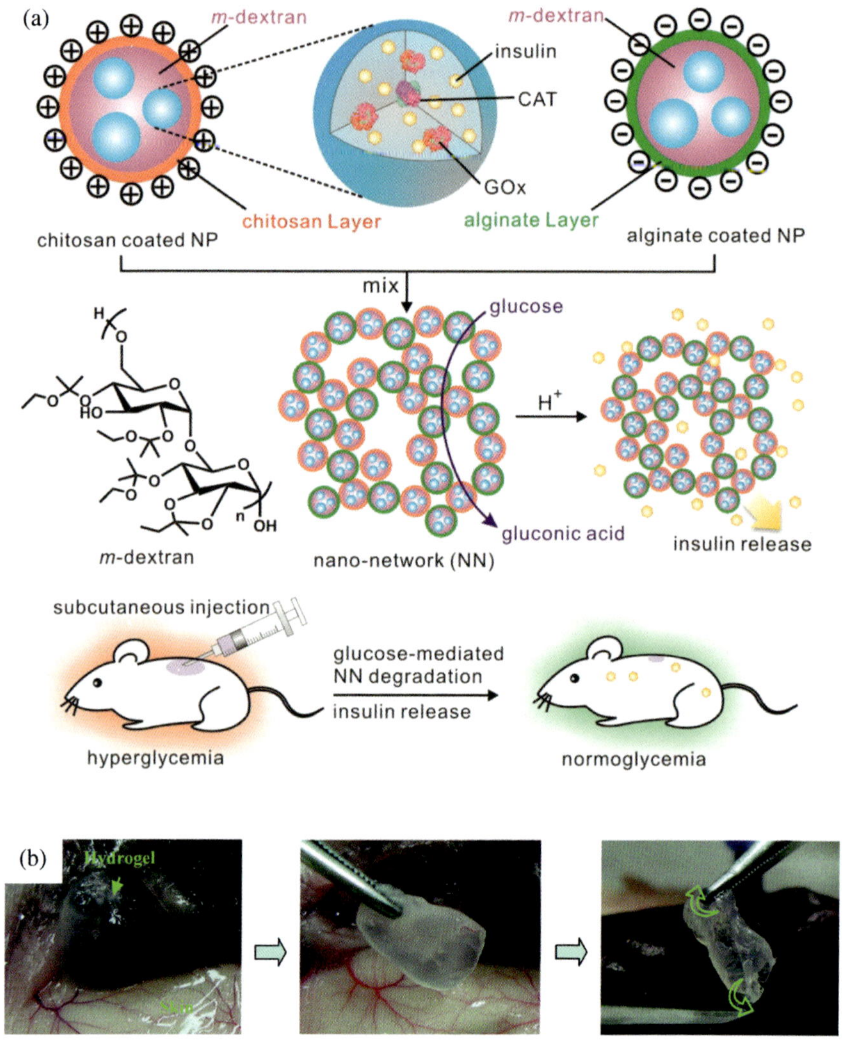


Figure 3.11 (a) Schematic of the loading, colloidal assembly, and injection of oppositely charged dextran nanoparticles for the responsive delivery of insulin under hyperglycemic conditions. Reprinted from [184] with permission from the American Chemical Society. (b) Injectable implantation of hydrophobically modified chitosan that forms a macrophase-separated network in response to a shift to physiological pH. Reprinted from [186] with permission from the Royal Society of Chemistry.

thermally triggered $\beta$-sheet formation in the silk domains. This method produced shear-thinning injectable gels that were also cytoprotective [185]. Hydrophobically modified chitosan can be prepared by attaching N-linked palmitoyl groups, forming viscoelastic liquids due to the associative nature of the pendant aliphatic chains below pH 6.5 [186, 187]. In these examples, the pH-responsive nature of the mildly substituted chitosan backbone was utilized to trigger macrophase separation upon injection into tissues at physiological pH, resulting in up to an order of magnitude increase in the shear elastic modulus.

Examples from this class of physical hydrogels that target biomedical applications generally have disordered but percolating structures and exhibit a porous microscale structure that is strongly dependent on the processing conditions. While it can be difficult to generalize the designs of these gels, that they can be formed with relatively simple processing (physical blending of substituents and occasionally mild chemical modifications) is advantageous for the rapid translation of injectable biomaterials to the clinic.

## 3.8 Perspectives and Conclusion

The design of injectable physical hydrogels will require new fundamental and applied knowledge in an inherently interdisciplinary field, with important consequences for our ability to address major gaps in the treatment of acute and chronic medical conditions. This review began with a physical description of non-covalently crosslinked hydrogels, discussed several common experimental techniques for studying their behavior, and provided an overview of the primary types of physical hydrogels being developed for injectable delivery. These classes of injectable physical hydrogels were considered in order of increasing length scale of interactions: associating polymers with sticker-like interactions, micellar gels of nanoscale aggregates, high aspect ratio nanofibers, and colloidal and macrophase-separated gels. Several examples from each class were discussed in the context of the mechanism by which they were delivered: viscous liquids that responsively self-assembled under physiological conditions, or shear-thinning solids that rapidly recovered and self-healed following injection. Important general design considerations that were addressed throughout this review were: (1) manipulating molecular design to promote intermolecular associations for improved mechanical robustness; (2) consideration of gel thermodynamics (as a function concentration, crosslinking density, association energy, etc.) to ensure the preparation of homogeneous formulations; and, (3) the structural origins of the dynamic mechanical response of physical gels.

Clearly, physically crosslinked, injectable hydrogels are emerging as important biomedical solutions to address critical treatment gaps in drug delivery and tissue engineering, repair, and regeneration [137, 188–191]. Indeed, non-covalent interactions are universal in complex soft matter, governing the structure and mechanics of synthetic systems like implantable hydrogels as well as hierarchically structured biological tissues. Exercising control over these transient, physical interactions is essential to control gelation and injectability for practical applications. Engineering viscoelasticity is also a natural approach to the development of complex systems that can better integrate with tissues at the molecular, cellular, and tissue level [115, 192]. Employing these non-covalent interactions appropriately will support the development of high performance soft biomaterials with emergent mechanical properties such as high toughness, extensibility, and resilience. From this perspective, there are many exciting opportunities for the development of injectable systems that can be molecularly programmed to mature into tissue-like implants on timescales relevant to tissue repair and regeneration. Conceivably, such structural and mechanical maturation may involve numerous timescales and even have multiple triggers for dynamical maturation, including chemical, biological, or mechanical signals. Of course, materials with this complex behavior would still need to be delivered to the patient as an injectable, shelf-stable formulation.

Numerous scientific and technological hurdles remain before such materials can be brought to the clinic. However, it is becoming clearer that the complex cellular-directed assembly of physically and chemically crosslinked biological tissues will provide niches for many different types of crosslinking mechanisms. Combinations of multiple physical interactions (such as associating polymers and micellar gels) and injectable mechanisms (shear-thinning and responsive self-assembly) can yield tough implants [9], and composites made from hybrid polymer–peptide conjugates that form specific interactions with biomolecules to produce hierarchically structured injectable systems [36]. Combining physical and chemical crosslinking is another approach to introducing temporal control into injectable formulations, coupling the benefits of rapidly setting, responsive micellar gelation with long-term stabilization from chemical crosslinking [38, 193–196]. Advances in the design and control of multi-crosslinked, hierarchically structured injectable hydrogels are likely to have a long-term impact on biomaterial development for minimally invasive surgical treatments.

While there is a clear case for increasing the complexity of injectable biomedical materials for addressing material performance objectives, there is also a strong argument to be made for attention to compositional simplicity in the development of products for near-term applications. As has been described throughout this review, the synthetic diversity of systems that achieve similar phenomenological behavior is quite large. The safety profile of a new biomedical material will be assessed in part by its immediate cytotoxicity, acute and chronic inflammatory behavior, and the decomposition and processing of its components. Experimental investigation of this latter issue is largely absent from the literature evaluating new physically injectable biomaterials. While improving, a complete understanding of matrix signals interactions that direct cell behavior is limited. Thus, gel designs that use biologically sourced, tissue-specific extracellular matrix components are being explored to balance the benefits of simplicity with the need to control cell behavior for specific applications [197].

Moving forward, it will be increasingly critical to take a simplifying approach when developing and applying physical hydrogels for biomedical applications. Indeed, common, commercially available, implantable soft biomaterials are made from a handful of synthetic and biologically derived polymers, with ease of production and mitigation of safety and regulatory risks being as important as potency in determining the viability of translational projects [198]. The developmental timeframe for biomaterials is long and costly, requiring multi-year studies for even relatively simple formulations [199]. With careful consideration of these important regulatory issues at the outset, new injectable biomaterials will continue to make impactful, near-term advances as ground-breaking, minimally invasive surgical tools. As management of chemical and physical complexity improves through simple and robust design strategies, high performance gel formulations are likely to dramatically change the way medicine approaches the repair and regeneration of damaged and diseased tissues.

## References

[1] Anseth, K.S. and Burdick, J.A. (2002). New directions in photopolymerizable biomaterials. *MRS Bulletin*, **27**, pp. 130–136.

[2] Toh, W.S., Lim, T.C., Kurisawa, M., and Spector, M. (2012). Modulation of mesenchymal stem cell chondrogenesis in a tunable hyaluronic acid hydrogel microenvironment. *Biomaterials*, **33**, pp. 3835–3845.

[3] Temenoff, J.S., Shin, H., Conway, D.E., Engel, P.S., and Mikos, A.G. (2003). *In vitro* cytotoxicity of redox radical initiators for cross-linking of oligo(poly(ethylene glycol) fumarate) macromers. *Biomacromolecules*, **4**, pp. 1605–1613.

[4] Aguado, B.A., Mulyasasmita, W., Su, J., Lampe, K.J., and Heilshorn, S.C. (2011). Improving viability of stem cells during syringe needle flow through the design of hydrogel cell carriers. *Tissue Eng. Part A*, **18**, pp. 806–815.

[5] Haines-Butterick, L. *et al.* (2007). Controlling hydrogelation kinetics by peptide design for three-dimensional encapsulation and injectable delivery of cells. *Proc. Nat. Acad. Sci. USA*, **104**, pp. 7791–7796.

[6] Olsen, B.D., Kornfield, J.A., and Tirrell, D.A. (2010). Yielding behavior in injectable hydrogels from telechelic proteins. *Macromolecules*, **43**, pp. 9094–9099.

[7] Shen, W., Zhang, K., Kornfield, J.A., and Tirrell, D.A. (2006). Tuning the erosion rate of artificial protein hydrogels through control of network topology. *Nat. Mater.*, **5**, pp. 153–158. Available at: http://www.nature.com/nmat/journal/v5/n2/suppinfo/nmat1573_S1.html.

[8] Tang, S., Glassman, M.J., Li, S., Socrate, S., and Olsen, B.D. (2014). Oxidatively responsive chain extension to entangle engineered protein hydrogels. *Macromolecules*, **47**, pp. 791–799.

[9] Glassman, M.J., Chan, J., and Olsen, B.D. (2013). Reinforcement of shear thinning protein hydrogels by responsive block copolymer self-assembly. *Adv. Funct. Mat.*, **23**, pp. 1182–1193.

[10] Tripathi, A., Tam, K.C., and McKinley, G.H. (2006). Rheology and dynamics of associative polymers in shear and extension: theory and experiments. *Macromolecules*, **39**, pp. 1981–1999.

[11] Petka, W.A., Harden, J.L., McGrath, K.P., Wirtz, D., and Tirrell, D.A. (1998). Reversible hydrogels from self-assembling artificial proteins. *Science*, **281**, pp. 389–392.

[12] Kirkland, S.E. *et al.* (2007). Thermoreversible hydrogels from RAFT-synthesized BAB triblock copolymers: steps toward biomimetic matrices for tissue regeneration. *Biomacromolecules*, **9**, pp. 481–486.

[13] Yamaguchi, N. *et al.* (2007). Growth factor mediated assembly of cell receptor-responsive hydrogels. *J. Am. Chem. Soc.*, **129**, pp. 3040–3041.

[14] Loveless, D.M., Jeon, S.L., and Craig, S.L. (2005). Rational control of viscoelastic properties in multicomponent associative polymer networks. *Macromolecules*, **38**, pp. 10171–10177.

[15] Glassman, M.J. and Olsen, B.D. (2013). Structure and mechanical response of protein hydrogels reinforced by block copolymer self-assembly. *Soft Matter*, **9**, pp. 6814–6823.

[16] Wong Po Foo, C.T.S., Lee, J.S., Mulyasasmita, W., Parisi-Amon, A., and Heilshorn, S.C. (2009). Two-component protein-engineered physical hydrogels for cell encapsulation. *Proc. Nat. Acad. Sci. USA*, **106**, pp. 22067–22072.

[17] Mulyasasmita, W., Cai, L., Hori, Y., and Heilshorn, S.C. (2014). Avidity-controlled delivery of angiogenic peptides from injectable molecular-recognition hydrogels. *Tissue Eng. Part A*, **20**, pp. 2102–2114.
[18] Kolesky, D.B. *et al.* (2014). 3D bioprinting of vascularized, heterogeneous cell-laden tissue constructs. *Adv. Mat.*, **26**(19), pp. 3124–3130.
[19] Dobrynin, A.V. (2004). Phase diagram of solutions of associative polymers. *Macromolecules*, **37**, pp. 3881–3893.
[20] Rubinstein, M. and Dobrynin, A.V. (1999). Associations leading to formation of reversible networks and gels. *Curr. Opin. Coll. Int. Sci.*, **4**, pp. 83–87.
[21] Semenov, A.N. and Rubinstein, M. (1998). Thermoreversible gelation in solutions of associative polymers. 1. Statics. *Macromolecules*, **31**, pp. 1373–1385.
[22] Winnik, M.A. and Yekta, A. (1997). Associative polymers in aqueous solution. *Curr. Opin. Coll. Int. Sci.*, **2**, pp. 424–436.
[23] Tanaka, F. (2011). *Polymer Physics: Applications to molecular association and thermoreversible gelation.* Cambridge University Press, Cambridge.
[24] Annable, T., Buscall, R., Ettelaie, R., and Whittlestone, D. (1993). The rheology of solutions of associating polymers: Comparison of experimental behavior with transient network theory. *J. Rheol.*, **37**, pp. 695–726.
[25] Graessley, W.W. (2003). *Polymeric liquids and networks.* Garland Science.
[26] Rubinstein, M. and Colby, R. (2003). *Polymer physics.* Oxford University Press, Oxford.
[27] Larson, R.G. (1999). *The structure and rheology of complex fluids*, Vol. 4. Oxford University Press, New York.
[28] Yan, C. and Pochan, D.J. (2010). Rheological properties of peptide-based hydrogels for biomedical and other applications. *Chem. Soc. Rev.*, **39**, pp. 3528–3540.
[29] Yount, W.C., Loveless, D.M., and Craig, S.L. (2005). Strong means slow: Dynamic contributions to the bulk mechanical properties of supramolecular networks. *Ange. Chem.*, **117**, pp. 2806–2808.
[30] Yount, W.C., Loveless, D.M., and Craig, S.L. (2005). Small-molecule dynamics and mechanisms underlying the macroscopic mechanical properties of coordinatively cross-linked polymer networks. *J. Am. Chem. Soc.*, **127**, pp. 14488–14496.
[31] Webber, R.E. and Shull, K.R. (2004). Strain dependence of the viscoelastic properties of alginate hydrogels. *Macromolecules*, **37**, pp. 6153–6160.
[32] Pham, Q., Russel, W., Thibeault, J., and Lau, W. (1999). Micellar solutions of associative triblock copolymers: The relationship between structure and rheology. *Macromolecules*, **32**, pp. 5139–5146.
[33] Semenov, A. and Rubinstein, M. (2002). Dynamics of entangled associating polymers with large aggregates. *Macromolecules*, **35**, pp. 4821–4837.
[34] Seitz, M.E., Burghardt, W.R., and Shull, K.R. (2009). Micelle morphology and mechanical response of triblock gels. *Macromolecules*, **42**, pp. 9133–9140.

[35] Pochan, D.J. *et al.* (2003). Thermally reversible hydrogels via intramolecular folding and consequent self-assembly of a *de novo* designed peptide. *J. Am. Chem. Soc.*, **125**, pp. 11802–11803.
[36] Matsusaki, M. *et al.* (2011). Physical and specific crosslinking of collagen fibers by supramolecular nanogelators. *Adv. Mat.*, **23**, pp. 2957–2961 .
[37] Russel, W.B. (1991). Concentrated colloidal dispersions. *MRS Bulletin*, **16**, pp. 27–31.
[38] Klouda, L. *et al.* (2011). Thermoresponsive, in situ cross-linkable hydrogels based on N-isopropylacrylamide: Fabrication, characterization and mesenchymal stem cell encapsulation. *Acta Biomater.*, **7**(4), pp. 1460–1467.
[39] Haque, M.A., Kurokawa, T., Kamita, G., and Gong, J.P. (2011). Lamellar bilayers as reversible sacrificial bonds to toughen hydrogel: hysteresis, self-recovery, fatigue resistance, and crack blunting. *Macromolecules*, **44**, pp. 8916–8924 .
[40] Temenoff, J.S. and Mikos, A.G. (2000). Review: tissue engineering for regeneration of articular cartilage. *Biomaterials*, **21**, pp. 431–440.
[41] Zhou, C., Hillmyer, M.A., and Lodge, T.P. (2012). Efficient formation of multicompartment hydrogels by stepwise self-assembly of thermoresponsive ABC triblock terpolymers. *J. Am. Chem. Soc.*, **134**, pp. 10365–10368.
[42] Koonar, I., Zhou, C., Hillmyer, M.A., Lodge, T.P., and Siegel, R.A. (2012). ABC triblock terpolymers exhibiting both temperature- and pH-sensitive micellar aggregation and gelation in aqueous solution. *Langmuir*, **28**, pp. 17785–17794.
[43] Ozbas, B., Kretsinger, J., Rajagopal, K., Schneider, J.P., and Pochan, D.J. (2004). Salt-triggered peptide folding and consequent self-assembly into hydrogels with tunable modulus. *Macromolecules*, **37**, pp. 7331–7337.
[44] Zhang, L., Furst, E.M., and Kiick, K.L. (2006). Manipulation of hydrogel assembly and growth factor delivery via the use of peptide–polysaccharide interactions. *J. Control. Release*, **114**, pp. 130–142.
[45] Tae, G., Kornfield, J.A., and Hubbell, J.A. (2005). Sustained release of human growth hormone from *in situ* forming hydrogels using self-assembly of fluoroalkyl-ended poly(ethylene glycol). *Biomaterials*, **26**, pp. 5259–5266.
[46] Shimada, T., Megley, K., Tirrell, M., and Hotta, A. (2011). Fluid mechanical shear induces structural transitions in assembly of a peptide–lipid conjugate. *Soft Matter*, **7**, pp. 8856–8861 .
[47] Wang, X., Kluge, J.A., Leisk, G.G., and Kaplan, D.L. (2008). Sonication-induced gelation of silk fibroin for cell encapsulation. *Biomaterials*, **29**, pp. 1054–1064.
[48] Yan, C. *et al.* (2010). Injectable solid hydrogel: mechanism of shear-thinning and immediate recovery of injectable $\beta$-hairpin peptide hydrogels. *Soft Matter*, **6**(20), pp. 5143–5156.
[49] Yan, C. *et al.* (2012). Injectable solid peptide hydrogel as a cell carrier: effects of shear flow on hydrogels and cell payload. *Langmuir*, **28**, pp. 6076–6087.
[50] Sharei, A. *et al.* (2013). A vector-free microfluidic platform for intracellular delivery. *Proc. Nat. Acad. Sci. USA*, **110**, pp. 2082–2087.

[51] McQueen, A., Meilhoc, E., and Bailey, J.E. (1987). Flow effects on the viability and lysis of suspended mammalian cells. *Biotech. Lett.*, **9**, pp. 831–836.
[52] Lutolf, M. and Hubbell, J. (2005). Synthetic biomaterials as instructive extracellular microenvironments for morphogenesis in tissue engineering. *Nat. Biotech.*, **23**, pp. 47–55.
[53] Bryant, S.J., Arthur, J.A., and Anseth, K.S. (2005). Incorporation of tissue-specific molecules alters chondrocyte metabolism and gene expression in photocrosslinked hydrogels. *Acta Biomater.*, **1**, pp. 243–252.
[54] Burdick, J.A. and Anseth, K.S. (2002). Photoencapsulation of osteoblasts in injectable RGD-modified PEG hydrogels for bone tissue engineering. *Biomaterials*, **23**, pp. 4315–4323.
[55] Nie, T., Akins Jr., R.E., and Kiick, K.L. (2009). Production of heparin-containing hydrogels for modulating cell responses. *Acta Biomater.*, **5**, pp. 865–875.
[56] Leight, J.L., Alge, D.L., Maier, A.J., and Anseth, K.S. (2013). Direct measurement of matrix metalloproteinase activity in 3D cellular microenvironments using a fluorogenic peptide substrate. *Biomaterials*, **34**, pp. 7344–7352.
[57] Kharkar, P.M., Kiick, K.L., and Kloxin, A.M. (2013). Designing degradable hydrogels for orthogonal control of cell microenvironments. *Chem. Soc. Rev.*, **42**, pp. 7335–7372.
[58] Tanaka, F. and Edwards, S. (1992). Viscoelastic properties of physically crosslinked networks. 1. Transient network theory. *Macromolecules*, **25**, pp. 1516–1523.
[59] Tanaka, F. and Edwards, S. (1992). Viscoelastic properties of physically crosslinked networks: Part 2. Dynamic mechanical moduli. *J. Non-Newt. Fluid Mech.*, **43**, pp. 273–288.
[60] Bird, R.B., Armstrong, R.C., and Hassager, O. (1987). *Dynamics of polymeric liquids. Vol. 1: Fluid mechanics*. Wiley-Blackwell.
[61] Helgeson, M.E., Moran, S.E., An, H.Z., and Doyle, P.S. (2012). Mesoporous organohydrogels from thermogelling photocrosslinkable nanoemulsions. *Nat. Mater.*, **11**, pp. 344–352.
[62] Winter, H.H. and Chambon, F. (1986). Analysis of linear viscoelasticity of a crosslinking polymer at the gel point. *J. Rheol.*, **30**, pp. 367–382.
[63] Winter, H.H. (1987). Can the gel point of a cross-linking polymer be detected by the G′ – G? crossover? *Poly. Eng. Sci.*, **27**, pp. 1698–1702.
[64] Dae Han, C. and Kim, J.K. (1993). On the use of time-temperature superposition in multicomponent/multiphase polymer systems. *Polymer*, **34**, pp. 2533–2539.
[65] Evans, R., Tassieri, M., Auhl, D., and Waigh, T.A. (2009). Direct conversion of rheological compliance measurements into storage and loss moduli. *Phys. Rev. E*, **80**.
[66] Ewoldt, R.H., Hosoi, A.E. and McKinley, G.H. (2007). Rheological fingerprinting of complex fluids using large amplitude oscillatory shear (LAOS) flow. *Ann. Trans.-Nordic Rheol. Soc.*, **15**, p. 3.

[67] Hyun, K. *et al.* (2011). A review of nonlinear oscillatory shear tests: Analysis and application of large amplitude oscillatory shear (LAOS). *Prog. Poly. Sci.*, **36**, pp. 1697–1753.
[68] Blackwell, B.C. and Ewoldt, R.H. (2014). A simple thixotropic–viscoelastic constitutive model produces unique signatures in large-amplitude oscillatory shear (LAOS). *J. Non-Newt. Fluid Mech.*, **208**, pp. 27–41.
[69] Ewoldt, R.H., Hosoi, A., and McKinley, G.H. (2008). New measures for characterizing nonlinear viscoelasticity in large amplitude oscillatory shear. *J. Rheol.*, **52**, pp. 1427–1458.
[70] Erk, K.A., Henderson, K.J., and Shull, K.R. (2010). Strain stiffening in synthetic and biopolymer networks. *Biomacromolecules*, **11**, pp. 1358–1363.
[71] Myers, K.M., Paskaleva, A.P., House, M., and Socrate, S. Mechanical and biochemical properties of human cervical tissue. *Acta Biomater.*, **4**(1), pp. 104–116.
[72] Sanabria-DeLong, N., Crosby, A.J., and Tew, G.N. (2008). Photo-cross-linked PLA-PEO-PLA hydrogels from self-assembled physical networks: mechanical properties and influence of assumed constitutive relationships. *Biomacromolecules*, **9**, pp. 2784–2791.
[73] Cui, J. *et al.* (2012). Synthetically simple, highly resilient hydrogels. *Biomacromolecules*, **13**(3), pp. 584–588.
[74] Parsons, E.M., Boyce, M.C., Parks, D.M., and Weinberg, M. (2005). Three-dimensional large-strain tensile deformation of neat and calcium carbonate-filled high-density polyethylene. *Polymer*, **46**, pp. 2257–2265.
[75] Yu, G., Yan, X., Han, C., and Huang, F. (2013). Characterization of supramolecular gels. *Chem. Soc. Rev.*, **42**, pp. 6697–6722.
[76] Whitmore, L. and Wallace, B.A. (2008). Protein secondary structure analyses from circular dichroism spectroscopy: methods and reference databases. *Biopolymers*, **89**, pp. 392–400.
[77] Sreerama, N. and Woody, R.W. (2000). Estimation of protein secondary structure from circular dichroism spectra: comparison of CONTIN, SELCON, and CDSSTR methods with an expanded reference set. *Anal. Biochem.*, **287**, pp. 252–260.
[78] Pelton, J.T. and McLean, L.R. (2000). Spectroscopic methods for analysis of protein secondary structure. *Anal. Biochem.*, **277**, pp. 167–176.
[79] Moerner, W. (2012). Microscopy beyond the diffraction limit using actively controlled single molecules. *J. Microscopy*, **246**, pp. 213–220.
[80] Moerner, W. (2007). New directions in single-molecule imaging and analysis. *Proc. Nat. Acad. Sci. USA*, **104**, pp. 12596–12602.
[81] Eiselt, P., Yeh, J., Latvala, R.K., Shea, L.D., and Mooney, D.J. (2000). Porous carriers for biomedical applications based on alginate hydrogels. *Biomaterials*, **21**, pp. 1921–1927.
[82] Smith, A.M. *et al.* (2008). Fmoc-diphenylalanine self assembles to a hydrogel via a novel architecture based on $\pi$–$\pi$ interlocked $\beta$-sheets. *Adv. Mater.*, **20**(1), pp. 37–41.

[83] Pochan, D.J., Pakstis, L., Ozbas, B., Nowak, A.P., and Deming, T.J. (2002). SANS and cryo-TEM study of self-assembled diblock copolypeptide hydrogels with rich nano- through microscale morphology. *Macromolecules*, **35**, pp. 5358–5360.

[84] Hule, R.A. *et al.* (2008). Correlations between structure, material properties and bioproperties in self-assembled $\beta$-hairpin peptide hydrogels. *Faraday Discuss.*, **139**, pp. 251–264.

[85] Zhou, M. *et al.* (2009). Self-assembled peptide-based hydrogels as scaffolds for anchorage-dependent cells. *Biomaterials*, **30**(13), pp. 2523–2530.

[86] Engler, A.J., Rehfeldt, F., Sen, S., and Discher, D.E. (2007). Microtissue elasticity: measurements by atomic force microscopy and its influence on cell differentiation, in Y.L. Wang and D.E. Discher (eds.), *Methods in Cell Biology*, Vol. 83. Academic Press, pp. 521–545.

[87] Li, Z. and Deming, T.J. (2010). Tunable hydrogel morphology via self-assembly of amphiphilic pentablock copolypeptides. *Soft Matter*, **6**, pp. 2546–2551.

[88] Erk, K.A. and Shull, K.R. (2011). Rate-dependent stiffening and strain localization in physically associating solutions. *Macromolecules*, **44**, pp. 932–939.

[89] Erk, K.A., Martin, J.D., Hu, Y.T., and Shull, K.R. (2012). Extreme strain localization and sliding friction in physically associating polymer gels. *Langmuir*, **28**, pp. 4472–4478.

[90] Tixier, T., Tabuteau, H., Carriere, A., Ramos, L., and Ligoure, C. (2010). Transition from "brittle" to "ductile" rheological behavior by tuning the morphology of self-assembled networks. *Soft Matter*, **6**, pp. 2699–2707.

[91] Berret, J.-F. and Serero, Y. (2001). Evidence of shear-induced fluid fracture in telechelic polymer networks. *Phys. Rev. Lett.*, **87**.

[92] Skrzeszewska, P.J. *et al.* (2010). Fracture and self-healing in a well-defined self-assembled polymer network. *Macromolecules*, **43**(7), pp. 3542–3548.

[93] Tabuteau, H., Mora, S., Ciccotti, M., Hui, C.-Y., and Ligoure, C. (2011). Propagation of a brittle fracture in a viscoelastic fluid. *Soft Matter*, **7**, pp. 9474–9483.

[94] Ligoure, C. and Mora, S. (2013). Fractures in complex fluids: the case of transient networks. *Rheol. Acta*, **52**, pp. 91–114.

[95] Stukalin, E.B., Cai, L.-H., Kumar, N.A., Leibler, L., and Rubinstein, M. (2013). Self-healing of unentangled polymer networks with reversible bonds. *Macromolecules*, **46**, pp. 7525–7541.

[96] Fox, J.D. and Rowan, S.J. (2009). Supramolecular polymerizations and main-chain supramolecular polymers. *Macromolecules*, **42**, pp. 6823–6835.

[97] Sun, T.L. *et al.* (2013). Physical hydrogels composed of polyampholytes demonstrate high toughness and viscoelasticity. *Nature Mater.*, **12**, pp. 932–937.

[98] Holten-Andersen, N. *et al.* (2014). Metal-coordination: using one of nature's tricks to control soft material mechanics. *J. Mater. Chem. B*, **2**, pp. 2467–2472.

[99] Feldman, K.E., Kade, M.J., Meijer, E.W., Hawker, C.J., and Kramer, E.J. (2009). Model transient networks from strongly hydrogen-bonded polymers. *Macromolecules*, **42**, pp. 9072–9081.

[100] Wang, C., Kopecek, J., and Stewart, R.J. (2001). Hybrid hydrogels cross-linked by genetically engineered coiled-coil block proteins. *Biomacromolecules*, **2**, pp. 912–920.

[101] Li, J., Lewis, C.L., Chen, D.L., and Anthamatten, M. (2011). Dynamic mechanical behavior of photo-cross-linked shape-memory elastomers. *Macromolecules*, **44**, pp. 5336–5343.

[102] Holten-Andersen, N. *et al.* (2011). pH-induced metal-ligand cross-links inspired by mussel yield self-healing polymer networks with near-covalent elastic moduli. *Proc. Nat. Acad. Sci. USA*, **108**(7), pp. 2651–2655.

[103] Rodell, C.B., Kaminski, A.L., and Burdick, J.A. (2013). Rational design of network properties in guest–host assembled and shear-thinning hyaluronic acid hydrogels. *Biomacromolecules*, **14**, pp. 4125–4134.

[104] Rubinstein, M. and Semenov, A.N. (1998). Thermoreversible gelation in solutions of associating polymers. 2. Linear dynamics. *Macromolecules*, **31**, pp. 1386–1397.

[105] Akagi, Y. *et al.* (2011). Examination of the theories of rubber elasticity using an ideal polymer network. *Macromolecules*, **44**(14), pp. 5817–5821.

[106] Asai, D. *et al.* (2012). Protein polymer hydrogels by *in situ*, rapid and reversible self-gelation. *Biomaterials*, **33**(21), pp. 5451–5458.

[107] Nejadnik, M.R. *et al.* (2013). Calcium-mediated secondary cross-linking of bisphosphonated oligo(poly(ethylene glycol) fumarate) hydrogels. *Macromol. Biosci.*, **13**(10), pp. 1308–1313.

[108] Fullenkamp, D.E., He, L., Barrett, D.G., Burghardt, W.R., and Messersmith, P.B. (2013). Mussel-inspired histidine-based transient network metal coordination hydrogels. *Macromolecules*, **46**, pp. 1167–1174.

[109] Sun, J.-Y. *et al.* (2012). Highly stretchable and tough hydrogels. *Nature*, **489**, pp. 133–136. Available at: http://www.nature.com/nature/journal/v489/n7414/abs/nature11409.html.

[110] Westhaus, E. and Messersmith, P.B. (2001). Triggered release of calcium from lipid vesicles: a bioinspired strategy for rapid gelation of polysaccharide and protein hydrogels. *Biomaterials*, **22**, pp. 453–462.

[111] Hori, Y., Winans, A.M., and Irvine, D.J. (2009). Modular injectable matrices based on alginate solution/microsphere mixtures that gel *in situ* and co-deliver immunomodulatory factors. *Acta Biomater.*, **5**, pp. 969–982.

[112] Rubinstein, M. and Dobrynin, A.V. (1997). Solutions of associative polymers. *Trends Poly. Sci.*, **5**, pp. 181–186.

[113] Shen, W., Lammertink, R.G.H., Sakata, J.K., Kornfield, J.A., and Tirrell, D.A. (2005). Assembly of an artificial protein hydrogel through leucine zipper aggregation and disulfide bond formation. *Macromolecules*, **38**, pp. 3909–3916.

[114] Xu, C. and Kopeček, J. (2008). Genetically engineered block copolymers: influence of the length and structure of the coiled-coil blocks on hydrogel self-assembly. *Pharm. Res.*, **25**, pp. 674–682.

[115] Kopeček, J. and Yang, J. (2012). Smart self-assembled hybrid hydrogel biomaterials. *Ang. Chem. Int. Ed.*, **51**, pp. 7396–7417.
[116] Mi, L., Fischer, S., Chung, B., Sundelacruz, S., and Harden, J.L. (2005). Self-assembling protein hydrogels with modular integrin binding domains. *Biomacromolecules*, **7**, pp. 38–47.
[117] Parisi-Amon, A., Mulyasasmita, W., Chung, C., and Heilshorn, S.C. (2013). Protein-engineered injectable hydrogel to improve retention of transplanted adipose-derived stem cells. *Adv. Health. Mater.*, **2**, pp. 428–432.
[118] Lu, H.D., Charati, M.B., Kim, I.L., and Burdick, J.A. (2012). Injectable shear-thinning hydrogels engineered with a self-assembling Dock-and-Lock mechanism. *Biomaterials*, **33**, pp. 2145–2153.
[119] Soranno, D.E., Lu, H.D., Weber, H.M., Rai, R., and Burdick, J.A. (2013). Immunotherapy with injectable hydrogels to treat obstructive nephropathy. *J. Biomed. Mater. Res. Part A*, **102**(7), pp. 2173–2180.
[120] Jiang, Z., You, Y., Deng, X., and Hao, J. (2007). Injectable hydrogels of poly($\epsilon$-caprolactone-co-glycolide)–poly($\epsilon$-ethylene glycol)–poly(caprolactone-co-glycolide) triblock copolymer aqueous solutions. *Polymer*, **48**, pp. 4786–4792.
[121] Ramos, L. and Ligoure, C. (2007). Structure of a new type of transient network: entangled wormlike micelles bridged by telechelic polymers. *Macromolecules*, **40**, pp. 1248–1251.
[122] Nakaya-Yaegashi, K., Ramos, L., Tabuteau, H., and Ligoure, C. (2008). Linear viscoelasticity of entangled wormlike micelles bridged by telechelic polymers: an experimental model for a double transient network. *J. Rheol.*, **52**, pp. 359–377.
[123] Krogstad, D.V. *et al.* (2013). Effects of polymer and salt concentration on the structure and properties of triblock copolymer coacervate hydrogels. *Macromolecules*, **46**(4), pp. 1512–1518.
[124] Pawar, G.M. *et al.* (2012). Injectable hydrogels from segmented PEG-bisurea copolymers. *Biomacromolecules*, **13**(12), pp. 3966–3976.
[125] Pham, Q., Russel, W., Thibeault, J., and Lau, W. (1999). Micellar solutions of associative triblock copolymers: Entropic attraction and gas-liquid transition. *Macromolecules*, **32**, pp. 2996–3005.
[126] Meng, X.-X. and Russel, W.B. (2006). Telechelic associative polymers: Interaction potential and high frequency modulus. *Journal of Rheology (1978-present)*, **50**(2), pp. 169–187.
[127] Weng, W., Li, Z., Jamieson, A.M., and Rowan, S.J. (2008). Control of gel morphology and properties of a class of metallo-supramolecular polymers by good/poor solvent environments. *Macromolecules*, **42**, pp. 236–246.
[128] Drzal, P.L. and Shull, K.R. (2003). Origins of mechanical strength and elasticity in thermally reversible, acrylic triblock copolymer gels. *Macromolecules*, **36**, pp. 2000–2008.
[129] Vasquez, P.A., McKinley, G.H., and Cook, L.P. (2007). A network scission model for wormlike micellar solutions: I. Model formulation and viscometric flow predictions. *J. Non-Newt. Fluid Mech.*, **144**, pp. 122–139.

[130] Cromer, M., Cook, L.P., and McKinley, G.H. (2009). Extensional flow of wormlike micellar solutions. *Chem. Eng. Sci.*, **64**, pp. 4588–4596.

[131] Germann, N., Cook, L.P., and Beris, A.N. (2014). Investigation of the inhomogeneous shear flow of a wormlike micellar solution using a thermodynamically consistent model. *J. Non-Newt. Fluid Mech.*, **207**, pp. 21–31.

[132] Christensen, T., Hassouneh, W., Trabbic-Carlson, K., and Chilkoti, A. (2013). Predicting transition temperatures of elastin-like polypeptide fusion proteins. *Biomacromolecules*, **14**, pp. 1514–1519.

[133] Vogt, A.P. and Sumerlin, B.S. (2009). Temperature and redox responsive hydrogels from ABA triblock copolymers prepared by RAFT polymerization. *Soft Matter*, **5**, pp. 2347–2351.

[134] Wright, E.R. and Conticello, V.P. (2002). Self-assembly of block copolymers derived from elastin-mimetic polypeptide sequences. *Adv. Drug Del. Rev.*, **54**, pp. 1057–1073.

[135] Zhang, Z. *et al.* (2011). Biodegradable and thermoreversible PCLA–PEG–PCLA hydrogel as a barrier for prevention of post-operative adhesion. *Biomaterials*, **32**(21), pp. 4725–4736.

[136] Matson, J.B. and Stupp, S.I. (2012). Self-assembling peptide scaffolds for regenerative medicine. *Chem. Comm.*, **48**, pp. 26–33.

[137] Webber, M.J., Berns, E.J., and Stupp, S.I. (2013). Supramolecular nanofibers of peptide amphiphiles for medicine. *Israel J. Chem.*, **53**, pp. 530–554.

[138] Cui, H., Webber, M.J., and Stupp, S.I. (2010). Self-assembly of peptide amphiphiles: From molecules to nanostructures to biomaterials. *Pept. Sci.*, **94**, pp. 1–18.

[139] Altunbas, A. and Pochan, D.J. (2012). Peptide-based and polypeptide-based hydrogels for drug delivery and tissue engineering, in T. Deming (ed.), *Peptide-Based Materials, Topics in Current Chemistry*, Vol. 310. Springer, Berlin–Heidelberg, Ch. 206, pp. 135–167.

[140] Dankers, P.Y.W. *et al.* (2012). Hierarchical formation of supramolecular transient networks in water: a modular injectable delivery system. *Adv. Mater.*, **24**(20), pp. 2703–2709.

[141] Stupp, S.I. and Palmer, L.C. (2013). Supramolecular chemistry and self-assembly in organic materials design. *Chem. Mater.*, **26**, pp. 507–518.

[142] Baral, A. *et al.* (2014). Assembly of an injectable noncytotoxic peptide-based hydrogelator for sustained release of drugs. *Langmuir*, **30**(3), pp. 929–936.

[143] Altunbas, A., Lee, S.J., Rajasekaran, S.A., Schneider, J.P., and Pochan, D.J. (2011). Encapsulation of curcumin in self-assembling peptide hydrogels as injectable drug delivery vehicles. *Biomaterials*, **32**, pp. 5906–5914.

[144] Cigognini, D. *et al.* (2011). Evaluation of early and late effects into the acute spinal cord injury of an injectable functionalized self-assembling scaffold. *PLoS ONE*, **6**(5), p. e19782.

[145] Huang, H. *et al.* (2011). Design of a shear-thinning recoverable peptide hydrogel from native sequences and application for influenza H1N1 vaccine adjuvant. *Soft Matter*, **7**, pp. 8905–8912.
[146] Hule, R.A., Nagarkar, R.P., Hammouda, B., Schneider, J.P., and Pochan, D.J. (2009). Dependence of self-assembled peptide hydrogel network structure on local fibril nanostructure. *Macromolecules*, **42**, pp. 7137–7145.
[147] Rieth, S., Baddeley, C., and Badjic, J.D. (2007). Prospects in controlling morphology, dynamics and responsiveness of supramolecular polymers. *Soft Matter*, **3**, pp. 137–154.
[148] Lowik, D.W.P.M. and van Hest, J.C.M. (2004). Peptide based amphiphiles. *Chem. Soc. Rev.*, **33**, pp. 234–245.
[149] Castelletto, V., Hamley, I.W., Cenker, C., and Olsson, U. (2010). Influence of salt on the self-assembly of two model amyloid heptapeptides. *J. Phys. Chem. B*, **114**, pp. 8002–8008.
[150] Guo, H. *et al.* (2013). The robust hydrogel hierarchically assembled from a pH sensitive peptide amphiphile based on silk fibroin. *Biomacromolecules*, **14**(8), pp. 2733–2738.
[151] Dong, H., Paramonov, S.E., Aulisa, L., Bakota, E.L., and Hartgerink, J.D. (2007). Self-assembly of multidomain peptides: Balancing molecular frustration controls conformation and nanostructure. *J. Am. Chem. Soc.*, **129**, pp. 12468–12472.
[152] Niece, K.L. *et al.* (2008). Modification of gelation kinetics in bioactive peptide amphiphiles. *Biomaterials*, **29**(34), pp. 4501–4509.
[153] MacKintosh, F., Käs, J., and Janmey, P. (1995). Elasticity of semiflexible biopolymer networks. *Phys. Rev. Lett.*, **75**, pp. 4425–4428.
[154] Storm, C., Pastore, J.J., MacKintosh, F.C., Lubensky, T.C., and Janmey, P.A. (2005). Nonlinear elasticity in biological gels. *Nature*, **435**, pp. 191–194.
[155] Ramachandran, S., Trewhella, J., Tseng, Y., and Yu, Y.B. (2006). Coassembling peptide-based biomaterials: Effects of pairing equal and unequal chain length oligopeptides. *Chem. Mater.*, **18**, pp. 6157–6162.
[156] Greenfield, M.A., Hoffman, J.R., Olvera de la Cruz, M., and Stupp, S.I. (2009). Tunable mechanics of peptide nanofiber gels. *Langmuir*, **26**, pp. 3641–3647.
[157] Pashuck, E.T., Cui, H., and Stupp, S.I. (2010). Tuning supramolecular rigidity of peptide fibers through molecular structure. *J. Am. Chem. Soc.*, **132**, pp. 6041–6046.
[158] Nagy, K.J., Giano, M.C., Jin, A., Pochan, D.J., and Schneider, J.P. (2011). Enhanced mechanical rigidity of hydrogels formed from enantiomeric peptide assemblies. *J. Am. Chem. Soc.*, **133**, pp. 14975–14977.
[159] Yucel, T., Micklitsch, C.M., Schneider, J.P., and Pochan, D.J. (2008). Direct observation of early-time hydrogelation in $\beta$-hairpin peptide self-assembly. *Macromolecules*, **41**, pp. 5763–5772.
[160] Larsen, T.H., Branco, M.C., Rajagopal, K., Schneider, J.P., and Furst, E.M. (2009). Sequence-dependent gelation kinetics of $\beta$-hairpin peptide hydrogels. *Macromolecules*, **42**, pp. 8443–8450.

[161] Schneider, J.P. *et al.* (2002). Responsive hydrogels from the intramolecular folding and self-assembly of a designed peptide. *J. Am. Chem. Soc.*, **124**(50), pp. 15030–15037.

[162] Kretsinger, J.K., Haines, L.A., Ozbas, B., Pochan, D.J. and Schneider, J.P. (2005). Cytocompatibility of self-assembled β-hairpin peptide hydrogel surfaces. *Biomaterials*, **26**, pp. 5177–5186.

[163] Giano, M.C., Pochan, D.J., and Schneider, J.P. (2011). Controlled biodegradation of self-assembling β-hairpin peptide hydrogels by proteolysis with matrix metalloproteinase-13. *Biomaterials*, **32**, pp. 6471–6477.

[164] Haines-Butterick, L.A., Salick, D.A., Pochan, D.J., and Schneider, J.P. (2008). *In vitro* assessment of the pro-inflammatory potential of β-hairpin peptide hydrogels. *Biomaterials*, **29**, pp. 4164–4169.

[165] Salick, D.A., Kretsinger, J.K., Pochan, D.J., and Schneider, J.P. (2007). Inherent antibacterial activity of a peptide-based β-hairpin hydrogel. *J. Am. Chem. Soc.*, **129**, pp. 14793–14799.

[166] Salick, D.A., Pochan, D.J., and Schneider, J.P. (2009). Design of an injectable β-hairpin peptide hydrogel that kills methicillin-resistant *Staphylococcus aureus*. *Adv. Mater.*, **21**, pp. 4120–4123.

[167] Haines, L.A. *et al.* (2005). Light-activated hydrogel formation via the triggered folding and self-assembly of a designed peptide. *J. Am. Chem. Soc.*, **127**(48), pp. 17025–17029.

[168] Collier, J.H. *et al.* (2001). Thermally and photochemically triggered self-assembly of peptide hydrogels. *J. Am. Chem. Soc.*, **123**(38), pp. 9463–9464.

[169] Lin, B.F. *et al.* (2012). pH-responsive branched peptide amphiphile hydrogel designed for applications in regenerative medicine with potential as injectable tissue scaffolds. *J. Mater. Chem.*, **22**, pp. 19447–19454 .

[170] Hartgerink, J.D., Beniash, E., and Stupp, S.I. (2011). Self-assembly and mineralization of peptide-amphiphile nanofibers. *Science*, **294**, pp. 1684–1688.

[171] Guilbaud, J.-B. *et al.* (2010). Enzymatic catalyzed synthesis and triggered gelation of ionic peptides. *Langmuir*, **26**(13), pp. 11297–11303.

[172] Toledano, S., Williams, R.J., Jayawarna, V., and Ulijn, R.V. (2006). Enzyme-triggered self-assembly of peptide hydrogels via reversed hydrolysis. *J. Am. Chem. Soc.*, **128**, pp. 1070–1071.

[173] Geisler, I.M. and Schneider, J.P. (2012). Evolution-based design of an injectable hydrogel. *Adv. Funct. Mater.*, **22**, pp. 529–537.

[174] Webber, M.J. *et al.* (2010). Development of bioactive peptide amphiphiles for therapeutic cell delivery. *Acta Biomater.*, **6**(1), pp. 3–11.

[175] Sinthuvanich, C., Haines-Butterick, L.A., Nagy, K.J., and Schneider, J.P. (2012). Iterative design of peptide-based hydrogels and the effect of network electrostatics on primary chondrocyte behavior. *Biomaterials*, **33**, pp. 7478–7488.

[176] Nowak, A.P., Breedveld, V., Pine, D.J., and Deming, T.J. (2003). Unusual salt stability in highly charged diblock co-polypeptide hydrogels. *J. Am. Chem. Soc.*, **125**, pp. 15666–15670.

[177] Pakstis, L.M. *et al.* (2003). Effect of chemistry and morphology on the biofunctionality of self-assembling diblock copolypeptide hydrogels. *Biomacromolecules*, **5**(2), pp. 312–318.
[178] Yang, C.-Y. *et al.* (2009). Biocompatibility of amphiphilic diblock copolypeptide hydrogels in the central nervous system. *Biomaterials*, **30**(15), pp. 2881–2898.
[179] Song, B. *et al.* (2012). Sustained local delivery of bioactive nerve growth factor in the central nervous system via tunable diblock copolypeptide hydrogel depots. *Biomaterials*, **33**(35), pp. 9105–9116.
[180] Branco, M.C., Pochan, D.J., Wagner, N.J., and Schneider, J.P. (2010). The effect of protein structure on their controlled release from an injectable peptide hydrogel. *Biomaterials*, **31**, pp. 9527–9534.
[181] Veiga, A.S. *et al.* (2012). Arginine-rich self-assembling peptides as potent antibacterial gels. *Biomaterials*, **33**(35), pp. 8907–8916.
[182] Choueiri, R.M., Klinkova, A., Thérien-Aubin, H., Rubinstein, M. and Kumacheva, E. Structural transitions in nanoparticle assemblies governed by competing nanoscale forces. *J. Am. Chem. Soc.*, **135**(28), pp. 10262–10265.
[183] Wang, H. *et al.* (2011). Oppositely charged gelatin nanospheres as building blocks for injectable and biodegradable gels. *Adv. Mater.*, **23**(12), pp. H119–H124.
[184] Gu, Z. *et al.* (2013). Injectable nano-network for glucose-mediated insulin delivery. *ACS Nano*, **7**(5), pp. 4194–4201.
[185] Gong, Z., Yang, Y., Ren, Q., Chen, X., and Shao, Z. (2012). Injectable thixotropic hydrogel comprising regenerated silk fibroin and hydroxypropylcellulose. *Soft Matter*, **8**, pp. 2875–2883.
[186] Chiu, Y.-L. *et al.* (2009). Rapidly in situ forming hydrophobically-modified chitosan hydrogels via pH-responsive nanostructure transformation. *Soft Matter*, **5**, pp. 962–965.
[187] Chiu, Y.-L. *et al.* (2009). pH-triggered injectable hydrogels prepared from aqueous N-palmitoyl chitosan: *In vitro* characteristics and *in vivo* biocompatibility. *Biomaterials*, **30**, pp. 4877–4888.
[188] Macaya, D. and Spector, M. (2012). Injectable hydrogel materials for spinal cord regeneration: a review. *Biomed. Mater.*, **7**.
[189] Radhakrishnan, J., Krishnan, U.M., and Sethuraman, S. (2014). Hydrogel based injectable scaffolds for cardiac tissue regeneration. *Biotech. Adv.*, **32**, pp. 449–461.
[190] Kretlow, J.D., Klouda, L., and Mikos, A.G. (2007). Injectable matrices and scaffolds for drug delivery in tissue engineering. *Adv. Drug Del. Rev.*, **59**, pp. 263–273.
[191] Kretlow, J.D., Young, S., Klouda, L., Wong, M., and Mikos, A.G. (2009). Injectable biomaterials for regenerating complex craniofacial tissues. *Adv. Mater.*, **21**, pp. 3368–3393.
[192] Jia, X. and Kiick, K.L. (2009). Hybrid multicomponent hydrogels for tissue engineering. *Macromol. Biosci.*, **9**, pp. 140–156.

[193] Ekenseair, A.K. *et al.* (2012). Structure-property evaluation of thermally and chemically gelling injectable hydrogels for tissue engineering. *Biomacromolecules*, **13**(9), pp. 2821–2830.
[194] Hacker, M.C., Klouda, L., Ma, B.B., Kretlow, J.D., and Mikos, A.G. (2008). Synthesis and characterization of injectable, thermally and chemically gelable, amphiphilic poly(N-isopropylacrylamide)-based macromers. *Biomacromolecules*, **9**, pp. 1558–1570.
[195] Temenoff, J.S. *et al.* (2003). Thermally cross-linked oligo(poly(ethylene glycol) fumarate) hydrogels support osteogenic differentiation of encapsulated marrow stromal cells *in vitro*. *Biomacromolecules*, **5**(1), pp. 5–10.
[196] Zhang, J. *et al.* (2011). Physically associated synthetic hydrogels with long-term covalent stabilization for cell culture and stem cell transplantation. *Adv. Mater.*, **23**(43), pp. 5098–5103.
[197] Medberry, C.J. *et al.* (2013). Hydrogels derived from central nervous system extracellular matrix. *Biomaterials*, **34**(4), pp. 1033–1040.
[198] Place, E.S., Evans, N.D., and Stevens, M.M. (2009). Complexity in biomaterials for tissue engineering. *Nat. Mater.*, **8**, pp. 457–470. Available at: http://www.nature.com/nmat/journal/v8/n6/suppinfo/nmat2441_S1.html.
[199] Berlemann, U. and Schwarzenbach, O. (2009). An injectable nucleus replacement as an adjunct to microdiscectomy: 2 year follow-up in a pilot clinical study. *Eur. Spine J.*, **18**, pp. 1706–1712.

# Chapter 4

# Injectable Hydrogels by Chemical Crosslinking

B. Balakrishnan*

*Department of Biosciences & Bioengineering, Indian Institute of Technology, Bombay, Mumbai, Maharashtra, India-400076*

## 4.1 Introduction

Development of injectable hydrogels (either through *in situ* gelling or shear-thinning) has attracted much attention recently due to its simple method of delivery and targeted action. Based on the method of preparation, the hydrogels are broadly classified as physically or chemically crosslinked. Physical crosslinking occurs mainly by self-assembling of polymers in the presence of various triggers like pH, temperature, and ions, as discussed in Chapter 3. Most of the shear-thinning hydrogels have physical crosslinks. Despite the advantages of physical crosslinking in terms of its reversible, self-assembling nature, which produces minimal inflammatory response in principle by avoiding the use of toxic crosslinking agents, these self-assembled domains have been found to disrupt upon mechanical stress more easily than covalent crosslinks. Chemical crosslinking, on the other hand, is irreversible in nature as covalent bonds are formed between the polymers. These networks can be made biodegradable by designing polymers and chemical reactions in such a way that they can undergo cleavage due to hydrolysis, or in response to enzymes. Small molecule crosslinkers like glutaraldehyde and carbodiimides used for making hydrogels have been found

*Author to whom correspondence should be addressed. Email: bijibal@rediffmail.com.

to be toxic and are not recommended for the preparation of *in situ* gelling systems. Attempts are being made to minimize the toxic effects of crosslinkers by the identification of new crosslinkers and crosslinking methods with minimal adverse effects. In the following sections, the commonly employed chemical reactions as well as the new concepts that can be exploited for *in situ* gelation are described in detail.

## 4.2 Chemical Crosslinking Methods for *In Situ* Gelation

A number of questions need to be addressed while preparing chemically crosslinked *in situ* gelling systems for various biomedical applications:

(a) Are the polymer precursor solutions and crosslinking agents/initiators (in case of photocrosslinking) biocompatible, sterilizable, and injectable?
(b) Can gelation occur at physiological temperature and pH?
(c) Is gelation time appropriate to fill the defect and form gel at the defect site?
(d) If gelation is slow, do the polymer precursor solutions have sufficient shelf life after mixing?
(e) Does the crosslinking reaction lead to change of pH or temperature, or formation of any by-products? If yes, is it within permissible limits?
(f) Is the *in situ* forming gel appropriate for a particular biomedical application in terms of mechanics, degradation profile, and bioactivity?
(g) Does the *in situ* forming gel integrate with the host tissue?

Proper understanding of crosslinking reaction mechanisms and molecules involved in the preparation of *in situ* gelling hydrogels is essential to answer all of these questions. Commonly utilized polymers for *in situ* gelation via chemical crosslinking are derivatives of synthetic polymers like PEG and poly(vinyl alcohol) (PVA), and biopolymers like hyaluronic acid (HA), dextran, alginate, chondroitin sulfate (CS), chitosan, cellulose, heparin, and collagen/gelatin. In this chapter, discussion is restricted to commonly employed chemical crosslinking methods (Figure 4.1) like Schiff base reaction, free radical-mediated crosslinking, Michael-type addition, disulfide crosslinking, and genipin crosslinking. Special mention is given to the click reaction concept used for *in situ* gelation and multiple crosslinking methods.

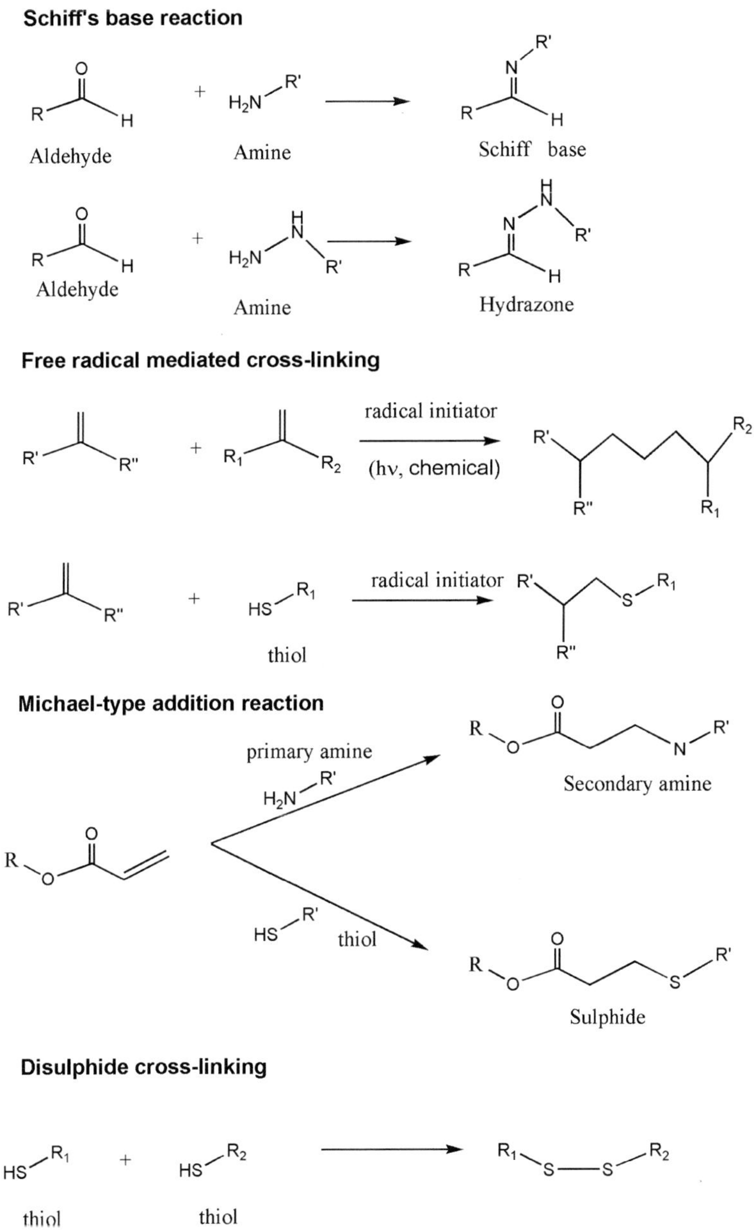


Figure 4.1 Common chemical reactions employed for the preparation of hydrogels *in situ*.

### 4.2.1 *Schiff base reaction*

Schiff base reaction is one of the widely accepted strategies for the preparation of hydrogels, particularly because of the mild reaction conditions and high reaction rates. It has been acknowledged as a simple and metal-free click reaction for preparation of biodegradable hydrogels having specific association mechanism [1]. The method involves the reaction between amino groups and aldehyde/ketone groups, leading to the formation of imine linkage between two molecules (Figure 4.1). The reaction has also been used to insert pH-sensitive linkages mainly for drug delivery applications.

#### 4.2.1.1 In situ *gelling hydrogels based on biopolymers*

Alginate, dextran, CS, and HA are some of the extensively used biopolymers to prepare injectable hydrogels exploiting the Schiff base reaction. In the case of polymers having *cis*-glycols, the introduction of aldehyde functionality by periodate oxidation is an easy approach, particularly for biopolymers. Macromolecular dialdehyde derivative thus formed can react with polymers having an amino functional group to form crosslinks. Gelatin, water-soluble derivatives of chitosan and adipic acid dihydrazide, or ethylene diamine derivatives of synthetic polymers as well as biopolymers are the other components extensively used.

Dextrans, linear polymers of glucose linked by $\alpha$-1,6-linkages with some branching by $\alpha$-1,3-linkages, have been used clinically to prevent vascular occlusion, as a plasma volume expander, and for anticoagulation [2]. Gelatin, which is a partially hydrolyzed collagen, also has a long history of medical use [3]. *In situ* crosslinking of partially oxidized dextran (Odex) with amino groups of gelatin was reported by Schacht *et al.* [4]. The gelation time was found to be dependent on the pH and ionic strength of the buffer, and was reported to be about 3 min in a phosphate buffer (pH 7). Though the hydrogel films [5] revealed dose- and incubation time-dependent cytotoxicity with keratinocytes, *in vivo* biocompatibility studies showed only moderate foreign body reaction around the subcutaneous implants [6]. The potential of these hydrogels as an injectable delivery system is yet to be explored, but they have been evaluated as preformed hydrogel films for controlled delivery of bioactive molecules.

Jayakrishnan and coworkers reported the preparation of injectable *in situ* gelling systems employing a Schiff base reaction between various oxidized polysaccharides and gelatin for a variety of biomedical applications [7–11]. Alginate, seaweed-derived anionic linear polysaccharides composed of 1,4-linked $\beta$-D-mannuronate and 1,4-linked $\alpha$-L-guluronate residues in varying proportions, was periodate oxidized to different degrees in heterogeneous reaction conditions (as a dispersion in (1:1) ethanol/water mixture). By employing oxidized alginate of appropriate molecular weight and degree of oxidation, hydrogels were prepared with gelatin instantaneously in the presence of a small amount of borax (0.1 M), due to Schiff base formation between the $\varepsilon$-amino groups of lysine or hydroxylysine side groups of gelatin and the available aldehyde groups of oxidized alginate. Gelation was found to be fast with type A gelatin, as more amino groups are available for Schiff base formation. Fast gelation in the presence of borax is attributed to the alkaline pH of the medium which facilitates the formation of the Schiff base as well as the ability of borax to complex with hydroxyl groups of polysaccharides. The properties of these gels are tunable wherein the gelling time could be decreased by increasing the concentration of oxidized alginate, gelatin, and borax, and degree of oxidation of alginate. Gelation occurred within 13 s for gels prepared using 20% solution of oxidized alginate having degree of oxidation 57% and 15% gelatin solution. The degree of crosslinking can be varied by changing the concentration of oxidized alginate and gelatin, and the hydrogels were found to be biocompatible and completely degradable in all the combinations. The injectability of the system was demonstrated by subcutaneous and intramuscular injection of the hydrogel using a double syringe applicator in which one syringe was filled with 20% solution of oxidized alginate having degree of oxidation 57% in 0.1 M borax solution, and the other with 15% solution of gelatin. Gel was explanted and was found to be intact, confirming the formation of the hydrogel *in situ* at the target site (Figure 4.2). Gel injected intramuscularly and subcutaneously was found to be completely degradable with minimal foreign body response. The prospect of the system as an injectable drug delivery vehicle and as a tissue engineering scaffold was established, respectively, by using primaquine as a model drug and by encapsulation of hepatocytes inside the gel matrix. Primary hepatocytes maintained viability and functionality within the matrix for a period of four weeks [7]. The efficacy of these gels as wound-healing agents, evaluated on a full thickness rat wound model, revealed complete wound re-epithelialization within 15 days (Figure 4.2). The gels were shown to perform more effectively than commercially available sodium alginate (Kaltostat®) dressing, where seven months of foreign body reaction was observed, whereas for these gels, foreign body reaction was found to subside within 15 days, demonstrating that the material was undergoing degradation on the wound bed and the degradation products

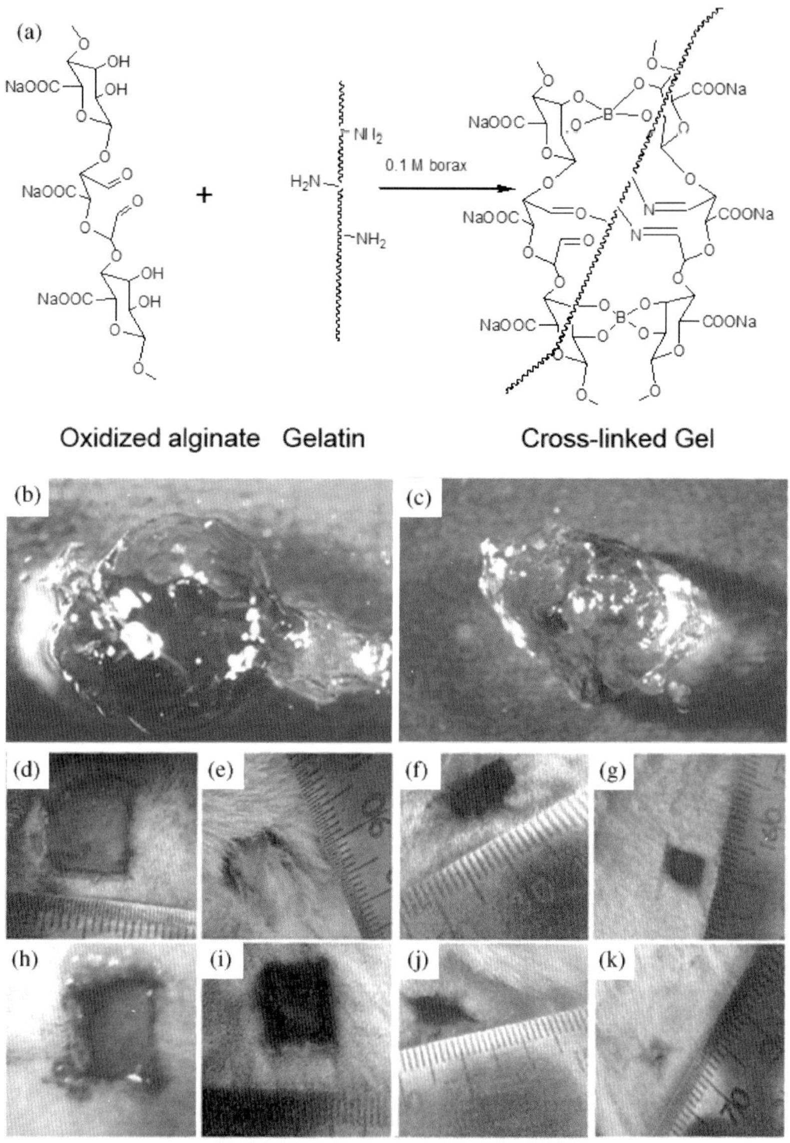


Figure 4.2 Reaction scheme for the preparation of oxidized alginate-crosslinked gelatin hydrogel via Schiff base reaction (a); Gross appearance of gel (b); Gel explanted from subcutaneous tissue after 1 hr of injection (c); Representative photographs of macroscopic appearance of 1×1 cm wound excised on rat (d), gel applied on wound (h), control wounds at 5 (e), 10 (f) and 15 days (g) and test wounds at 5 (i), 10 (j), and 15 days (k). Reprinted from [7 & 9] with permission from Elsevier.

were not inducing any adverse reaction [8]. The wound-healing efficacy of the gels was improved by the incorporation of N′-2′-*O*-dibutyryl adenosine cyclic AMP, and was found to achieve complete healing within ten days [9]. Gentamycin-loaded gels showed antibacterial property against gram negative *P. aeruginosa* and gram positive *S. aureus*, which could be useful in preventing wound infections [10]. The gel has also demonstrated potential as an injectable, biomimetic adhesive scaffold for cartilage tissue regeneration [12].

Oxidized alginate has further been used to make *in situ* forming hydrogels with *N,O*-carboxymethyl chitosan, exhibiting tunable gelling time by varying oxidation degree of alginate. Hydrogels demonstrated non-cytotoxicity towards NIH-3T3 cells and controlled delivery of loaded bovine serum albumin, initially by diffusion followed by a degradation-dependent mechanism indicating its potential application as a drug delivery and tissue engineering matrix [13]. Partially oxidized alginate has been used to prepare *in situ* forming gels with hydroxypropyl chitosan, and was evaluated for corneal endothelium reconstruction [14]. Partially oxidized carboxymethyl cellulose has also been found to undergo Schiff base reaction with gelatin in the presence of borax, leading to instantaneous hydrogel formation (within 15 s). Hydrogel has been found to be biodegradable and induced negligible reactive oxygen species generation and minimal immunological response. The gel demonstrated good adhesion onto cartilage surface, and promoted migration and viability of primary murine chondrocytes while maintaining the genetic profile [15]. However, hydrogels prepared using oxidized CS and gelatin were found not to be *in situ* gelling [11]. When oxidized gum arabic was used to make imine linkages with an anti-malarial drug (primaquine) via Schiff base reaction, it formed hydrogel within 5 min and was found to be useful as an injectable, biodegradable drug depot with long-term controlled delivery [16]. *In situ* gelling systems were also prepared using water-soluble chitosan hydrochloride and Odex as a surgical tissue adhesive by Schiff base reaction between aldehyde groups of Odex and amino groups of chitosan [17]. Gelation was very fast (within 5 s) for these systems. The gels were porous in morphology and non-toxic towards L929 mouse fibroblast cell line. Bonding strength, measured as the load required for the glued rat skin to peel off by adding standard weights, was found to be in the range 220–370 $gf/cm^2$ for all concentrations of chitosan, degree of oxidations of dextran, and all setting times tried, which is much higher than the values obtained for fibrin glue (45–50 $gf/cm^2$). The burst strength of these gels at a setting time of 5 min was shown to be within the range 230–370 mm of Hg indicating efficacy as a tissue sealant. Efficacy evaluation of the gel on a rabbit liver parenchymal injury model by making a liver lobe edge resection of approximately 1.5 cm length and liver lobe circular excision of approximately 1 cm diameter, revealed that the gel could effectively control the hemostasis of liver injury. The tissue response elicited was comparable with the FDA-approved, commercially available tissue adhesive Bioglue™. Evaluation of chitosan hydrochloride/dextran dialdehyde glue on a sheep lung injury model under normal physiological as well as coagulopathic conditions showed complete sealing of the site with no observed air leak in any of the animals, whereas the incisions sealed by control Bioglue™ displayed persistent air leak. The test gel demonstrated superior performance with similar tissue response. Evaluation of sealing efficacy of this tissue sealant as an adjunct to sutures in aortic incision probed on a pig aortic injury model demonstrated effective sealing of the aortic incisions. The sealing was intact in all the cases for two weeks, with no incidences of aneurisms. The endoluminal surface was clean without any thrombus and neointimal formation was observed across the incision site [18].

Preparation of *in situ* gelling hydrogels using Odex and other water-soluble derivatives of chitosan such as N-carboxyethyl chitosan (CEC) and thiolated chitosan has also been reported. The gels with N-carboxyethyl chitosan were fast gelling under body temperature and pH. Tunable gelling time, swelling, and degradation can be achieved by varying the degree of oxidation of dextran. The gels were cytocompatible and were shown to enhance wound healing [19]. Triple-penetrating network hydrogels prepared by Zhang *et al.* [20], employing Odex, teleostean, and N-CEC, exhibited combined characteristics of the three individual binary hydrogel formulations (i.e., Odex/CEC, Odex/teleostean, and CEC/teleostean) with respect to the rheological profile and mechanical properties. This triple inter penetrating network showed an improved degradation pattern compared to binary formulations and has been evaluated for the controlled delivery of rosiglitazone. The same research group [21] has tried to make an *in situ* gelling double network structure using thiolated chitosan and Odex without using any extraneous crosslinking agents. In addition to Schiff base formation, thiolated chitosan can undergo disulfide intracrosslinking, resulting in fast gelation and formation of an interpenetrating network structure. These gels were superior to hydrogels prepared from Odex and N-CEC in terms of mechanical properties, and exhibited slow degradation with very mild tissue response [21]. Odex was further explored to make *in situ* gelling systems for antifungal therapy with carboxymethyl cellulose-hydrazide via formation of hydrazone linkages. Briefly, Amphotericin-B was rendered water soluble by conjugation to Odex, demonstrated antifungal activity against *C. albicans*, and fast gelling with carboxymethyl cellulose hydrazide. Controlled release of the drug from the gels for 11 days showed antifungal activity, but the antifungal activity was absent when the drug suspension was just entrapped within the gel. As both gels exhibited fungal-killing properties upon contact, it can be envisaged that injectable systems of these types, having soluble or insoluble drug formulations, could be useful for the treatment of local fungal infections, with or without concurrent systemic therapy [22]. Schiff base reaction has also been utilized to make *in situ* gelling hydrogels using Odex and polymers such as polyallylamine (PAA), oligochitosan, and glycol chitosan. The Odex-PAA combination was found to be the most pro-hemostatic among these, as indicated by a decrease in clotting time and an increase in clot strength [23]. Hydrazide-modified carboxymethyldextran and Odex, or oxidized carboxymethylcellulose were also found to undergo Schiff base reaction, leading to *in situ* gelation. Gelation time was concentration-dependent in both cases. Gels having Odex shrank after gelling while that with oxidized carboxymethyl cellulose swelled and reduced the formation of peritoneal adhesions [24].

HA is a non-sulfated glycosaminoglycan (GAG) found in the extracellular matrix (ECM) of many soft connective tissues, and is made up of alternating units of D glucuronic acid and *N*-acetyl-D-glucosamine, linked together via alternating $\beta$-1,4 and $\beta$-1,3 glycosidic bonds. This is synthesized at the inner wall of the plasma membrane by the enzyme HA synthase and is extruded out to the ECM without any modification. Considering the potential of HA in contributing towards the mechanical integrity of the ECM and regulation of various cellular processes by binding with cell receptors such as CD44 and RHAMM, it has been modified and exploited for the preparation of injectable hydrogels for various biomedical applications. For example, periodate oxidized HA has been utilized to form *in situ* gelling hydrogel (within 1 min) with HA-adipic hydrazide. It has also been employed to make *in situ* gelling hydrogel with N-succinyl chitosan for cartilage tissue engineering. These gels exhibited tunable gelling time, compressive moduli, and degradation profile. Gels were reported to have supported the adhesion of encapsulated bovine articular chondrocytes [25]. The same gel in conjugation with glucose oxidase and catalase was exploited to make *in situ* gelling, biodegradable, and glucose-responsive hydrogels. Enzyme-triggered conversion of glucose can lead to a reduction of microenvironment pH, resulting in enhanced swelling and hydrolysis of Schiff linkages, leading to the release of insulin, which is foreseen to have potential in adipose tissue engineering [26]. Partially oxidized HA has also been shown to form gel (within 3–8 min) with adipic hydrazide [27]. Though the gel was biocompatible and promoted nucleus pulposus cellular expression of type II collagen and aggrecan, the viscoelastic properties were inadequate for clinical use. To improve the viscoelastic properties in order to be adequate to bear and transfer load, gelatin was conjugated to the partially oxidized HA and used to form Schiff base with adipic hydrazide. Complex shear modulus similar to native tissue (∼11–14 kPa for hydrogels, 11.3 kPa for native NP) was obtained for the resulting gels. These gels were found to promote adhesion, viability, proliferation, and gene expression (collagen type II, aggrecan, SOX-9, HIF-A) of NP cells [28]. Oxidized heparin has also been employed to make injectable hydrogels with water-soluble chitosan having lactose groups to immobilize fibroblast growth factor (FGF-2). The hydrogel demonstrated biodegradability *in vivo* with almost complete degradation within about 20 days after subcutaneous injection. FGF-2 incorporated gel induced significant neovascularization and fibrous tissue formation near the injected site, indicating controlled release of biologically active FGF-2 molecules from the hydrogel, and subsequent induction of vascularization [29].

#### 4.2.1.2 In situ *gelling systems based on synthetic polymers*

Derivatives of synthetic polymers have also been explored to make *in situ* gelling systems by means of Schiff base reaction with complementary polymers. Polyvinyl alcohol (PVA) functionalized to have aldehyde and hydrazide groups through carbamate linkages was found to undergo self-crosslinking via Schiff base formation to form hydrogel instantaneously at neutral pH and room temperature. The hydrogel supported the viability of murine neuroblastoma N2a cells for four days of culture. Detailed evaluation has still to be carried out in order to establish the potential of the system for biomedical applications [30]. In another study, propionaldehyde, butyraldehyde, or 2-oxoethyl succinate-functionalized PEG was reacted with a peptide-based dendritic crosslinker ($Lys_3Cys_4$) to form a wound-specific corneal adhesive. Gelation took place within minutes to form elastic transparent hydrogels. Degradation was found to be dependent on the type of bond formed during crosslinking, either thiazolidine or pseudoproline, and ranged from days to months [31]. PEG amines and Odex have been used to make *in situ* gelling adhesive hydrogels that gels within seconds. With an increase in aldehyde/amino ratio, bulk elastic moduli and gelation time varied from $16.1 \pm 0.8$ to $161.0 \pm 10.8$ kPa, and $6.7 \pm 2.5$ to $69.2 \pm 7.5$ s, respectively, indicating a significant network response to aldehyde concentration. Elastic moduli and gelation time demonstrated formation of fast-gelling, stiffer networks with an increase in CHO:$NH_2$ ratio. However, the bulk properties became insensitive to CHO:$NH_2$ at higher aldehyde content (CHO:NH2 >3), indicating that the extent and kinetics of network formation approach saturation at critical concentrations of dextran aldehyde [32].

Though Schiff base reaction is easy to translate for clinical applications because of its simple and easy method of execution, the low stability of imine linkages and thus formed hydrogels are limiting factors for those applications in which long term stability is required.

### 4.2.2 *Free radical-mediated crosslinking*

With the advances in synthetic chemistry, many new multifunctional monomers and macromers have been synthesized with a view of making injectable hydrogels by crosslinking via free radical-initiated polymerization. Free radical polymerization is characterized by three different stages:

(1) Initiation, in which free radicals are generated by means of initiators and subsequent reaction of the radicals with monomer molecules by opening the double bond to form new radicals;
(2) Propagation, where monomer radicals are successively reacted to form long chains and continuously propagate the reactive radical center; and,
(3) Termination, where chain growth stops.

Initiation can be induced by means of thermal, photo, and redox initiators, and among them, photo and redox free radical-initiated polymerizations are employed for making injectable hydrogels as both can be conducted under physiological conditions. This section discusses the various free radical polymerization methods used for making injectable hydrogels for biomedical applications.

#### 4.2.2.1 *Crosslinking via redox polymerization*

Crosslinking of polymers via redox polymerization is one of the free radical-mediated reactions in which radicals are generated via an oxidation reduction reaction. In this method, crosslinking of acrylate functionalized polymers is performed by employing a redox pair initiator system which generates radicals during oxidation reduction reaction. The most commonly used redox pair initiator systems for biomedical applications are sodium persulfate (NaPS), ammonium persulfate (APS) (oxidant)/ascorbic acid (AA), or $N$,$N$,$N'$,$N'$-tetramethylethylene diamine (TEMED) (reducer). The advantage of the APS/TEMED system is that when combined in equimolar amounts, the system remains neutral, while the APS/AA system is strongly acidic [33]. The concentration of redox initiators must be optimized before it is used for making *in situ* gelling systems *in vivo*. The benefits of redox polymerization include low activation energy, very short induction period, radical generation at reasonable rates over a very wide range of temperatures, and aqueous reaction medium.

Attempts have been made to create oligo(poly(ethylene glycol) fumarate (OPF) hydrogels as injectable carriers for marrow stromal cells (MSCs) in orthopedic tissue engineering applications [33]. Two different persulfate oxidizing agents, APS and NaPS with three AA-based reducing agents, AA, sodium ascorbate (Asc), and magnesium ascorbate-2-phosphate (Asc-2) and their combinations, were evaluated to establish the relationship between pH, exposure time, and cytotoxicity for rat MSCs. Regardless of the reagent used, lower concentration (10 mM) was associated with small pH change (decreased by 1.5 units) and corresponding high (90%) cellular viability. However, at 100 mM and above, MSC viability was significantly reduced in the case of APS/AA or NaPS/Asc compared to NaPS/Asc-2 system, although gelation time of OPF was more for NaPS/Asc-2 than that for the APS/AA system. These studies indicate that final pH and/or radical formation have a significant impact on MSC viability and detailed investigation must be carried out if OPF systems are to be employed as injectable hydrogels for cell encapsulation studies.

In another study where a water-soluble crosslinking system consisting of PEG diacrylate (PEG-DA), APS, and AA was used to make injectable poly(propylene fumarate-co-ethylene glycol) hydrogels. By changing the PEG block length of the poly(PF-co-EG) and the concentration of the crosslinking system, the equilibrium water content and the sol fraction of the hydrogels could be varied from $57.1 \pm 0.3$ to $79.7 \pm 0.2\%$ and

2.5 ± 0.0 to 3.33 ± 5.4%, respectively. The onset of gelation times was also shown to vary from 1.1 ± 0.1 to 4.3 ± 0.2 min. All the hydrogels demonstrated tensile strength and tensile moduli in the range 61.7 ± 18.2 to 401.3 ± 67.5 kPa and 0.4 ± 0.0 to 3.3 ± 0.3 MPa, respectively. Endothelial cells were found to attach to the hydrogels suggesting that injectable poly(propylene fumarate-co-ethylene glycol) hydrogels formed by employing this water-soluble crosslinking system could be useful for tissue engineering applications [34]. The same research group tested an APS/TEMED redox initiator system for the preparation of injectable OPF hydrogels in the presence of a co-crosslinker, $N-N'$-methylenebisacrylamide (Figure 4.3), wherein gelation occurred within 4 min. These gels could entrap DNA physically and maintained sustained release for a period of 45–62 days, depending upon the molecular weight of PEG used for OPF preparation, suggesting the hydrogels have potential for gene delivery [35].

Biopolymer-based injectable hydrogels were prepared by radical polymerization of methacrylic and lactic acid-grafted chitosan, initiated by the redox system APS/TEMED. Hydrogel formed after 8 mins was found to maintain normal morphology of chondrocytes through a culture duration of 12 days [36]. Redox-initiated crosslinking consisting of AA and NaPS was

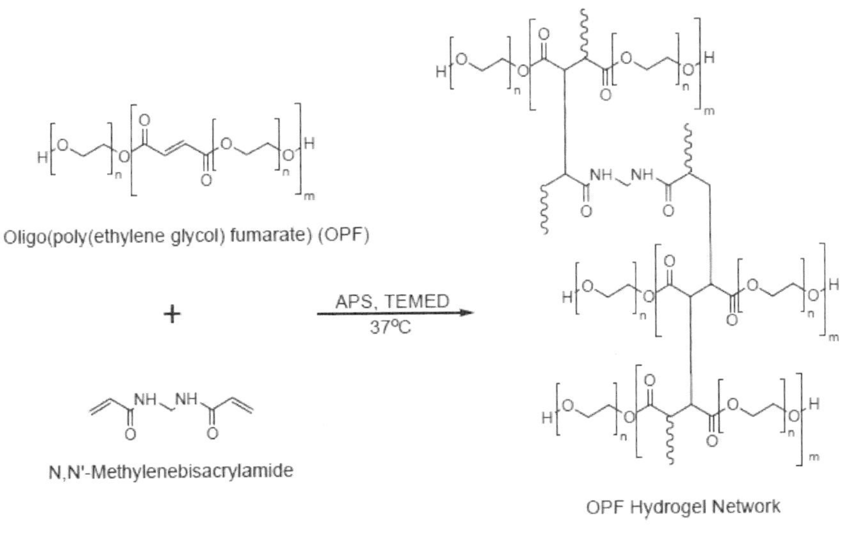


Figure 4.3 Crosslinking reaction between OPF and $N-N'$ methylenebisacrylamide using APS/TEMED redox initiator to form *in situ* gelling hydrogel. Reprinted from [35] with permission from Elsevier.

used to make *in situ* forming hydrogel based on chitooligosaccharides after functionalization with methacrylate groups. By changing the concentration of the oligomer and redox pair, degree of acetylation of chitooligomer, and the methacrylation degree of the modified oligomer, gelation time was found to be easily modified between 1.5 and 60 min for these biodegradable gels, which could be used as injectable as well as preformed hydrogels for drug delivery and tissue engineering applications [37]. PEG extended with oligo-lactide and end-capped with acryloyl groups was crosslinked in phosphate-buffered saline in the presence of a redox initiator system, APS/ TEMED, to make injectable hydrogels suitable for tissue engineering. Gelation time was found to decrease with an increase in concentration of TEMED and gelling time of about 10 min was achieved by adjusting the concentration to 1.2 $\mu$L/g. A temperature increase of only 1.1°C was observed, indicating that the formation of hydrogel *in situ* might not lead to significant harm to embedded cells and surrounding tissues. The modulus of hydrogel was higher compared to the physically crosslinked counterpart, Pluronics®. Degradation rate was tunable with a change in degree of crosslinking, and length and composition of oligoester blocks [38]. Composite injectable hydrogels, made with hydroxyapatite particles as fillers employing redox crosslinking using PEG-diacrylate as crosslinker and APS/TEMED as initiator system, has been evaluated for bone tissue engineering. Time sweep rheology analysis showed that gelation time decreased with an increase in concentration of PEGMC and the fastest gelation (212 s) occurred in the case of PEGMC(0.8)/HA. The stiffness of these gels was unaffected by the presence of HA, but was controlled by the PEGMC matrix. It is envisaged that just as calcium ions of apatite crystals in bones are associated with citrate molecules, the HA particles will also have interaction with citric acid units attached to PEGMC chains, which ultimately would be beneficial for use in bone tissue engineering [39].

Injectable hydrogel prepared by redox-initiated crosslinking of methacrylate derivative of HA (MeHA) using an APS/TEMED initiator has been evaluated as a bulking agent to reduce left ventricular (LV) wall stress created following myocardial infarction (MI). Hydrogels were prepared using MeHA macromer (4 wt%) having different degree of methacrylation (30% or 60%) using varying concentration of APS (5–12.5 mM)/TEMED (56.25 mM) in such a way that they have similar degradation and tissue distribution upon injection, but have differential moduli (~8 versus ~43 kPa) and gelation (Figure 4.4). These hydrogels were injected into a clinically relevant ovine MI model to evaluate the associated beneficial effect of intramyocardial hydrogel injection on the remodeling response based on hydrogel mechanics. Injection of all hydrogels significantly increased the wall thickness in the apex and basilar infarct regions compared with the control infarct. However, the higher-modulus (High MeHA) treatment group had a smaller infarct area and reductions in normalized end-diastolic and end-systolic volumes with better functional outcomes (cardiac output and ejection fraction) than the low-modulus (Low MeHA) and control infarct groups (Figure 4.4), indicating that the modulus of hydrogel plays a major role in the treatment of MI. This fundamental information can be used

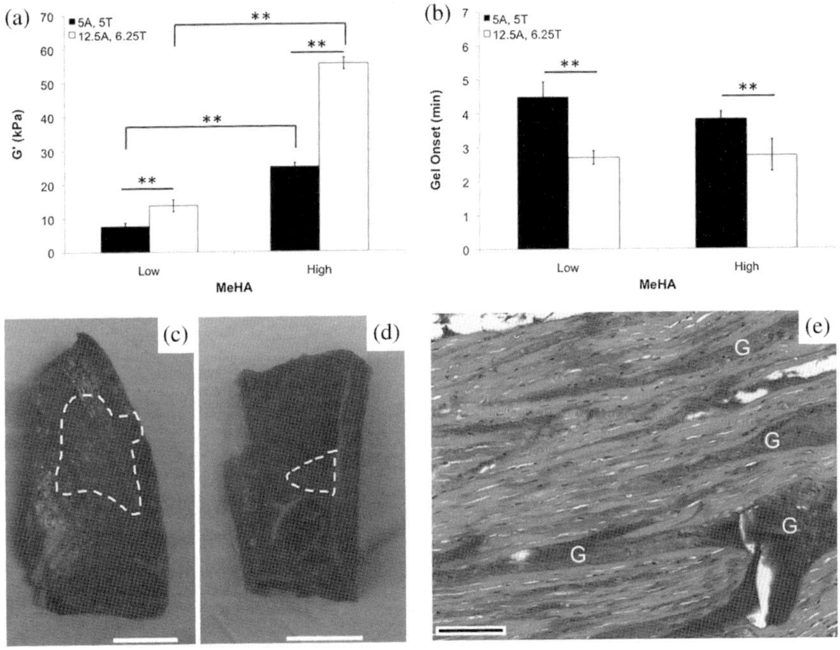


Figure 4.4 MeHA hydrogel mechanical properties and gelation time behavior. (a) Storage modulus and (b) gelation onset time for various macromer (MeHA Low or High) and initiator [5.0 mM/5.0 mM or 12.5 mM/6.25 mM APS (A)/TEMED (T)] combinations. n = 3 per group. Data are presented as mean ± SD. $^{**}P < 0.01$. (c and d) Explanted tissue with 4% by weight MeHA High (outlined with dashed line) gelled with (c) 5.0 mM APS/5.0 mMTEMED or (d) 12.5 mM APS/6.25 mMTEMED, showing differences in gel distribution based on initiator concentrations. (Scale bars, 10 mm.) (e) Representative H&E-stained image of the cardiac tissue at the apex of a MeHA High gel (labeled "G")-treated infarct 24 h post-injection, demonstrating integration with tissue. (Scale bar, 100 $\mu$m.) Reprinted from [40] with permission from PNAS.

in the rational design of therapeutic materials for treatment of this condition [40]. Further, to evaluate the effect of degradability of hydrogels in the treatment of MI, two hydrolytically degradable hydroxyethyl methacrylated HA (HeMA-HA) hydrogels and stable MeHA hydrogels with low and high degradation profile but similar initial mechanics were prepared and evaluated. It was demonstrated that the stable hydrogels maintained myocardial wall thickness in the apical and basilar regions more efficiently (low MeHA: apical: 6.5 mm, basilar: 7 mm; high MeHA: apical: 7.0 mm basilar: 7.2 mm) than the hydrolytically degrading (low HeMA-HA: apical: 3.5 mm, basilar: 6.0 mm; high HeMA-HA: apical: 4.1 mm, basilar: 6.1 mm) hydrogels, but both hydrogel groups outperformed the controls. Both treatments resulted in increased vessel formation compared to the control, but with stable MeHA hydrogels, LV volume reduction was effective for eight weeks whereas degradable hydrogels were effective only for two weeks [41]. Further, hydrogel-carrying polylactide microspheres ($\sim$50 $\mu$m diameter) with a tunable degradation profile were also evaluated. Injection of fast degrading HeMA-HA hydrogels containing 75 $\text{mgmL}^{-1}$ microspheres to the infarcted LV wall of sheep was shown to reduce the LV dilation, enhanced myocardial wall thickness and the presence of vessels in the apical infarct region to $\sim$35 and $\sim$60%, respectively, compared to empty hydrogels [42].

As the properties of hydrogels prepared by employing redox initiators are highly dependent on the initiator concentration and since some of these initiators have demonstrated severe biocompatibility issues, wide applicability of this method for biomedical applications is restricted. To substantiate this, there are reports to show the cytotoxicity of the chitosan methacrylate-based hydrogels made by the APS/TEMED system compared to those made using photoinitiator Irgacure 2959 at the same gelation time and in the presence of cells [43].

#### 4.2.2.2 *Crosslinking via photopolymerization*

Photocrosslinking is another type of free radical polymerization method to make injectable hydrogels, where reaction is initiated via photoinitiators which decompose upon exposure to UV or visible light, leading to the formation of free radicals. Since UV/visible light can be applied *in vivo* in a minimally invasive manner by means of laparoscopic devices, catheters, or transdermal illumination, photopolymerization is a suitable method for the preparation of *in situ* gelling hydrogels [44]. This method is superior to other radical polymerization methods mainly because of their temporal and spatial control, leading to control over polymerization exotherms and time of gelation. This can be used for the fabrication of even complex structures using lasers or masks. The basic principle of this method is the introduction of polymerizable vinyl groups onto the polymer using certain compounds (e.g., methacrylol chloride, glycidyl methacrylate, methacrylic anhydride) and exposure to UV/visible light in the presence of some acceptable initiators (e.g., camphorquinone, Irgacure 2959 (I2959)) to form hydrogels. Further, copolymerization of polymers with acrylic monomers can also be performed by this method resulting in preparation of *in situ* gelling systems. The concentration and type of the photoinitiators should be controlled in such a way that the photoirradiation will not produce any unwanted reactions like tissue necrosis due to heat generation and cell toxicity.

HA has been exploited extensively for the preparation of *in situ* gelling hydrogels via photocrosslinking. Methacrylate derivative of HA (2% solution) has been found to readily fill irregularly shaped defects *in vivo* and crosslink rapidly upon exposure to light, resulting in the formation of a three-dimensional hydrogel of considerable mechanical integrity. These hydrogels promoted the retention of the chondrocytic phenotype and cartilage matrix synthesis by encapsulated chondrocytes *in vitro* and accelerated healing in an *in vivo* osteochondral defect model [45]. Burdick *et al.* [46] used these MeHA hydrogels (initiator-I2959; long wavelength UV; intensity $\sim 4\,\mathrm{mW/cm^2}$; irradiation for 10 min) to encapsulate primary auricular swine chondrocytes and implanted subcutaneously on nude mice. Neocartilage formation studied for a period of eight weeks showed comparable results with PEG dimethacrylate (PEGDM) hydrogels, which are extensively studied as photocrosslinked chondrocyte carriers. More uniformly distributed GAG formation, one of the main functionalities of chondrocytes, was seen in MeHA hydrogels than in non-degradable PEGDM hydrogels (Figure 4.5). Hydrogel prepared by photocrosslinking of HA with PEG-diacrylate encapsulated with mesenchymal cells and TGF-$\beta$-3 showed chondrocyte differentiation [47]. Hydrolytically degradable MeHA hydrogels prepared by incorporating repeating units of $\alpha$-hydroxy esters (e.g., lactic acid) between the HA and the polymerizing methacrylate moiety (initiator, I2959; UV 365 nm; intensity $\sim 2\,\mathrm{mW/cm^2}$; irradiation time, 10 min) were found to undergo complete degradation into HA, lactic acid, and chains of poly(methacrylic acid) even without hyaluronidase within nine days [48]. *In situ* forming HA hydrogels with degradable crosslinks, prepared by photocrosslinking of HEMA-HA macromer having reactive methacrylate group and hydrolyzable ester groups between the methacrylate functionality and

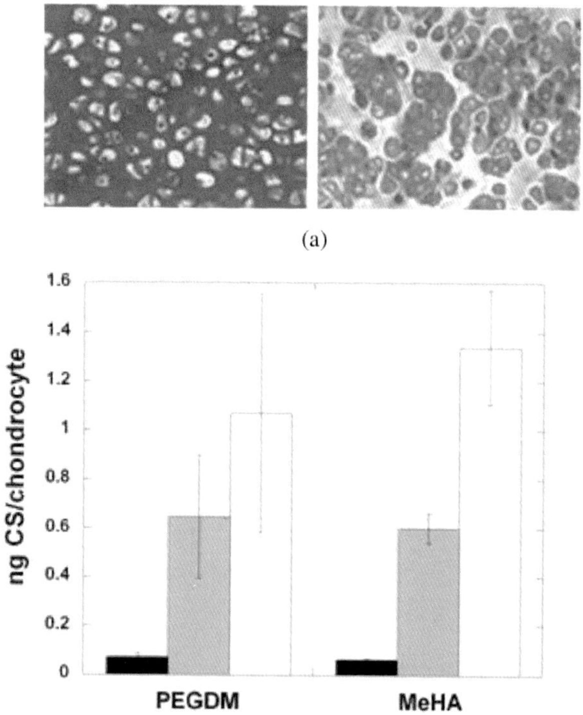


(a)

(b)

Figure 4.5 Neocartilage formation by swine chondrocytes photoencapsulated in HA and PEG hydrogels. (a) Histological sections (stained for GAG, bar = 100 $\mu$m) of auricular chondrocytes photoencapsulated in 2% 350 kDa MeHA (left) and 10% PEGDM (right) hydrogels 6 weeks after implantation in nude mice. (b) Glycosaminoglycan content (reported as ng CS/chondrocyte) immediately after photoencapsulation (black) and after 4 (gray) and 8 (white) weeks of auricular chondrocyte culture in nude mice for 2% 350 kDaMeHA and 10 wt% PEGDM. Reprinted from [46] with permission from the American Chemical Society.

HA backbone, have been shown to enhance the engraftment of circulating bone marrow derived cells (BMCs) in the myocardium following MI. These hydrogels maintained sustained release of recombinant stromal cell-derived factor-1alpha (rSDF-1$\alpha$) and HA to the injured myocardium and stimulated BMC chemotaxis up to four-fold *in vitro* through CXCR4 and CD44 receptor signaling, respectively. Gel containing rSDF-1$\alpha$ enhanced BMC homing to the remodeling myocardium significantly better than delivery of only rSDF-1$\alpha$ to the heart following experimental MI in mice [49]. Hybrid hydrogels having tunable physical and biological properties prepared by photocrosslinking of methacrylate derivatives of HA and gelatin exhibited enhanced cell spreading compared to MeHA hydrogels [50].

A methacrylate derivative of CS and alginate also has been employed to make injectable hydrogels for cartilage tissue engineering. Photocrosslinkable hydrogels from CS, prepared by polymerization of methacrylate derivative of CS and its copolymerization with PEG diacrylate (PEODA) in the presence of UV photo initiator I2959 using UV light source (intensity $\sim$10 mW/cm$^2$) at a wavelength of 365 nm, were found to have good mechanical strength [51]. Methacrylate derivatives of alginate, synthesized by reaction with 2-aminoethyl methacrylate (AEMA) were photocrosslinked in the presence of I2959 initiator with UV light at 365 nm at intensities 8–20 mW/cm$^2$ for 8 min to form hydrogels [52]. AEMA imparts biodegradable linkages within the hydrogels which are susceptible to hydrolysis. Attempts were also made to make the hydrogel biodegradable by partially oxidizing alginate prior to methacrylation. Physical properties of these hydrogels could be tuned by varying the degree of alginate oxidation and weight concentration of OMA, in addition to varying the degree of alginate methacrylation. Photocrosslinked hydrogel prepared by using alginate with degree of oxidation 9% (OMA-9) and 14% (OMA-14) (2 w/v %) lost 73.9 and 100%, respectively, of their mass in DMEM by 3 weeks, indicating that the rate of oxidation could control the degradation of the hydrogels. Swelling ratio of OMA-14 was found to be more than that of OMA-9 (Figure 4.6) [53]. Photocrosslinked alginate hydrogels containing interpenetrating HA chains demonstrated stem cell chondrogenesis, high viability with metabolic activity comparable to calcium crosslinked alginate hydrogels for up to ten days of culture, and up-regulation of chondrogenic gene markers indicating the potential of photocrosslinked, injectable alginate as an injectable delivery system for cell-based cartilage repair therapies [54].

Macromers synthesized by coupling acrylated PEG to the side chain of the biodegradable polyphosphoester poly(6-aminohexyl propylene phosphate) (PPE-HA) were also found to undergo photocrosslinking (8 min at 365 nm UV light at $\sim$8–20mW/cm$^2$ in the presence of initiator (I2959)) to give elastic biodegradable injectable hydrogels. Bone marrow-derived goat mesenchymal stem cells maintained viability when encapsulated within these gels and exhibited mineralization when cultured in osteogenic medium for three weeks [55]. Methacrylation of glycol chitosan upon photocrosslinking in the presence of I2959 on exposure to UV source (320–480 nm) at

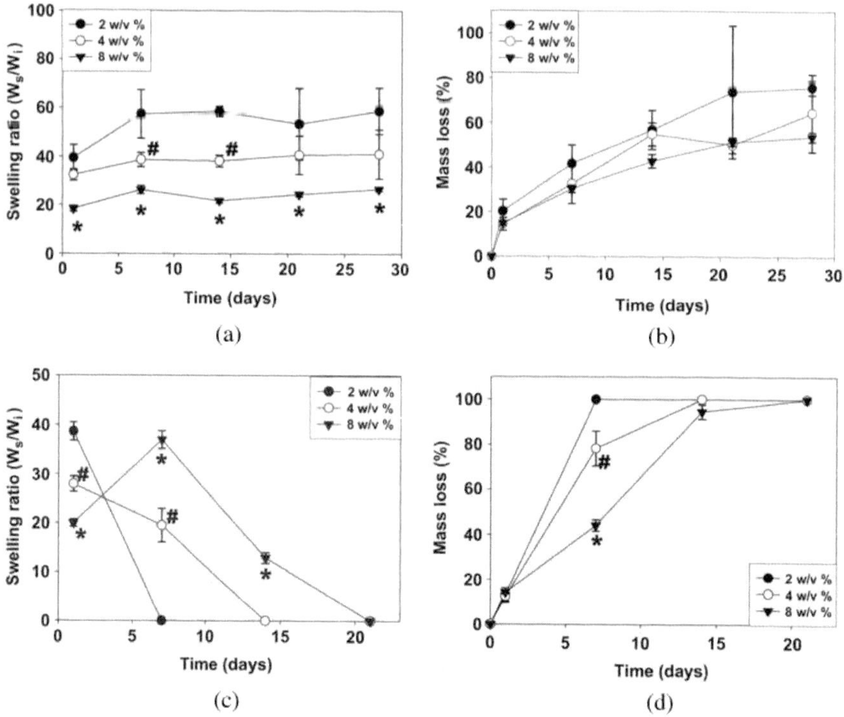


Figure 4.6 Swelling ratios of photocrosslinked (a) OMA-9 and (b) OMA-14 hydrogels and mass loss of (c) OMA-9 and (d) OMA-14 hydrogels in DMEM over time. Values represent mean ± standard deviation ($N = 3$). $^*p < 0.05$ compared with 2 and 4 w/v%, and $^\#p < 0.05$ compared with 2 w/v %. Reprinted from [53] with permission from Elsevier.

intensities ranging from 25 to 102 mW/cm$^2$ for 9 min yielded biocompatible hydrogels which undergo degradation in the presence of lysozyme. The elastic moduli obtained for these hydrogels (0.44 MPa) were lower than that for human tibial articular cartilage (1–10 MPa), indicating the need for further modification of the system to improve the mechanical strength [56]. Photocrosslinkable gelatin hydrogels were also prepared in a similar way by derivatization with methacrylic acid, followed by UV irradiation (365 nm) in the presence of I2959 for 20 min. It was found that the hydrogel supported chondrocyte growth and maintained chondrocyte phenotype under *in vitro* culture conditions for 12 days [57]. Photocrosslinking has also been exploited to stabilize injectable hydrogels prepared using a self-healing Dock-and-Lock system having shear-thinning properties. Peptides were modified to undergo secondary crosslinking through light-initiated radical polymerization of methacrylate functional groups to tune gel mechanics and erosion kinetics. These secondary crosslinked hydrogels demonstrated retention of self-adhesive property as well as enhanced modulus [58].

A comparative study on the characteristics of photo- and redox-initiated crosslinking to make injectable hydrogels using citric acid-derived water-soluble and biodegradable polymers has been performed. PEG maleate citrates (PEGMCs) were synthesized by reacting citric acid with PEG 200 and maleic acid (Figure 4.7). PEGMCs were shown to undergo *in situ* crosslinking into biodegradable elastomeric polyester hydrogels using photo- as well as redox-crosslinking methods. While photocrosslinking was in the presence of a crosslinker, acrylic acid, and photoinitator, 2,2′-azobis(2-methyl propionamidine) dihydrochloride, redox-crosslinking was in the presence of 0.026 M APS and 0.11 M TEMED. The citric acid derived polymeric approach is fascinating in the sense that it can preserve the carboxylate and hydroxyl groups, the highly favored pendant chemistries for functionalization of biomaterials with drugs or biomolecules, for crosslinking. The sol content of these hydrogels was found to decrease with an increase in the concentration of crosslinkers, but increased with concentration of maleic acid. However, it was envisaged that the trans-conformation of maleic acid would encourage more coiled pre-polymer chains when a higher number of maleate units was incorporated in the middle of chains which can undergo intramolecular crosslinking in addition to interchain crosslinking, and the intracrosslinked chain could be leached out of the network by the solvent. Tensile studies have shown that those polymers having higher content of citric acid and relatively less maleic acid display enhanced elongation and ultimate tensile strength and decreased initial modulus. Further, it was found that degradation of these hydrogels was controlled by the concentration of maleic acid and crosslinker as well as the type of crosslinking. Photocrosslinked hydrogels are fast degrading compared to redox-crosslinked ones where only 70% of the mass was degraded after 30 days. These hydrogels were evaluated as a protein delivery vehicle and the release of protein was shown to be slow from those gels having higher citrate content due to hydrogen bonding [59].

Though the photocrosslinking method has many advantages over other crosslinking methods in terms of control of gelation and fabrication of complex three-dimensional structures, the degradability of the resulting hydrogels and potential toxicity remain concerns. Researchers are trying to make them biodegradable by introducing hydrolytically

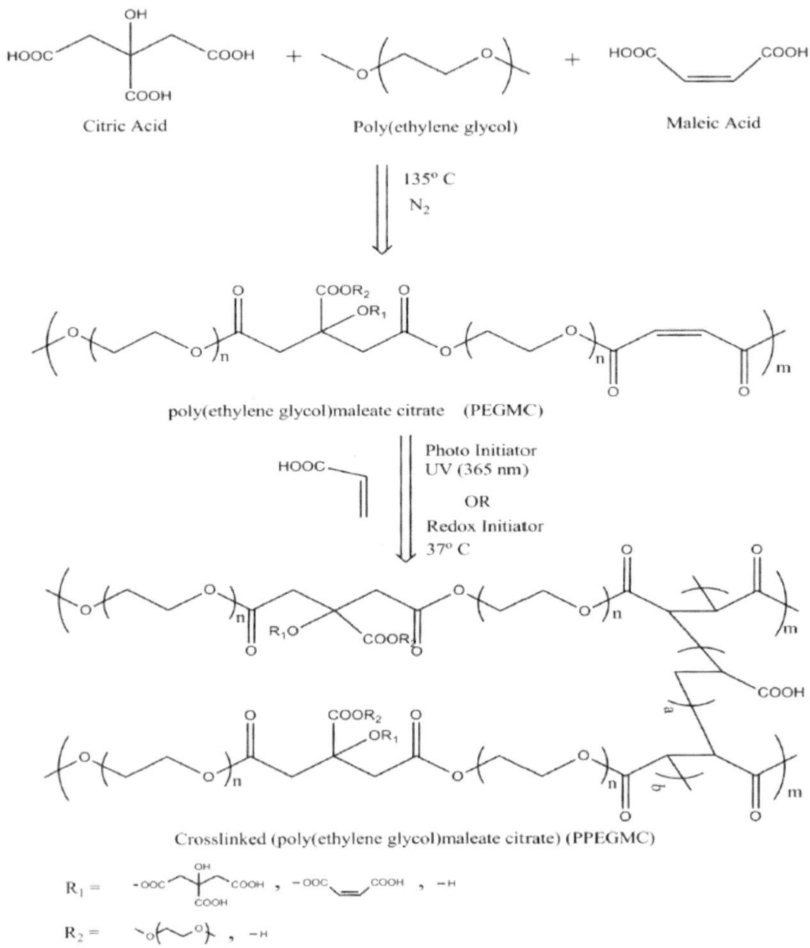


Figure 4.7 Schematic representation of PPEGMC synthesis. Reprinted from [59] with permission from Elsevier.

labile spacers. Radical chemistry, photoinitiator efficiency (the ability to produce radicals), and exposure time can dramatically affect cell viability. Photoinitiators, which may have toxic effects and prolonged irradiation time, leading to a local temperature increase that can subsequently harm neighboring cells and tissues, is one of the other limiting factors. In many cases, photopolymerization has not proven suitable for injectable *in vivo* use, in spite of the development of transdermal photoactivation methodology [44]. Hence, attempts have been made to make use of visible light photoinitiating conditions to encapsulate cells in order to minimize the potentially damaging effects of UV irradiation. Collagen has demonstrated photocrosslinking when exposed to visible light (440–500 nm, 0.5 W/cm$^2$) in the presence of riboflavin (type II photoinitiator) (gelling time 40 s). Implantation of chondrocytes-encapsulated, photocrosslinked collagen hydrogels demonstrated hyaline type cartilage formation and integration of newly formed cartilage within six weeks, suggesting potential as an injectable cartilage tissue engineering scaffold [60, 61]. Injectable hydrogels were also prepared using methacrylated glycol chitosan (MeGC) and HA by photocrosslinking with a riboflavin photoinitiator under visible light (400–500 nm, 300 mW/cm$^2$, irradiation for 40 s). The hydrogels were shown to be biocompatible, and the presence of HA in the crosslinked chitosan network significantly improved cell proliferation and cartilaginous ECM production by encapsulated chondrocytes, indicating the high potential of these composite hydrogels as carriers of cells and bioactive molecules for the treatment of cartilage damage [62].

Thiol-ene photopolymerization is an alternative approach in which the toxic side effects of irradiation can be minimized by reducing the concentration of initiators and irradiation time. Because of the resistant nature of this reaction towards oxygen inhibition, it can rapidly form hydrogel even at low initiator concentrations and with less irradiation time [63]. Besides being bio-orthogonal and biocompatible, the advantage of thiol-ene photopolymerization is spatial and temporal control over reactions. Multi-arm PEG-norbornene (PEG-NB) macromers crosslinked by matrix metalloproteinase (MMP) cleavable peptides flanked with bis-cysteines via step-growth thiol-ene click reaction (Michael addition) has been found to be more homogeneous with a higher conversion rate compared to chain growth-polymerized gels with similar crosslinking density. However, Michael reaction is spontaneous and therefore the temporal and spatial control will be forfeited and disulfide formation is a possible side reaction [64]. A thiol-ene hydrogel system prepared by photopolymerization of a 4-arm PEG-NB macromer and bis-cysteine-terminated and chymotrypsin-sensitive peptide sequence for the encapsulation of pancreatic MIN6 $\beta$-cells has shown more minimal cellular damage, improved gel fraction, mesh sizes, and mechanical properties than PEG-diacrylate hydrogels having equal functionality [65]. Copolymerization of thiolated gelatin with PEG-diacrylate using photoinitiator Irgacure upon exposure to long wavelength UV ($\lambda$ max = 365 nm, intensity at 100 mW/cm$^2$) for 2 mins yielded hydrogel with tunable mechanical properties and long-term cell anchorage points, thus supporting cell attachment and growth in the three-dimensional environment [66].

### 4.2.3 *Michael-type addition reaction*

The Michael-type addition reaction presents another method for *in situ* formation of polymeric hydrogels. The reaction involves addition of nucleophiles (called 'Michael donor') to an $\alpha,\beta$-unsaturated carbonyl bond attached to an electron-withdrawing group (called 'Michael acceptor') without the use of any radical initiators. Among the Michael addition reactions, the reaction of thiols to form a carbon–sulfur bond represents a key reaction in biosynthesis. The reaction can occur under physiological conditions (temperature and pH) within minutes and has been extensively employed in developing *in situ* gelling systems.

#### 4.2.3.1 In situ *gelling systems based on synthetic polymers*

Hubbel and coworkers [67] were the first group who utilized the Michael-type addition reaction for the formation of degradable PEG hydrogels suitable for the delivery of protein drugs, wherein an aqueous solution of PEG-multiacrylate having albumin particles was mixed with an aqueous solution of PEG-dithiol to rapidly form hydrogel. The release mechanism of protein is controlled by dissolution of protein in the presence of PEG and diffusion of protein through the gel. During crosslinking, the protein remains unaffected, and 65% of the total incorporated protein exhibited a zero-order release kinetics for a period of four days. Studies on the role of thiolates as the reactive species in a Michael-type addition reaction between thiol-containing peptides and PEG-diacrylates revealed that the reactivity is influenced by the sign and number of the neighboring charges and not by the position of charges, which formed the basis of a rational design of peptides that can be used to make polymer/peptide or protein conjugates via a Michael-type addition reaction [68]. Based on this, Hubbel group prepared various non-interactive and bio-interactive hydrogel networks, using branched, end-functionalized PEG chains and bioactive thiol-bearing peptides as structural building blocks [69, 70]. Variants of two clinically relevant forms of vascular endothelial growth factors (VEGF), namely VEGF121 and VEGF165, were also explored to make PEG-peptide bio-interactive hydrogel networks (gelling time 15 min) to support endothelial cell migration and formation of vascular tissue [71]. Adhesive and plasmin-sensitive hydrogels prepared by

Michael-type addition of PEG-bis-vinylsulfone with RGD (adhesive peptide) and GCYKNRCGYKNRCG-NH2 (plasmin-sensitive peptide having three cystein residues) were able to better support cellular infiltration compared to those lacking either of these two peptides. Bone morphogenetic protein-2 (BMP-2) incorporated within the hydrogel was found to induce bone healing in a rat model [72]. A PEG-based thiol containing polymer, prepared by copolymerization of 2-mercaptosuccinic acid with diamino-PEG, was crosslinked with PEG divinyl sulfone to form *in situ* hydrogel (Figure 4.8). The reaction was pH-dependent (rapid between pH 6 and 8) and the hydrogel encapsulated with bovine serum albumin demonstrated complete release after 20 days, indicating that the encapsulated protein remains unaffected during crosslinking and was simply physically trapped within the hydrogel. Gels remained intact at the site of injection even after six months, indicating their suitability as a long-term protein delivery depot [73]. The Michael-type addition reaction of thiol-capped PVA chains and the vinyl moiety contained in a grafted methacrylic PVA yielded hydrogel (after 60 min) suitable for surgical applications entailing *in situ* vitreous replacement [74]. There are also reports showing preparation of *in situ* forming hydrogels by the step-growth polymerization of maleimide- and thiol-terminated PEG macromers via Michael-type addition [75].

#### 4.2.3.2 In situ *gelling systems based on biopolymers*

Biopolymers have also been used to make *in situ* gelling hydrogels based on Michael-type addition reaction. Injectable hydrogels have been prepared using thiolated HA, $\alpha,\beta$-unsaturated esters (diacrylates and dimethacrylate), and amides (diacrylamides and dimethylacrylamide) of PEG. Rapid gelation with no temperature rise was observed for PEG-diacrylate. Gelation time was found to increase in the order PEG-diacrylate $>$ PEG-dimethacrylate $>$ PEG-acrylamide $>$ PEG-diacrylamide, as the electrophilicity of double bond decreases in the same manner [76]. An acellular hydrogel matrix comprised of the fibronectin functional domains (FNfds) coupled to a hyaluronate backbone was also prepared via this reaction to stimulate wound repair. Briefly, cysteine-tagged FNfds were first coupled to a homobifunctional PEG derivative followed by coupling to thiol-modified HA to obtain a crosslinked hydrogel matrix, which besides being biocompatible induced rapid and group recruitment of stromal fibroblasts which were absent in unmodified and RGD-tethered hydrogels [77]. A Michael-type

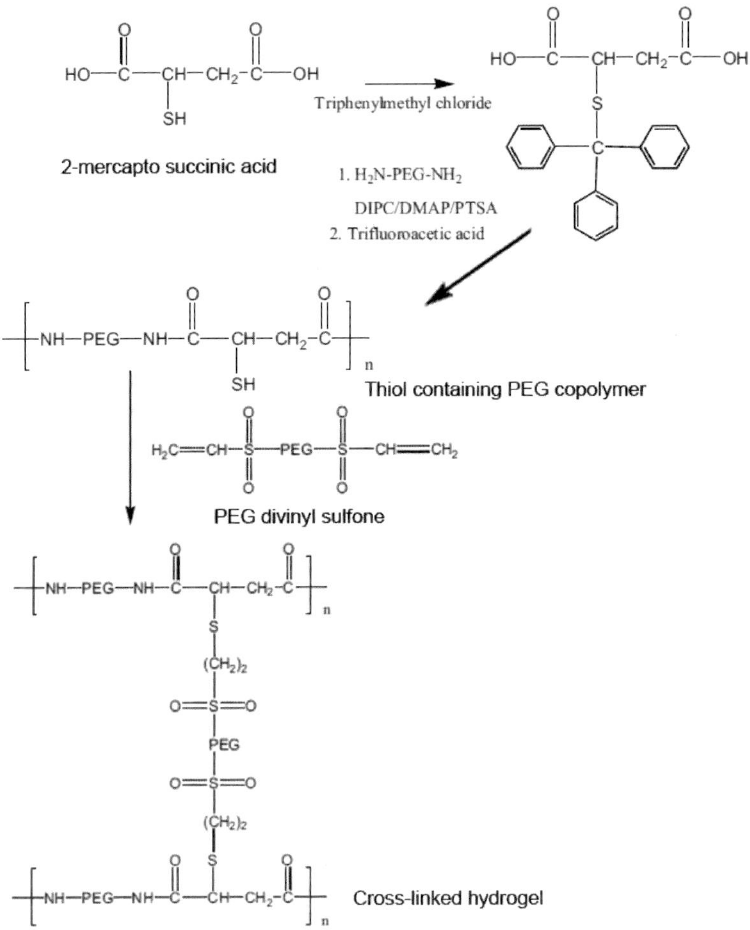


Figure 4.8 Schematic representation of preparation of PEG-based hydrogel by Michael-type addition reaction. Reprinted from [73] with permission from Elsevier.

addition reaction between acrylated HA and tetrathiolated PEG also resulted in the formation of hydrogel within 10 min under physiological conditions. Acrylation of HA was done by the coupling of adipic acid dihydrazide with HA, followed by acrylation with *N*-acryloxysuccinimide. These hydrogels demonstrated tunable gelation time, compression, and elastic modulus by varying the concentration of acrylated HA and the molecular weight of HA. Hydrogel based on 50-kDa HA were completely degradable by 100U/ml hyaluronidase *in vitro*. Human mesenchymal cells encapsulated in BMP-2-loaded gels demonstrated 55% enhanced cellular viability *in vitro* and expression of osteocalcin and mature bone formation *in vivo*, substantiating the usefulness of hydrogel for tissue regeneration [78, 79]. Thiolated HA and 4-arm PEG-vinyl sulfone yielded hydrogel with tunable gelation time. The gelation time decreased with an increase in the concentration of polymer precursors and molecular weight (45–185 KDa) of HA used. The shortest gelation time obtained was 1 min. The degradation time of hydrogels was found to increase (up to 15 days) with the molecular weight of HA, its degree of substitution, and concentration of polymer. Gels were evaluated for cartilage tissue engineering and demonstrated an increase in chondrocyte proliferation during the first week of culture [80]. Nano-structured HA hydrogel was prepared by a combination of atom transfer radical polymerization (ATRP) and Michael-type addition reaction, in which one component was biodegradable PEG monomethyl ether methacrylate-co-poly(hydroxylethyl methacrylate) nanogel with pendent hydroxyl groups prepared via electron transfer ATRP and subsequently acrylated, and the other component was thiolated HA. Gelation took place within a few minutes after mixing the precursor solutions resulting in a nano-structured gel, with a porous three-dimensional structure and uniform distribution of nanogels [81].

Thiol-functionalized dextrans (14 $M_n$ or 31 kDa) with degrees of substitution ranging from 12 to 25 were prepared to form hydrogel *in situ* using tetra-acrylated PEG or a dextran vinyl sulfone conjugate. Thiolation of dextran was performed by reaction with cystamine after activation of the hydroxyl groups of dextran with 4-nitrophenyl chloroformate. Highly elastic and biodegradable hydrogels with tunable moduli were formed instantaneously when thiolated dextran reacted with tetracrylated PEG or dextran divinyl sulfone. Degradation was fast (three to seven weeks) for the thiolated dextran/dextrandivinylsulfone system, whereas dex-SH/PEG-4-Acr hydrogels took about 7 to over 21 weeks, depending on the degree of substitution, concentration, and dextran molecular weight [82]. The degradation profile of these hydrogels was influenced by the presence of encapsulated chondrocytes. Chondrocytes remained viable within the matrix for three weeks [83]. Tetrathiolated PEG has also been used to crosslink dextran vinyl sulfone to make *in situ* gelling hydrogels [82]. The properties of these hydrogels (degradation and modulus) were also controlled by the degree of vinyl sulfone substitution and dextran molecular weight. Degradation of these hydrogels and the protein release profile from them could be adjusted by introducing spacers like ethyl or propyl groups between the thioether and the ester bond of the dextran vinyl sulfone conjugates [84]. Thiolated glycol chitosan and water soluble oligo(acryloyl carbonate)-b-PEG-b-oligo(acryloyl carbonate) triblock copolymers have also been used to make injectable hydrogels via a Michael-type addition reaction. Robust hydrogels with tunable gelation time and storage modulus were formed rapidly upon mixing aqueous solutions of precursor polymer solutions at remarkably low polymer concentrations of 1.5–4.5 wt% under physiological conditions. The gels have a microporous structure with low swelling and slow degradation (over six months) under physiological pH and temperature. Specific bioactivity can be rendered to these hydrogels due to the availability of multiple acryoyl groups which can undergo conjugation with thiol-containing biomolecules. For example, glycol chitosan hydrogels were made cell adhesive by modification with Gly-Arg-Gly-Asp-Cys (GRGDC), which well supported adhesion and growth of both MG63 osteoblast and L929 fibroblast cells, indicating great potential in the development of injectable cell-specific bioactive extracellular matrices for tissue engineering [85]. *In situ* chitosan-PEG hydrogel was also prepared by a two-step reaction in which initially 2-carboxyethyl acrylate molecules were grafted to the primary amine functional groups in chitosan, followed by a Michael-type addition reaction with thiol end groups in the PEG. Gelation took place at 10 min after mixing the precursor liquid solutions [86]. A heparin-based hydrogel for maintenance of functional hepatocytes was prepared by a Michael-type addition reaction of thiolated heparin and acrylated PEG by raising the temperature from 25 to 37°C. Cells entrapped within the hydrogel were viable and maintained high levels of albumin and urea synthesis after three weeks in culture. Hepatocyte growth factor (HGF)-incorporated hydrogel demonstrated controlled release of GF molecules and only 40% was released after 30 days in culture. These HGF-containing hydrogels exhibited significantly higher levels of albumin and urea synthesis compared to bare hydrogel, demonstrating the potential of heparin-based hydrogels in encapsulation and maintenance of difficult-to-culture primary hepatocytes, as matrices for *in vitro* differentiation of hepatocytes or stem cells, and as vehicles for transplantation of these cells [87].

One of the disadvantages of the Michael-type addition reaction is its susceptibility to premature crosslinking of the thiol-functionalized precursor polymers via oxidative disulfide crosslinking, which limits the shelf life of precursor polymer solutions. Furthermore, the reaction cannot be employed to make *in situ* gelation under acidic conditions, which is seen particularly in the case of chronic inflammation.

### 4.2.4 Disulfide crosslinking

It was mentioned in the previous section that some of the thiolated polymers can undergo self-crosslinking with time, resulting in disulfide bonds between polymer chains. This property has been utilized for the preparation of injectable hydrogels. One typical example includes formation of disulfide linkages between the thiol groups of cysteine residues found in proteins. Various thiolated derivatives of chitosan such as chitosan-4-thiobutylamidine [88], chitosan-thioglycolic acid [89] and chitosan-thioethylamidine [90] have been prepared for making intra- or intermolecular disulfide crosslinks at pH above 5 [91]. However, the viscosity increase of all these derivatives was not high enough to make *in situ* gelling hydrogels. Chitosan coupled with thioglycolic acid, which takes 40 min for gelation, has been shown to undergo faster gelation in the presence of oxidizing agents like hydrogen peroxide, sodium periodate, APS, and sodium hypochlorite (within 20 min). Gels prepared by using ammonium persulfate were found to be cytotoxic, while other hydrogels maintained cellular viability of more than 50%. The potential of these hydrogels as drug delivery depots for respiratory, gastrointestinal, and reproductive tract has been identified [92]. Disulfide-crosslinked HA hydrogels gained much attention because of their dual degradability, as they can undergo cleavage enzymatically by hyaluronidase and chemically by biological reductants such as glutathione (GSH) and cysteine. However, for these hydrogels prepared by air-oxidation or in the presence of strong oxidants such as gaseous oxygen and dimethyl sulfoxide, the control and monitoring of the progress of gelation was found to be difficult, and they often required a prolonged reaction time, limiting their use as injectable systems [93].

*In situ* forming PEG hydrogels have been prepared utilizing disulfide bonds wherein 8-arm-PEG-SH was crosslinked in the presence of either $H_2O_2$ or 8-arm-PEG-S-TP (thiopyridyl-terminated 8-arm-PEG) resulting in the formation of hydrogel networks denoted as $H_2O_2$ hydrogels and S-TP hydrogels, respectively (Figure 4.9). Gelation of 8-arm-PEG-SH with 8-arm-PEG-S-TP (within 10 s) was shown to be faster than that with $H_2O_2$. While both intra- and intermolecular disulfide linkages are possible with $H_2O_2$, only intermolecular sulfide linkages are possible with 8-arm-PEG-S-TP. The hydrogels demonstrated a low degree of swelling, good mechanical strength, and provided sustained doxycycline release for

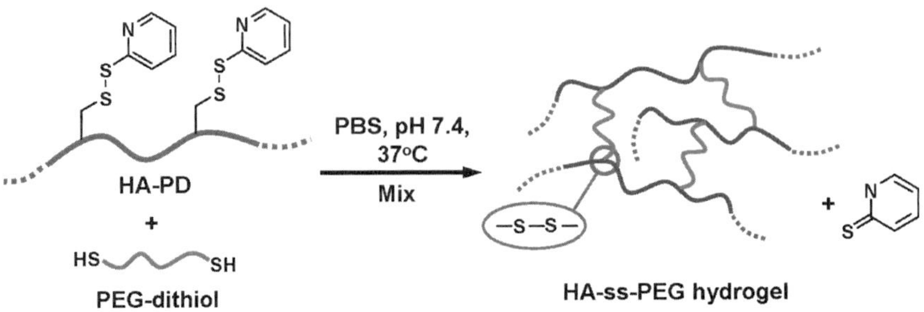


Figure 4.9 Schematic representation of $H_2O_2$ hydrogels and S-TP hydrogels via disulfide bridges. Reprinted from [94] with permission Elsevier.

up to ten days and improved *in vivo* wound-healing efficacy results on nitrogen mustard-exposed skin compared to untreated skin and skin treated with placebo hydrogels [94].

Another strategy has been employed to make disulfide crosslinked HA hydrogels with PEG dithiol as crosslinkers without using oxidizing agents. Here, HA was couple with 1,6-diaminohexane using carbodiimide chemistry to yield amine-derivatized HA, followed by reaction with *N*-Succinimidyl 3-(2-pyridyldithio)propionate to yield HA with pyridyl disulfide groups. The gel was used to encapsulate a chromogenic pyridine-2-thione (Figure 4.10). Gelling time determined by rheological time sweep analysis was within 4–5 mins. Degradation of these gels in the presence of a hyaluronidase enzyme was found to be restricted to surface, which can be attributed to the poor diffusion of the large enzyme molecules (55 kDa) within the hydrogels. However, the same gel at a GSH concentration of 50 mM demonstrated fast cleavage of sufficient number of crosslinks, resulting in the loss of polymer mass and structural integrity without gaining any mass. As GSH exists at different concentrations in extracellular space (2–20 $\mu$M), cytoplasm (0.5–10 mM), and tumor tissues (four-fold more), these dual responsive degradation properties of the disulfide-crosslinked HA hydrogels could find unique opportunities of controlling the release of drugs under different enzymatic and reductive conditions [95].

Not much research has been conducted in the area of disulfide crosslinking for the development of the injectable hydrogels. The use of cytotoxic oxidizing agents and prolonged gelation time may be the limiting factors.

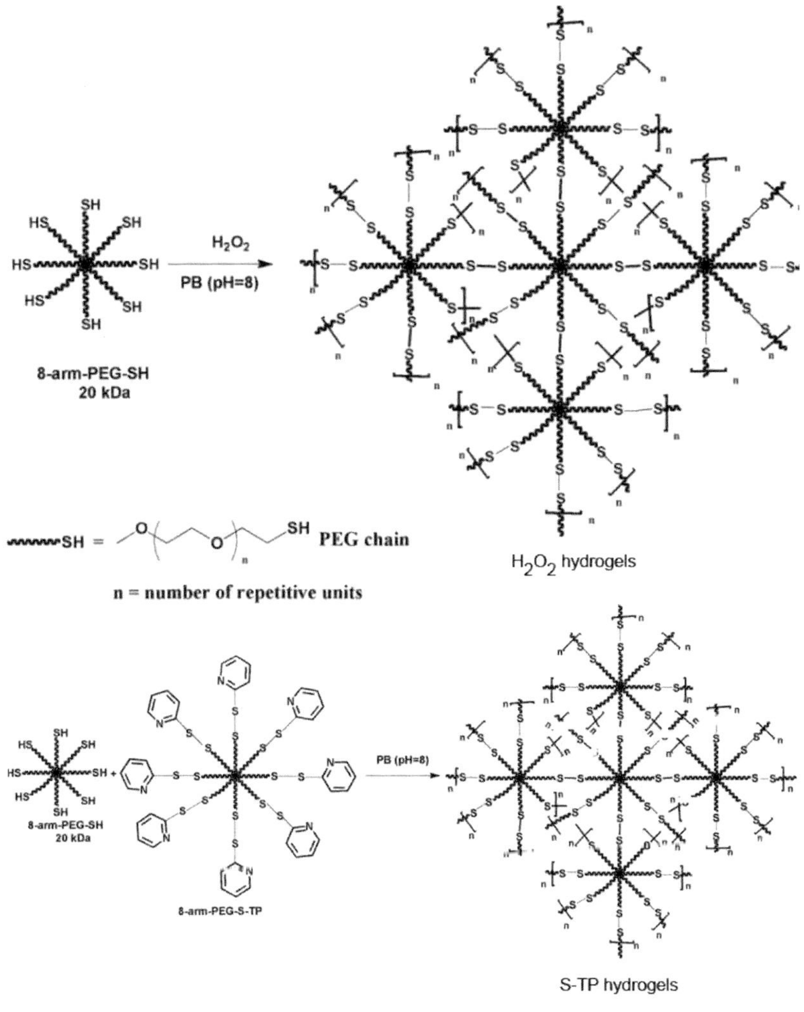


Figure 4.10 Synthesis of disulfide crosslinked HA hydrogel via the thiol–disulfide exchange reaction between HA functionalized with pyridyl disulfide and PEGdithiolate, releasing the chromogenic small molecule, pyridine-2-thione. Reprinted from [95] with permission from the American Chemical Society.

### 4.2.5 *Genipin crosslinking*

Genipin, a naturally occurring iridoid glycoside having $C_{11}H_{14}O_5$ molecular structure is extracted from gardenia fruit (*Gardenia jasminoides ELLIS*).

Being less toxic, it has been used as an alternative to glutaraldehyde for repairing tissues [96]. Genipin undergoes spontaneous crosslinking with protein, collagen [97], gelatin [98], and chitosan [99]. and has been used to make hydrogels for various biomedical applications. The crosslinking mechanism of genipin is still under investigation. One plausible suggestion for the mechanism is formation of a stable bond through an intra/intermolecular crosslink structure, as can be seen in Figure 4.11 [100].

Toxicity evaluation with chondrocytes and osteoblasts demonstrated the dose-dependent toxicity of genipin. It has been recommended that the concentration of genipin should be within 0.5 mM, and above which genipin can induce cell apoptosis via reactive oxygen species production [101]. *In situ* gelation (within 2 mins) was observed by genipin (0.15%) crosslinking of a chitosan solution neutralized with glycerol-phosphate disodium salt. Oscillatory time sweep analysis of these gels demonstrated an increase in storage modulus with time, followed by a faster reduction, exhibiting a discontinuity at around 130 min and subsequently achieving the structural network steady state. This sudden decrease has been attributed to the syneresis properties of the crosslinked gel, causing the gel to shrink [102]. Gel was rapidly formed upon *in vivo* injection and localized at the injection site, and remained in position for at least one week [103]. Genipin has also been used to crosslink multi-arm (8-arm and 4-arm) amino-terminated PEG to form injectable hydrogels. Gelation was slower with 8-arm PEG compared to that with 4-arm. Though the gelation time was found to decrease with the increase in concentration of genipin, in all the combinations tried the lowest gelling time obtained was 1.3 h, which is rather high compared to other *in situ* gelling systems. Enhanced adhesion of human adipose-derived stem cells was observed when seeded onto the 4-arm compared to 8-arm PEG hydrogels, suggesting the potential for use of genipin crosslinked multi-arm PEG hydrogels as injectable scaffolds in a variety of tissue engineering applications [104]. Further, genipin has been used as an *in situ* covalent crosslinker for collagen gels which continued to crosslink even after injection, as monitored by noting the fluorescence [105]. These gels have been found to promote viability of mesenchymal stem cells. Gels incorporated with lipid microtubules containing FGF decreased the cytotoxic effects of genipin towards the astrocytes and promoted cellular infiltration towards the gel, which could be useful for the treatment of spinal cord injuries [106]. Extensive work has not yet been reported on genipin crosslinking, mainly because of its toxicity issues.

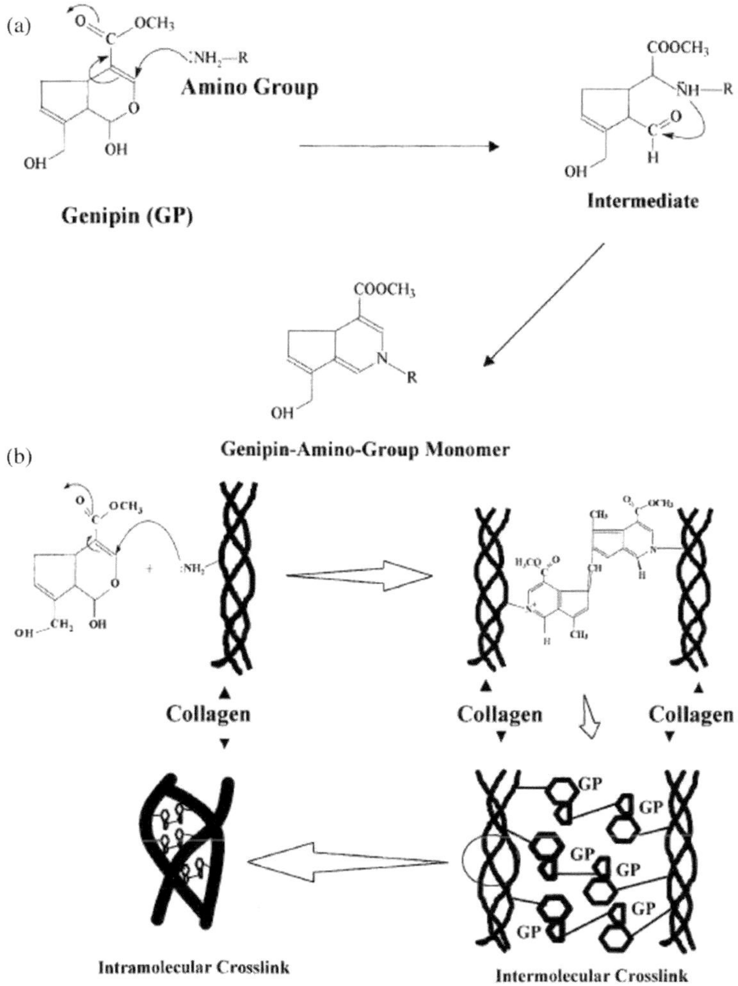


Figure 4.11 (a) Presumable reaction mechanism of genipin with aminogroups. (b) Schematic representation of presumable fixation mechanism of tissues with genipin. Reprinted from [96] with permission from Elsevier.

### 4.2.6 *Click reaction concept for* in situ *gelation*

Click reaction, according to the original definition by Sharpless, is a reaction which meets following criteria:

(a) Proceeds with a high thermodynamic driving force to allow for high yields and reasonable reaction times with no problematic side products.

(b) Simple method of purification of the products without using any chromatographic techniques [107].

Common click reactions, used for regenerative medicine has been reviewed elsewhere [108]. One of the most common click reactions is azide-alkyne cyclo addition, where an organic azide reacts with an alkyne to form a triazole ring in the presence of a Cu(I) catalyst (Figure 4.12). However, this is not recommended for the preparation of *in situ* forming hydrogels because of toxicity concerns over the use of copper. Attempts were made to find a copper-free azide-alkyne click reaction and resulted in the development of the strain-promoted azide-alkyne [3 + 2] cycloaddition between strained fluorinated cyclooctynes and azides. The angle distortion of the cyclooctyne accelerates the reaction, giving it suitability for physiological conditions without the need for the catalyst, making it appropriate for *in situ* crosslinking [109]. Though this reaction has been employed for the crosslinking and surface modification of dendrimers and polymeric nanostructures, it has not been extensively explored for the preparation of *in situ* gelling systems.

Other than cycloadditions, nucleophilic substitutions, radical additions, Michael-type additions, and Schiff base reaction can also be considered as click reactions (see previous sections). The Diels–Alder reaction is another promising click reaction that involves [4 + 2] cycloaddition between diene and dienophile, and is accelerated by the presence of water. The reaction has been used to make HA-based hydrogels employing furan modified HA and dimaleimide derivative of PEG in aqueous solution in the absence of additional crosslinking agents and catalysts. The mechanical and degradation properties were found to be tunable with the furan:maleimide ratio. However, slow gelation limits its use as an injectable hydrogel [110]. Native chemical ligation, a widely employed click reaction for a variety of peptide ligations and for biomolecular modification of surfaces, also has been utilized for making *in situ* gelling using branched PEG having $N$-hydroxysuccinimide-activated oxo-esters and $N$-cysteine endgroups under physiological conditions without using any catalysts and exogenous thiols. Gelation occurred rapidly and reaction proceeded through an oxo-ester-mediated native chemical ligation pathway, which involved thiol capture to form a thioester intermediate, followed by an $S$-to-$N$ acyl rearrangement to yield an amide crosslink. The gelation time was pH and temperature dependent. Further, the hydrogels showed tissue-adhesive properties and demonstrated lap shear adhesion strength of

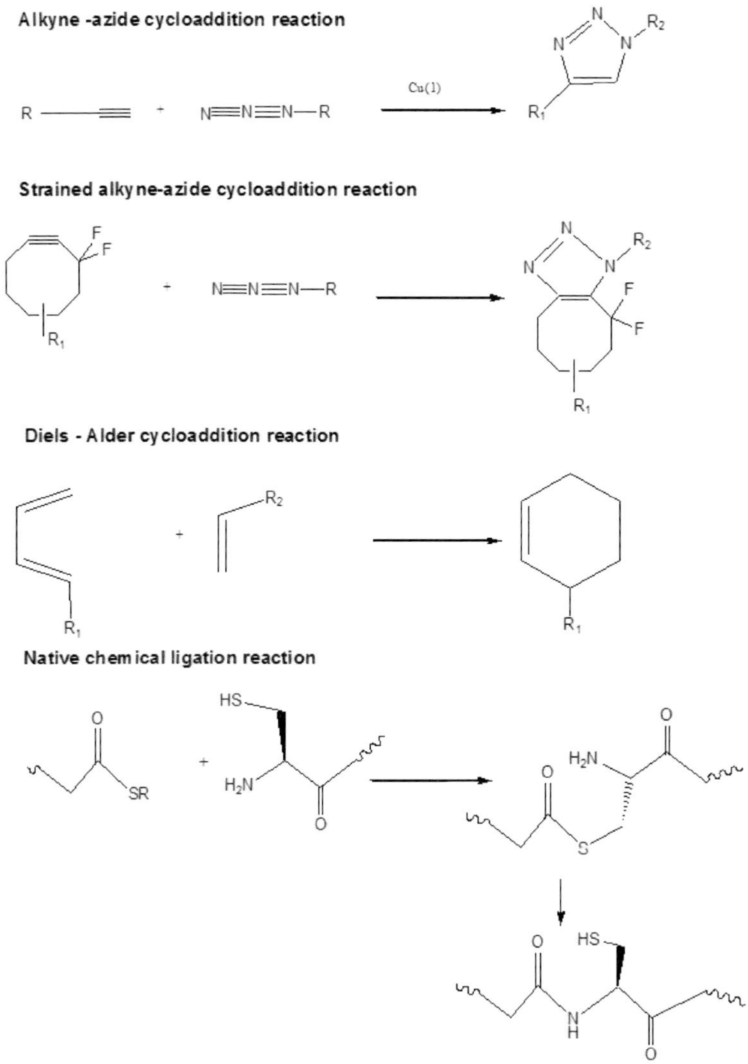


Figure 4.12 Common click reactions used for preparation of injectable hydrogels.

46 kPa. The hydrogels exerted minimal inflammatory response upon subcutaneous injection, making them an attractive candidate for *in vivo* applications such as wound healing, tissue repair, drug delivery, and tissue engineering [111].

### 4.2.7 *Other methods*

*In situ* forming hydrogel by condensation reaction between a multifunctional PEG *N*-hydroxysuccinimide ester and poly(allylamine hydrochloride) has been evaluated for hemorrhage control. The hydrogel was shown to provide a combination of rapid gelation and swelling properties suitable for hemorrhage control [112]. Preparation of a series of PVA-based multifunctional crosslinkers for making *in situ* gelling systems with oxidized HA has been reported. In one approach, thiol, cysteine 1,2-aminothiol, and aminooxy side chains were incorporated into PVA by direct coupling of the protected nucleophilic functionalities to PVA's hydroxyl groups via carbamate linkages followed by acidic deprotection. For this, the hydroxyl groups of PVA were initially activated with 1,1′-carbonyldiimidazole followed by a reaction with appropriate amine containing molecules. The reaction was performed in such a way that the degree of substitution did not exceed 5% in order to ensure water solubility. In the second approach, PVA was amino-derivatized to a low degree (3%) and used to quantitatively react with different *N*-hydroxysuccinimide esters of carboxylic acids containing maleimide, R-iodoacetyl, or acrylate thiophilic functionalities. Amino-derivatized PVA was also explored for further functionalization with semicarbazide-terminated side chains. These new PVA derivatives were evaluated for their ability to undergo *in situ* crosslinking reactions with partially oxidized HA and other PVA derivatives carrying electrophilic functionalities. Crosslinking was due to Michael-type addition and Schiff base reaction, respectively, forming hydrogels with thiol containing PVA derivatives and aldehyde derivatives of HA. Though multifunctional PVA crosslinkers were shown to give short gelation times, wide use of this strategy could be limited due to the poor aqueous solubility of the derivatives, as it requires initial dissolution in DMSO followed by multiple aqueous extractions [113]. The organophosphorous crosslinker, $\beta$-[tris(hydroxymethyl)phosphino]propionic acid (THPP) has been used to chemically crosslink elastin-like polypeptides rapidly in aqueous solution to form hydrogels via Mannich-type condensation of hydroxymethylphosphines with primary and secondary amines of amino acids. Hydrogels based on crosslinked ELP[KV2F-64] remained intact at physiological temperature and promoted viability of human chondrocytes up to 12 weeks. Great potential is envisaged for THPP crosslinkers because of the aqueous reaction condition, cytocompatibility of the crosslinker, ability of rapid gelation within 1 min, and subsequent crosslinking over hours to form hydrogels with mechanical properties similar to that of cartilage. However, the stability and mechanical properties under *in vivo* conditions are yet to be evaluated in detail [114]. Five homo-bifunctional PEG derivatives examined as crosslinking agents for thiol-modified HA and gelatin demonstrated as having the order of reactivity: maleimide $>$ iodoacetate $>$ bromoacetate $>$ iodoacetamide $>$ acrylate $\gg$ bromoacetamide, with rates increasing exponentially with increasing pH. Gelation time varied from less than 1 min to 2 h and the addition of the crosslinkers showed minimal cytotoxicity toward primary human dermal fibroblasts [115].

### 4.2.8 *Multiple crosslinking methods*

As well as using one single method for crosslinking, multiple methods of crosslinking have been attempted for the preparation of injectable hydrogels. Shear-thinning hydrogels have used secondary photocrosslinking to make stable gels with tunable mechanics and erosion [58]. In another study, a two-step crosslinking (initially Schiff base reaction between Odex and amino gelatin, followed by UV light-induced radical reaction of 4-arm(ethylene glycol) -acrylate) has been employed for the secondary network. The reaction proceeded under physiological conditions to form an IPN hydrogel which supported the attachment and spreading of pre-osteoblasts (MC3T3-E1), envisaging that this hierarchically designed hydrogel could be a promising candidate for bone or cartilage tissue engineering applications [116]. Multiple crosslinking methods have also been employed to improve the mechanical properties of physically crosslinked gels.

## 4.3 Conclusions

Injectable hydrogels have gained much appreciation recently with their intrinsic advantages like ease of administration and filling of geometrically complex target sites, targeted site delivery, and improved patient compliance. Advances in synthetic chemistry have led to the identification of various chemical reactions suitable for the development of injectable *in situ* forming hydrogels for various biomedical applications. Several issues like the toxicity of polymer precursors, their sterilizability and injectability, the time required for gelation, stability of gels formed *in situ*, and other physiochemical properties suitable for a particular application need to be addressed during the development of injectable hydrogels. Among the chemically crosslinked *in situ* forming hydrogels, those formed via Schiff base reaction, free radical-mediated crosslinking, Michael-type addition reaction, disulfide crosslinking, and genipin crosslinking are more widely accepted for biomedical applications. Researchers are also trying to combine the advantages of both physical as well as chemical crosslinking to make mechanically robust injectable hydrogels. The new concept of click reaction has revolutionized this area of research by the introduction of reactions like strain-promoted alkyne-azide cycloaddition, Diels–Alder cycloaddition, and native chemical ligation which were rarely employed in this field previously. The success of development of the injectable hydrogels relies on how effectively these can be translated for clinical use. Injectable hydrogels with similar composition and concentration of chemical constituents of ECM have also been developed. However, the design of biomimetic injectable hydrogels with proper temporal and spatial complexity and bioactive functions of ECM that control cellular behavior is still a major challenge for biomedical applications like tissue engineering and regenerative medicine.

## References

[1] Sun, J. and Tan, H. (2013). Alginate-based biomaterials for regenerative medicine applications. *Materials*, **6**(4), pp. 1285–1309.

[2] Mehvar, R. (2000). Dextrans for targeted and sustained delivery of therapeutic and imaging agents. *J. Control. Release*, **69**, pp. 1–25.

[3] Olsen, D. *et al.* (2003). Recombinant collagen and gelatin for drug delivery. *Adv. Drug Deliv. Rev.*, **55**, pp. 1547–1567.

[4] Schacht, E., Bogdanov, B., van den Bulcke, A., and de Rooze, N. (1997). Hydrogels prepared by crosslinking of gelatin with dextran dialdehyde. *React. Funct. Polym.*, **33**, pp. 109–116.

[5] Draye, J.-P. *et al.* (1998). *In vitro* release characteristics of bioactive molecules from dextran dialdehyde cross-linked gelatin hydrogel films. *Biomaterials*, **19**, pp. 99–107.

[6] Draye, J.-P. *et al.* (1998). *In vitro* and *in vivo* biocompatibility of dextran dialdehyde cross-linked gelatin hydrogel films. *Biomaterials*, **19**, pp. 1677–1687.

[7] Balakrishnan, B. and Jayakrishnan, A. (2005). Self-cross-linking biopolymers as injectable *in situ* forming biodegradable scaffolds. *Biomaterials*, **26**, pp. 3941–3951.

[8] Balakrishnan, B., Mohanty, M., Umashankar, P.R., and Jayakrishnan, A. (2005). Evaluation of an *in situ* forming hydrogel wound dressing based on oxidized alginate and gelatin. *Biomaterials*, **26**, pp. 6335–6342.

[9] Balakrishnan, B., Mohanty, M., Fernandez, A.C., Mohanan, P.V., and Jayakrishnan, A. (2006). Evaluation of the effect of incorporation of dibutyryl cyclic adenosine monophosphate in an *in situ* forming hydrogel wound dressing based on oxidized alginate and gelatin. *Biomaterials*, **27**, pp. 1355–1361.

[10] Balakrishnan, B., Jayakrishnan, A., Kumar, S.S.P., and Nandkumar, M.A. (2012). Anti-bacterial properties of an *in situ* forming hydrogel based on oxidized alginate and gelatin loaded with gentamycin. *Trends Biomater. Artif. Organs*, **26**(3), pp. 139–145.

[11] Dawlee, S., Sugandhi, A., Balakrishnan, B., Labarre, D., and Jayakrishnan, A. (2005). Oxidized chondroitin sulfate-cross-linked gelatin matrixes: a new class of hydrogels. *Biomacromolecules*, **6**, pp. 2040–2048.

[12] Balakrishnan, B., Joshi, N., Jayakrishnan, A., and Banerjee, R. (2014). Self-crosslinked oxidized alginate/gelatin hydrogel as injectable, adhesive biomimetic scaffolds for cartilage regeneration. *Acta Biomater.*, **10**, pp. 3650–3663.

[13] Li, X. *et al.* (2012). A covalently crosslinked polysaccharide hydrogel for potential applications in drug delivery and tissue engineering. *J. Mater. Sci. Mater. Med.*, **23**, pp. 2857–2865.

[14] Liang, Y. *et al.* (2011). An *in situ* formed biodegradable hydrogel for reconstruction of the corneal endothelium. *Colloids Surf. B Biointerfaces*, **82**, pp. 1–7.

[15] Balakrishnan, B., Joshi, N., Jayakrishnan, A., and Banerjee, R. (2013). Borate aided Schiff base formation yields *in situ* gelling hydrogels for cartilage regeneration. *J. Mater. Chem. B.* **1**, pp. 5564–5577.

[16] Nishi, K.K. and Jayakrishnan, A. (2007). Self-gelling primaquine-gum arabic conjugate: an injectable controlled delivery system for primaquine. *Biomacromolecules*, **8**, pp. 84–90.

[17] Jayakrishnan, A., Labarre, D., Alexandre, L., Balakrishnan, B., and Umashankar, P.R. (2008). Biocompatible and biodegradable biopolymer matrix. *US Patent App. 12/746,625.* (Manuscript under preparation.)

[18] Jayakrishnan, A., Labarre, D., Alexandre, L., Balakrishnan, B., and Umashankar, P.R. (2010). Biocompatible and biodegradable biopolymer matrix. *EP Patent 2,231,134.*

[19] Weng, L., Romanov, A., Rooney, J., and Chen, W. (2009). Non-cytotoxic, *in situ* gelable hydrogels composed of N-carboxyethyl chitosan and oxidizeddextran. *Biomaterials*, **29**, pp. 3905–3913.

[20] Zhang, H., Qadeer, A., and Chen, W. (2012). *In situ* gelable interpenetrating double network hydrogel formulated from binary components: Thiolated chitosan and oxidized dextran. *Biomacromolecules*, **12**, pp. 1428–1437.

[21] Zhang, H., Qadeer, A., Mynarcik, D., and Chen, W. (2012). Delivery of rosiglitazone from an injectable triple interpenetrating network hydrogel composed of naturally derived materials. *Biomaterials*, **32**, pp. 890–898.

[22] Hudson, S.P., Langer, R., Fink, G.R., and Kohane, D.S. (2010). Injectable *in situ* cross-linking hydrogels for local antifungal therapy. *Biomaterials*, **31**, pp. 1444–1452.

[23] Peng, H.T. and Shek, P.N. (2009). Development of *in situ*-forming hydrogels for hemorrhage control. *J. Mater. Sci. Mater. Med.*, **20**, pp. 1753–1762.

[24] Ito, T., Yeo, Y., Highley, C.B., Bellas, E., and Kohane, D.S. (2007). Dextran-based *in situ* cross-linked injectable hydrogels to prevent peritoneal adhesions. *Biomaterials*, **28**, pp. 3418–3426.

[25] Tan, H., Chu, C.R., Payne, K.A., and Marra, K.G. (2009). Injectable *in situ* forming biodegradable chitosan-hyaluronic acid based hydrogels for cartilage tissue engineering. *Biomaterials*, **30**, pp. 2499–2506.

[26] Tan, H., Rubin, J.P., and Marra, K.G. (2010). Injectable *in situ* forming biodegradable chitosan-hyaluronic acid based hydrogels for adipose tissue regeneration. *Organogenesis*, **6**, pp. 173–180.

[27] Su, W.Y., Chen, Y.C., and Lin, F.H. (2010). Injectable oxidized hyaluronic acid/adipic acid dihydrazide hydrogel for nucleus pulposus regeneration. *Acta Biomater.*, **6**, pp. 3044–3055.

[28] Chen, Y.C., Su, W.Y., Yang, S.H., Gefen, A., and Lin, F.H. (2013). *In situ* forming hydrogels composed of oxidized high molecular weight hyaluronic acid and gelatin for nucleus pulposus regeneration. *Acta Biomater.*, **9**, pp. 5181–5193.

[29] Fujita, M. *et al.* (2004). Vascularization *in vivo* caused by the controlled release of fibroblast growth factor-2 from an injectable chitosan/non-anticoagulant heparin hydrogel. *Biomaterials*, **25**, pp. 699–706.

[30] Ossipov, D.A., Brännvall, K., Forsberg-Nilsson, K., and Hilborn, J. (2007). Formation of the first injectable poly(vinyl alcohol) hydrogel by mixing of functional PVA precursors. *J. Appl. Polym. Sci.*, **106**(1), pp. 60–70.

[31] Oelker, A.M., Berlin, J.A., Wathier, M., and Grinstaff, M.W. (2011). Synthesis and characterization of dendron cross-linked PEG hydrogels as corneal adhesives. *Biomacromolecules*, **12**(5), pp. 1658–1665.

[32] Shazly, T.M., Artzi, N., Boehning, F., and Edelman, E.R. (2008). Viscoelastic adhesive mechanics of aldehyde-mediated soft tissue sealants. *Biomaterials*, **29**, pp. 4584–4591.

[33] Temenoff, J.S., Shin, H., Conway, D.E., Engel, P.S., and Mikos, A.G. (2003). In vitro cytotoxicity of redox radical initiators for cross-linking of oligo(poly(ethylene glycol) fumarate) macromers. *Biomacromolecules*, **4**, pp. 1605–1613.

[34] Shung, A.K., Behravesh, E., Jo, S., and Mikos, A.G. (2003). Crosslinking characteristics and cell adhesion to an injectable poly(propylene fumarate-co-ethylene glycol) hydrogel using a water-soluble crosslinking system. *Tissue Eng.*, **9**, pp. 243–254.

[35] Kasper, F.K. *et al.* (2005). *In vitro* release of plasmid DNA from oligo(poly(ethylene glycol) fumarate) hydrogels. *J. Control. Release*, **104**, pp. 521–539.

[36] Hong, Y. *et al.* (2007). Covalently crosslinked chitosan hydrogel: Properties of *in vitro* degradation and chondrocyte encapsulation. *Acta Biomater.*, **3**, pp. 23–31.

[37] Boesel, L.F., Reis, R.L., and Román, J.S. (2009). Innovative approach for producing injectable, biodegradable materials using chitooligosaccharides and green chemistry. *Biomacromolecules*, **10**, pp. 465–470.
[38] Zhu, W. and Ding, J. (2005). Synthesis and characterization of a redox-initiated, injectable, biodegradable hydrogel. *J. Applied Polym. Science.*, **99**, pp. 2375–2383.
[39] Jiao, Y. *et al.* (2012). A rheological study of biodegradable injectable PEGMC/HA composite scaffolds. *Soft Matter*, **8**, pp. 1499–1507.
[40] Ifkovits, J.L. *et al.* (2010). Injectable hydrogel properties influence infarct expansion and extent of postinfarction left ventricular remodeling in an ovine model. *Proc. Nat. Acad. Sci. USA*, **107**, pp. 11507–11512.
[41] Tous, E. *et al.* (2012). Tunable hydrogel-microsphere composites that modulate local inflammation and collagen bulking. *Acta Biomater.*, **8**, pp. 3218–3227.
[42] Tous, E. *et al.* (2011). Influence of injectable hyaluronic acid hydrogel degradation behavior on infarction-induced ventricular remodeling. *Biomacromolecules*, **12**, pp. 4127–4135.
[43] Hu, X. and Gao, C. (2007). Photoinitiating polymerization to prepare biocompatible chitosan hydrogels. *J. Appl. Polym. Sci.*, **110**, pp. 1059–1067.
[44] Elisseeff, J. *et al.* (1999). Transdermal photopolymerization for minimally invasive implantation. *Proc. Nat. Acad. Soc. USA*, **96**(6), pp. 3104–3107.
[45] Erickson, I.E. *et al.* (2009). Macromer density influences mesenchymal stem cell chondrogenesis and maturation in photocrosslinked hyaluronic acid hydrogels. *Osteoarthr. Cart.*, **17**, pp. 1639–1648.
[46] Burdick, J.A., Chung, C., Jia, X., Randolph, M.A., and Langer, R. (2005). Controlled degradation and mechanical behavior of photopolymerized hyaluronic acid networks. *Biomacromolecules*, **6**, pp. 386–391.
[47] Sharma, B., Williams, C.G., Khan, M., Manson, P., and Elisseeff, J.H. (2007). *In vivo* chondrogenesis of mesenchymal stem cells in a photopolymerized hydrogel. *Plast. Reconstr. Surg.*, **119**, pp. 112–120.
[48] Sahoo, S., Chung, C., Khetan, S., and Burdick, J.A. (2008). Hydrolytically degradable hyaluronic acid hydrogels with controlled temporal structures. *Biomacromolecules*, **9**, pp. 1088–1092.
[49] Purcell, B.P., Elser, J.A., Mu, A., Margulies, K.B., and Burdick, J.A. (2012). Synergistic effects of SDF-1$\alpha$ chemokine and hyaluronic acid release from degradable hydrogels on directing bone marrow derived cell homing to the myocardium. *Biomaterials*, **33**, pp. 7849–7857.
[50] Camci-Unal, G., Cuttica, D., Annabi, N., Demarchi, D., and Khademhosseini, A. (2013). Synthesis and characterization of hybrid hyaluronic acid-gelatin hydrogels. *Biomacromolecules*, **14**, pp. 1085–1092.
[51] Li, Q. *et al.* (2004). Photocrosslinkable polysaccharides based on chondroitin sulfate. *J. Biomed. Mater. Res.*, **68**, pp. 28–33.
[52] Jeon, O., Bouhadir, K.H., Mansour, J.M., and Alsberg, E. (2009). Photocrosslinked alginate hydrogels with tunable biodegradation rates and mechanical properties. *Biomaterials*, **30**, pp. 2724–2734.

[53] Jeon, O., Alt, D.S., Ahmed, S.M., and Alsberg, E. (2012). The effect of oxidation on the degradation of photocrosslinkable alginate hydrogels. *Biomaterials*, **33**, pp. 3503–3514.
[54] Coates, E.E., Riggin, C.N., and Fisher, J.P. (2013). Photocrosslinked alginate with hyaluronic acid hydrogels as vehicles for mesenchymal stem cell encapsulation and chondrogenesis. *J. Biomed. Mater. Res. A.*, **101**, pp. 1962–1970.
[55] Li, Q. *et al.* (2006). Biodegradable and photocrosslinkable polyphosphoester hydrogel. *Biomaterials*, **27**, pp. 1027–1034.
[56] Amsden, B.G., Sukarto, A., Knight, D.K., and Shapka, S.N. (2007). Methacrylated glycol chitosan as a photopolymerizable biomaterial. *Biomacromolecules.*, **8**, pp. 3758–3766.
[57] Hu, X., Ma, L., Wang, C., and Gao, C. (2009). Gelatin hydrogel prepared by photo-initiated polymerization and loaded with TGF-beta1 for cartilage tissue engineering. *Macromol. Biosci.*, **9**, pp. 1194–1201.
[58] Lu, H.D., Soranno, D.E., Rodell, C.B., Kim, I.L., and Burdick, J.A. (2013). Secondary photocrosslinking of injectable shear-thinning Dock-and-Lock hydrogels. *Adv. Healthcare Mater.*, **2**, pp. 1028–1036.
[59] Gyawali, D. *et al.* (2010). Citric acid-derived *in situ* crosslinkable biodegradable polymers for cell delivery. *Biomaterials*, **31**, pp. 9092–9105.
[60] Ibusuki, S. *et al.* (2007). Photochemically cross-linked collagen gels as three-dimensional scaffolds for tissue engineering. *Tissue Eng.*, **13**, pp. 1995–2001.
[61] Randolph, M.A., Ibusuki, S., Redmond, R.W., Kochevar, I.E., and Gill, T.J. (2007). Photo-crosslinking collagen gel for tissue engineered cartilage. *Eur. Cell Mater.*, **13**, p. 12.
[62] Hu, J. *et al.* (2012). Visible light crosslinkable chitosan hydrogels for tissue engineering. *Acta Biomater.*, **8**, pp. 1730–1738.
[63] McCall, J.D. and Anseth, K.S. (2012). Thiolene photopolymerizations provide a facile method to encapsulate proteins and maintain their bioactivity. *Biomacromolecules*, **13**, pp. 2410–2417.
[64] Fairbanks, B.D. *et al.* (2009). A versatile synthetic extracellular matrix mimic via thiol-norbornene photopolymerization. *Adv. Mater.*, **21**, pp. 5005–5010.
[65] Lin, C.C., Raza, A., and Shih, H. (2011). PEG hydrogels formed by thiol-ene photo-click chemistry and their effect on the formation and recovery of insulin-secreting cell spheroids. *Biomaterials*, **32**, pp. 9685–9695.
[66] Fu, Y. *et al.* (2012). 3D cell entrapment in crosslinked thiolated gelatin-poly(ethylene glycol) diacrylate hydrogels. *Biomaterials*, **33**, pp. 48–58.
[67] Elbert, D.L., Pratt, A.B., Lutolf, M.P, Halstenberg, S., and Hubbell, J.A. (2001). Protein delivery from materials formed by self-selective conjugate addition reactions. *J. Control. Release*, **76**, pp. 11–25.
[68] Lutolf, M.P., Tirelli, N., Cerritelli, S., Cavalli, L., and Hubbell, J.A. (2001). Systematic modulation of Michael-type reactivity of thiols through the use of charged amino acids. *Bioconjug. Chem.*, **12**, pp. 1051–1056.

[69] Lutolf, M.P. and Hubbell, J.A. (2003). Synthesis and physiochemical characterization of end-linked poly(ethylene glycol)-co-peptide hydrogels formed by Michael-type addition. *Biomacromolecules*, **4**, pp. 713–722.

[70] Lutolf, M.P. *et al.* (2003). Repair of bone defects using synthetic mimetics of collagenous extracellular matrices. *Nat. Biotechnol.*, **21**, pp. 513–518.

[71] Zisch, A.H. *et al.* (2003). Cell-demanded release of VEGF from synthetic, biointeractive cell ingrowth matrices for vascularized tissue growth. *FASEB J.*, **17**, pp. 2260–2282.

[72] Pratt, A.B., Weber, F.E., Schmoekel, H.G., Müller, R., and Hubbell, J.A. (2004). Synthetic extracellular matrices for *in situ* tissue engineering. *Biotechnol. Bioeng.*, **86**, pp. 27–36.

[73] Qiu, B. *et al.* (2003). A hydrogel prepared by *in situ* cross-linking of a thiol-containing poly(ethylene glycol)-based copolymer: a new biomaterial for protein drug delivery. *Biomaterials*, **24**, pp. 11–18.

[74] Tortora, M., Cavalieri, F., Chiessi, E., and Paradossi, G. (2007). Michael-type addition reactions for the *in situ* formation of poly(vinyl alcohol)-based hydrogels. *Biomacromolecules*, **8**, pp. 209–214.

[75] Fu, Y. and Kao, W.J. (2011). *In situ* forming poly(ethylene glycol)-based hydrogels via thiol-maleimide Michael-type addition. *J. Biomed. Mater. Res. A.*, **98**, pp. 201–211.

[76] Shu, X.Z., Liu, Y., Palumbo, F.S., Luo, Y., and Prestwich, G.D. (2004). *In situ* crosslinkable hyaluronan hydrogels for tissue engineering. *Biomaterials*, **25**(7–8), pp. 1339–1348.

[77] Ghosh, K., Ren, X.D., Shu, X.Z., Prestwich, G.D., and Clark, R.A. (2006). Fibronectin functional domains coupled to hyaluronan stimulate adult human dermal fibroblast responses critical for wound healing. *Tissue Eng.*, **12**, pp. 601–613.

[78] Kim, J. *et al.* (2007). Bone regeneration using hyaluronic acid-based hydrogel with bone morphogenic protein-2 and human mesenchymal stem cells. *Biomaterials*, **28**, pp. 1830–1837.

[79] Kim, J. *et al.* (2009). Characterization of low-molecular-weight hyaluronic acid-based hydrogel and differential stem cell responses in the hydrogel microenvironments. *J. Biomed. Mater. Res. A.*, **88**, pp. 967–975.

[80] Jin, R. *et al.* (2010). Synthesis and characterization of hyaluronic acid-poly(ethylene glycol) hydrogels via Michael addition: An injectable biomaterial for cartilage repair. *Acta Biomater.*, **6**, pp. 1968–1972.

[81] Bencherif, S.A., Washburn, N.R., and Matyjaszewski, K. (2009). Synthesis by AGET ATRP of degradable nanogel precursors for *in situ* formation of nanostructured hyaluronic acid hydrogel. *Biomacromolecules.*, **10**, pp. 2499–2507.

[82] Hiemstra, C., Aa, L.J., Zhong, Z., Dijkstra, P.J., and Feijen, J. (2007). Rapidly *in situ*-forming degradable hydrogels from dextran thiols through Michael addition. *Biomacromolecules*, **8**(5), pp. 1548–1556.

[83] Hiemstra, C., Zhong, Z., van Steenbergen, M.J., Hennink, W.E., and Feijen, J. (2007). Release of proteins and basic fibroblast growth factor from

*in situ* forming degradable dextran hydrogels. *J. Control. Release*, **122**, pp. 71–78.

[84] Jukes, J.M. *et al.* (2010). A newly developed chemically crosslinked dextran-poly(ethyl glycol) hydrogel for cartilage tissue engineering. *Tissue Eng. Part A.*, **16**, pp. 565–573.

[85] Yu, Y. *et al.* (2011). Novel injectable biodegradable glycol chitosan-based hydrogels crosslinked by Michael-type addition reaction with oligo(acryloyl carbonate)-b-poly(ethylene glycol)-b-oligo(acryloyl carbonate) copolymers. *J. Biomed. Mater. Res. A.*, **99**, pp. 316–326.

[86] Kim, M.S., Choi, Y.J., Noh, I., and Tae, G. (2007). Synthesis and characterization of *in situ* chitosan-based hydrogel via grafting of carboxyethyl acrylate. *J.Biomed. Mater. Res. A.*, **83**, pp. 674–682.

[87] Kim, M., Lee, J.Y., Jones, C.N., Revzin, A., and Tae, G. (2010). Heparin-based hydrogel as a matrix for encapsulation and cultivation of primary hepatocytes. *Biomaterials*, **31**, pp. 3596–3603.

[88] Bernkop-Schnürch, A., Hornof, M., and Zoidl, T. (2003). Thiolated polymers–thiomers: synthesis and *in vitro* evaluation of chitosan–2-iminothiolane conjugates. *Int. J. Pharm.*, **260**, pp. 229–237.

[89] Hornof, M.D., Kast, C.E., and Bernkop-Schnürch, A. (2003). *In vitro* evaluation of the visco-elastic properties of chitosan–thioglycolic acid conjugates. *Eur. J. Pharm. Bio-pharm.*, **55**, pp. 185–190.

[90] Krauland, A.H., Hoffer, M.H., and Bernkop-Schnürch, A. (2005). Viscoelastic properties of a new *in situ* gelling thiolated chitosan conjugate. *Drug Dev. Ind. Pharm.*, **31**, pp. 885–893.

[91] Kast, C. and Bernkop-Schnürch, A. (2001). Thiolated polymers–thiomers: Development and *in vitro* evaluation of chitosan–thioglycolic acid conjugates. *Biomaterials*, **22**, pp. 2345–2352.

[92] Sakloetsakun, D., Hombach, J.M., and Bernkop-Schnürch, A. (2009). *In situ* gelling properties of chitosan–thioglycolic acid conjugate in the presence of oxidizing agents. *Biomaterials*, **30**, pp. 6151–6157.

[93] Shu, X.Z., Liu, Y., Luo, Y., Roberts, M.C., and Prestwich, G.D. (2002). Disulfide cross-linked hyaluronan hydrogels. *Biomacromolecules*, **3**, pp. 1304–1311.

[94] Anumolu, S.S. *et al.* (2011). Doxycycline hydrogels with reversible disulfide cross links for dermal wound healing of mustard injuries. *Biomaterials.*, **32**, pp. 1204–1217.

[95] Choh, S.Y., Cross, D., and Wang, C. (2011). Facile synthesis and characterization of disulfide-cross-linked hyaluronic acid hydrogels for protein delivery and cell encapsulation. *Biomacromolecules*, **12**, pp. 1126–1136.

[96] Chang, Y. *et al.* (2005). Cell-free xenogenic vascular grafts fixed with glutaraldehyde or genipin: *In vitro* and *in vivo* studies. *J. Biotechnol.*, **120**, pp. 207–219.

[97] Jeng, L., Olsen, B.R., and Spector, M. (2012). Engineering endostatin-expressing cartilaginous constructs using injectable biopolymer hydrogels. *Acta Biomater.*, **8**, pp. 2203–2212.

[98] Bigi, A., Cojazzi, G., Panzavolta, S., Roveri, N., and Rubini, K. (2002). Stabilization of gelatin films by crosslinking with genipin. *Biomaterials*, **23**, pp. 4827–4832.
[99] Mwale, F. *et al.* (2005). Biological evaluation of chitosan salts cross-linked to genipin as a cell scaffold for disk tissue engineering. *Tissue Eng.*, **11**, pp. 130–140.
[100] Chang, Y., Tsai, C.C., Liang, H.C., and Sung, H.W. (2001). Reconstruction of the right ventricular outflow tract with a bovine jugular vein graft fixed with a naturally occurring crosslinking agent (genipin) in a canine model. *J. Thorac. Cardiovasc. Surg.*, **122**, pp. 1208–1218.
[101] Wang, C., Lau, T.T., Loh, W.L., Su, K., and Wang, D.A. (2011). Cytocompatibility study of a natural biomaterial crosslinker-genipin with therapeutic model cells. *J. Biomed. Mater. Res. B Appl. Biomater.*, **97**, pp. 58–65.
[102] Moura, M.J., Figueiredo, M.M., and Gil, M.H. (2007). Rheological study of genipin cross-linked chitosan hydrogels. *Biomacromolecules*, **8**, pp. 823–829.
[103] Moura, M.J., Faneca, H., Lima, M.P., Gil, M.H., and Figueiredo, M.M. (2011). *In situ* forming chitosan hydrogels prepared via ionic/covalent co-cross-linking. *Biomacromolecules*, **12**, pp. 3275–3284.
[104] Tan, H., DeFail, A., Rubin, J.P., Chu, C.R., and Marra, K.G. (2010). Novel multi-arm PEG-based hydrogels for tissue engineering. *J. Biomed. Mater. Res. A.*, **92**(3), pp. 979–987.
[105] Macaya, D., Ng, K.K., and Spector, M. (2011), Injectable collagen–genipin gel for the treatment of spinal cord injury: *in vitro* studies. *Adv. Funct. Mater.*, **21**(24), pp. 4788–4797.
[106] Macaya, DJ., Hayakawa, K., Arai, K., and Spector, M. (2013). Astrocyte infiltration into injectable collagen-based hydrogels containing FGF-2 to treat spinal cord injury. *Biomaterials*, **34**, pp. 3591–3602.
[107] Kolb, H.C., Finn, M.G., and Sharpless, K.B. (2001). Click chemistry: diverse chemical function from a few good reactions. *Angew. Chem. Int. Ed. Engl.*, **40**, pp. 2004–2021.
[108] Nimmo, C.M. and Shoichet, M.S. (2011). Regenerative biomaterials that "click": simple, aqueous-based protocols for hydrogel synthesis, surface immobilization, and 3D patterning. *Bioconjug. Chem.*, **22**, pp. 2199–2209.
[109] Agard, N.J., Prescher, J.A., and Bertozzi, C. R. (2004). A strain-promoted [3 + 2] azide-alkyne cycloaddition for covalent modification of biomolecules in living systems. *J. Am. Chem. Soc.*, **126**, pp. 15046–15047.
[110] Nimmo, C.M., Owen, S.C., and Shoichet, M.S. (2011). Diels–Alder click cross-linked hyaluronic acid hydrogels for tissue engineering. *Biomacromolecules*, **12**, pp. 824–830.
[111] Strehin, I., Gourevitch, D., Zhang, Y., Heber-Katz, E., and Messersmith, P.B. (2013). Hydrogels formed by oxo-ester mediated native chemical ligation. *Biomater. Sci.*, **1**, pp. 603–613.
[112] Peng, H.T., Blostein, M.D. and Shek, P.N. (2009). Experimental optimization of an *in situ*-forming hydrogel for hemorrhage control. *J. Biomed. Mater. Res. B Appl. Biomater.*, **89**, pp. 199–209.

[113] Ossipov, D.A., Piskounova, S., and Hilborn, J. (2008). Poly(vinyl alcohol) cross-linkers for *in vivo* injectable hydrogels. *Macromolecules*, **41**(11), pp. 3971–3982.

[114] Lim, D.W., Nettles, D.L., Setton, L.A., and Chilkoti, A. (2007). Rapid cross-linking of elastin-like polypeptides with (hydroxymethyl)phosphines in aqueous solution. *Biomacromolecules*, **8**, pp. 463–470.

[115] Vanderhooft, J.L., Mann, B.K., and Prestwich, G.D. (2007). Synthesis and characterization of novel thiol-reactive poly(ethylene glycol) cross-linkers for extracellular-matrix-mimetic biomaterials. *Biomacromolecules*, **8**(9), pp. 2883–2889.

[116] Geng, X., Mo, X., Fan, L., Yin, A., and Fang, J. (2012). Hierarchically designed injectable hydrogel from oxidized dextran, amino gelatin and 4-arm poly(ethylene glycol)-acrylate for tissue engineering application. *J. Mater. Chem.*, **22**, pp. 25130–25139.

## Chapter 5

# Injectable Hydrogels by Enzymatic Crosslinking

S. V. Gohil

*Department of Orthopaedic Surgery, UConn Health, Farmington, CT*
*New England Musculoskeletal Institute, UConn Health, Farmington, CT*
*Institute for Regenerative Engineering, UConn Health, Farmington, CT*

C. W. Hanna

*Department of Orthopaedic Surgery, UConn Health, Farmington, CT*

L. S. Nair*

*Department of Orthopaedic Surgery, UConn Health, Farmington, CT*
*Institute for Regenerative Engineering, UConn Health, Farmington, CT*
*Department of Materials Science and Engineering*
*University of Connecticut, Storrs, CT*
*New England Musculoskeletal Institute, UConn Health, Farmington, CT*
*Department of Biomedical Engineering, UConn Health, Farmington, CT*

### 5.1 Introduction

Until recently, the tissue regeneration approaches typically relied on three essential components: scaffolds, designed to maintain the cells in a three-dimensional environment; cells, to ultimately form the new tissue; and, signals or biological cues, such as growth factors to guide the gene expression and extra cellular matrix (ECM) production by the implanted or host cells (as discussed in Chapter 1) [1, 2]. The convergence of tissue engineering with

*Author to whom correspondence should be addressed. Email: nair@uchc.edu.

advanced biomaterials science, stem cell science and developmental biology led to the emergence of 'regenerative engineering', which is defined as 'the integration of materials science and tissue engineering with stem and developmental cell biology and regenerative medicine toward the regeneration of complex tissues, organs, or organ systems' [3].

Recent developments in custom scaffold preparation for cell/growth factor delivery, chemical and biological modification of scaffolds for improved cell adhesion, and bioactivity, along with the use of new stem cell sources have resulted in significant advances in the field [4]. However, maintaining the long-term viability and functions of encapsulated cells is still a challenge. Furthermore, obtaining complete spatiotemporal control on the release of bioactive signals and protein-based growth factors, while preventing their denaturation, is yet to be achieved.

Hydrogels have emerged as a promising candidate for cell- and growth factor-based regenerative engineering strategies due to their unique tissue-mimetic properties such as the ability to allow gas/nutrient diffusion and aqueous three-dimensional environments [5, 6]. As discussed in Chapter 2, hydrogels are highly hydrated three-dimensional networks composed of crosslinked hydrophilic polymers with a wide range of pharmaceutical, biomedical, and tissue engineering applications. They are capable of delivering entrapped therapeutics *in vivo* in a spatiotemporally controlled, sustained fashion [7, 8]. Additionally, they can be fabricated under physiological conditions, enabling gelation in the presence of tissue, cells, and bioactive molecules. The high water content and porosity of hydrogels provide a tissue-like micro environment for cells, promoting facile transport and thereby permitting long-term cell survival. The gel chemistry can further be systematically varied to introduce biological cues, creating tailored hydrogel niches that promote or suppress targeted cell interactions or functions [9]. *In situ* crosslinkable, injectable hydrogels are especially desirable clinically for minimally invasive applications using endoscopic or percutaneous procedures [10]. In essence, hydrogels provide a tailorable microenvironment that can support cell survival and function.

The crosslinks required for hydrogel fabrication can be generated by: (a) physical association processes (Chapter 3) such as ion-induced gelation, thermogelation, inclusion complex formation, stereocomplex formation, and complementary binding; and (b) chemical processes (Chapter 4) such as redox/photopolymerization, and *via* classic organic reactions [11]. Among these approaches, the potential to exploit enzyme-catalyzed reactions as biological stimulus to trigger hydrogel assembly holds several advantages.

This is because the enzymes (a) are uniquely chemo-, regio-, and enantioselective, thus have high specificity; (b) work under mild conditions (aqueous, pH 5–8, 37°C) without unwanted side reactions or cytotoxicity and do not require the use of reagents or crosslinkers which are toxic to cells; and, (c) play key roles as selective catalysts in cell pathways and disease states [12, 13]. These enzymatic hydrogels can be formed by covalent crosslinking of macromolecular polymers or by the self-assembly of smaller molecules [14]. Enzymatically triggered self-assembly of small molecules into hydrogels, which can take place *in vitro*, *in vivo*, extra- or intracellularly has recently attracted much attention. If self-assembly occurs in an aqueous medium, the resulting gel is referred to as 'supramolecular hydrogel' and the small molecule is referred to as 'hydrogelator' [15]. Supramolecular hydrogels have been utilized for a variety of applications ranging from drug delivery, controlling cell fate, and tissue engineering applications [16–19]. This chapter summarizes the different types of enzyme-based crosslinking methods for hydrogel preparation with a focus on the origin and mechanism of action, along with their applications in the field of regenerative engineering.

## 5.2 Commonly Used Enzyme Systems for Hydrogel Preparation

### 5.2.1 *Transglutaminase*

#### 5.2.1.1 *Origin and mechanism*

Transglutaminases belong to a family of enzymes which catalyze post-transcriptional modification of proteins via the formation of isopeptide bonds. Protein crosslinking can occur either through $\varepsilon$-($\gamma$-glutamyl)lysine bonds or via incorporation of primary amines at selected peptide-bound glutamine residues [20]. These crosslinked proteins are important in proper blood clotting and wound healing [21], as well as skin and hair maintenance [22]. Furthermore, transglutaminases have been found to play an important role in cellular functions such as both pro- and anti-apoptosis [23], as well as attachment and migration [24]. Mammalian transglutaminases require the binding of $Ca^{2+}$ for their activity, as this binding disrupts the four domains found in these proteins, exposing the active site [20, 25]. Therefore, the microenvironment in which transglutaminases are found plays an important role in their activity [26]. Recently, research on transglutaminases has increased dramatically due to the discovery of the enzyme's role in diseases such as celiac disease [27] and various neurodegenerative conditions [28].

In mammals, eight transglutaminases have been described, but only six have been characterized [20]. For the crosslinking of hydrogels, three major transglutaminases have been studied: Factor XIIIa (FXIIIa), transglutaminase-2 (TG2), and microbial transglutaminase (mTG). FXIIIa, which is important for proper blood clotting and wound healing [21] is formed through the proteolytic cleaving of Factor XIII in the presence of thrombin [20]. TG2, also commonly known as 'tissue transglutaminase', is a multifunctional protein capable of acting as a transglutaminase, a G protein, a kinase, a protein disulfide isomerase, or an adaptor protein [25, 26]. mTG was initially discovered for the purpose of producing transglutaminase on an industrial scale [29]. Not only did this discovery enable the mass production of this enzyme, but mTG was found to be active independent of the presence of $Ca^{2+}$. Initial research into mTG focused on the use of this enzyme in food processing [30], but it has since been expanded into other fields, including regenerative engineering [31].

The general mechanism of transglutaminase-induced polymer crosslinking is shown in Scheme 5.1. For protein-based polymers such as gelatin or collagen, transglutaminases catalyze the transamidation reaction between glutamyl and lysyl side groups resulting in polymer crosslinking through the formation of $\varepsilon$-($\gamma$-glutamyl) lysine isopeptide side-chain bridges [32]. Briefly, the glutamine residue first binds to the active site of the enzyme by

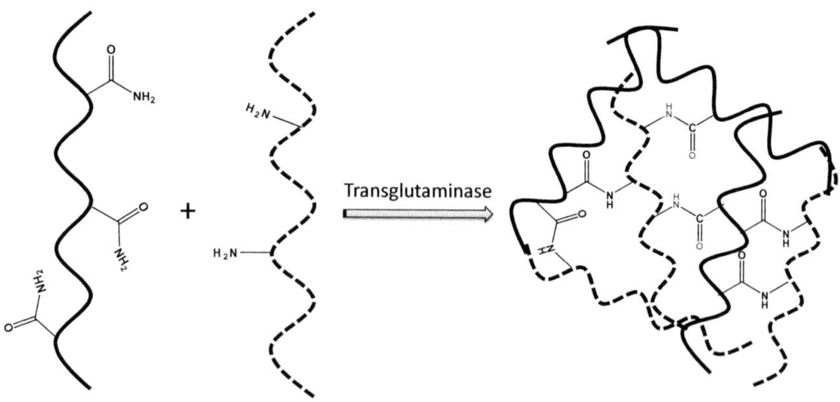


Scheme 5.1 Mechanism of transglutaminase mediated crosslinking process.

forming a $\gamma$-glutamylthiolester with a cysteine residue, releasing ammonia. The second substrate, which contains lysine, then binds to the active site and attacks the thiolester bond. This results in the formation of the $\varepsilon$-($\gamma$-glutamyl) lysine bond, while additionally re-establishing the enzyme into its original form [33].

#### 5.2.1.2 *Applications*

Transglutaminase-mediated crosslinking has been investigated for protein based natural polymers such as collagen and gelatin due to the presence of glutamine and lysine groups in the polymer chains. In general, transglutaminases can support the formation of isopeptide bonds via crosslinking by $\varepsilon$-($\gamma$-glutamyl) lysine bonds or through covalent conjugation of polyamines, lipid esterification, or glutamine residue deamination [20]. Table 5.1 shows examples of polymeric hydrogels prepared via transglutaminase-mediated crosslinking.

mTG is the most widely explored transglutaminase for enzymatic crosslinking due to its calcium independent activity and commercial availability. mTG has been used to develop gelatin hydrogels wherein the mTG-crosslinked gels were found to form more slowly, but were stronger and permanent when compared to mushroom tyrosinase-crosslinked hydrogels [34]. Also, the addition of chitosan to the hydrogels resulted in an increase in gel formation speed and strength, owing to the increased availability of amino groups for crosslinking. The study also demonstrated the ability of mTG-crosslinked hydrogels to encapsulate *E. Coli* cells, where the encapsulated cells could be subsequently released in the presence of protease K [35]. mTG-crosslinked gelatin hydrogels have shown to support 3T3 fibroblast growth *in vitro*, with mTG displaying no cytotoxic effects up to concentrations of 0.05% [36]. Concentrations below cytotoxic levels were able to dose dependently increase the mechanical properties of the hydrogels, although the gels were too weak for load-bearing applications.

In an attempt to create an artificial ECM, De Colli *et al.* [37] developed hydrogels of hyaluronic acid, chondroitin sulfate, and gelatin (G-HA-CS). As both hyaluronic acid and chondroitin sulfate do not possess the functional groups necessary form transglutaminase mediated crosslinking, they were functionalized with the dipeptide glycine-lysine (GK). The glutamic residues of gelatin could then react with the lysine residues of GK-modified hyaluronic acid/chondroitin sulfate to form $\varepsilon$-($\gamma$-glutamyl) lysine crosslinks in the presence of transglutaminase. The study demonstrated

Table 5.1 Transglutaminase-crosslinked hydrogels and their applications.

| Polymer | Remarks | Applications | Refs. |
|---|---|---|---|
| Gelatin | Reversible, *in situ* entrapment of *E. coli* | Potential applications in microfluidic biosensor systems | [35] |
| Gelatin | Effect of enzyme concentration on crosslinking and mechanical properties evaluated | Fibroblast culture, possible dermal applications | [36] |
| Hydrolyzed collagen fraction | Pronounced adhesion to abdominal tissue, inhibits aversive fibroblast and prostate carcinoma cell growth | Sustained drug release, neuritotrophic | [38] |
| Casein | Enzyme-assisted gelation with tunable characteristics | Prolonged drug release | [42] |
| Poly(ethylene glycol) | Proteolytically sensitive hydrogels with fibrin-like biomolecular characteristics | Drug/growth factor delivery for tissue engineering | [43, 44] |
| Collagen | Endostatin-expressing cartilaginous constructs | Articular cartilage regeneration, cell encapsulation and culture | [39] |
| Gelatin and gly-coaminoglycans | High internal phase emulsion technique | Hepatocyte culture | [37] |
| PEG–peptide conjugates | Injectable, *in situ* formed, biodegradable | Tissue adhesive | [45] |

that the hydrogels may be effective for liver tissue regeneration, as primary rat hapatocytes and C3A cells could be encapsulated within the hydrogels. The encapsulated hepatocytes showed good viability, large cell aggregates, satisfactory cell–cell contacts, and production of high levels of albumin and urea (Figure 5.1). mTG-crosslinked gelatin hydrogels have also been investigated *in vitro* as a drug delivery vehicle to facilitate neuronal regeneration following radical prostatectomy due to carcinoma [38]. These hydrogels were found to be capable of suppressing carcinoma cell growth when loaded with the chemotherapeutic drug cytosine arabinocide, while supporting *in vitro* axonal regeneration. Furthermore, the hydrogels exhibited fast gelation and were capable of attaching to rat abdominal tissue when crosslinked *in situ*.

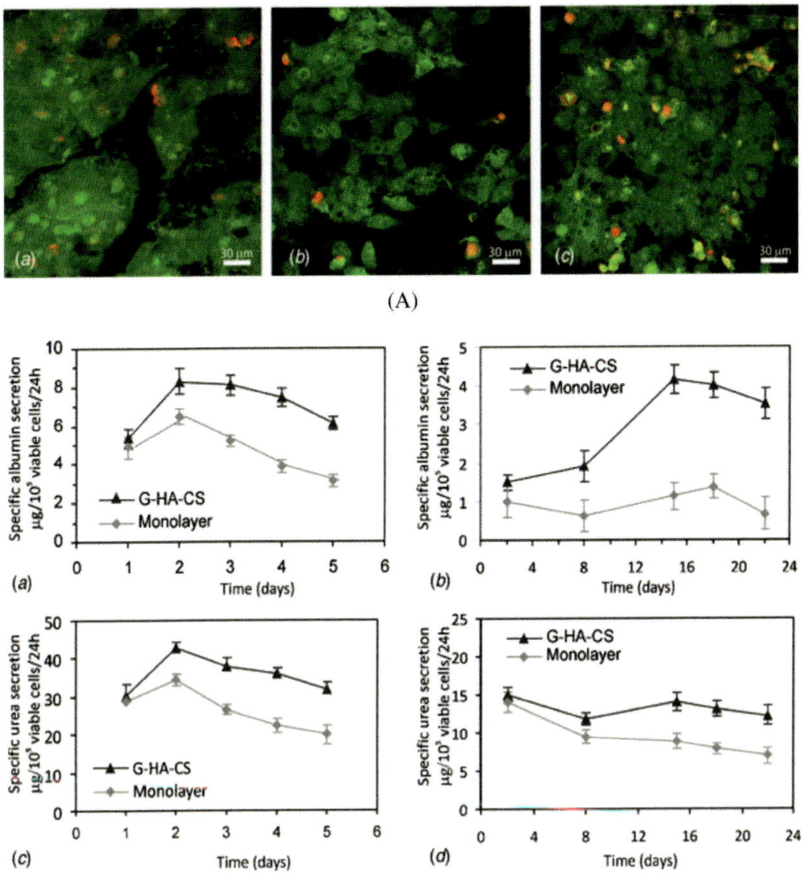


(B)

Figure 5.1 (A) Live/dead staining. Confocal micrographs of SYTO-10/DEAD-Red double-stained primary hepatocytes 4 days post-seeding (a) and of C3A cells at 4 days (b) and 15 days (c) post-seeding within G-HA-CS scaffold. Most cells in aggregates are stained only fluorescent green (viable). (B) Normalized albumin secretion (a), (b) and urea production (c), (d) by primary hepatocytes (a), (c) and C3A cells (b), (d) on both G-HA-CS scaffold and monolayer polystyrene plates. Reproduced from [37] with permission from IOP publishing.

Different crosslinkers, including TG2 and mTG, were utilized to develop injectable collagen-based hydrogels as a delivery vehicle for goat mesenchymal stem cells transfected to overexpress endostatin. The study demonstrated the ability of the system to undergo covalent crosslinking *in vivo* and the potential to serve as a tissue engineered cartilage construct by suppressing angiogenesis via endostatin overexpression [39]. Davis *et al.* [40] utilized guinea pig TG2 and recombinant human TG2 for crosslinking a new class of high molecular weight recombinant protein polymers. The protein polymers were designed to have evenly spaced transglutaminase substrates (lysine or glutamine) for crosslinking into a hydrogel with a range of viscoelastic properties. The hydrogels displayed tunable mechanical, swelling, and degradation properties, based on the frequency of glutamine and lysine amino acid repeats in a given protein's base peptide sequence. Furthermore, these hydrogels were able to support mouse 3T3 and primary human fibroblast cell growth, in both two- and three-dimensional cultures.

Guo *et al.* [41] utilized a two-step process for fabricating hydrogels from a globular protein 'soy glycinin', with potential applications in drug delivery. Briefly, soy glycinin was mixed with an anionic polysaccharide, dextran sulfate, and heated to partially unfold the protein. This led to the formation of a highly charged electrostatic complex which was subsequently crosslinked using mTG to obtain a transparent hydrogel. Larger amounts of dextran sulfate and lower ionic strength were favorable for increasing the transparency of the gels. Casein hydrogels crosslinked in the presence of mTG have also been developed [42]. Use of higher concentrations of mTG led to the formation of stronger hydrogels with reduced gelation time. Gelation time was also found to be temperature dependent and could be described by an Arrhenius plot with apparent activation energy of 95.4 kJ/mol. Thus, an increase in the reaction temperature favored the enzyme assisted gelation of the aqueous casein system. This may be attributed to the fact that higher temperatures accelerated the molecular motion and thus increased the chance of inter- or intramolecular crosslinking of casein under the catalysis action of mTG. The feasibility to modulate the drug release profile from the hydrogel by changing mTG concentration was demonstrated by encapsulating a model drug, vitamin-B12.

Transglutaminases have also been utilized for developing hydrogels from synthetic polymers. For instance, FXIIIa-crosslinked cell instructive poly(ethylene) glycol (PEG) scaffolds were developed for tissue engineering. Matrix metalloproteinases (MMP) degradable lysine and glutamine sequences were grafted to 8-arm PEG macromers, and subsequently crosslinked with FXIIIa [43]. Furthermore, cell adhesion ligands and vascular endothelial growth factor (VEGF) were grafted to the scaffold. This resulted in a scaffold which promoted cell attachment while also providing controlled growth factor release, as the gel degradation ensued [44]. Primary stromal cells were found to invade and proteolytically remodel these networks *in vitro* and *in vivo* due to the presence of MMP-sensitive linker peptides. These hybrid networks could be used as true alternatives for fibrin as well as provisional drug delivery platforms in tissue engineering [44]. Hu *et al.* [45] designed multifunctional PEG-based polymers containing covalently bound peptide substrates of transglutaminase enzyme, for use as tissue adhesives. The adhesive strength of guinea pig skin bonded using these hydrogels and that of a commercial fibrin tissue adhesive were found to be approximately similar. In summary, the transglutaminase-mediated crosslinking provides a versatile, natural method of preparing hydrogels with tunable mechanical and drug release characteristics for a wide variety of applications.

### 5.2.2 *Peroxidases*

#### 5.2.2.1 *Origin and mechanism*

Peroxidase enzymes play an important role in the human body. Human peroxidases are responsible for consuming hydrogen peroxide ($H_2O_2$), which is harmful to mammalian cells, and produce antibacterial products to facilitate oral health. Furthermore, human peroxidases such as myeloperoxidase, eosinophil peroxidase, lactoperoxidase, and thyroid peroxidase play important roles in immunity, biosynthesis of hormones, and pathogenesis [46–48].

Horseradish peroxidase (HRP) is a well-characterized enzyme obtained from the roots of horseradish. This enzyme belongs to the superfamily of plant peroxidases [49]. Of the multiple isoenzymes found in horseradish, HRP is the most abundant, and has been the focus of scientific study [50]. HRP is a 44-kDa enzyme comprised of 308 amino acid residues, and contains a heme group, two calcium atoms, and eight glycosylated aspargi-nine carbohydrate linkage sites [49, 50]. The heme group appears at the active site of the protein while the calcium atoms are found both distal and proximal to the heme plane, linked to the heme group through a series of hydrogen bonds [50]. Loss of these calcium ions leads to a decrease in both the enzyme activity and thermostability of HRP. HRP and its site-directed mutagenesis have been studied for use in bioanalytical applications, diagnostic kits, natural-product and fine-chemical synthesis, and medical diagnostics [49, 50]. Recently, this enzyme has been studied for the synthesis of enzymatically crosslinked injectable hydrogels.

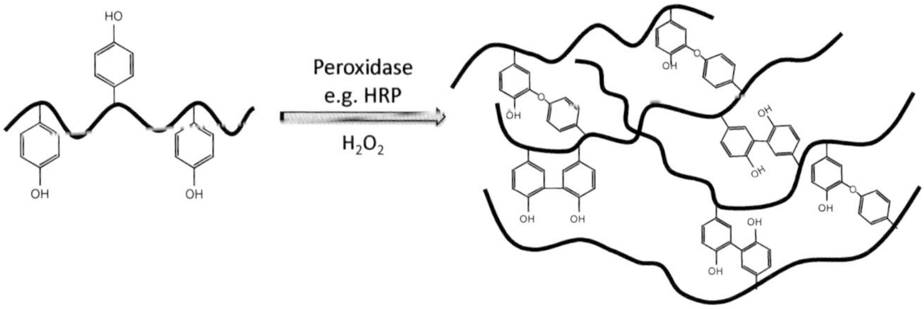


Scheme 5.2 Mechanism of peroxidase mediated crosslinking process.

In the presence of $H_2O_2$, HRP catalyzes the formation of radical species by the oxidation of two phenolic substrate molecules with concomitant reduction of $H_2O_2$ to water. First, $H_2O_2$ reacts with the heme group at the active site of HRP, resulting in an oxoferryl center and a porphyrin-based cation radical [50]. This complex is a powerful reducing agent, and undergoes two single electron oxidation reactions in the presence of a phenolic oxidizing agent to return to its ground state. The reaction results in the generation of two water molecules. Subsequently, the two phenolic radical species may undergo various spontaneous reactions including radical couplings, which mainly lead to the formation of different dityrosyl-type crosslinks, as shown in Scheme 5.2.

#### 5.2.2.2 *Applications*

HRP has been used extensively to form enzymatically crosslinked hydrogels for tissue engineering. Numerous systems utilizing phenol modified proteins, polysaccharides [51, 52] as well as synthetic polymers such as PEG [53, 54], have been crosslinked by HRP for use as tissue engineered scaffolds. Table 5.2 shows a list of polymers used for the fabrication of HRP-crosslinked hydrogels and their applications. Here, we highlight studies that used HRP/$H_2O_2$-crosslinked hydrogels for engineering specific tissues.

Gelatin is the most widely explored protein-based polymer for fabrication of peroxidase crosslinkable hydrogels. HPA-modified gelatin conjugates (Gtn-HPA) were utilized to fabricate hydrogels with tunable stiffness, for controlling the proliferation rate of human fibroblasts (HFF-1) in both two- and three-dimensional culture environments [55]. Here, the stiffness of the hydrogels was varied by changing the $H_2O_2$ and Gtn-HPA concentrations. It was found that the proliferation rate of HFF-1

Table 5.2 Peroxidase-crosslinked hydrogels and their applications.

| **Polymer** | **Remarks** | **Applications** | **Refs.** |
|---|---|---|---|
| Gelatin | Hydroxyphenyl propionic acid (HPA)-modified hydrogels with tunable stiffness | Cellular delivery, cartilage regeneration | [55], [56] |
| Gelatin | HPA and tyramine modified for increased mechanical strength | Osteogenic differentiation of mesenchymal stem cells | [57] |
| Gelatin | Tyramine modification by carbodiimide coupling | Injectable hydrogel for osteoblast delivery | [5] |
| Amylopectin derivative and Collagen | Phenol modified amylopectin, serum degradable, can be fabricated into spherical/tubular constructs | Cell encapsulation and release | [61] |
| Poly(ethylene glycol) | 3,4-Dihydroxyphenylalanine modification | Potential bioadhesive for medical purposes | [54] |
| Gelatin and Poly(ethylene glycol) | Tyramine conjugates synthesized using p-nitrophenylchloroformate as coupling reagent | Injectable matrix for tissue engineering and regenerative medicine | [53] |
| Dextran | Chondrocytes maintain cell viability and round morphology | Cartilage tissue engineering | [62] |
| Dextran and hyaluronic acid | Polysaccharide hybrids resembling proteoglycan structure | Cartilage tissue engineering | [62] |
| Hyaluronic acid | Dexamethasone-loaded hydrogels | Rheumatoid arthritis treatment | [66] |
| Alginic–hyaluronic acid composite | Composite approach for modulating mechanical properties of gel | Cellular delivery | [69] |
| Alginate, gelatin, and albumin | Tunable gelation, mechanical properties and cell adhesiveness | Scaffold for tissue engineering and carrier for drug delivery | [52] |
| Carboxymethyl cellulose | Cell sheets could be harvested using cellulase enzyme within 5 min | Cell sheet engineering | [51] |
| Dextran and heparin | Tunable swelling and mechanical properties | Scaffold for cartilage engineering | [63] |
| Gum tragacanth | Tyramine-modified, polysaccharide based gel with gelation time $<30$ seconds | Potential use for biomedical application | [64] |

using Gtn-HPA hydrogels was strongly dependent on the hydrogel stiffness. The HFF-1 exhibited higher proliferation rate under two-dimensional conditions, when the stiffness of the hydrogel was increased. In contrast, the HFF-1 cultured inside the hydrogel (i.e., three-dimensional culture) remained non-proliferative for 12 days before a stiffness-dependent proliferation profile was shown. Furthermore, the proliferation rate was also found to decrease with an increase in the stiffness of the hydrogel in a three-dimensional culture environment, unlike in a two-dimensional environment. The tunable stiffness of Gtn-HPA hydrogels was also found to modulate chondrocyte cellular functions with a dramatic impact on cartilage tissue histogenesis and repair in a rabbit model [56]. Chondrocytes cultured within a Gtn-HPA hydrogel of medium stiffness ($G' = 1000\,Pa$) showed the highest level of sulfated glycosaminoglycan production, as well as the highest ratio of Col-II to Col-I gene expression *in vitro*. In an osteochondral defect repair model, the defects implanted with chondrocytes in hydrogels of medium stiffness showed markedly more hyaline cartilage formation, smoother surface, and better integration with adjacent cartilage, compared to defects treated with hydrogels of low or high stiffness.

Another study investigated the feasibility of further modulating the gel properties by increasing the phenol content. The phenol content of Gtn-HPA hydrogels was enhanced by conjugation of tyramine to the Gtn-HPA, which resulted in Gtn-HPA-Tyr hydrogels [57]. Since gelatin has both amino and carboxyl groups, phenol containing HPA and tyramine can be coupled to gelatin using the amine and carboxyl groups. The modulus of the gels was found to vary with $H_2O_2$ concentration, although Gtn-HPA-Tyr hydrogels displayed higher storage moduli than Gtn-HPA hydrogels. Furthermore, human mesenchymal stem cells (hMSCs) encapsulated in Gtn-HPA-Tyr hydrogels with a storage modulus higher than 20,000, Pa, expressed increased levels of the osteogenic markers osteocalcin (OC) and RUNX2. hMSCs encapsulated in these hydrogels also expressed alkaline phosphatase and calcium accumulation. Therefore, the tyramine-substituted hydrogels (Gtn-HPA-Tyr) were found to be appropriate for bone tissue engineering, due to their enhanced mechanical properties.

Several studies have investigated tyramine-modified gelatin hydrogels as cell delivery vehicles. Injectable, enzymatically crosslinked tyramine-modified gelatin gels were developed and utilized as osteoblast delivery vehicles using murine preosteoblast MC3T3-E1 cells [5]. These injectable gels were found to support the adhesion, spreading, and three-dimensional growth of MC3T3-E1 cells. The study also showed the feasibility to activate extracellular signal-regulated kinase phosphorylation and active $\beta$-catenin expression, when cultured in the presence of inductive molecules such as insulin and LiCl. A recent study demonstrated the potential of this crosslinking approach for developing cell-encapsulated spherical hydrogel particles [58]. The spherical particles were prepared by dropping HepG2 mammalian cell suspended in aqueous tyramine-conjugated gelatin solution into an aqueous solution containing 1 mM $H_2O_2$. The sphericity of the hydrogel particles was found to decrease with decreasing concentrations of tyramine-modified gelatin, HRP, and $H_2O_2$, whereas the hydrogel particle membrane layer thickness could be controlled by altering the incubation time in $H_2O_2$ solution.

Park *et al.* [53] developed an *in situ* crosslinkable gelatin-poly(ethylene glycol)-tyramine (GPT) hydrogel as a novel injectable material for tissue engineering. The gelation times were found to reduce with increasing HRP concentration, whereas the storage moduli increased with increasing $H_2O_2$ concentration. The *in vitro* and *in vivo* studies demonstrated the excellent biocompatibility of the gels. The GPT gels were further conjugated with a SVVYGLR peptide, a potent angiogenic factor, to enhance endothelial cell activities and neo-vascularization [59]. *In vitro* cell studies using human umbilical vein endothelial cells (HUVECs) and *in vivo* subcutaneous injection studies demonstrated that the *in situ* conjugation of SVVYGLR sequences into phenol residues of GPT hydrogels enhanced the activity of HUVECs *in vitro* and stimulated the formation of new blood vessels in the hydrogel matrices *in vivo.*

In another study, injectable hydrogels were fabricated from bovine lactoferrin, using peroxidase-mediated crosslinking [60]. Lactoferrin is a 78 kDa glycoprotein and has shown to decrease osteoclastogenesis and increase osteoblast proliferation, survival, and differentiation. The enzymatically crosslinked lactoferrin hydrogels were found to present a conducive microenvironment to support the viability and osteogenic differentiation of mouse stromal cells. The cells encapsulated in bovine lactoferrin (bLF) gel showed the expression of active $\beta$-catenin and phosphorylated Akt, two proteins known to be significantly up-regulated by lactoferrins.

In addition to proteins, HRP has been exploited to induce the gelation of various polysaccharides and glycosaminoglycans. For instance, tyramine-substituted amylopectin hydrogels (AP-Ph) were developed, formed into microparticles and microfibers, and were shown to be degradable in serum while maintaining encapsulated endothelial cell viability [61]. When microfibers were embedded into a collagen matrix, incubation in serum

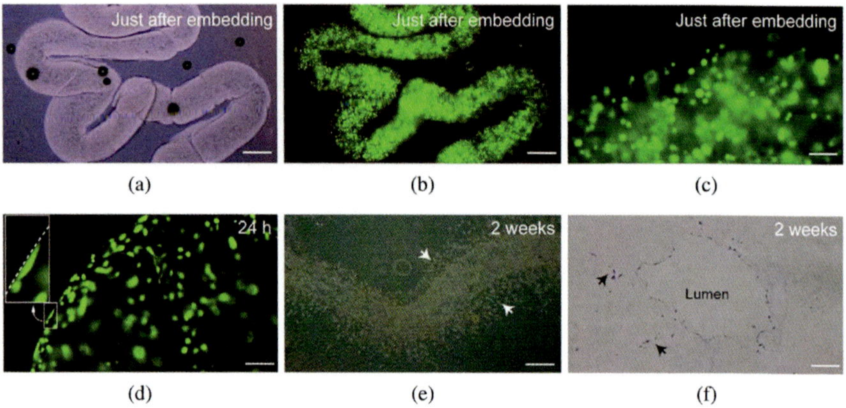


Figure 5.2 ((a): Bright field; (b), (c): Fluorescence field) Cell-enclosing Tyramine substituted amylopectin gel (AP-Ph) fibers immediately after embedding in collagen gel, (d) cells released from AP-Ph gel fiber and elongated on surface of lumen, (e) tubular construct after two weeks of culture, and (f) cross-section of the tubular construct stained with hematoxylin and eosin. The area indicated by the white rectangle in photomicrograph (d) is magnified at the surface of the cavity, showing the cell elongated on the surface (dotted line). The arrows in photomicrographs (e) and (f) indicate that the cells migrated into ambient collagen gel. Bars in photomicrographs (a), (b), and (e): 500 $\mu$m; in photographs (c), (d), and (f): 100 $\mu$m. Reprinted from [61] with permission from John Wiley and Sons.

released the encapsulated endothelial cells which spontaneously formed tubular structures within the collagen sponge. The study demonstrated the ease with which this system could be used to spontaneously generate vascular networks (Figure 5.2). Tyramine-substituted dextran hydrogels have been prepared via enzymatic crosslinking with $H_2O_2$ in the presence of HRP for cartilage tissue engineering [62]. Hydrogel strength and gelation time were found to be dependent on dextran concentration and extent of tyramine substitution. These hydrogels were found to support bovine chondrocyte viability while maintaining their round morphology. Furthermore, chondrocytes produced a matrix rich in glycosaminoglycans and collagen type II, demonstrating the potential of these hydrogels for cartilage regeneration. Jin *et al.* [63] developed injectable hydrogels from a mixture of dextran–tyramine conjugate and heparin–tyramine conjugate, and tested different weight ratios for cartilage regeneration. The gels fabricated from a 50:50 weight ratio of dextran–tyramine conjugate and heparin–tyramine conjugate showed the highest cell viability and proliferation. The hydrogel also enhanced the production of chondroitin sulfate and collagen, when compared to dextran–tyramine hydrogels.

In another study, hydrogels developed from carboxylmethylcellulose modified with phenolic groups (CMC-Ph) were tested as a substrate for cell sheet technology [51]. The gel showed good cell adhesion and proliferation of L929 fibroblast cells. The cell sheet could easily be harvested from the confluent monolayer after about 5 min of digestion of the gel using cellulase (5 U/mL). Tavakol *et al.* [64] utilized Iranian gum tragacanth, a heterogeneous highly branched anionic polysaccharide obtained from the exudate of *Astragalus fluccosus*, to develop injectable, *in situ* forming hydrogels for biomedical applications. At concentrations $\leq$0.1% (w/v), the gels showed no toxicity towards hMSCs and Caco-2 cells.

Lih *et al.* [65] utilized *in situ* curable chitosan-poly(ethylene glycol)-tyramine (CPT) hydrogels as efficient tissue adhesives, wherein PEG modified with tyramine was grafted onto chitosan backbone to enhance its solubility and to crosslink into three-dimensional networks. The hydrogels showed adhesiveness ranging from 3- to 20-fold that of fibrin glue (Greenplast®). When applied to a rat liver defect or a rat skin incision model, the gel showed excellent hemostatic properties (within 5 s) and wound-healing effects. Histological results demonstrated the superior healing effects of CPT hydrogels when compared to suture, fibrin glue and cyanoacrylate, with complete recovery by two weeks post-implantation.

Dexamethasone (DMT)-loaded injectable tyramine modified hyaluronic acid (HA-Tyr) hydrogels were developed for the treatment of rheumatoid arthritis [66]. These gels were synthesized by amide bond formation between carboxyl groups of HA and amine groups of tyramine, followed by HRP mediated crosslinking in the presence of $H_2O_2$. Intra-articular injection of these drug-loaded hydrogels was found to successfully treat rheumatoid arthritis in a collagen-induced arthritis model in rats with almost complete recovery of damaged cartilage within the tested time period of four weeks. Further, Kurisawa *et al.* utilized sequential injection of hydrogen peroxide containing HA–Tyr solution and HRP for *in vivo* fabrication of biodegradable hydrogels via enzyme-induced oxidative coupling [67]. The same group reported the preparation of a multi functional hydrogel by coupling phenol containing synthetic peptides. [Arg-Gly-Asp (RGD) peptide bearing two phenol moieties (phenol$_2$-poly(ethylene glycol)-RGD)] with phenol moieties of HA-Tyr, via HRP-mediated crosslinking reaction [68]. HUVECs cultured on or within the RGD-modified hydrogels showed significantly different adhesion behavior, from non-adherence on the HA-Tyr hydrogel to strong adherence on hydrogels modified with phenol$_2$-poly(ethylene glycol)-RGD. The altered cell adhesion behavior led to improved cell proliferation, migration, and formation of capillary-like network in the hydrogel *in vitro* and formation of functional vasculature after two weeks of subcutaneous implantation *in vivo*. Tyramine substituted hyaluronic acid and alginic acid hydrogels with mechanical properties dependent on the proportion of hyaluronic to alginic acid have also been reported for cartilage tissue engineering [69]. Both alginic and hyaluronic acid hydrogels supported ATDC-5 pre-chondrocyte viability and proliferation.

In conclusion, peroxidase-mediated crosslinking has been successfully utilized for developing injectable hydrogels from a wide range of chemically diverse polymers of natural as well as synthetic origin.

### 5.2.3 *Tyrosinase and laccase*

#### 5.2.3.1 *Origin and mechanism*

Tyrosinase, also known as polyphenol oxidase, is an enzyme which catalyzes the formation of *o*-quinones from phenols and *o*-diphenols. It is found in plants, where it is responsible for the browning of fruits and vegetables, as well as in animals, where it plays an important role in melanin production. Therefore, tyrosinase inhibitors have been extensively studied in food and cosmetic industries to prevent food browning and hyper pigmentation [70]. Studies have shown that positive tyrosinase immunostaining may help in diagnosing melanocytic lesions [71]. It has been reported that the consumption of mushrooms rich in tyrosinase and melanin, might affect the anti-melanoma humoral immune response [72]. Additionally, tyrosinases have been used in biosensors, such as gold nanoparticle-modified glassy carbon electrodes [73] and zinc oxide nanorod microarrays-modified nanocrystalline diamond electrodes [74], for the detection of phenolic compounds.

Mushroom tyrosinase has been extensively studied since it is highly homologous to mammalian tyrosinases, commercially available, and inexpensive [70, 75]. Although tyrosinases have been heavily used in the food and cosmetic industries, these enzymes, particularly those of mushroom origin, have only recently begun to be investigated for the development of enzymatically crosslinked injectable hydrogels. Tyrosinases catalyze the oxidation of either monophenols or *o*-diphenols into *o*-quinones through two distinct reactions, Cycle I and Cycle II respectively, both of which require oxygen [76]. In cycle I, tyrosinase passes through four states to accomplish oxidation. The monophenol is oxidized as it binds to a single copper atom in the binucleate cuprous active site of the tyrosinase enzyme. This molecule is then reduced into an *o*-diphenol as it detaches, leaving tyrosinase with a single oxygen atom at its binucleate cuprous reactive site. The *o*-diphenol is subsequently oxidized once more as it binds to both copper atoms at the active site of the tyrosinase enzyme. Finally, this molecule is reduced to form an *o*-quinone and a water molecule. The addition of oxygen returns the binucleate cuprous active site of tyrosinase to its fully oxygenated state. Cycle I can experience a lag phase when exogenous *o*-diphenol or a similar reductant is not available [75].

On the other hand, cycle II boasts five states. Here, the *o*-diphenol is oxidized as it binds to copper atoms at the binucleate cuprous active site. Next, it is reduced into an *o*-quinone, removing oxygen from the binucleate cuprous active site and producing a water molecule. Another *o*-diphenol is then oxidized as it binds to the cuprous active site before it is reduced once more into a second *o*-quinone product. This final step removes the second oxygen atom from the binucleate cuprous active site and produces another water molecule. Once again, the addition of oxygen restores tyrosinase's active site to its fully oxygenated state. Therefore, each turn of cycle II produces two *o*-quinones and two water molecules. Cycle II is not subject to the lag phase which may be experienced in cycle I due to the fact that *o*-diphenols are in excess, as they are the original substrate [75]. Hence, tyrosinases can act as mono-phenolases catalyzing hydroxylation of monophenolic compounds to *o*-diphenols as well as di-phenolases catalyzing subsequent oxidation of formed *o*-diphenols to respective *o*-quinones, with concomitant reduction of molecular oxygen to water. Surface-exposed tyrosyl side chains of proteins or phenol modified polymer chains can also be converted into *o*-quinones. These *o*-quinones can react spontaneously, mainly via 1,4-additions with amino groups of other polymeric chains or with side chains of lysine, tyrosine, and cysteine residues to form covalent crosslinks, as shown in Scheme 5.3 [32].

Laccase is a multicopper enzyme belonging to the blue multicopper oxidase family, mainly of fungal, bacterial, and plant origin. This enzyme has also been found in some insects. Laccase has various functions, including participation in lignin biosynthesis, plant pathogenicity, and the degradation of plant cell walls, insect sclerotization, bacterial melanization, and melanin-related virulence for humans. It catalyzes the single electron

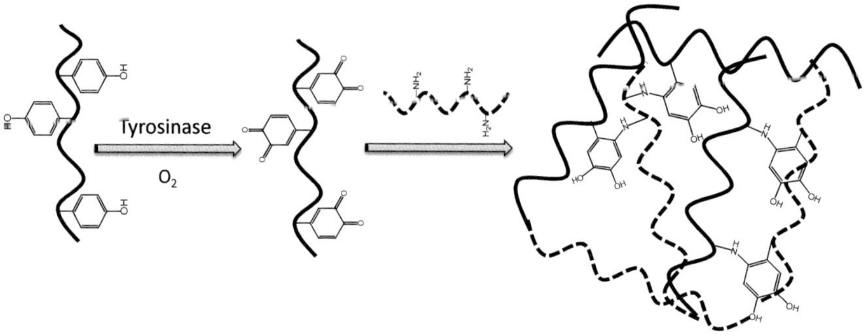


Scheme 5.3 Mechanism of tyrosinase-mediated crosslinking process.

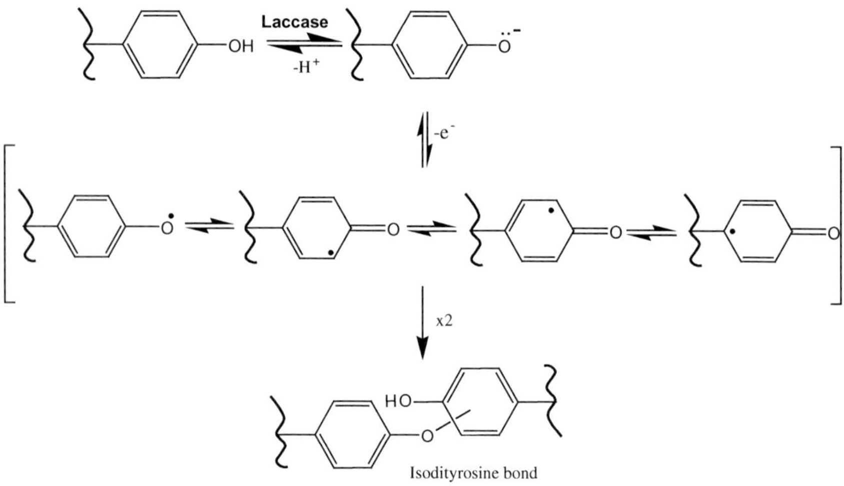


Scheme 5.4 Mechanism of laccase-mediated crosslinking process.

oxidation of aromatic compounds such as phenols, anilines, and benzothiols, with concomitant reduction of molecular oxygen to water as follows [77]:

$$4\,Substrate\,(Reduced) + O_2 + 4H^+ + 4e^- \xrightarrow{Laccase} 2H_2O + 4\,Substrate\,(Oxidized).$$

In general, laccase and peroxidases abstract a hydrogen atom from phenolic compounds and aromatic amines leading to radicals, while tyrosinase-catalyzed reactions result in quinone formation. As shown in Scheme 5.4, the crosslinking of tyrosine-containing peptides by laccase proceeds via tyrosyl radicals that form primarily isodityrosine and also a small number of dityrosine bonds [77]. Laccase is also capable of inducing the formation of disulfide bonds by the oxidation of cysteines into cystine in the presence of ferulic acid [78]. The laccase-generated radicals can be further oxidized to quinones or undergo various coupling reactions, as described earlier. *o*-Quinones generated from low molecular weight phenolic compounds by laccases or peroxidases can further react with amino, sulfhydryl, thioether, phenolic, indole, and imidazole groups of proteins, leading to crosslinking [79].

#### 5.2.3.2 *Applications*

Tyrosinase-crosslinked hydrogels have been studied for various biomedical applications (Table 5.3). Choi *et al.* [80] reported bio-inspired, tissue-adhesive gelatin hydrogels with good elastic and hemostatic properties, prepared by the tyrosinase-mediated synthesis of L-3,4-dihydroxyphenylalanine (L-DOPA) in gelatin followed by $Fe^{3+}$ ion crosslinking (DOPA-$Fe^{3+}$gelatin hydrogel). Here, the tyrosine residues in human gelatin were converted into DOPA by enzymatic reaction with tyrosinase. Upon the addition of $Fe^{3+}$ ions, the DOPA-modified gelatin formed a sticky hydrogel within seconds through complexation between the DOPA molecules and $Fe^{3+}$ ions. The

Table 5.3 Hydrogel fabricated by tyrosinase/laccase-mediated crosslinking and their applications.

| Polymer | Remarks | Applications | Refs. |
|---|---|---|---|
| Silk fibroin/ chitosan | Biocompatible, no use of chemical crosslinkers | Scaffold fabrication | [82] |
| Chondroitin sulfate | Variable gelation and degradation times | Sustained drug delivery and tissue engineering | [81] |
| Gelatin | DOPA-modified, $Fe^{3+}$ ion-crosslinked | Tissue adhesive, hemostatic | [80] |
| Collagen | Increased denaturation temperature and reduced degradation by microbial collagenase, use of non-toxic crosslinkers | Cell implantation for tissue engineering | [83] |
| Chitosan and gelatin | Laccase-assisted coupling between polyphenols of *H. virginiana* and 1° $NH_2$ groups of chitosan and gelatin | Bioactive, antibacterial wound dressing for chronic wound healing | [84] |

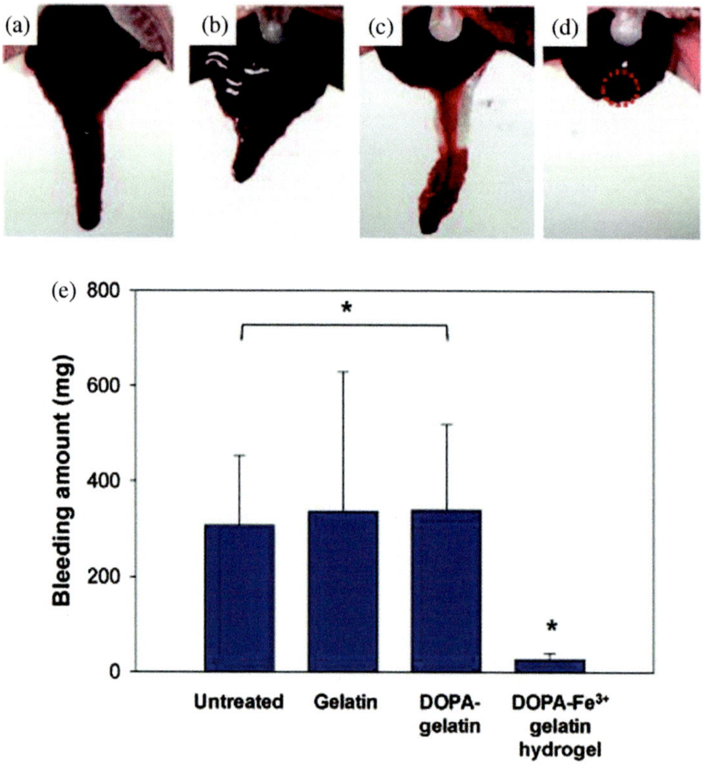


Figure 5.3 The evaluation of hemostatic efficiency of the DOPA–$Fe^{3+}$ gelatin hydrogel in a bleeding liver model of rats. Bleeding was induced using an 18-G needle, and 100 μL of DOPA–$Fe^{3+}$ gelatin hydrogel was immediately applied to the bleeding site. Macroscopic images were taken 1 min after pricking with a needle (no treatment, a) and 1 min after applying gelatin (b), DOPA–gelatin (c) and DOPA–$Fe^{3+}$ gelatin hydrogel (d) to the bleeding livers of rats. The red circle denotes the sealing of the defect by the DOPA–$Fe^{3+}$ gelatin hydrogel (e) blood loss by weight from the damaged livers. Data are shown as the mean ± standard deviation ($n = 5$) with significance at $^*p < 0.01$. Reprinted from [80] with permission from the Royal Society of Chemistry.

hemostatic ability of the DOPA-$Fe^{3+}$ gelatin hydrogels was explored using a rat model of liver hemorrhage. Shortly after the injection of DOPA-$Fe^{3+}$ gelatin hydrogel, the bleeding was arrested in the hemorrhaging site of the liver (Figure 5.3).

Chen *et al.* [34] compared the ability of mushroom tyrosinase and mTG to catalyze gel formation from gelatin and chitosan. mTG led to the formation of strong and permanent gels from gelatin solutions, even in the absence of chitosan. Gels formed in the presence of chitosan were stronger and the gelation was faster. In contrast to mTG, tyrosinase-catalyzed reactions did not lead to gel formation unless chitosan was present. Tyrosinase-catalyzed gelatin-chitosan gels were observed to be transient and considerably weaker than mTG-catalyzed gels. The gel strength (i.e., elastic modulus) of tyrosinase-catalyzed gelatin-chitosan gels peaked at 5h, after which the gels broke, as shown by rheological analysis. The strength of both mTG-catalyzed and tyrosinase-catalyzed gels could be adjusted by altering the gelatin and chitosan compositions.

Tyrosinases have also been used to crosslink tyramine-substituted chondroitin sulfate to form hydrogels for tissue engineering and drug delivery [81]. Gelation was found to be dependent on polymer concentration, degree of tyramine substitution, and tyrosinase concentration. The gels were completely biodegradable by chondroitinase and they supported fibroblast viability. Furthermore, the gels were able to maintain a sustained release profile of a small molecule (methylene blue), over an extended time period. Tyrosinase has also been used to develop silk fibroin-chitosan hydrogels for tissue engineering applications [82]. The crosslinking was confirmed through viscosity measurement, UV spectroscopy, light scattering, differential scattering calorimetry, and adhesive shear strength.

Collagen and collagen peptide fragments of defined molecular weight have also been successfully crosslinked in the presence of tyrosinases [83]. The tensile strength of the resultant constructs can be enhanced in the presence of added phenolic compounds like caffeic acid, chlorogenic acid, and catechine. Furthermore, hydrogel degradation with mammalian collagenase was reduced in the presence of these phenolic compounds. This resistance to degradation was attributed to the fact that enzymatic crosslinking of collagen partially leads to the formation of an insoluble gel-like material, especially in the presence of phenolic molecules. Similar trends were observed for gelatin and gelatin hydrolysate, where these proteinaceous compounds were able to be crosslinked in the presence of tyrosinases, and the addition of phenolic compounds enhanced the degree of crosslinking [79].

Laccase-catalyzed oxidation has been utilized for the successful fabrication of gelatin–chitosan composite hydrogels in the presence of plant polyphenols obtained from *Hamamelis virginiana* (C/G-PC) [84]. These hydrogels were investigated as bioactive hydrogel dressings. Rheological studies revealed that the mechanical properties of the hydrogels were a function of the enzymatic reaction time and precursor concentrations. Furthermore, the hydrogels were found to inhibit the growth of gram-negative *Pseudomonas aeruginosa* and gram-positive *Staphylococcus aureus*, the two most commonly found bacteria in chronic wounds. The antibacterial capacity of the C/G-PC hydrogels was attributed to the combined action of the intrinsic antibacterial activity of chitosan (interactions between its cationic amino groups and the negatively charged microbial cell membranes, resulting in the alteration of cell membrane permeability) and the covalently attached as well as released phenolic compounds.

Collagen peptides have also been crosslinked in the presence of laccases, derived from two different mushrooms, *Trametes hirsuta* and *Trametes versicolor* [83]. The *Trametes versicolor* laccase exhibited higher oxygen consumption, indicating that it was easier for this enzyme to access substrates inside collagen coils. Furthermore, incubation of collagen with laccases in the presence of phenolic molecules led to gels with higher tensile strength as well as reduced collagenase degradation due to higher degree of crosslinking. Similarly, these mushroom laccases were used to crosslink gelatin [79]. The potential of crosslinking gelatine and gelatine hydrolysate by laccase was shown by size exclusion chromatography (SEC) and sodium dodecyl sulfate polyacrylamide gel electrophoresis (SDS-PAGE). Gelatine hydrolysate showed higher crosslinking, likely due to less steric hindrance for laccase in accessing proper amino acid substrates. The addition of small phenolic compounds to the reaction mixtures led to a further increase in the degree of crosslinking, as described earlier. Laccase has been used to crosslink globular proteins as well [85]. Laccase was able to crosslink bovine serum albumin (BSA), $\beta$-lactoglobulin, and $\alpha$-casein to different degrees as well as through different pathways. The protein structure was therefore found to play a significant role in the crosslinking reaction. Furthermore, the presence of ferulic acid was found to be important for the crosslinking reaction.

3,4-Dihydroxyphenylalanine (DOPA) residues, which are known to impart adhesive and curing capabilities to mussel adhesive proteins, have been grafted to branched PEG and crosslinked in the presence of mushroom tyrosinase [54]. Polymer crosslinking time was found to be dependent on the PEG architecture, polymer concentration, and enzyme concentration. It has been suggested that these hydrogels may be useful for medical applications. Overall, tyrosinase- and laccase-assisted crosslinking methods have great potential for the development of hydrogels for a wide range of biomedical applications.

### 5.2.4 *Lysyl oxidase*

#### 5.2.4.1 *Origin and mechanism*

Lysyl oxidase is a copper-containing amine oxidase, which oxidizes primary amines into reactive aldehydes. Additionally, four lysyl oxidase-like proteins have been identified. Lysyl oxidase and lysyl oxidase-1 are proteins which are secreted as zymogens, and subsequently cleaved into their active forms in the ECM. Lysyl oxidase has been found to play an important role in the proper development of the respiratory and cardiovascular systems in mice. Mice lacking the lysyl oxidase gene showed significantly less elastin fibers compared to the wild-type mice in the distal airways, bronchi, bronchioles, and trachea [86]. No difference in collagen content was found, but collagen in mice lacking lysyl oxidase gene was observed to be loose, disperse, and unorganized. Furthermore, mice were found to have deficient crosslinking of collagen and elastin in blood vessels, leading to a lack of resilience and elasticity, which ultimately resulted in a high prevalence of aortic aneurisms [87].

Lysyl oxidase requires two cofactors for catalysis, namely, a covalently bound lysine tyrosinylquinone (LTQ) cofactor and a copper ion [32]. Lysyl oxidase catalyzes protein crosslinking via the formation of anhydrolysinonorleucine crosslinks or through an aldol condensation reaction [88]. In the presence of oxygen and water, lysyl oxidase can oxidize the primary $\varepsilon$-amino group at the end of an accessible lysine residue into a reactive aldehyde. This aldehyde group may then react with the unmodified $\varepsilon$-amino group of another lysine residue to form a Schiff base product, as shown in Scheme 5.5. Alternatively, this reactive aldehyde may react with another aldehyde on a modified lysine residue to undergo aldol condensation reaction.

#### 5.2.4.2 *Applications*

Table 5.4 provides examples of polymers used for hydrogel fabrication by lysyl oxidase-mediated crosslinking, along with their applications. As lysyl oxidase stabilizes the ECM through the chemical crosslinking of collagen and elastin, it was used to improve the mechanical strength of ECM for application in tissue engineered vascular constructs [89]. Rat aortic smooth muscle cells, transfected with lysyl oxidase cDNA, and subsequently seeded in collagen gels were found to produce similar levels of ECM proteins compared to non-transfected cell controls. However, increased ECM matrix

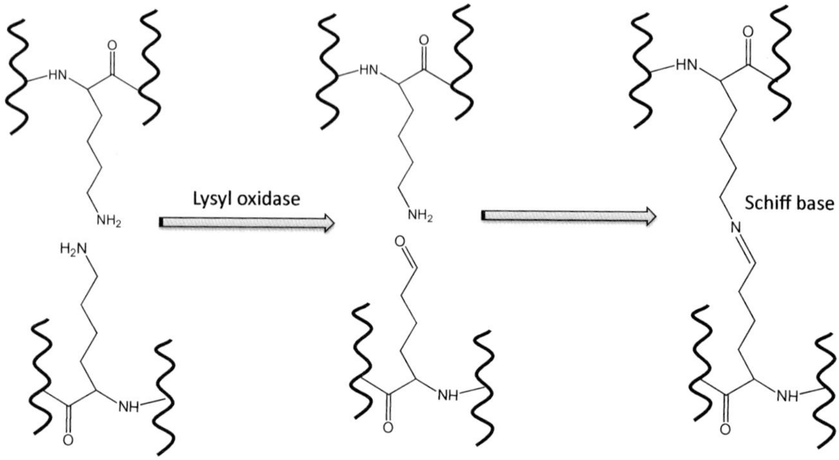


Scheme 5.5 Mechanism of lysyl oxidase mediated crosslinking process.

Table 5.4 Lysyl oxidase-crosslinked hydrogels and their applications.

| Polymer/ Monomer | Remarks | Applications | Refs. |
|---|---|---|---|
| Multidomain peptides containing lysine residues | Nanofibrous peptide hydrogel | Cell encapsulation | [92] |
| Tropoelastin | Two-component system | Study of elastin assembly *in vitro* | [91] |
| Collagen | Lysyl oxidase-producing gene activated matrices, enhanced mechanical strength of wound site | Dermal wound healing | [90] |
| Collagen | Lysyl oxidase transfected smooth muscle cells seeded in gels, improved mechanical properties of constructs | Engineering tissue with enhanced mechanical property | [89] |

crosslinking was found in the transfected cell-seeded scaffolds as evidenced by increased desmosine (a product of lysyl oxidase-mediated crosslinking of elastin) concentrations and mechanical strength. The study demonstrated the ability of lysyl oxidase to improve the strength of tissue engineered scaffolds.

A recent study utilized overexpression of lysyl oxidase to increase collagen and elastin crosslinking for improving the mechanical strength of ECM during dermal would healing [90]. Briefly, a gene-activated matrix (GAM), composed of collagen gels loaded with lysyl oxidase transgene plasmid DNA (pDNA) complexed with polyethylenimine (PEI), was utilized for the localized and sustained transfection of cells involved in the healing process. The hydrogels successfully transfected fibroblasts *in vitro* and significantly enhanced the mechanical strength of the wound site, when compared to non-DNA-loaded scaffolds in a rat dermal wound model. This study again demonstrated the use of lysyl oxidase gene therapy for improving ECM crosslinking and wound healing.

Purified lysyl oxidase from yeast (*Pichia pastoris*) has also been utilized to facilitate the oxidation and subsequent crosslinking of recombinant human tropoelastin. The crosslinked human tropoelastin formed an elastin-like polymer that is elastic, showed hydrogel-like behavior, and contained typical elastin crosslinks including lysinonorleucine, allysine aldol, and desmosine [91]. Additionally, lysyl oxidase has been used to crosslink self-assembled peptide hydrogels [92]. Self-assembled peptide hydrogels containing lysine residues grew mechanically stronger upon incubation in culture media, while those with glutamate instead of lysine did not. This phenomenon was explained by the presence of lysyl oxidase in the culture media. These lysine containing self assembled peptide hydrogels can also be crosslinked using plasma amine oxidase, a similar oxidative enzyme. Therefore, lysyl oxidase is a promising enzyme for the development of enzymatically crosslinked hydrogels for tissue engineering.

### 5.2.5 *Phosphatase*

#### 5.2.5.1 *Origin and mechanism*

A phosphatase is an enzyme belonging to the hydrolase class that removes a phosphate group from its substrate by hydrolyzing phosphoric acid monoesters into a phosphate ion and a molecule with a free hydroxyl group. Alkaline phosphatase, as the name implies, works best at an alkaline pH with optimum *in vitro* activity at a pH of 10. Most human tissues contain alkaline phosphatase; kidney, liver, bone, intestine, reticuloendothelial tissue, and placenta are particularly rich sources [93]. Acid phosphatases, on the other hand, can catalyze the hydrolysis of orthophosphate monoesters under acidic conditions [94]. Yang *et al.* [95] reported

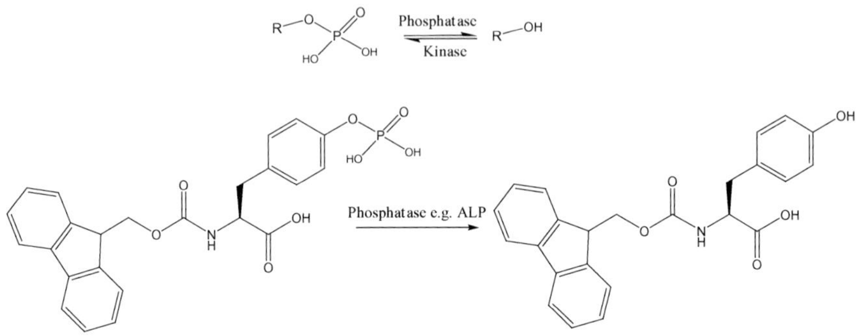


(a)

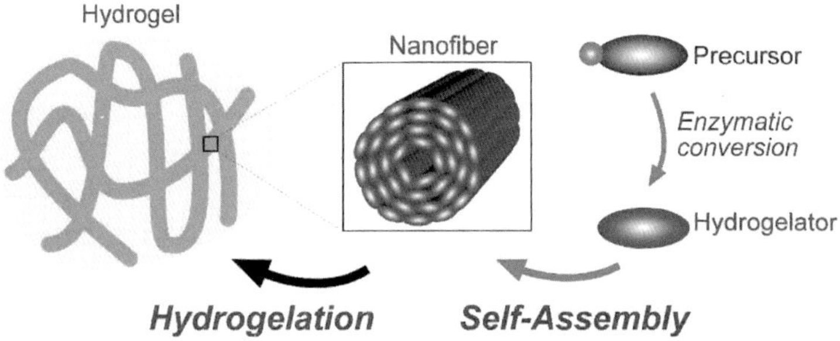


(b)

Scheme 5.6 (a) Use of phosphatase/kinase switch for controlled and reversible hydrogelation and mechanism of phosphatase-mediated hydrogelation process using a small molecule 'hydrogelator'; (b) Enzymatic conversion of a precursor small molecule into a hydrogelator leads to formation of nanofibrous supramolecular hydrogels. Reprinted from [15] with permission from the American Chemical Society.

alkaline phosphatase-induced gelation of *N*-fluorenylmethyloxycarbonyl tyrosine phosphate under basic conditions by dephosphorylation of $PO_4^{3-}$, which results in the conversion of the ionic group into a neutral group. The neutralization leads to a change in balance of hydrophobicity/hydrophilicty of the precursor, yielding a small molecule hydrogelator, which could self-assemble to form supramolecular hydrogels. Scheme 5.6 shows the mechanism of the phosphatase-mediated hydrogelation process. The enzymatic conversion of a precursor small molecule into a hydrogelator leads to changes in hydrophobicity/hydrophilicity, resulting in reorganization into $\beta$-sheet features due to hydrogen bonding. This is followed by self-assembly into nanofibrous supramolecular hydrogels. Furthermore, controlled and reversible hydrogelation can be achieved in these systems by use of enzyme switches. For instance, a phosphatase/kinase enzyme pair has been utilized for the fabrication of supramolecular hydrogels wherein the kinase catalyzes the attachment of phosphate groups to a protein/small molecule/hydrogelator (phosphorylation), and phosphatase removes them (dephosphorylation) [96].

### 5.2.5.2 *Applications*

Phosphatases have been widely explored for the generation of supramolecular hydrogels for tissue engineering (Table 5.5). A series of phosphatase-catalyzed supramolecular hydrogels that self-assemble into nanofibers have been studied for this purpose. Self-assembled supramolecular hydrogels have been shown to form from Nap-FFGEY (naphthalene–Phe-Phe-Gly-Glu-Tyr) hydrogelator [96]. A kinase/phosphatase switch was used to control the phosphorylation and dephosphorylation of the hydrogelator, thus regulating the formation of supramolecular hydrogels. Briefly, treating with phosphatase would lead to hydrogel formation, whereas addition of a kinase would induce a gel–sol phase transition in the presence of adenosine triphosphates (ATPs). Here, the Tyr residue is converted into tyrosine phosphate to form a more hydrophilic molecule, Nap-FFGEY-P(O)$(OH)_2$. The hydrogelator was found to be biocompatible, and when the precursor was injected subcutaneously in mice, supramolecular hydrogels were formed within 1 h. Another study reported grafting of Taxol (an apoptotic drug used for the

Table 5.5 Phosphatase-crosslinked hydrogels and their applications.

| **Polymer/Monomer** | **Remarks** | **Applications** | **Refs.** |
|---|---|---|---|
| $\beta$-amino acid derivatives | Self-assembled nanofibers, stable, supramolecular hydrogels | Potential in tissue engineering, drug delivery, and wound healing | [16] |
| Napthalene-FFGEY | Kinase/phosphatase enzyme switch regulated gelation | Novel biomaterial | [96] |
| Fluorenylmethyloxycarbonyl, tyrosine phosphate | Self-assembled matrices for calcium phosphate mineralization | Hybrid biocomposites | [17] |
| Short peptides | Gel properties similar to collagen | Tissue engineering, design of novel biomaterials | [18] |

treatment of cancer), to hydrogelators for creating supramolecular hydrogels [19]. Taxol activity was found to be conserved during the hydrogelation process, demonstrating the ability to use this technology for the development of hydrogels as drug delivery vehicles.

In an attempt to generate collagen-like self-assembled supramolecular hydrogels, a series of naphthalene-modified peptide sequences were studied, changing only a single amino acid residue to differentiate each peptide [18]. The best hydrogel was determined to be the peptide sequence incorporating lysine due to its superior mechanical properties as well as its non-cytotoxicity towards mouse fibroblasts. Self-assembled supramolecular hydrogels have also been prepared by the alkaline phosphatase-catalyzed dephophorylation of *N*-FMOC-tyrosine-phosphate [17]. Hydrogelators formed were found to self-assemble into nanofibrous structures, and when mineralized with calcium phosphate, the resulting hydrogels showed a drastic increase in mechanical strength. Such hydrogels may hold promise as scaffolds for bone tissue engineering.

$\beta$-amino acid derivative hydrogelators have also been used to form supramolecular hydrogels in the presence of acid phosphatase [16]. The storage modulus of the hydrogels and their gelation time was found to be dependent on the concentration of acid phosphatase used. Furthermore, $\beta$-amino acid derivative supramolecular hydrogels were found to be more stable *in vivo* when compared to hydrogels formed with $\alpha$-amino acid derivatives.

### 5.2.6 *Metalloproteinases*

A metalloproteinase is a protease enzyme whose catalytic mechanism involves a metal. They are further classified into two subgroups: endopeptidases, which cleave peptide bonds of non-terminal amino acids (i.e., within the molecule), and exopeptidases, which cleave the terminal or the penultimate peptide bond, thereby releasing a single amino acid or dipeptide from the peptide chain. MMPs are zinc-dependent endopeptidases capable of degrading all kinds of ECM proteins and are known to play a major role in various cellular functions. The peptide cleavage capability of MMPs has been utilized to form hydrogelators which can self-assemble to form degradable supramolecular hydrogels. Yang *et al.* [97] synthesized peptide sequences of Phe-Phe-Phe-Phe-Cys-Gly-Leu-Asp-Asp (FFFFCGLDD), which are cleavable by MMP-9 between the glycine and leucine residues and can subsequently self-assemble to form supramolecular hydrogels. The hydrogels were found to be elastic with nanofibrous structure. It was suggested that these hydrogels may hold great promise for cancer therapy and drug delivery, where the MMPs produced by cancerous tissue could initiate gelation and facilitate the sustained release of therapeutic drugs. Similarly, supramolecular hydrogels from the precursor *N*-Palmitoyl-Gly-Gly-Gly-His-Gly-Pro-Leu-Gly-Leu-Ala-Arg-Lys-$NH_2$ have been formed through the enzymatic cleavage of the -Pro-Leu-Gly-Leu- residue sequence by MMP-7 [98]. The precursor solution was found to form micelles, but following the enzymatic cleavage, nanofibrous supramolecular hydrogels were formed. These hydrogels were also suggested for the treatment of cancers overexpressing MMPs. Therefore, self-assembled supramolecular hydrogels catalyzed by MMP cleavage of peptide precursors hold great promise for drug delivery and possibly tissue engineering.

Thermolysin is the most stable member of a family of metalloproteinases produced by *Bacillus* such as *Bacillus thermoproteolyticus rokko*. It requires one zinc ion for enzyme activity and four calcium ions for structural stability and is widely used for peptide bond formation through reversed hydrolysis. Self-assembled supramolecular hydrogels of FMOC-peptide sequences have been created through thermolysin catalysis [99]. The process was found to be fully reversible and driven by equilibrium. Guilbaud *et al.* [100] developed thermolysin-triggered tetrapeptide (FEFK, F:Phe, E:Glu, K:Lys)-derived hydrogels with high shear moduli > 1 MPa for cell encapsulation (Figure 5.4). The mechanism of gelation was found to involve formation of

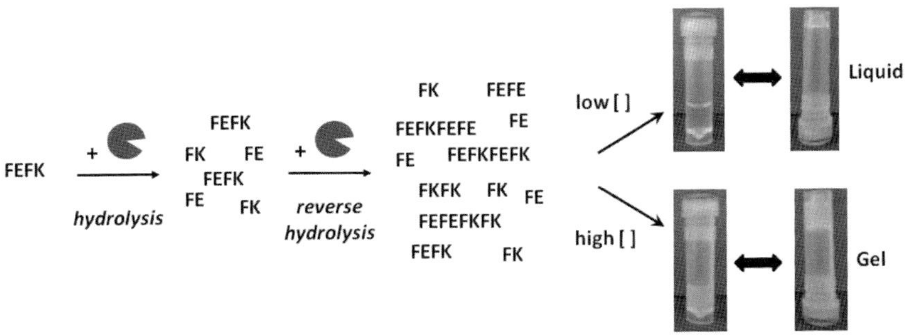


Figure 5.4 Schematic representation of the thermolysin-catalyzed synthesis of peptide based hydrogels. For $C_0 < 100$ mg $mL^{-1}$ (low), the amount of octapeptides synthesized is lower than the critical gelation concentration and no gelation is observed. For $C_0 \geq 100$ mg $mL^{-1}$ (high), the amount of octapeptides synthesized is higher than the critical gelation concentration and gelation of the sample is observed. Reprinted from [101] with permission from the American Chemical Society.

longer octapeptide sequences such as FEFEFKFK and FEFKFEFK. These octapeptides self-assemble under appropriate conditions into $\beta$-sheet rich fibers that subsequently associate/entangle to form hydrogels [101, 102].

### 5.2.7 *DNA ligase and polymerase*

DNA is an attractive material for the generation of nanostructures including hydrogels. This is due to the availability of a plethora of enzymes which can modify DNA with high accuracy, thereby providing a toolbox to construct nanometer-scaled objects. Um *et al.* [103] reported the self-assembly of branched DNA molecules, possessing a complementary sticky end whose sequences were palindromic, via ligase-catalyzed reactions. These branched DNA molecules were able to hybridize to and ligate with each other via T4 DNA ligase, serving as both monomers and crosslinkers (Figure 5.5a). Self-assembly leads to the formation of a large-scale, three-dimensional structure with hydrogel properties. The same group later utilized the DNA hydrogel consisting of genes as part of the gel scaffolding for producing functional proteins (termed 'the P-gel system' or 'P-gel') in the absence of any living cells with an efficiency of about 300 times higher than solution-based systems [104]. Here, the X-shaped DNA (X-DNA) were ligated with linear plasmids (genes) and placed in a polydimethylsiloxane (PDMS) micromold along with T4 DNA ligase. Subsequently, protein was expressed simply by incubating the P-gel micropads with cell lysates for a specific time period (Figures 5.5b–d).

Lee *et al.* [105] also utilized a T4 ligase (polymerase) enzyme to elongate DNA chains, weaving them non covalently into a meta-hydrogel with potential applications in drug release, cell therapy, electric switches, and flexible circuits. Taq DNA polymerase was also utilized for the enzymatic synthesis of rigid, branched DNA building blocks, to generate DNA networks by self-assembly [106]. The enzyme used each arm of a Y-shaped 3-armed DNA construct as the primer strand and incorporated the branched construct into large assemblies during polymerase chain reaction, to form functional hydrogels. Qi *et al.* [107] reported the use of DNA as programmable, sequence-specific "glue" to self assemble PEG hydrogel cubes with edge length ranging from 30 microns to 1 millimeter in the presence of Phi29 DNA polymerase. Using these hydrogel cuboids displaying face-specific DNA glues, diverse structures including dimers, extended chains, and open network structures were created by submerging them in aqueous solution followed by gentle agitation. Similarly, using a liquid-liquid interface made

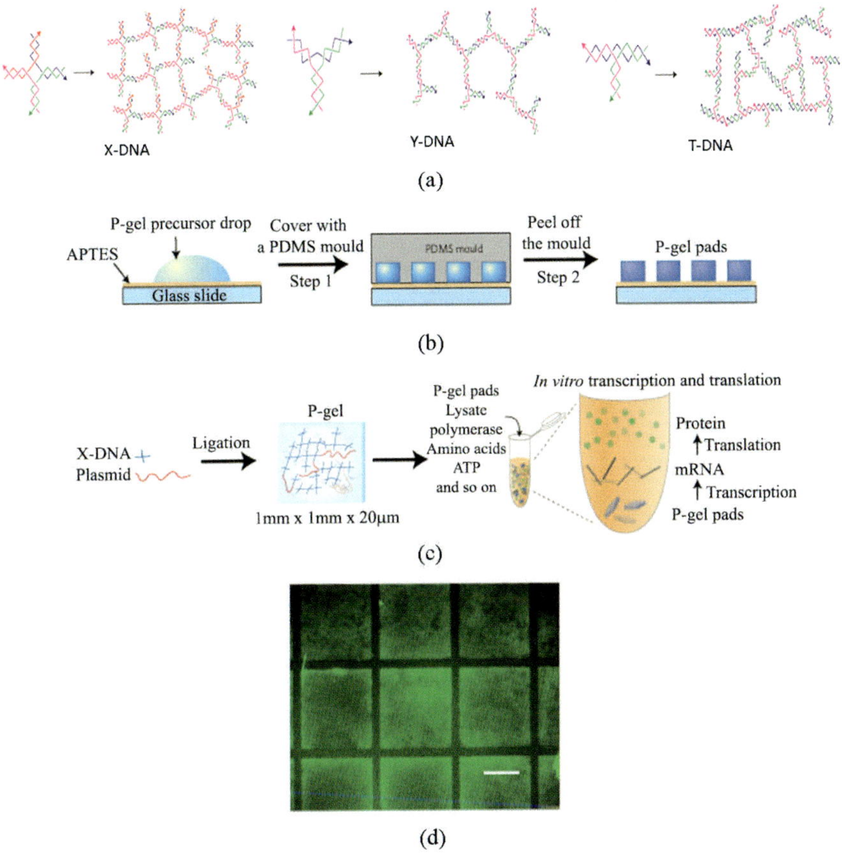


Figure 5.5 (a) X-, Y- and T-DNA (branched DNA monomers) serve as crosslinkers to form networked gels. (b) A schematic diagram illustrating the formation of P-gel micropads (side-view). The P-gel precursor drop, which contains X-DNA, genes, and T4 DNA ligase, was confined within a PDMS mold with precisely defined dimensions: 1.0 mm × 1.0 mm × 20 $\mu$m (20 nL) to control and adjust the exact gene amount within the P-gel scaffolding and also to obtain a higher surface-to-volume ratio of P-gel on an APTES-coated glass slide. Extra solution was able to flow out owing to the surface modification (Step 1). After gelation, the PDMS mould was peeled off from the substrate (Step 2), and P-gel micropads were formed using the pre-designed dimensions. (c) A schematic diagram of the gelation process through enzymatic crosslinking and cell-free expression with P-gel pads. (d) A fluorescent image of the P-gel pads after staining with SYBR I. Each micropad was separated by 100 $\mu$m spacers (scale bar: 500 $\mu$m). Adapted from [103, 104] with permission from Macmillan Publishers Ltd.

of PBS and Fluorinert FC-40 liquid, dimers, chains of fixed length, square shapes or T junctions were also obtained. DNA-directed self-assembly of shape-controlled hydrogel is expected to open new doors to address the challenges of building complex, self-assembled, three-dimensional structures for diverse applications in materials science and especially in the field of biomaterials.

## 5.3 Conclusions

*In situ* crosslinking provides an added advantage for hydrogel fabrication by allowing gelation under physiological conditions, and *in situ* drug/protein/cellular encapsulation thereby eliminating post-loading step, and by avoiding cytotoxic and denaturing conditions. Enzymatic crosslinking can be utilized to develop complex, bioactive hydrogels which can be programmed into two- and three-dimensional networks, not only by crosslinking various natural or synthetic polymers/monomers but also by self-assembly of small peptides and DNA. These advances in the design and fabrication of enzymatically crosslinkable hydrogels will serve as a stepping stone for the development of novel biomaterials for regenerative engineering applications.

## References

[1] Persidis, A. (1999). Tissue engineering. *Nat. Biotechnol.*, **17**(5), pp. 508–510.

[2] Ikada, Y. (2006). Challenges in tissue engineering. *J. R. Soc. Interface*, **3**(10), pp. 589–601.

[3] Verma, S. *et al.* (2011). Biodegradable polymers for emerging clinical use in tissue engineering, in A.J. Domb, N. Kumar, and A. Ezra (eds.), *Biodegradable Polymers in Clinical Use and Clinical Development*. John Wiley & Sons, Inc., Hoboken, NJ, US, pp. 555–619.

[4] Howard, D. *et al.* (2008). Tissue engineering: strategies, stem cells and scaffolds. *J. Anat.*, **213**(1), pp. 66–72.

[5] Amini, A.A. and Nair, L.S. (2012). Enzymatically cross-linked injectable gelatin gel as osteoblast delivery vehicle. *J. Bioact. Compat. Polym.*, **27**(4), pp. 342–345.

[6] Nicodemus, G.D. and Bryant, S.J. (2008). Cell encapsulation in biodegradable hydrogels for tissue engineering applications. *Tissue Eng. Part B Rev.*, **14**(2), pp. 149–165.

[7] Ashley, A.A. and Nair, L.S. (2012). Injectable hydrogels for bone and cartilage repair. *Biomed. Mater.*, **7**(2), p. 024105.

[8] Nguyen, M.K. and Alsberg, E. (2014). Bioactive factor delivery strategies from engineered polymer hydrogels for therapeutic medicine. *Prog. Polym. Sci.*, **39**(7), pp. 1235–1265.
[9] Nuttelman, C.R. *et al.* (2008). Macromolecular monomers for the synthesis of hydrogel niches and their application in cell encapsulation and tissue engineering. *Prog. Polym. Sci.*, **33**(2), pp. 167–179.
[10] Amini, A.A. and Nair, L.S. (2012). Injectable hydrogels for bone and cartilage repair. *Biomed. Mater.*, **7**(2), p. 024105.
[11] Ko, D.Y. *et al.* (2013). Recent progress of *in situ* formed gels for biomedical applications. *Prog. Polym. Sci.*, **38**(3–4), pp. 672–701.
[12] Toledano, S. *et al.* (2006). Enzyme-triggered self-assembly of peptide hydrogels via reversed hydrolysis. *J. Am. Chem. Soc.*, **128**(4), pp. 1070–1071.
[13] Teixeira, L.S. *et al.* (2012). Enzyme-catalyzed crosslinkable hydrogels: emerging strategies for tissue engineering. *Biomaterials*, **33**(5), pp. 1281–1290.
[14] Hahn, M.E. and Gianneschi, N.C. (2011). Enzyme-directed assembly and manipulation of organic nanomaterials. *Chem. Commun.*, **47**, pp. 11814–11821.
[15] Yang, Z., Liang, G., and Xu, B. (2008). Enzymatic hydrogelation of small molecules. *Acc. Chem. Res.*, **41**(2), pp. 315–326.
[16] Yang, Z. *et al.* (2007). *In vitro* and *in vivo* enzymatic formation of supramolecular hydrogels based on self-assembled nanofibers of a beta-amino acid derivative. *Small*, **3**(4), pp. 558–562.
[17] Schnepp, Z.A.C., Gonzalez-McQuire, R., and Mann, S. (2006). Hybrid biocomposites based on calcium phosphate mineralization of self-assembled supramolecular hydrogels. *Adv. Mater.*, **18**(14), pp. 1869–1872.
[18] Hu, Y. *et al.* (2010). Supramolecular hydrogels inspired by collagen for tissue engineering. *Org. Biomol. Chem.*, **8**(14), pp. 3267–3271.
[19] Gao, Y. *et al.* (2009). Enzyme-instructed molecular self-assembly confers nanofibers and a supramolecular hydrogel of taxol derivative. *J. Am. Chem. Soc.*, **131**(38), pp. 13576–13577.
[20] Griffin, M., Casadio, R., and Bergamini, C.M. (2002). Transglutaminases: nature's biological glues. *Biochem. J.*, **368**(Pt 2), pp. 377–396.
[21] Richardson, V.R. *et al.* (2013). Substrates of Factor XIII-A: roles in thrombosis and wound healing. *Clin. Sci. (Lond.)*, **124**(3), pp. 123–137.
[22] Greenberg, C.S., Birckbichler, P.J., and Rice, R.H. (1991). Transglutaminases: multifunctional cross-linking enzymes that stabilize tissues. *FASEB J.*, **5**(15), pp. 3071–3077.
[23] Fesus, L. and Szondy, Z. (2005). Transglutaminase 2 in the balance of cell death and survival. *FEBS Lett.*, **579**(15), pp. 3297–3302.
[24] Akimov, S.S. and Belkin, A.M. (2001). Cell surface tissue transglutaminase is involved in adhesion and migration of monocytic cells on fibronectin. *Blood*, fibronectin. *Blood*, **98**(5), pp. 1567–1576.
[25] Fesus, L. and Piacentini, M. (2002). Transglutaminase 2: an enigmatic enzyme with diverse functions. *Trends Biochem. Sci.*, **27**(10), pp. 534–539.

[26] Park, D., Choi, S.S., and Ha, K.S. (2010). Transglutaminase 2: a multi-functional multi-functional protein in multiple subcellular compartments. *Amino Acids*, 619–631.
[27] Di Sabatino, A. *et al.* (2012). The function of tissue transglutaminase in celiac disease. *Autoimmun. Rev.*, **11**(10), pp. 746–753.
[28] Grosso, H. and Mouradian, M.M. (2012). Transglutaminase 2: biology, relevance to neurodegenerative diseases and therapeutic implications. *Pharmacol. Ther.*, **133**(3), pp. 392–410.
[29] Ando, H. *et al.* (1989). Purification and characteristics of a novel transglutaminase derived from microorganisms. *Agric. Biol. Chem.*, **53**(10), pp. 2613–2617.
[30] Yokoyama, K., Nio, N., and Kikuchi, Y. (2004). Properties and applications of microbial transglutaminase. *Appl. Microbiol. Biotechnol.*, **64**(4), pp. 447–454.
[31] Zhu, Y. and Tramper, J. (2008). Novel applications for microbial transglutaminase beyond food processing. *Trends Biotechnol.*, **26**(10), pp. 559–565.
[32] Heck, T. *et al.* (2013). Enzyme-catalyzed protein crosslinking. *Appl. Microbiol. Biotechnol.*, **97**(2), pp. 461–475.
[33] Lorand, L. and Graham, R.M. (2003). Transglutaminases: crosslinking enzymes with enzymes with pleiotropic functions. *Nat. Rev. Mol. Cell Biol.*, **4**(2), pp. 140–156.
[34] Chen, T. *et al.* (2003). Enzyme-catalyzed gel formation of gelatin and chitosan: potential for *in situ* applications. *Biomaterials*, **24**(17), pp. 2831–2841.
[35] Chen, T. *et al.* (2003). Enzymatic methods for *in situ* cell entrapment and cell release. *Biomacromolecules*, **4**(6), pp. 1558–1563.
[36] Broderick, E.P. *et al.* (2005). Enzymatic stabilization of gelatin-based scaffolds. *J. Biomed. Mater. Res. Part B: Appl. Biomater.*, **72B**, pp. 37–42.
[37] De Colli, M. *et al.* (2012). A biomimetic porous hydrogel of gelatin and glycosaminoglycans cross-linked with transglutaminase and its application in the culture of hepatocytes. *Biomed. Mater.*, **7**(5), p. 055005.
[38] Franke, K. *et al.* (2013). Prostate carcinoma cell growth-inhibiting hydrogel supports axonal regeneration *in vitro*. *Neurosci. Lett.*, **541** pp. 248–252.
[39] Jeng, L., Olsen, B.R., and Spector, M. (2012). Engineering endostatin-expressing cartilaginous constructs using injectable biopolymer hydrogels. *Acta Biomater.*, **8**(6), pp. 2203–2212.
[40] Davis, N.E. *et al.* (2010). Modular enzymatically crosslinked protein polymer hydrogels for *in situ* gelation. *Biomaterials*, **31**(28), pp. 7288–7297.
[41] Guo, J., Zhang, Y., and Yang, X.-Q. (2012). A novel enzyme cross-linked gelation method for preparing food globular protein-based transparent hydrogel. *Food Hydrocoll.*, **26**(1), pp. 277–285.
[42] Song, F. *et al.* (2010). Novel casein hydrogels: formation, structure and controlled drug release. *Colloids Surf. B Biointerfaces*, **79**(1), pp. 142–148.
[43] Ehrbar, M. *et al.* (2007). Biomolecular hydrogels formed and degraded via site-specific enzymatic reactions. *Biomacromolecules*, **8**(10), pp. 3000–3007.

[44] Ehrbar, M. *et al.* (2007). Enzymatic formation of modular cell-instructive fibrin analogs for tissue engineering. *Biomaterials*, **28**(26), pp. 3856–3866.
[45] Hu, B.-H. and Messersmith, P.B. (2005). Enzymatically cross-linked hydrogels and their adhesive strength to biosurfaces. *Orthod. Craniofac. Res.*, **8**, pp. 145–149.
[46] Klebanoff, S.J. *et al.* (2013). Myeloperoxidase: a front-line defender against phagocytosed microorganisms. *J. Leukoc. Biol.*, **93**(2), pp. 185–198.
[47] Jong, E.C., Henderson, W.R., and Klebanoff, S.J. (1980). Bactericidal activity of eosinophil peroxidase. *J. Immunol.*, **124**(3), pp. 1378–1382.
[48] Furtmüller, P.G. *et al.* (2006). Active site structure and catalytic mechanisms of human peroxidases. *Arch. Biochem. Biophys.*, **445**(2), pp. 199–213.
[49] Zakharova, G.S., Uporov, I.V., and Tishkov, V.I. (2011). Horseradish peroxidase: modulation of properties by chemical modification of protein and heme. *Biochemistry*, **76**(13), pp. 1391–4401.
[50] Veitch, N.C. (2004). Horseradish peroxidase: a modern view of a classic enzyme. *Phytochem.*, **65**(3), pp. 249–259.
[51] Sakai, S., Ogushi, Y., and Kawakami, K. (2009). Enzymatically crosslinked carboxymethylcellulose-tyramine conjugate hydrogel: cellular adhesiveness and feasibility for cell sheet technology. *Acta Biomater.*, **5**(2), pp. 554–559.
[52] Sakai, S. *et al.* (2010). *In situ* simultaneous protein-polysaccharide bioconjugation and hydrogelation using horseradish peroxidase. *Biomacromolecules*, **11**(5), pp. 1370–1375.
[53] Park, K.M. *et al.* (2011). *In situ* cross-linkable gelatin-poly(ethylene glycol)-tyramine hydrogel via enzyme-mediated reaction for tissue regenerative medicine. *J. Mater. Chem.*, **21**, pp. 13180–13187.
[54] Lee, B.P., Dalsin, J.L., and Messersmith, P.B. (2002). Synthesis and gelation of DOPA-modified poly(ethylene glycol) hydrogels. *Biomacromolecules*, **3**(5), pp. 1038–1047.
[55] Wang, L. S., Chung, J. E., and Kurisawa, M. (2012). Controlling fibroblast proliferation with dimensionality-specific response by stiffness of injectable gelatin hydrogels. *J. Biomater. Sci. Polym. Ed.*, **23**, pp. 1793–1806.
[56] Wang, L.S. *et al.* (2014). Modulation of chondrocyte functions and stiffness-dependent cartilage repair using an injectable enzymatically crosslinked crosslinked hydrogel with tunable mechanical properties. *Biomaterials* **35**(7), pp. 2207–2217.
[57] Wang, L.S. *et al.* (2012). Enzymatically cross-linked gelatin-phenol hydrogels with a broader stiffness range for osteogenic differentiation of human mesenchymal stem cells. *Acta Biomater.*, **8**(5), pp. 1826–1837.
[58] Sakai, S. *et al.* (2014). Horseradish peroxidase-mediated encapsulation of mammalian cells in hydrogel particles by dropping. *J. Microencap.*, **31**(1), pp. 100–104.
[59] Park, K.M. *et al.* (2012). *In situ* SVVYGLR peptide conjugation into injectable gelatin-poly(ethylene glycol)-tyramine hydrogel via enzyme-mediated reaction for enhancement of endothelial cell activity and neo-vascularization. *Bioconj. Chem.*, **23**(10), pp. 2042–2050.

[60] Amini, A.A. *et al.* (2014). Enzymatically cross-linked bovine lactoferrin as injectable hydrogel for cell delivery. *Tissue Eng. Part A*, **20**(21–22), pp. 2830–2839.
[61] Sakai, S. *et al.* (2012). Rapidly serum-degradable hydrogel templating fabrication of spherical tissues and curved tubular structures. *Biotechnol. Bioeng.*, **109**(11), pp. 2911–2919.
[62] Jin, R. *et al.* (2010). Enzymatically-crosslinked injectable hydrogels based on biomimetic dextran–hyaluronic acid conjugates for cartilage tissue engineering. *Biomaterials*, **31**(11), pp. 3103–3113.
[63] Jin, R. *et al.* (2011). Chondrogenesis in injectable enzymatically crosslinked heparin/dextran hydrogels. *J. Control. Release*, **152**(1), pp. 186–195.
[64] Tavakol, M. *et al.* (2014). Synthesis and characterization of an enzyme mediated *in situ* forming hydrogel based on gum tragacanth for biomedical applications. *Iran J. Biotech.*, **12**(1), p. e15811.
[65] Lih, E. *et al.* (2012). Rapidly curable chitosan–PEG hydrogels as tissue adhesives for hemostasis and wound healing. *Acta Biomater.*, **8**(9), pp. 3261–3269.
[66] Kim, K.S. *et al.* (2011). Injectable hyaluronic acid-tyramine hydrogels for the treatment of rheumatoid arthritis. *Acta Biomater.*, **7**(2), pp. 666–674.
[67] Kurisawa, M. *et al.* (2005). Injectable biodegradable hydrogels composed of hyaluronic acid–tyramine conjugates for drug delivery and tissue engineering. *Chem. Commun.*, pp. 4312–4314.
[68] Wang, L.S. *et al.* (2014). Enzymatic conjugation of a bioactive peptide into an injectable hyaluronic acid-tyramine hydrogel system to promote the formation of formation of functional vasculature. *Acta. Biomater.*, **10**(6), pp. 2539–2550.
[69] Ganesh, N. *et al.* (2013). Enzymatically cross-linked alginic-hyaluronic acid composite hydrogels as cell delivery vehicles. *Int. J. Biol. Macromol.*, **55**, pp. 289–294.
[70] Chang, T.S. (2009). An updated review of tyrosinase inhibitors. *Int. J. Mol. Sci.*, **10**(6), pp. 2440–2475.
[71] Boyle, J.L. *et al.* (2002). Tyrosinase expression in malignant melanoma, desmoplastic melanoma, and peripheral nerve tumors. *Arch. Pathol. Lab. Med.*, **126**(7), pp. 816–822.
[72] Dordic, M. *et al.* (2012). Immunity to melanin and to tyrosinase in melanoma patients, and in people with vitiligo. *BMC Complement. Altern. Med.*, **12**, p. 109.
[73] Sanz, V.C. *et al.* (2005). Development of a tyrosinase biosensor based on gold nanoparticles-modified glassy carbon electrodes: Application to the measurement of a bioelectrochemical polyphenols index in wines. *Anal. Chim. Acta*, **528**(1), pp. 1–8.
[74] Zhao, J., Wu, D., and Zhi, J. (2009). A novel tyrosinase biosensor based on biofunctional ZnO nanorod microarrays on the nanocrystalline diamond electrode for detection of phenolic compounds. *Bioelectrochemistry*, **75**(1), pp. 44–49.
[75] Seo, S.Y., Sharma, V.K., and Sharma, N. (2003). Mushroom tyrosinase: recent prospects. *J. Agric. Food Chem.*, **51**(10), pp. 2837–2853.

[76] De, A., Mandal, S., and Mukherjee, R. (2008). Modeling tyrosinase activity. Effect of ligand topology on aromatic ring hydroxylation: an overview. *J. Inorg. Biochem.*, **102**(5–6), pp. 1170–1189.
[77] Mattinen, M.-L. *et al.* (2005). Laccase-catalyzed polymerization of tyrosine-containing peptides. *FEBS J.*, **272**(14), pp. 3640–3650.
[78] Selinheimo, E. *et al.* (2007). Elucidating the mechanism of laccase and tyrosinase in wheat bread making. *J. Agric. Food Chem.*, **55**(15), pp. 6357–6365.
[79] Jus, S. *et al.* (2012). Enzymatic cross-linking of gelatine with laccase and tyrosinase. *Biocatal. Biotransform.*, **30**(1), pp. 86–95.
[80] Choi, Y.C. *et al.* (2014). Human gelatin tissue-adhesive hydrogels prepared by enzyme-mediated biosynthesis of DOPA and $Fe^{3+}$ ion crosslinking. *J. Mater. Chem.*, **2**, pp. 201–209.
[81] Jin, R., Lou, B., and Lin, C. (2013). Tyrosinase-mediated *in situ* forming hydrogels from biodegradable chondroitin sulfate–tyramine conjugates. *Polym. Int.*, **62**(3), pp. 353–361.
[82] Kang, G.D. *et al.* (2004). Silk fibroin/chitosan conjugate crosslinked by tyrosinase. *Macromol. Res.*, **12**(5), pp. 534–539.
[83] Jus, S. *et al.* (2011). Cross-linking of collagen with laccases and tyrosinases. *Mater. Sci. Eng. C*, **31**(5), pp. 1068–1077.
[84] Rocasalbas, G. *et al.* (2013). Laccase-assisted formation of bioactive chitosan/gelatin hydrogel stabilized with plant polyphenols. *Carbohydr. Polym.*, **92**(2), pp. 989–996.
[85] Steffensen, C.L. *et al.* (2008). Cross-linking proteins by laccase-catalyzed oxidation: importance relative to other modifications. *J. Agric. Food Chem.*, **56**(24), pp. 12002–12010.
[86] Maki, J.M. *et al.* (2005). Lysyl oxidase is essential for normal development and function of the respiratory system and for the integrity of elastic and collagen collagen fibers in various tissues. *Am. J. Pathol.*, **167**(4), pp. 927–936.
[87] Maki, J.M. *et al.* (2002). Inactivation of the lysyl oxidase gene Lox leads to aortic aneurysms, cardiovascular dysfunction, and perinatal death in mice. *Circulation*, **106**(19), pp. 2503–2509.
[88] Lucero, H.A. and Kagan, H.M. (2006). Lysyl oxidase: an oxidative enzyme and effector of cell function. *Cell. Mol. Life Sci.*, **63**(19–20), pp. 2304–2316.
[89] Elbjeirami, W.M. *et al.* (2003). Enhancing mechanical properties of tissue-engineered constructs via lysyl oxidase crosslinking activity. *J. Biomed. Mater. Res A*, **66**(3), pp. 513–521.
[90] Lau, Y.K., Gobin, A.M., and West, J.L. (2006). Overexpression of lysyl oxidase to increase matrix crosslinking and improve tissue strength in dermal wound healing. *Ann. Biomed. Eng.*, **34**(8), pp. 1239–1246.
[91] Mithieux, S.M. *et al.* (2005). A model two-component system for studying the architecture of elastin assembly *in vitro*. *J. Struct. Biol.*, **149**(3), pp. 282–289.
[92] Bakota, E.L. *et al.* (2011). Enzymatic cross-linking of a nanofibrous hydrogel. *Biomacromolecules*, **12**(1), pp. 82–87.

[93] Vroon, D.H. and Israili, Z. (1990). Alkaline phosphatase and gamma glutamyltransferase, in H.K. Walker, W.D. Hall, and J.W. Hurst (eds.), *Clinical Methods: the history, physical, and laboratory examinations*, 3rd Edn. Butterworths, Boston.
[94] Bull, H. *et al.* (2002). Acid phosphatases. *Mol. Pathol.*, **55**(2), pp. 65–72.
[95] Yang, Z. *et al.* (2004). Enzymatic formation of supramolecular hydrogels. hydrogels. *Adv. Mater.*, **16**(16), pp. 1440–1444.
[96] Yang, Z. *et al.* (2006). Using a kinase/phosphatase switch to regulate a supramolecular hydrogel and forming the supramolecular hydrogel *in vivo. J. Am. Chem. Soc.*, **128**(9), pp. 3038–3043.
[97] Yang, Z., Ma, M., and Xu, B. (2009). Using matrix metalloprotease-9 (MMP-9) to trigger supramolecular hydrogelation. *Soft Matter*, **5**, pp. 2546–2548.
[98] Koda, D. *et al.* (2010). Proteinase-mediated drastic morphological change of peptide–amphiphile to induce supramolecular hydrogelation. *Chem. Commun.*, **46**, pp. 979–981.
[99] Williams, R.J. *et al.* (2009). Enzyme-assisted self-assembly under thermodynamic control. *Nat. Nanotechnol.*, **4**(1), pp. 19–24.
[100] Szkolar, L. *et al.* (2014). Enzymatically triggered peptide hydrogels for 3D cell encapsulation and culture. *J. Peptide Sci.*, **20**(7), pp. 578–584.
[101] Guilbaud, J.B. *et al.* (2013). Effect of enzyme concentration of the morphology and properties of enzymatically triggered peptide hydrogels. *Biomacromolecules*, **14**(5), pp. 1403–1411.
[102] Guilbaud, J.B. *et al.* (2010). Enzymatic catalyzed synthesis and triggered gelation of ionic peptides. *Langmuir*, **26**(13), pp. 11297–11303.
[103] Um, S.H. *et al.* (2006). Enzyme-catalysed assembly of DNA hydrogel. *Nat. Mater.*, **5**(10), pp. 797–801.
[104] Park, N. *et al.* (2009). A cell-free protein-producing gel. *Nat. Mater.*, **8**(5), pp. 432–437.
[105] Lee, J.B. *et al.* (2012). A mechanical metamaterial made from a DNA hydrogel. *Nat. Nanotechnol.*, **7**(12), pp. 816–820.
[106] Buβkamp, H. *et al.* (2014). A new building block for DNA network formation by self-assembly and polymerase chain reaction. *Beilstein J. Org. Chem.*, **10**, pp. 1037–1046.
[107] Qi, H. *et al.* (2013). DNA-directed self-assembly of shape-controlled hydrogels. *Nat. Commun.*, **4**, p. 2275.

# Chapter 6

# Injectable Hydrogels as Tissue Adhesives

N. Oliva
*Institute for Medical Engineering and Science, Massachusetts Institute of Technology, Cambridge, MA*

N. Artzi*
*Institute for Medical Engineering and Science, Massachusetts Institute of Technology, Cambridge, MA*
*Department of Medicine, Brigham and Women's Hospital Harvard Medical School, Cambridge, MA*

## 6.1 Introduction

About 3,500 years ago, Romans and Greeks described adhesives made of a variety of materials such as egg whites, blood, or milk. However, it was not until 1845 that a surgeon used a rubber adhesive applied to strips of fabric: the invention of the surgical tape. Since then, a number of adhesive products have been approved and commercialized, including a variety of tapes and other adhesives such as cyanoacrylates. They represent a dramatic improvement over existing techniques, especially with regard to healing times and biocompatibility. However, there is still room for improvement: most medical adhesives need to be used in combination with sutures or tapes and their function is limited in most cases to pure mechanics aiming to keep both parts of the wound together. Seizing the opportunity for improvement in the field of adhesive technology, from the development of

*Author to whom correspondence should be addressed. Email: nartzi@mit.edu.

new formulations that help improve healing, to release of anti-inflammatory drugs, anti-fungal treatments, or anesthetics, will dramatically improve patients' clinical outcome and quality of life.

## 6.2 Adhesives

An adhesive may be defined as a material which, when applied to two distinct surfaces, brings them together and resists separation [1]. Adhesives can be classified in a number of ways, the most used being by origin of raw adhesive components and by method of adhesion.

Adhesives can be made from synthetic or natural sources. Among the synthetic materials, poly(ethylene glycol) (PEG) is the most widely used as a backbone polymer, together with others such as poly(vinyl alcohol) and poly(lactic acid). Synthetic hydrogels can be engineered and formed reproducibly at a relative low cost and with tunable properties. Natural hydrogels such as collagen, hyaluronic acid, or chondroitin sulfate (CS) can provide high biocompatibility and promote healing, but are often more expensive and difficult to manufacture.

Adhesion type is categorized as reactive or non-reactive, based on whether it hardens by chemical reaction or not. Examples of non-reactive adhesives include pressure-sensitive adhesives (PSA), contact adhesives, and hot adhesives. PSAs are unique in that they form a strong bond with only finger pressure. The adhesion mechanism behind them relies on the adhesive being soft enough to wet-out the adherent surface and form bonds that can be weak, such as van der Waals bonds, or strong, such as ionic or hydrogen bonds. These adhesives are designed in a way that balances flow to wet-out the surface, and resistance to flow to counteract shear stress [2]. A contact adhesive is capable of adhering to itself, bringing together any two surfaces it is attached to. This type of adhesive is applied to both surfaces and allowed to harden, after which they are brought together to impart adhesion. The mechanism behind it is composed of a number of effects, which are not well understood. By infrared spectroscopy, Zheleva found the existence of physical interactions, chemical grafting of polymers, and chemical interactions between the components in the adhesive [3]. Hot adhesives are thermoplastic materials characterized by an increase in cohesive strength as the temperature decreases [4]. These materials provide adhesion by solidification mechanisms.

Reactive adhesives, meanwhile, are those whose mechanisms of adhesion involve chemical reactions between their components. They can be classified as single-component or multicomponent depending on the number of parts involved in the reaction. Single-component adhesives, such as acrylate-based adhesives, harden by chemical reaction upon external stimuli including humidity, exposure to oxygen, heat, UV light or salt. Multicomponent adhesives can also react upon the addition of a catalyst or following a spontaneous reaction between the functional groups in each component upon mixing.

## 6.3 Tissue Adhesives for Medical Applications

Surgical sealants are often used in conjunction with sutures to seal incisions. Sutures are point closure devices that do not provide a liquid- or airtight seal, especially critical in disease states that impart compromised performance or following imperfect tissue apposition for suturing. Application of an adhesive sealant on top of the suture line prevents this passage through the sutures.

Mehdizadeh and Yang defined tissue adhesives and sealants as a "group of liquid or semi-liquid compounds that can be applied to a tissue incision for the purpose of closing wounds, adhering to soft tissues, and for hemostasis" [5]. They also outlined the ideal properties a tissue adhesive should have, including the following:

- Low toxicity, ease of preparation and sterilizability;
- Low viscosity prior to application and rapid solidification on the target tissue;
- Maintaining tissue adhesion post-application;
- Improve hemostasis, healing, and regeneration;
- Degradability at a rate suitable for the application;
- Non-toxic degradation by-products; and,
- Cost-effective.

### 6.3.1 *Commercially available surgical adhesives*

Cyanoacrylates are known for their high adhesion strength. Their nitrile and ester groups assist in forming a strong adhesive bond, but degrade into histotoxic substances, most notably formaldehyde. As such, cyanoacrylates are predominately used to close external incisions or wounds, with less prominent internal use. A primary method employed to allow internal use by minimizing toxicity is to slow degradation, providing the body with more time to process the toxins before they increase [6]. As with many external adhesives, cyanoacrylates can be formulated into a PSA [7] and thus comply with many of the restrictions stated for dermal adhesives.

Fibrin glues (such as Tisseel and Crosseal) are considered to be the natural counterpart of cyanoacrylates for their ability to induce hemostasis and seal tissues. Made out of thrombin and fibrinogen, fibrin glues are highly biocompatible and have minimal toxicity, albeit with limited adhesion strength. Although fibrin sealants have only been commercially available since 1998, their utility as a hemostatic agent was first discovered in 1909 by Bergel [8]. Due to its origin, many growth factors bind to fibrin, including vascular endothelial growth factors (VEGF), transforming growth factor-$\beta$1 (TGF-$\beta$1), and basic fibroblastic growth factor (bFGF) [9–11]. Currently there are a few on-label uses that include hemostasis, colonic sealing, and skin graft attachment. Fibrin glue has also been shown to serve as a scaffold for drug and gene delivery, increasing its therapeutic value [12–17].

The limitations of fibrin glue and cyanoacrylates have prompted much research into finding alternatives; other sealants such as collagen-based sealants and natural and synthetic hydrogels have been formulated, each with their own unique traits.

CoSeal is a commercially available, rapidly gelling tissue sealant made of tetra-succinimidyl PEG and tetra-thiol PEG [18, 19]. It is mainly used in cardiac surgery sealing, such as aortic root and aortic arch repair, aortic reconstruction, coronary artery bypass graft (CABG), or left ventricular assist device (LVAD) implantation. While having suitable hemostatic and sealing properties, CoSeal is a brittle, non-elastic sealant with relatively low tensile strength.

BioGlue is a surgical adhesive also used in cardiac surgery composed of bovine serum albumin and glutaraldehyde. This bioadhesive can crosslink through amine-aldehyde chemistry and adhere both to the tissue through chemical bonds and to the graft through mechanical bonding within the interstices of the graft matrix. Even though it has high adhesion strength, studies have shown significant toxicity associated with the release of glutaraldehyde from the tissue adhesive [20].

Azadani *et al.* compared a number of commercially available sealants used in cardiac surgery (CoSeal, Crosseal, Tisseel, and BioGlue) in terms of elasticity, hemostatic efficacy, biodegradability, and associated complications [21]. BioGlue showed outstanding elasticity and hemostatic properties compared to its counterparts, but its high stiffness when applied to aortic anastomoses caused aortic mobility restrictions that led to stenosis [22].

Even though all currently available medical adhesives have some benefits (adhesion, hemostatic efficacy, low toxicity, etc.), there is still not one that meets and satisfies all the requirements for an optimal tissue surgical adhesive. This is the reason and the motivation that has led a number of groups in the biomaterial research field to develop improved medical adhesives in the last decade.

### 6.3.2 *Classification of medical adhesives based on their properties*

Adhesives for medical applications can be either classified by their degradability, mode of application and type of adhesive–tissue interactions (the latter is developed extensively in the next section on hydrogels).

#### 6.3.2.1 *Degradability*

Biodegradable or resorbable polymers have gained a lot of attention in recent years as these materials can be degraded by their host [23] and be eliminated after performing their action, thus reducing the risk of chronic inflammation. This classification can be applied to any biomaterial, regardless of it being adhesive. However, adhesive material is a great example of a medical device that is critical to have and yet is only temporarily required, and hence it is emphasized in this section.

Table 6.1 summarizes the most widely used biodegradable polymers, classified as synthetic and natural polymers, and their mechanism of degradation [24]. These mechanisms need to be studied thoroughly for each system and application to enable predictive degradation rate in light of the specific environment. For instance, a polymer that is enzymatically cleaved by a particular protein present in a specific disease may be suitable for the triggered release of a drug to combat such disease, while a non-resorbable material is adequate for long-term uses, such as hip or knee replacements. This table highlights a variety of degradation mechanisms: hydrolysis, enzymatic, microbial thermo-oxidation, photo-oxidation, and disintegration. The most important and widely used in adhesive development for medical applications are hydrolysis and enzymatically degradable polymers.

Table 6.1 Summary of resorbable polymers and their degradation mechanism. With permission from "An overview of Degradable Polymers; Degradable polymers and materials: Principles and Practice", 2nd Edition" by Cheng, H.N. (2012). American Chemical Society.

| **Type** | **Examples** | **Mechanism(s)** |
|---|---|---|
| 1. *Synthetic polymers* | | |
| a. Condensation polymers | Polyesters, polyamides, polyureas, polyurethanes, polyanhydrides | Hydrolysis, microbial or enzymatic action |
| b. Addition polymers | Addition of pro-oxidant or photo-sensitizer. Use of biodegradable or sensitized comonomers | Thermo-oxidation, photo-oxidation |
| c. Water-soluble polymers (WSP) | Poly(vinyl alcohol), PEG | Microbial action |
| d. Blends (with natural polymers or WSP) | Starch-based blends, PLA-based blends | Disintegration, case of oxidation and degradation |
| 2. *Natural polymers (biopolymers or agro-polymers)* | | |
| a. Polysaccharides | Starch (& its derivatives), cellulose (& its derivatives), chitin, alginate, hyaluronic acid | |
| b. Proteins | Soy protein, zein, casein, wheat gluten, gelatin. Also derivatives | Hydrolysis, microbial or enzymatic action |
| c. Lipics | Tryglycerides (& its derivatives) | |
| d. Semi-natural polymers | (i) Agro-monomers, e.g., lactic acid, 1,3-propanediol (ii) Microbial polymers, e.g., PHA | |
| e. Blends of natural polymers | Starch-PLA, starch-PHA, cellulosic esters-PHA | |

Hydrolytic degradation involves the breakdown of the adhesive network by the action of water. Some examples include networks formed of polyesters, polyanhydrides, and some polysaccharides, among others. The degradation rate in these cases is dictated by the bond half-life in aqueous medium, which varies from 0.1 hours for polyanhydrides to over 3 years in the case of polyesters [25]. However, these values may be altered by environmental factors, such as pH or the water uptake capability of the adhesive (the latter is critical to consider when designing swellable adhesive hydrogels).

Enzymatically degradable adhesives are those whose components are susceptible to being cleaved by enzymes, such as polyamides, cellulose, collagen, or hyaluronic acid-based materials. This type of degradation mechanism is enzyme concentration-dependent and hence the material will degrade at a rate determined by its local environment. As mentioned, high enzymatic activity due to a pathological condition can be used to trigger degradation of the adhesive and thus release of encapsulated drugs or factors in the material.

#### 6.3.2.2 *Mode of application*

Adhesive materials can be classified based on their mode of application as injectable or non-injectable materials. In the former, a single-component or multicomponent adhesive in a liquid state is cured *in situ* upon injection. Non-injectable materials can be divided into implants or tapes. In implants, the bulk of the material is pre-cured *a priori* and is then surgically implanted. Tapes also consist of pre-cured adhesive materials on a plastic, woven, or other non-degradable material platform that aids the application and protects the adhesive. Tapes are typically used for external applications such as skin wounds, while implantable materials are most commonly used in surgical applications.

When used surgically, injectable materials present the advantage of requiring minimally invasive procedures for their implantation, which can be done through a syringe needle or catheter into a cavity [26], unlike implantable materials that require open surgery, which increases tissue trauma, risk of infection, and overall morbidity [27]. As the curing process of injectable materials takes place directly on the surface of the target organ, the liquid components allow for an intimate interface with more interactions, which in turn improves the adhesive and mechanical properties. This is critical for uneven or curved surfaces, which are present in most organs and are accentuated in pathological states such as inflammation or neoplasia. However, injectable materials pose certain challenges in their design. First, there is a limited range of chemical reactions that can take place spontaneously and in a reasonable time frame in physiological conditions, and these reactions should not harm or modify the tissue to which they adhere (tissue surface chemistry will be analyzed in detail later in the section on hydrogel adhesives). Another challenge for *in situ* material polymerization is the toxicity of the monomers as they are in direct contact with the cells and can be uptaken or cause tissue crosslinking similar to the process observed in tissue fixation using formaldehyde or glutaraldehyde. An alternative approach to reducing toxicity is to use macromolecules with distant surface functional groups as reactive units instead of small monomeric units to form the adhesive. These macromolecules are less likely to be uptaken by cells or penetrate as deeply to the tissue due to their size, and the right distribution of chemical groups throughout their chains would prevent tight tissue crosslinking [28].

Injectable adhesives can also be classified based on whether they are hydrophobic or hydrophilic. Hydrophobic adhesives, such as the glue developed by Lang *et al.*, are water-resistant adhesives that act as a barrier [29]. Hydrophilic or hydrogel adhesives present the advantage of mimicking biological tissues and being extremely biocompatible. However, elasticity and barrier function tend to be compromised with respect to hydrophobic adhesives.

## 6.4 Current View on Adhesive Hydrogels

Hydrogels are three-dimensional, hydrophilic, polymeric networks that can store large amounts of water or biological fluids, a property that makes them highly biocompatible [30–32]. These materials resemble natural tissues due to the large amounts of structural water and their soft consistency [31, 32]. They also have the ability to swell (absorb and retain water in their structure). The swelling ratio is determined by their initial degree of crosslinking, which is controlled by material formulation, and can be modified on demand as a response to certain stimuli, such as temperature [33, 34], pH [35–39], light [40, 41], magnetic field [42, 43], electric current [44, 45], or antigens [40, 46]. Control over crosslinking density defines the mesh size, and hence the rate of passive diffusion, which are key parameters in developing a controlled drug delivery system [47].

A large number of hydrogels conjugated to cell adhesive molecules have been developed in the field of tissue engineering in recent years [48–52]. Their purpose was mainly to serve as biomimetic materials for three-dimensional cell growth. The use of hydrogels as tissue adhesives rather than cell adhesives is not as widely reported, but is rapidly emerging. This section focuses on novel strategies for tissue adhesives, the underlying mechanisms of adhesion, and their applications in the medical field.

### 6.4.1 *Classification by chemical composition (cohesion and adhesion chemistry)*

When designing adhesives for medical applications, the tissue–material interactions must be considered. These interactions can have a mechanical or chemical character.

#### 6.4.1.1 *Mechanical interlocking*

The mechanical interlocking mechanism of adhesion is based on the infiltration of the adhesive material into the pores and crevices of the adhered surface, such as bone [5]. A macroscopic example that illustrates mechanical interlocking and helps visualize the mechanism is Velcro; the hooks latch onto the loops and physically hold the two materials together like an adhesive would. In the medical field, amalgam tooth fillings are believed to adhere to teeth by means of mechanical interlocking. However, adhesion of these materials to smooth surfaces has challenged the actual role of mechanical interlocking mechanism on adhesion.

Recently, bio-inspired adhesives based on mechanical interlocking have been developed. Pang *et al.* utilized the wing locking mechanism of beetles to synthesize tilted micro- and nanohair arrays capable of interlocking, generating shear adhesions of about $60\,\mathrm{N/cm^2}$ [53]. Similarly, inspired by the endoparasite *Pomphorhynchus laevis*, Yang *et al.* developed a swellable microneedle system that adheres to soft tissues by interlocking the tips with minimal tissue penetration, thus reducing tissue damage usually associated with sutures or staples. The tips then swell in the dermis of the adherent tissue, providing the interlocking adhesion [54].

#### 6.4.1.2 *Chemical interactions*

Chemical reactions usually provide stronger adhesion strength as the materials bind to tissue surfaces. Primary bonds (covalent, ionic, or metallic), unlike secondary bonds (hydrogen bonding or van der Waals), form strong connections between atoms, making the adhesive less dependent on assisting mechanisms. This strong bond can also potentially damage surrounding tissue, necessitating careful choice of basic material components and reactive groups. As tissue–material interactions must be controlled, simply providing an excess number of reactive groups to react with tissue functional groups may impart toxicity. Hence, material formulation must be titrated to tissue type and state. It is possible to consume internally excess groups by a second material component, thus increasing cohesion while eliminating toxicity [55].

Secondary bonds such as van der Waals forces and hydrogen bonding can be sufficient to establish adequate adhesion in some applications despite their weak intermolecular force. Increasing material surface area for interaction with tissue surfaces may facilitate stronger adhesion. A common example of van der Waals forces as the primary mechanism is the gecko. This lizard is known to use its thousands of hair-like setae and spatulas to increase surface area contact in lieu of surface tension, and in this way is able to potentially support 130 kg. Attempts to replicate this form of dry adhesion have proven quite difficult due to the small size of the setae and spatulas, 5 $\mu$m and 0.2 $\mu$m in length, respectively [56]. The gecko, however, has inspired much research, which has been successful on a larger scale and is moving toward medical use [57].

Table 6.2 provides a summary of current approaches to the formation of adhesive hydrogels based on internal crosslinking chemistry and tissue adhesive chemistry. The number of chemical reactions in the formation of adhesive hydrogels is limited to basically five types of chemistry: Schiff base reaction, amide formation, cathecol coupling, thiol-maleimide addition, and free-radical reaction. These groups are most commonly used since the formation of adhesive hydrogel requires hydrophilic building blocks that are capable of reacting spontaneously at physiological pH to form a hydrogel. Moreover, the material must contain chemical groups capable of reacting with the tissues to provide adhesion. These restraints greatly limit the repertoire of chemical reactions available to engineer such materials.

#### 6.4.1.2.1 Schiff base formation

A Schiff base is the imine product of the condensation of an amine and an aldehyde, with the release of a molecule of water (Reaction 6.1). This bond is labile and can be cleaved in aqueous solution by hydrolysis [78].

[Reaction 6.1]

Table 6.2 Summary of current adhesive hydrogel approaches and their classification based on the components forming the material, the internal chemistry to confer cohesion, and the tissue chemistry to achieve adhesion.

| **Component 1** | **Component 2** | **Internal chemistry** | **Tissue chemistry** | **Refs.** |
|---|---|---|---|---|
| Bovine serum albumin | Glutaraldehyde | Schiff base | Schiff base | [20] |
| Oxidized dextran | Polylysine | Schiff base | Schiff base | [58–60] |
| Oxidized dextran | PEG-amine | Schiff base | Schiff base | [28, 61–63] |
| Oxidized dextran | PAMAM Dendrimer | Schiff base | Schiff base | [55] |
| CS-NHS | Bone marrow | Amide formation | Amide formation | [64, 65] |
| CS-NHS | Bone marrow | Amide formation | Amide formation | [66] |
| HA-NHS | Whole blood | Amide formation | Amide formation | [67] |
| PPO-PEO-cathecol copolymer | | Catechol coupling | $\pi-\pi$ interactions | [68] |
| PPO-PEO-cathecol copolymer | | Cathecol coupling | Cathecol reaction | [69] |
| PEG-Ala-Ala-DOPA | | Cathecol coupling | $\pi-\pi$ interactions | [70] |
| MAPs-DOPA | | Cathecol coupling | Cathecol/$\pi-\pi$ reaction | [71] |
| EPL-PEG-MAL | CSS | Thiol-maleimide addition | Disulfide bonds | [72] |
| Tetra-succinimidyl PEG | Tetra-thiol PEG | Thiol-succinimidyl addition | Amide formation | [18, 19] |
| CS-aldehyde-metacrylate | PEG-acrylate | Free radical | Schiff base | [73, 74] |
| N-isopropylacrylamide | CS | Free radical | Free radical | [75] |
| PEG-diacrylate | | Free radical | Free radical | [76] |
| Poly(AAc-co-MBA) | | Free radical | Free radical | [77] |

A number of groups have taken advantage of this reaction mechanism to form adhesive hydrogels in which the bulk of the material is formed by a polymeric network with Schiff bases at the crosslinking points [28, 55, 58, 59, 61–63, 67]. The adhesion relies on excess aldehydes reacting with amines on the surface of tissues, provided mainly by proteins forming

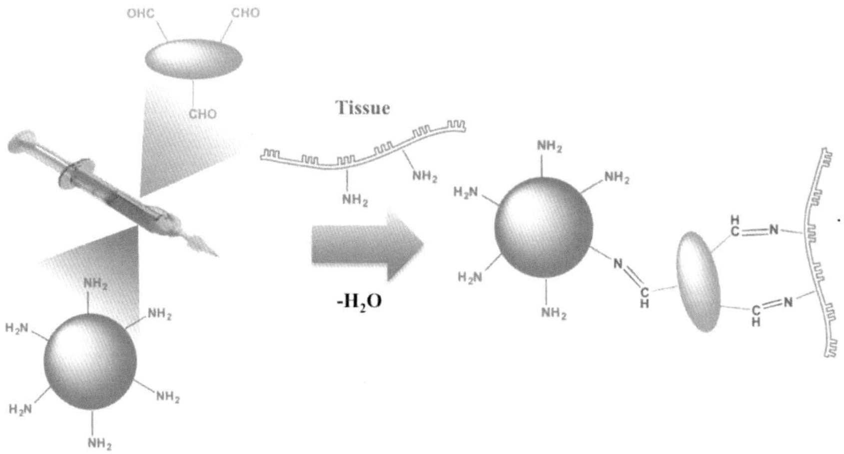


Figure 6.1 Schematic representation of the adhesion and cohesion mechanisms underlying amine-aldehyde adhesive hydrogels. Reprinted from [55] with permission from the American Chemical Society.

the extracellular matrix around the cells. Figure 6.1 shows the general mechanism of adhesion–cohesion in this type of materials.

Aldehydes provided by the first material component react both with tissue amines to impart adhesion and with amines provided by the second material component to impart cohesion. Hence, the mechanical and adhesive properties of the material can be tuned simultaneously by altering the aldehyde to amine ratio. This ratio can be modified by either altering the overall solid content of the aldehyde and amine-bearing polymers, or by changing the number of chemical groups available in each polymer. This is critical, as it may define not only the mechanical properties of the final adhesive, but also whether or not it will form a hydrogel or a stiff plastic-like polymer. For instance, when PAMAM dendrimer generation 5 (128 amines on the surface) at a solid content of 12.5% (w/v) was mixed with dextran-aldehyde (50% oxidation) at concentrations between 10 and 25% (w/v), the resulting product was a white, plastic-like product. However, if 75% of the amines were oxidized to alcohols, leaving the PAMAM dendrimer generation 5 with only 32 amines on the surface, the resulting material was a hydrogel (Figure 6.2) [55]. This proved that the reaction kinetics is critical to achieve a hydrogel structure, therefore the importance of controlling the ratios of the chemical groups.

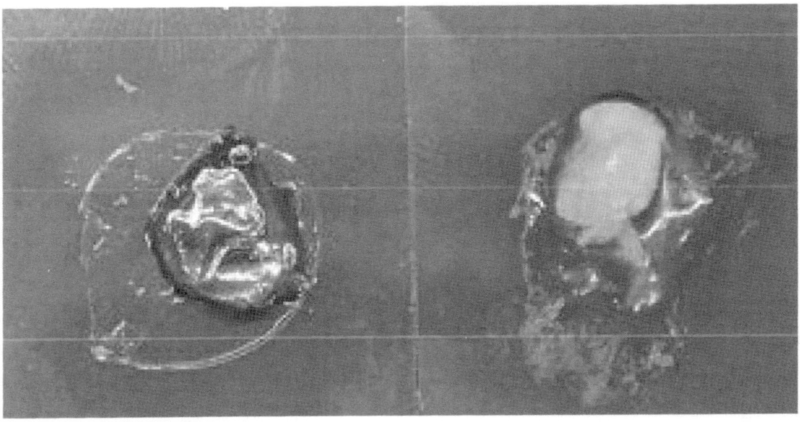

Figure 6.2 Images of dendrimer-dextran hydrogel (left) and non-hydrogel polymeric plastic (right). Reprinted from [55] with permission from the American Chemical Society.

Another important parameter is the number of amines available on the tissue, as this number differs from one tissue to another [55, 61]. This variability will also dictate adhesion and cohesion parameters given a specific hydrogel formulation. Adjusting the number of amine reactive groups for each target tissue based on tissue amine density is critical to ensuring optimal adhesion strength and performance. This will be further discussed in a later section. The aldehyde:amine ratio is further critical in determining the appropriate gelation time to be slow enough to enable proper application and fast enough so it cures on a specific target tissue without flowing to adjacent organs.

6.4.1.2.2 Amide formation

Another approach to adhesive hydrogel formation is the use of N-hydroxysuccinimide (NHS)-activated carboxylic acid-bearing components and amine-bearing components to form a stable amide bond (Reaction 6.2).

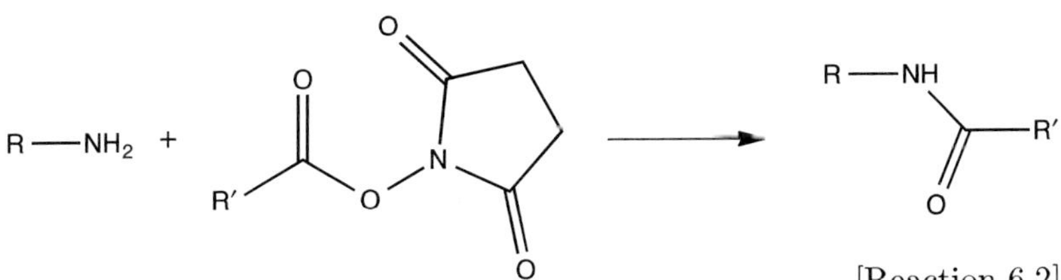


[Reaction 6.2]

NHS can react with carboxylic acids to form the activated carboxylate, which in turn can spontaneously react with amines to form the cohesive bulk of the adhesive hydrogel. Some examples of natural polymers used for this purpose are chondroitin sulfate or hyaluronic acid, which have been used in their NHS-activated form to generate adhesive hydrogels with amine-bearing natural components such as whole blood or bone marrow for tissue engineering purposes [64–66].

The adhesion of this type of hydrogel relies on the reaction of the NHS ester-bearing compounds with the amines on the surface of tissues. Following the same principles, there is a competition between the amines on the tissue and those on the amine-bearing component to react with the NHS esters, thus dictating adhesion and cohesion.

While these adhesive hydrogels are also suitable as injectable hydrogels, they have the major drawback of being unstable in aqueous media prior to gelation. The NHS-activated esters are readily cleaved in water; therefore they must be kept dry at all times and be resuspended immediately before use. Once they react with amines on the second component or on the tissue surface, the amine bond formed is highly stable, with half-life values of 83,000 years [25]. In these materials, enzymatic degradation of natural polymers such as hyaluronic acid or chondroitin sulfate accounts for their biodegradability.

#### 6.4.1.2.3 Catechol coupling

Over the past decade, biomimetic materials have gained attention in the biomedical world in an attempt to mimic and reproduce mechanisms employed by animals and plants in nature. In the area of adhesive hydrogels, mussel-inspired adhesives are without doubt the most representative example of the biomimetic approach [68–71]. By studying the composition of the proteins present in the byssal threads of mussels, a number of synthetic polymers have been exploited, such as PPO-PEO copolymers, with incorporated DOPA-like structures, mainly catechol, into the network to achieve the adhesion mechanics of mussels (Figure 6.3).

There are two types of crosslinking mechanisms behind this type of hydrogels: oxidation-induced DOPA quinone-mediated covalent bond, and $Fe^{3+}$-mediated coordinative non-covalent bond (Figure 6.4). Kim *et al.* studied both mechanisms using a natural backbone of mussel adhesive proteins (MAPs). In the first mechanism, sodium periodate is typically used as an oxidant to induce covalent crosslinking via quinone coupling. In the

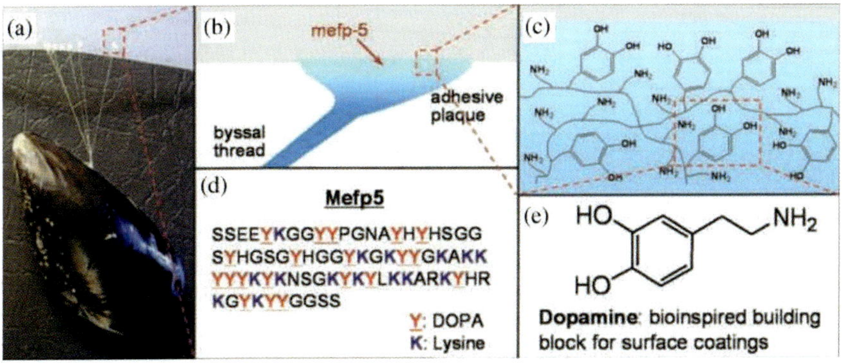


Figure 6.3 Mussel-inspired adhesive hydrogel containing dopamine as a building block to mimic the DOPA- and lysine-rich adhesive plaque of byssal threads in mussels. Reprinted from [70] with permission.

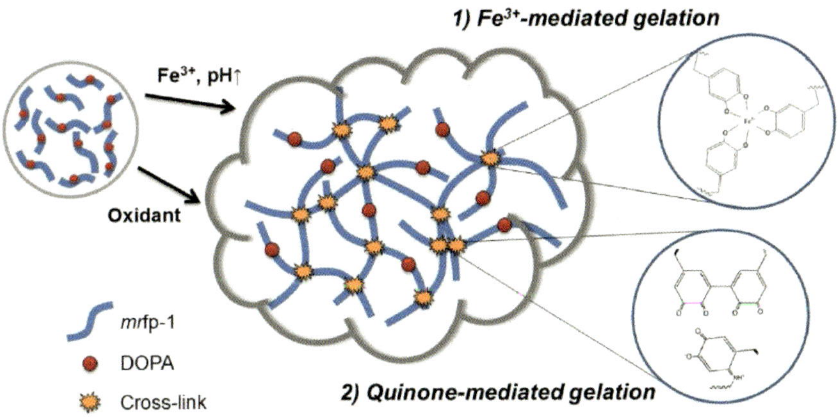


Figure 6.4 Schematic representation of the two crosslinking mechanisms of catechol-based adhesive hydrogels: $Fe^{3+}$-mediated or quinone-mediated. Reprinted from [71] with permission from the American Chemical Society.

second mechanism, the simple addition of $FeCl_3$ at high pH is sufficient to induce complexation of the DOPA with the iron cation. Given the nature of these bonds, oxidative crosslinking leads to an irreversible, covalent bond, thus with minimal swelling capability, while the coordination crosslinking renders a self-healing, swellable hydrogel [71].

Adhesion chemistry is based on the interactions of DOPA with biological tissues in two different manners — covalent and cation–$\pi$ or $\pi$–$\pi$ interactions. The covalent interaction relies on quinone-mediated crosslinking with amine or thiol groups on the surface of tissues, while $\pi$–$\pi$ are due to the electron distribution of the DOPA aromatic ring and are of weaker nature than covalent bonds. For this reason, quinone-crosslinked hydrogels have stronger adhesion to tissues than those $Fe^{3+}$-coordinated, which solely rely on $\pi$ interactions, with 200 versus 140 kPa, respectively [71].

While these adhesive hydrogels have outstanding mechanical properties and represent an excellent example of how nature can guide material design, they suffer the major drawback of having long gelation times ranging from 1 to 2 hours, precluding them from being directly injected to the wound site.

6.4.1.2.4 Maleimide-thiol addition

A less common but nonetheless interesting approach is the use of thiol-bearing components to perform a Michael addition on maleimide-bearing components (Reaction 6.3).

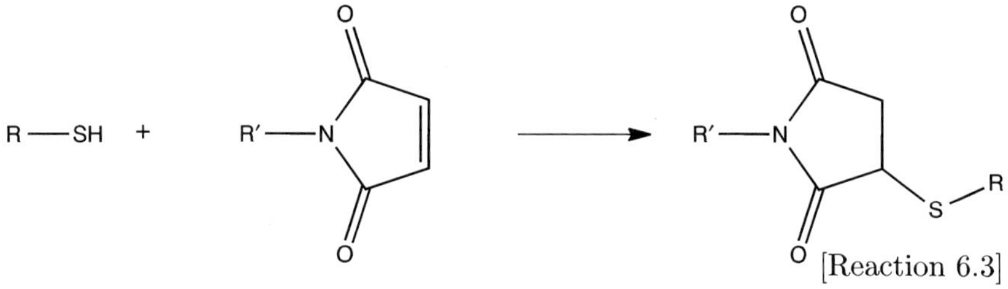


[Reaction 6.3]

Due to the structural similarities between maleimide and NHS, this reaction is very similar to the amide formation through NHS-esters in various aspects. Maleimide can be readily cleaved in an aqueous environment; therefore the maleimide-bearing component must be synthesized and immediately dried out and stored in anhydrous conditions to preserve its functionality. Maleimide also has high reactivity towards amines, as well as hydroxyls; thus, similarly to NHS, the second component of the adhesive hydrogel or the solvent used must be amine and alcohol free. However, these secondary reactions take place at a slower rate than that with thiols, making it possible to form the hydrogel before being degraded or reacted with other chemical groups. Maleimide secondary reactions with all the aforementioned groups (thiols, amines, and hydroxyls), take place mainly with extracellular matrix proteins and are responsible for providing tissue adhesive strength [78].

When the optimal polymer backbones and concentrations of thiol and maleimide groups are used, these gels are formed *in situ* on a short timescale (between 5 and 35 s) thus making them suitable as injectable hydrogels [72].

#### 6.4.1.2.5 Free-radical reaction

A free-radical reaction involves highly reactive unpaired electrons. Typically, polymers containing potential free-radical groups, such as acrylate moieties, can remain unreactive until an initiator is triggered to generate the first free radicals. The initiator can be triggered by a number of parameters such as heat [75] or UV light [73, 74, 77]. When engineering adhesive hydrogels, UV light is the most widely used crosslinking trigger as it allows for controlled crosslinking density, and non-penetrating radiation of high energy. Even though UV radiation poses the risk of DNA mutation in the irradiated cells, radiation times are typically too short to cause any deleterious genetic abnormalities and light can be focused to a specific point of interest, reducing damage to surrounding tissues.

A recent approach to photopolymerized adhesive hydrogels using UV light has been developed and widely studied for the treatment of cartilage and bone defects [73]. This novel material is based on a composite of chondroitin sulfate functionalized with methacrylate and aldehyde groups and PEG diacrylate (PEGDA). The chondroitin sulfate adhesive layer is applied to the defect with a brush and allowed to react with the tissue through the aldehyde groups and the excess washed away. The defect is then filled with PEGDA mixed with the photoinitiator and exposed to light at 365 nm (Figure 6.5). The acrylate moieties in the PEGDA are then photopolymerized to form the bulk hydrogel and simultaneously react with methacrylate groups on the CS adhered to the tissue. In this novel approach, the CS layer avoids damage or penetration of the cartilage, while the bulk of the hydrogel imparts optimal mechanical properties. The applications in tissue engineering of these hydrogels as cellular implants for cartilage and bone regeneration are discussed later in this section.

PEGDA adhesive hydrogels have also been used in combination with silicate nanoparticles (laponite) to dramatically improve the mechanical properties of the adhesive hydrogel [76]. Acrylate moieties can interact with silicate nanoparticles through hydrogen bonding and ion–dipole bonding to further physically crosslink the hydrogel's three-dimensional network, conferring improved adhesion strength compared to PEG only, as well as improved mechanical properties.

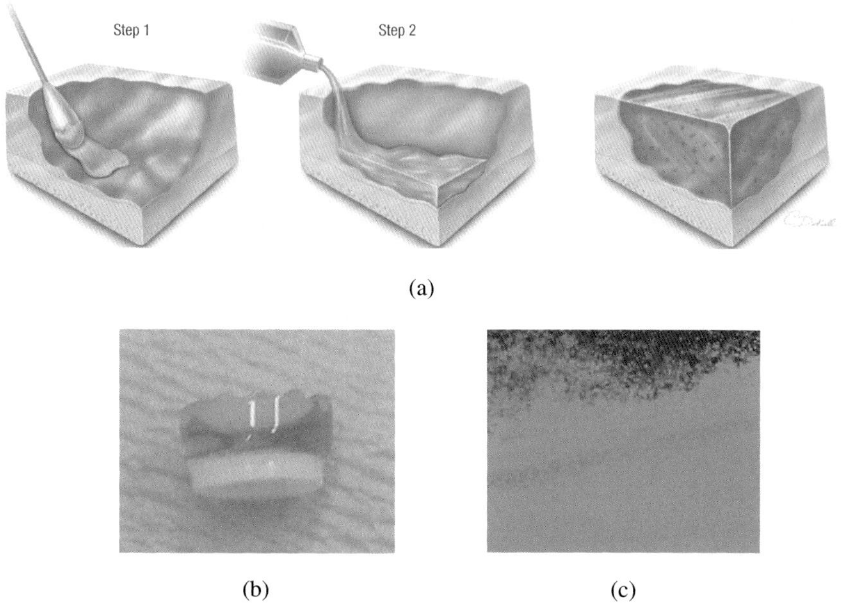


Figure 6.5 Application steps of dual-layer adhesive hydrogel — first layer adheres to the tissue through aldehyde groups and to the bulk of the material through acrylate moieties. PEGDA is used as hydrogel filler for the defect. Reprinted from [73] with permission.

From the last two examples it becomes clear that the field of adhesive hydrogels is evolving towards the combination of different reaction mechanisms to confer optimal mechanical and adhesive properties, while maintaining *in vivo* biocompatibility. This amalgamation of mechanisms is representative of adhesives in nature, where they do not purely rely on one chemical reaction with the substrate, but rather on a combination.

It's important to keep in consideration that the classification of adhesives following the type of material–tissue interactions is somehow artificial. Tissue porosity will confer some degree of mechanical interlocking to all adhesive formulations, and any adhesive based on an interlocking mechanism will have some sort of chemical interactions with the surrounding tissue (van der Waals and hydrogen bonds, at least), although these forces may not be sufficient for some applications. Cohen *et al.* presented a qualitative model describing the effects of different formulation parameters on the adhesion mechanisms (interlocking and chemical) using their adhesive hydrogel based on carbodiimide-crosslinked gelatin and alginate [79].

Nonetheless, there is usually one predominant mechanism to which most of the adhesion strength can be attributed.

The type of interaction with the substrate will determine not only adhesion strength to a particular tissue, but also a number of critical factors such as reversibility, degradation rate or toxicity. More importantly, biological tissues have a transient nature – they can undergo physical and chemical changes over time (with age) or under pathological states. For this reason, it is critical to examine adhesive–tissue interactions and study tissue surface modifications that could potentially affect material performance.

## 6.4.2 *Classification by application*

### 6.4.2.1 *Wound sealant*

The field of wound sealants is in desperate need of new materials capable of overcoming the limitations of the currently available medical sealants. While cyanoacrylate-based sealants have optimal mechanical properties, they are also highly toxic, thus not suitable for sealing internal wounds. On the other side of the spectrum, fibrin glues are highly biocompatible, but their adhesion strength to tissues is very weak.

Recent efforts in this field have led to the development of tunable adhesives based on amine-aldehyde chemistry [28, 55, 59–63]. As explained above (Section 6.4.1.2.1), these systems rely on a competitive reaction of the aldehydes on one of the components of the adhesive with amines present on the second component and on tissue surfaces. The oxidized dextran serves as the aldehyde-bearing component. Dextran is a polysaccharide composed of glucose rings naturally synthesized by lactic acid bacteria, such as *Leuconostoc mesenteroides* and *Streptococcus mutans*. The polysaccharide backbone makes it highly biocompatible and antithrombotic, and its bacterial origin makes it non-cleavable by human enzymes. Aldehydes present in small molecules, such as formaldehyde or glutaraldehyde, are highly toxic due to their ability to crosslink and penetrate the tissue. However, when these functional groups are scattered on a high molecular weight polymer, such as dextran, they are able to react with the tissue, forming Schiff bases without crosslinking proteins or penetrating deep into tissues, thus decreasing the toxicity associated with aldehyde reactivity.

A range of amine-bearing polymers has been used to achieve adhesive hydrogels: polylysine [59, 60], 8-arm PEG-amine [28, 61–63], or PAMAM dendrimers [55]. Amine-aldehyde adhesive hydrogels have been used as models to pioneer the field of adhesive–tissue interactions. Eight-arm PEG-amine in combination with dextran aldehyde was used to show that adhesion to tissues is not only a function of the adhesive formulation used (i.e., aldehyde groups provided to react with the tissue), but also a function of the target tissue amine density. When the same adhesive formulation is applied to different organs (heart, lung, liver, and intestine) the resulting adhesion strength varies in a correlative manner with the changes in amine density on the respective tissue surfaces: the more amines available on the tissue to react with the adhesive, the higher the adhesion strength [61]. This correlation was also found and quantified applying a single dendrimer-dextran hydrogel formulation to the three regions of the small intestine (jejunum, duodenum, and ileum), proving that even subtle changes in amine density within the same organ affect material performance [55]. Degradation kinetics was also organ-specific as a result of variations in the relative concentrations of tissue amines that affect the number of material aldehydes available for internal reaction. Therefore, hydrogel formulation must be tuned to match the target organ properties (in this case, amine density) in order to achieve optimal material performance.

The concepts learned from these studies resulted in a paradigm shift in adhesive material development. The understanding that materials are not static entities that can be formulated at the bench and used indiscriminately led to tissue- and state-specific adhesive material design and optimization.

#### 6.4.2.2 *Tissue engineering*

Adhesive hydrogels have been of great importance in advancing the field of cartilage and bone engineering. However, it is challenging to fill these defects with materials that replicate the complex native structure of cartilage and bone, and attempts to implant cartilage tissue in these defects show poor integration with the surrounding tissue, making regeneration challenging. Three systems capable of integrating into cartilage and bone and trigger its regeneration are highlighted herein. Wang *et al.* developed a photocrosslinkable hydrogel based on CS-aldehyde-methacrylate and PEG-diacrylate, also described above. Production of proteoglycans and collagen around the cells at the interface with the tissue explant was shown. *In vivo* results also showed proteoglycan deposition throughout the hydrogel and at the interface with the cartilage, proving integration of the newly formed tissue with the existing one [73]. Experiments in large animals and human trials showed safety, clinical applicability of this novel hydrogel for cartilage repair and inhibition of osseous overgrowth and fibrous scar tissue formation at the defect site, together with an improvement in defect tissue fill [74].

Simson *et al.* developed a system described in the section above based on CS functionalized with NHS and the host's bone marrow [65, 66]. The bulk network of the material served as a scaffold for fibrochondrocytes, which showed increased viability, proliferation, and migration as a function of bone marrow content in the adhesive. Meanwhile, CS improved adhesive strength and matrix production from fibrochondrocytes in their model of meniscal fusion. The increase in adhesion strength is readily explained by an increase in NHS reactive groups that can then be available to react with the tissue, while the CS itself drives the increased matrix production [65]. However, it was recognized that the system missed the biological cues present in natural tissues that, in turn, would drive further fibrochondrocyte proliferation and matrix deposition. For this reason, the adhesive hydrogel formulation was modified by adding bone morphogenetic protein-2 (BMP-2). The *in vitro* results showed that adding BMP-2 increased collagen production after three weeks with respect to a control hydrogel without BMP-2. While *in vivo* results showed no increased collagen or sulfated GAGs deposition with respect to the control, they did observe higher production of collagen type-II than type-I, consistent with hyaline cartilage [66].

Given the fact that the host's bone marrow is used as a scaffold, which is a natural component found in the host, this material is highly biocompatible. Chang *et al.* developed a similar, naturally derived, autologous, biodegradable approach for stem cell retention in the myocardium. The adhesive hydrogel was composed of HA functionalized with NHS and lysed whole blood. While HA serves as a harbor for cells through interaction with HA receptors on the cell surface, lysed whole blood – obtained from the patient – presents the benefit of avoiding rejection and pathogen transmission. Moreover, blood itself provides a source of growth factors and adhesion motifs. This type of adhesive hydrogel follows enzymatic degradation of the proteins in the lysed blood by proteases released by cells, showing minimal degradation after four weeks by hydrolysis alone. The system was used as a scaffold for stem cells and applied in a model of myocardial infarction, where it showed stem cell survival and proliferation, as well as increased ventricular systolic function after epicardial application [67].

This is only a glimpse into a growing field of the use of adhesive hydrogels as tissue engineering scaffolds. Data so far have been promising and will give rise to a number of products with clinical applicability. The use of adhesive hydrogels improves cell viability, allows for tunable formulations with embedded biological cues and adds the distinct advantage of being capable of adhering to the target tissue, increasing integration with surrounding tissues and improving regeneration.

#### 6.4.2.3 *Sustained release system*

A less studied area in the field of adhesive hydrogels is their use as a sustained release systems of drugs, genetic material, or antibacterial agents. While hydrogels by themselves have been widely used for this purpose, not much research has been conducted in incorporating these principles into the adhesive hydrogel field.

A biomimetic hydrogel inspired by mussels was designed to release antibacterial agents [69]. In this approach, they use a silver salt to oxidize the catechol groups and thus induce crosslinking of the network to form the adhesive hydrogel. The silver cations form nanoparticles upon oxidation that are entrapped in the bulk of the adhesive and released in a sustained manner. They showed a linear release of silver over time that had antimicrobial effect while having a non-cytotoxic effect on 3T3 fibroblasts. A major drawback of this system is the long gelation time required (10–60 mins depending on the ratio of silver to catechol), which makes it unsuitable for injection.

The capabilities of adhesive hydrogels to serve as scaffolds for sustained drug release systems are immense albeit underexplored. Local delivery may reduce side effects, enhance drug bioavailability and when embedded in adhesive scaffolds can facilitate controlled release. Beyond drug delivery, local release of genetic material would dramatically reduce its degradation by DNAses and RNAses before reaching the target site. In general, local release of any agent (antimicrobial, genetic material, or drugs) would greatly reduce clearance by the kidneys and liver, dramatically reducing the required dose to combat a specific disease.

## 6.5 Tissue Adhesive Characterization Techniques Using Fluorescent Probes

Let us now discuss fluorescence-based techniques to examine key parameters in the development of adhesive hydrogels *in vitro*, *ex vivo*, and *in vivo*, using the amine-aldehyde adhesive systems as a model.

### 6.5.1 *Adhesion strength*

A key requirement for biomedical adhesive hydrogels is adequate adhesion to tissues. Typically, uniaxial mechanical tests are used for this purpose using gelatin-coated surfaces to mimic a tissue extracellular matrix or excised tissues by themselves. Both of these approaches present advantages and limitations. On the one hand, coating surfaces with gelatin provides control and repeatability in the system by tuning the technique used to deposit the protein over the glass slide, thus giving consistent and reliable measurements of adhesion with low standard deviation. However, it also implies reducing a complex system like animal tissues to one component, as well as assuming that every tissue will have a particular collagen content. This system may not be representative of actual tissues and it poses the risk of failure when the adhesive is used *in vivo*. On the other hand, while excised tissues are representative of the *in vivo* system that the adhesive will incur, working with excised tissues limits the experiment duration owing to the short viability time of the samples (about 4 h from excision) and the large standard deviation associated with method intrinsic variability.

It is possible to use fluorescent probes to quantify adhesion strength to biological tissues. By tagging the adhesive and applying it to excised biological tissues, one can observe the tissue–material interface and assess the degree of adhesion. PAMAM dendrimer, where 1% of dendrimer molecules were tagged with fluorescein succinimidyl ester, was mixed with different concentrations of dextran aldehyde and applied to excised jejunum to form gels *in situ*. The tissues were then snap-frozen in liquid nitrogen, cryosectioned, and the tissue stained with propidium iodide. Figure 6.6 shows sections of jejunum (red) with adhesive formulations of increasing dextran content [55]. The higher the dextran aldehyde content, the denser (fewer, smaller pores) the interface with the tissue is, meaning more anchoring points. The fluorescence at the interface with the tissue linearly correlates with the uniaxial adhesion strength tests performed using an Instron mechanical tester. This indicates that, for this particular system, adhesion strength can be determined by fluorescence microscopy using the linear correlation obtained.

As discussed in previous sections, different organs have different properties, such as amine density on the surface available to react with

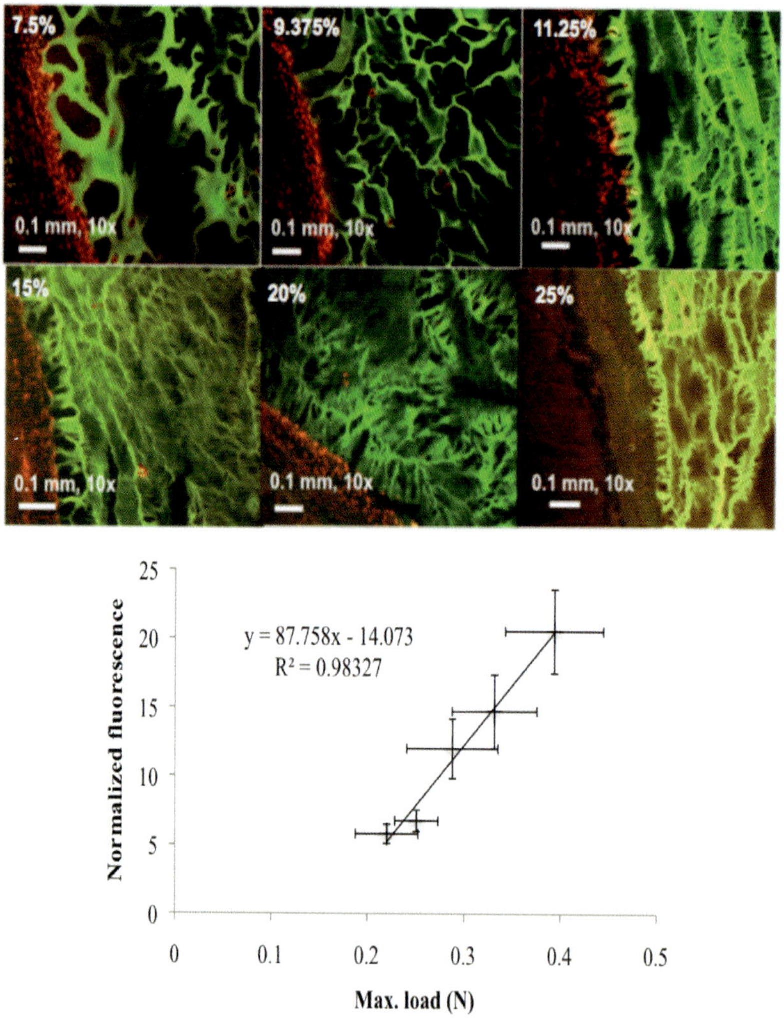


Figure 6.6 Fluorescence microscopy images of dendrimer-dextran interface with the surface of the small intestine (left) and the correlation of the interfacial intensity with the adhesion strength (right). Reprinted from [55] with permission from the American Chemical Society.

aldehydes provided by the adhesive. In order to create an *in vitro* system representative of the *in vivo* while saving animal tissues, it needs to provide similar amine density to that present *in vivo*. To study tissue amine density, tissues can be incubated with aldehyde-coated fluorescent microspheres,

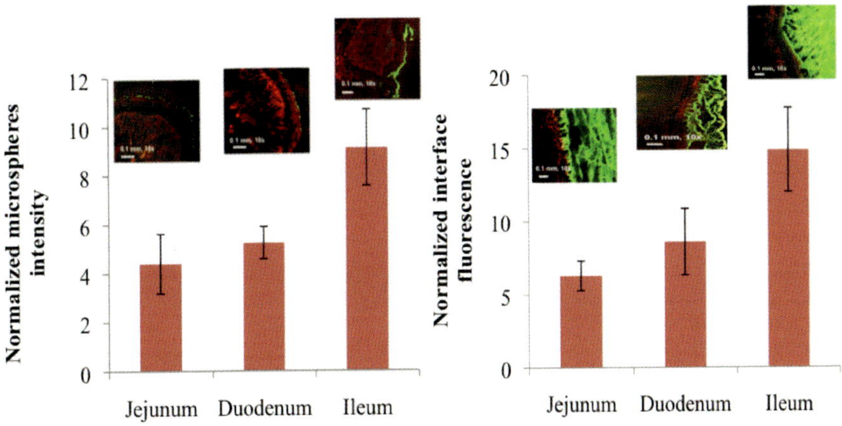


Figure 6.7 Fluorescence intensity of green fluorescent aldehyde-coated microspheres on the surface (left) and of fluorescently labeled dendrimer-dextran at the interface on the three regions of the small intestine. The data has a correlation coefficient of $R^2 = 0.99$. Reprinted from [55] with permission from the American Chemical Society.

which will react with amines on tissue surfaces and can then be quantified by optical techniques. Figure 6.7 shows cryosections of the three regions of the small intestine stained with propidium iodide (red) after incubation with fluorescent microspheres (green). Similarly to the data shown previously (Section 6.4.2.1), the ileum shows a denser, more fluorescent surface, indicating a higher amine density, which correlates with higher adhesion strength, as previously discussed [55].

Matching the gelatin amine density adsorbed on a surface to that of live tissues would allow for an *in vitro* system representative of the *in vivo* environment. This will enable a dramatic reduction in the number of animals required to develop and optimize the adhesive material formulation.

### 6.5.2 *In vitro degradation*

The degradation profile of a biomaterial is a key characteristic that must be properly addressed and studied *in vitro* before materials are implanted *in vivo*. Typically, material degradation is measured as the current weight normalized to initial weight. When it comes to hydrogels, however, weighing the material does not only account for degradation but also for swelling. In other words, it is not possible to assess solely material degradation using gravimetric analysis since the material absorbs water from its surroundings as it degrades, especially in the first few hours to days. An

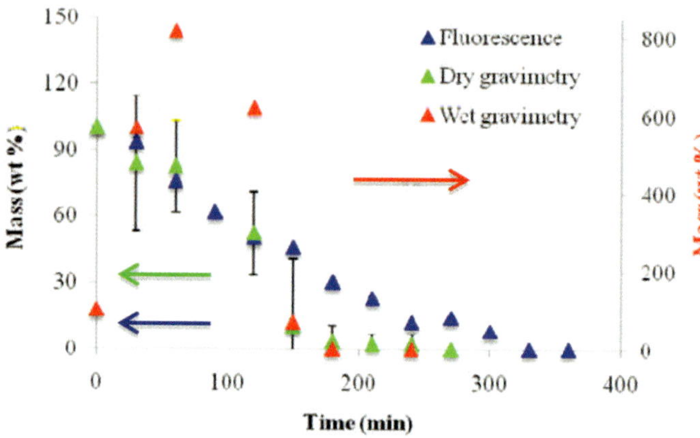


Figure 6.8 Hydrogel degradation tracked with three different techniques — fluorescence (blue triangles), dry gravimetry (green triangles), and wet gravimetry (red triangles). The fluorescence method allows for a more sensitive tracking with a smaller sample size. Reprinted from [80] with permission.

alternative approach would be to dry the material at each time point to eliminate the effect of water. However, this would be a destructive assay necessitating many samples, in addition to the limited ability to accurately measure the low weight of the dry material following water evaporation. A novel technique using fluorescent probes allows for sensitive degradation tracking while eliminating the swelling effect (Figure 6.8). By tagging the material, it is possible to follow material degradation and erosion over time using fluorescence techniques with only a few replicates for each sample [80].

### 6.5.3 In Vivo *degradation*

It has already been established that the *in vivo* environment is more complex than the simplified PBS- or culture media-based *in vitro* settings. A number of factors account for those differences, such as aqueous volume surrounding the adhesive hydrogel, enzymes present in those fluids and their concentration, and even more challenging to mimic: the effect of the immune system. Hence, it is essential that materials will be examined *in vivo* and their degradation studied relative to specific environments. Current approaches involve the implantation or injection of materials, usually in the subcutaneous space of mice or rats, where they are kept for a period of time, after which the animal is euthanized and the implants excised for further study, such as weighing to determine the percentage of material degraded. To minimize the number of animals used, implant degradation is followed only intermittently and not continuously. Critical information can then be overlooked, especially in the first 24 hours where material swelling and rapid release of unbound components occur. Using fluorescent probes to follow degradation allows for continuous and non-invasive tracking of material erosion, while avoiding the need for animal sacrifice at each time point. The procedure consists of the implantation of a fluorescently labeled adhesive hydrogel in the subcutaneous space of mice, which can then be scanned frequently using optical imaging, such as *in vivo* imaging system (IVIS). Degradation is easily measured by converting the loss of fluorescence to mass loss. Scanning the same specimen at different time points dramatically reduces sample-to-sample error and standard deviation [80].

More importantly, the information acquired from the *in vivo* experiments can be correlated with that obtained *in vitro*, conferring predictive capabilities to the system. *In vitro–in vivo* correlation between PEG-amine and dextran aldehyde degradation profiles for six different formulations allowed a correlation to be established that could then be used to predict the *in vivo* degradation rate of any other given formulation by performing solely the *in vitro* experiment (Figure 6.9) [80].

While *in vivo* experimentation is necessary to establish the properties of a biomaterial and ensure its functionality and safety, there is a need to identify determinant parameters controlling material *in vivo* behavior. This novel fluorescence technique allows for that to happen without a significant increase in the number of animals used for this purpose.

### 6.5.4 *Cellular uptake and toxicity*

Another important aspect of biomaterial design is the assessment of toxicity. There are a number of kits, such as live/dead assays or lactate dehydrogenase assays, which can quantify the amount of cell death associated with degradation byproducts. These assays, while extremely sensitive, do not provide insight into the mechanism contributing to the toxicity. Fluorescence assays provide the capability to observe morphological cellular changes associated with material dose and help implement modifications to reduce such toxicity. For instance, PAMAM dendrimers have been shown to impart toxicity in a dose-dependent manner above a certain threshold,

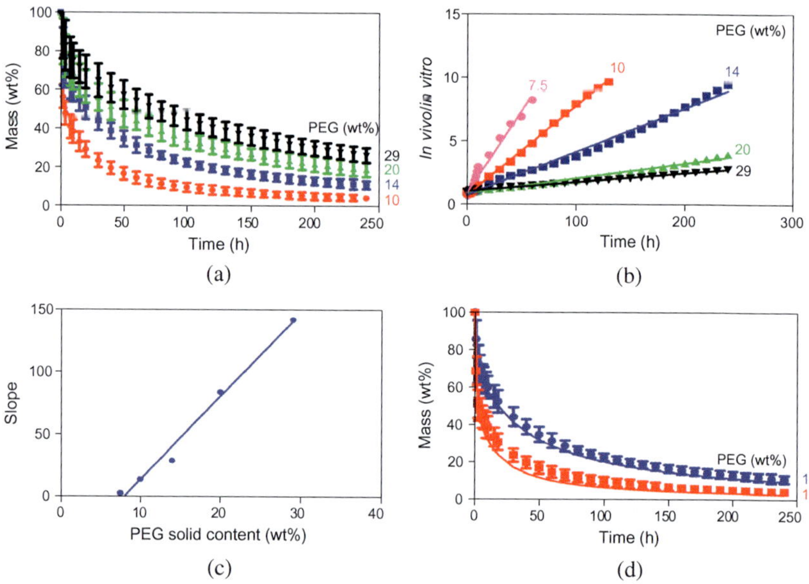


Figure 6.9 *In vivo–in vitro* correlation of material degradation (b and c) allows for the creation of predictive *in vivo* models from *in vitro* data (d – points are empirically accumulated and lines are model predictions). Reprinted from [80] with permission.

as their surface amines can create holes in the cellular membrane. Neutralizing those amines with other chemical groups that reduce or block their uptake can significantly improve material cytocompatibility. An example of this approach would be the oxidation of the amines to alcohols [55]. Figure 6.10 shows breast cancer cells (MDA-MB-468) treated with fluorescently labeled PAMAM dendrimer generation 5 with either 128 or 32 amines on the surface. While a cytotoxicity assay would only give relative numbers of toxicity, this assay allows the visualization of the cellular processes: dendrimers with 128 amines on the surface penetrate the cell imparting high toxicity, while the dendrimer with 32 surface amines reduces significantly cell uptake. Also, the dendrimer with 32 amines can be observed as discrete fluorescent points, indicating that the molecules are inside the endosomes, thus the uptake mechanism cannot happen through passive diffusion mechanisms, but rather through other specific endocytic processes.

The combination of existing cytotoxicity assays with the novel fluorescent technique will advance the understanding of cell–material interaction and help rationalize material optimization in light of the underlying cellular processes.

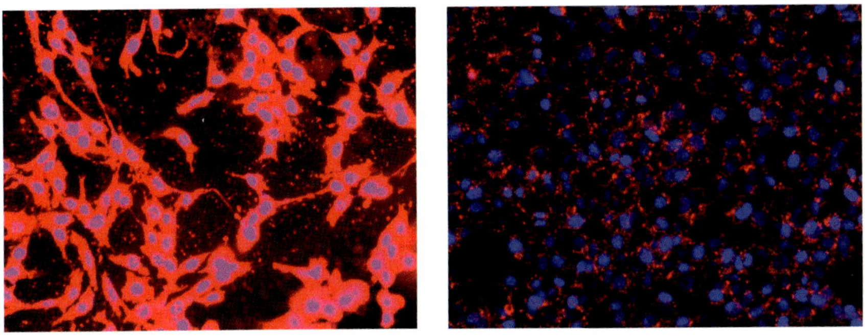

Figure 6.10 Texas Red®-dendrimer-treated breast cancer cells (MDA–MB–468). When the cells were incubated with generation 5 with 100% amines on the surface, high and uncontrolled uptake (intense red fluorescence) and toxicity (irregular cell shape and presence of debride) can be observed. Toxicity was decreased and uptake controlled in an endosomal manner when 75% of the amines were oxidized to alcohols. (Data not published.)

## 6.6 Summary and Implications

This chapter presents the emerging and rapidly growing field of adhesive hydrogels, while emphasizing the vital importance of studying the contextual performance of adhesive materials with respect to the specific tissue microenvironment they will encounter. Tissue type and state including potential pathological processes should be studied. Contextual material performance requires studying tissue and material properties in parallel to gain mechanistic insight and predictive (pre)clinical outcomes.

Adhesive materials have been used in a wide range of applications with notable advantages compared to their non-adhesive counterparts. Use of adhesive hydrogels in bone and cartilage improved tissue integration and regeneration capabilities that were otherwise challenging to achieve using non-adhesive materials. Local and sustained release of a number of therapeutic agents (drugs, genes, antibiotics, etc.) present remarkable advantages over systemic therapy by avoiding side effects and increasing compounds' bioavailability.

We are at the advent of a novel field of smart adhesive materials that have already demonstrated high potential to solve unmet clinical needs and the capability of being translated into the clinic, potentially leading to medical breakthroughs. Combination products that exploit the adhesive properties of such hydrogels to locally deliver factors from embedded cells, or deliver drugs and/or small molecules can lead to innovations in materials design that were previously unimaginable.

## References


[1] Kinloch, A.J. (1987). Adhesion and adhesives: Science and technology. London; New York: Chapman and Hall. x, p. 441.

[2] Benedek, I. (2004). Pressure-Sensitive Adhesives and Applications: Taylor & Francis.

[3] Zheleva, D. (2013). Mechanisms Of Interaction Between The Components In Adhesive Compositions Based On Chloroprene Rubber. *Journal of Chemical Technology and Metallurgy*. **48**(5), p. 8.

[4] Tse MFH, G. R. and Tathgur, A. (2006). Adhesion and Failure Mechanisms of a Model Hot Melt Adhesive Bonded to Polypropylene. *The Journal of Adhesion*. **61**, 25.

[5] Mehdizadeh, M. and Yang, J. (2013). Design strategies and applications of tissue bioadhesives. Macromolecular Bioscience. **13**(3), pp. 271–288.

[6] Dumitriu, S. (1994). Polymeric biomaterials. New York: Marcel Dekker, x, p. 845.

[7] Maska, R. (1972). Pressure-sensitive adhesives of cyanoacrylate polymers. Google Patents.

[8] Bergel, S. (1909). Wirkungen des Fibrins. Dtschr Med Wochenschr. **35**, 32.

[9] Catelas, I., Dwyer, J.F. and Helgerson, S. (2008). Controlled release of bioactive transforming growth factor beta-1 from fibrin gels *in vitro*. *Tissue Engineering Part C, Methods*, **14**(2), pp. 119–128.

[10] Liang, M.S. and Andreadis, S.T. (2011). Engineering fibrin-binding TGF-beta1 for sustained signaling and contractile function of MSC based vascular constructs. *Biomaterials*. **32**(33), pp. 8684–8693.

[11] Becker, J.C., Domschke, W. and Pohle, T. (2004). Biological *in vitro* effects of fibrin glue: fibroblast proliferation, expression and binding of growth factors. *Scandinavian Journal of Gastroenterology*, **39**(10), pp. 927–932.

[12] Anai, S., Hide, T., Takezaki, T., Kuroda, J., Shinojima, N., Makino, K., *et al.* (2014). Antitumor effect of fibrin glue containing temozolomide against malignant glioma. *Cancer Science*, **105**(5), pp. 583–591.

[13] Nie, S.P., Wang, X., Qiao, S.B., Zeng, Q.T., Jiang, J.Q., Liu, X.Q., *et al.* (2010). Improved myocardial perfusion and cardiac function by controlled-release basic fibroblast growth factor using fibrin glue in a canine infarct model. *Journal of Zhejiang University Science B*. **11**(12), pp. 895–904.

[14] Wen, Q., Zhou, C., Luo, W., Zhou, M. and Ma, L. (2014). Pro-osteogenic effects of fibrin glue in treatment of avascular necrosis of the femoral head *in vivo* by hepatocyte growth factor-transgenic mesenchymal stem cells. *Journal of Translational Medicine*, **12**(1), p. 114.

[15] Spicer, P.P. and Mikos, A.G. (2010). Fibrin glue as a drug delivery system. *Journal of Controlled Release: Official Journal of the Controlled Release Society*, **148**(1), pp. 49–55.

[16] Schillinger, U., Wexel, G., Hacker, C., Kullmer, M., Koch, C., Gerg, M., *et al.* (2008). A fibrin glue composition as carrier for nucleic acid vectors. *Pharmaceutical Research*, **25**(12), pp. 2946–2962.
[17] Aoyagi, Y., Kuroda, M., Asada, S., Bujo, H., Tanaka, S., Konno, S., *et al.* (2011). Fibrin glue increases the cell survival and the transduced gene product secretion of the ceiling culture-derived adipocytes transplanted in mice. *Experimental & Molecular Medicine*, **43**(3), 161–167.
[18] Wallace, D.G., Cruise, G.M., Rhee, W.M., Schroeder, J.A., Coker, G.T., Maroney, M.M. (2001). Rapid gelling biocompatible polymer composition. Google Patents.
[19] Wallace, D.G., Cruise. G.M., Rhee, W.M., Schroeder, J.A., Prior, J.J., Ju, J., *et al.* (2001). A tissue sealant based on reactive multifunctional polyethylene glycol. *Journal of Biomedical Materials Research*, **58**(5), 545–555.
[20] Furst, W. and Banerjee, A. (2005). Release of glutaraldehyde from an albumin-glutaraldehyde tissue adhesive causes significant *in vitro* and *in vivo* toxicity. *The Annals of Thoracic Surgery*, **79**(5), 1522–1528; discussion 9.
[21] Azadani, A.N., Matthews, P.B., Ge, L., Shen, Y., Jhun, C.S., Guy, T.S., *et al.* (2009). Mechanical properties of surgical glues used in aortic root replacement. *The Annals of Thoracic Surgery*, **87**(4), pp. 1154–1160.
[22] LeMaire, S.A., Schmittling, Z.C., Coselli, J.S., Undar, A., Deady, B.A., Clubb, F.J., Jr., *et al.* (2002). BioGlue surgical adhesive impairs aortic growth and causes anastomotic strictures. *The Annals of Thoracic Surgery*, **73**(5), pp. 1500–1505; discussion 6.
[23] Ulery, B.D., Nair, L.S. and Laurencin, C.T. (2011). Biomedical Applications of Biodegradable Polymers. *Journal of Polymer Science Part B, Polymer Physics*, **49**(12), pp. 832–864.
[24] Cheng, H.N. (2012). An Overview of Degradable Polymers. Degradable Polymers and Materials: Principles and Practice (2nd Edition). ACS Symposium Series. 1114: *American Chemical Society*. p. xiii–xiv.
[25] Gopferich, A. (1996). Mechanisms of polymer degradation and erosion. *Biomaterials*, **17**(2), pp. 103–114.
[26] Shoichet, M.S., Zahir, T., Ballios, B., Van, D.K.D. and Cooke, M. (2012). Injectable polymer composition for use as a cell delivery vehicle. Google Patents.
[27] Unterman, S.A., Marcus, N.A., Elisseeff, J.H. (2011). Injectable Polymers. Biodegradable Polymers in Clinical Use and Clinical Development: John Wiley & Sons, Inc. p. 631–664.
[28] Artzi, N., Shazly, T., Crespo, C., Ramos, A.B., Chenault, H.K., Edelman, E.R. (2009). Characterization of star adhesive sealants based on PEG/dextran hydrogels. *Macromolecular Bioscience*, **9**(8), 754–765.
[29] Lang, N., Pereira, M.J., Lee, Y., Friehs, I., Vasilyev, N.V., Feins, E.N., *et al.* (2014). A blood-resistant surgical glue for minimally invasive repair of vessels and heart defects. *Science Translational Medicine*, **6**(218), p. 218ra6.
[30] Langer, R. and Peppas, N.A. (2003). Advances in biomaterials, drug delivery, and bionanotechnology. *AIChE Journal*, **49**(12), pp. 2990–3006.

[31] Peppas, N.A., Bures, P., Leobandung, W. and Ichikawa, H. (2000). Hydrogels in pharmaceutical formulations. European journal of pharmaceutics and biopharmaceutics : official journal of Arbeitsgemeinschaft fur Pharmazeutische Verfahrenstechnik eV **50**(1), pp. 27–46.

[32] Peppas, N.A., Huang, Y., Torres-Lugo, M., Ward, J.H. and Zhang, J. (2000). Physicochemical foundations and structural design of hydrogels in medicine and biology. *Annual Review of Biomedical Engineering*, **2**, pp. 9–29.

[33] Jeong, B., Bae, Y.H., Lee, D.S. and Kim, S.W. (1997). Biodegradable block copolymers as injectable drug-delivery systems. *Nature*, **388**(6645), pp. 860–862.

[34] Park, T.G. (1999). Temperature modulated protein release from pH/temperature-sensitive hydrogels. *Biomaterials*, **20**(6), pp. 517–521.

[35] Siegel, R.A., Falamarzian, M., Firestone, B.A. and Moxley, B.C. (1988). pH-Controlled release from hydrophobic/polyelectrolyte copolymer hydrogels. Journal of Controlled Release. **8**(2), pp. 179–182.

[36] Brannon-Peppas, L. and Peppas, N.A. (1990). Dynamic and equilibrium swelling behaviour of pH-sensitive hydrogels containing 2-hydroxyethyl methacrylate. *Biomaterials*, **11**(9), 635–644.

[37] Peppas, N.A., Keys, K.B., Torres-Lugo, M. and Lowman, A.M. (1999). Poly(ethylene glycol)-containing hydrogels in drug delivery. *Journal of Controlled Release: Official Journal of the Controlled Release Society*, **62**(1–2), pp. 81–87.

[38] Risbud, M.V., Hardikar, A.A., Bhat, S.V., Bhonde, R.R. (2000). pH-sensitive freeze-dried chitosan-polyvinyl pyrrolidone hydrogels as controlled release system for antibiotic delivery. *Journal of Controlled Release: Official Journal of the Controlled Release Society*, **68**(1), pp. 23–30.

[39] Chen, S.C., Wu, Y.C., Mi, F.L., Lin, Y.H., Yu, L.C. and Sung, H.W. (2004). A novel pH-sensitive hydrogel composed of N,O-carboxymethyl chitosan and alginate cross-linked by genipin for protein drug delivery. *Journal of Controlled Release: Official Journal of the Controlled Release Society*, **96**(2), pp. 285–300.

[40] Al-Aribe, K. and Knopf, G.K. (2010). Photoresponsive hydrogel microvalve activated by bacteriorhodopsin proton pumps.

[41] Charati, M.B., Lee, I., Hribar, K.C. and Burdick, J.A. (2010). Light-sensitive polypeptide hydrogel and nanorod composites. *Small*, **6**(15), pp. 1608–1611.

[42] Szabo, D., Czako-Nagy, I.I., Zrinyi, M. and Vertes A. (2000). Magnetic and Mossbauer Studies of Magnetite-Loaded Polyvinyl Alcohol Hydrogels. *Journal of Colloid and Interface Science*, **221**(2), pp. 166–172.

[43] Liu, T.Y., Hu, S.H., Liu, T.Y., Liu, D.M. and Chen, S.Y. (2006). Magnetic-sensitive behavior of intelligent ferrogels for controlled release of drug. *Langmuir: The ACS Journal of Surfaces and Colloids*, **22**(14), pp. 5974–5978.

[44] Liu, G.Z., X; Tang, T. (2003). Stimuli-Responsive of Gelatin Hydrogel Under Direct Current Electric Field. *Acta Polymerica Sinica*, **1**(3), p. 5.

[45] Hou, C., Zhang, Q., Li, Y. and Wang, H. (2012). Graphene–polymer hydrogels with stimulus-sensitive volume changes. *Carbon*, **50**(5), pp. 1959–1965.

[46] Miyata, T., Asami, N., Uragami, T. (1999). A reversibly antigen-responsive hydrogel. *Nature*, **399**(6738), pp. 766–769.

[47] Hamidi, M., Azadi, A., Rafiei, P. (2008). Hydrogel nanoparticles in drug delivery. *Advanced Drug Delivery Reviews*, **60**(15), pp. 1638–1649.

[48] Mann, B.K., Gobin, A.S., Tsai, A.T., Schmedlen, R.H. and West, J.L. (2001). Smooth muscle cell growth in photopolymerized hydrogels with cell adhesive and proteolytically degradable domains: synthetic ECM analogs for tissue engineering. *Biomaterials*, **22**(22), pp. 3045–3051.

[49] Mann, B.K. and West, J.L. (2002). Cell adhesion peptides alter smooth muscle cell adhesion, proliferation, migration, and matrix protein synthesis on modified surfaces and in polymer scaffolds. *Journal of Biomedical Materials Research*, **60**(1), pp. 86–93.

[50] Luo, Y., Shoichet, M.S. (2004). A photolabile hydrogel for guided three-dimensional cell growth and migration. *Nature Materials*, **3**(4), pp. 249–253.

[51] Lutolf, M.P., Raeber, G.P., Zisch, A.H., Tirelli, N., Hubbell, J.A. (2003). Cell-Responsive Synthetic Hydrogels. *Advanced Materials*, **15**(11), pp. 888–892.

[52] Schukur, L., Zorlutuna, P., Cha, J.M., Bae, H. and Khademhosseini, A. (2013). Directed differentiation of size-controlled embryoid bodies towards endothelial and cardiac lineages in RGD-modified poly(ethylene glycol) hydrogels. *Advanced Healthcare Materials*, **2**(1), pp. 195–205.

[53] Pang, C., Kim, S.M., Rahmawan, Y. and Suh, K.Y. (2012). Beetle-inspired bidirectional, asymmetric interlocking using geometry-tunable nanohairs. *ACS applied materials & interfaces*, **4**(8), pp. 4225–4230.

[54] Yang, S.Y., O'Cearbhaill, E.D., Sisk, G.C., Park, K.M., Cho, W.K., Villiger, M., *et al.* (2013). A bio-inspired swellable microneedle adhesive for mechanical interlocking with tissue. *Nature Communications*, **4**, p. 1702.

[55] Oliva, N., Shitreet, S., Abraham, E., Stanley, B., Edelman, E.R. and Artzi, N. (2012). Natural tissue microenvironmental conditions modulate adhesive material performance. *Langmuir: The ACS Journal of Surfaces and Colloids*, **28**(43), pp. 15402–15409.

[56] Autumn, K. and Peattie, A.M. (2002). Mechanisms of Adhesion in Geckos. *Integrative and Comparative Biology*, **42**(6), 1081–1090.

[57] Mahdavi, A., Ferreira, L., Sundback, C., Nichol, J.W., Chan, E.P., Carter, D.J., *et al.* (2008). A biodegradable and biocompatible gecko-inspired tissue adhesive. *Proceedings of the National Academy of Sciences of the United States of America*, **105**(7), 2307–2312.

[58] Togo, Y., Takahashi, K., Saito, K., Kiso, H., Huang, B., Tsukamoto, H., *et al.* (2013). Aldehyded Dextran and ε-Poly(L-lysine) Hydrogel as Nonviral Gene Carrier. *Stem Cells International*, **2013**, p. 5.

[59] Kazusa, H., Nakasa, T., Shibuya, H., Ohkawa, S., Kamei, G., Adachi, N., *et al.* (2013). Strong adhesiveness of a new biodegradable hydrogel glue, LYDEX, for use on articular cartilage. *Journal of Applied Biomaterials & Functional Materials* **11**(3), pp. e180–e186.

[60] Hyon, S.H., Nakajima, N., Sugai, H. and Matsumura, K. (2013). Low cytotoxic tissue adhesive based on oxidized dextran and epsilon-poly-L-lysine. *Journal of Biomedical Materials Research Part A*.

[61] Artzi, N., Shazly, T., Baker, A.B., Bon, A. and Edelman, E.R. (2009). Aldehyde-amine chemistry enables modulated biosealants with tissue-specific adhesion. *Adv Mater.*, **21**(32–33), pp. 3399–3403.

[62] Artzi, N., Zeiger, A., Boehning, F., bon Ramos, A., Van Vliet, K. and Edelman, E.R. (2011). Tuning adhesion failure strength for tissue-specific applications. *Acta Biomaterialia*, **7**(1), pp. 67–74.

[63] Shazly, T.M., Artzi, N., Boehning, F. and Edelman, E.R. (2008). Viscoelastic adhesive mechanics of aldehyde-mediated soft tissue sealants. *Biomaterials*, **29**(35), pp. 4584–4591.

[64] Simson, J., Crist, J., Strehin, I., Lu, Q. and Elisseeff, J.H. (2013). An orthopedic tissue adhesive for targeted delivery of intraoperative biologics. *Journal of Orthopaedic Research: Official Publication of the Orthopaedic Research Society*, **31**(3), pp. 392–400.

[65] Simson, J.A., Strehin, I.A., Allen, B.W. and Elisseeff, J.H. (2013). Bonding and fusion of meniscus fibrocartilage using a novel chondroitin sulfate bone marrow tissue adhesive. *Tissue Engineering Part A*, **19**(15–16), pp. 1843–1851.

[66] Simson, J.A., Strehin, I.A., Lu, Q., Uy, M.O. and Elisseeff, J.H. (2013). An adhesive bone marrow scaffold and bone morphogenetic-2 protein carrier for cartilage tissue engineering. *Biomacromolecules*, **14**(3), pp. 637–643.

[67] Chang, C.Y., Chan, A.T., Armstrong, P.A., Luo, H.C., Higuchi, T., Strehin, I.A., *et al.* (2012). Hyaluronic acid-human blood hydrogels for stem cell transplantation. *Biomaterials*, **33**(32), pp. 8026–8033.

[68] Barrett, D.G., Bushnell, G.G., Messersmith, P.B. (2013). Mechanically robust, negative-swelling, mussel-inspired tissue adhesives. *Advanced Healthcare Materials*, **2**(5), pp. 745–755.

[69] Fullenkamp, D.E., Rivera, J.G., Gong, Y.K., Lau, K.H., He, L., Varshney, R., *et al.* (2012). Mussel-inspired silver-releasing antibacterial hydrogels. *Biomaterials*, **33**(15), pp. 3783–3791.

[70] Lee, H., Dellatore, S.M., Miller, W.M. and Messersmith, P.B. (2007). Mussel-inspired surface chemistry for multifunctional coatings. *Science*, **318**(5849), pp. 426–430.

[71] Kim, B.J., Oh, D.X., Kim, S., Seo, J.H., Hwang, D.S., Masic, A., *et al.* (2014). Mussel-mimetic protein-based adhesive hydrogel. *Biomacromolecules* **15**(5), 1579–1585.

[72] Zhou, Y., Nie, W., Zhao, J., Yuan, X. (2013). Rapidly *in situ* forming adhesive hydrogel based on a PEG-maleimide modified polypeptide through Michael addition. *Journal of Materials Science Materials in Medicine*, **24**(10), pp. 2277–2286.

[73] Wang, D.A., Varghese, S., Sharma, B., Strehin, I., Fermanian, S., Gorham, J., *et al.* (2007). Multifunctional chondroitin sulphate for cartilage tissue-biomaterial integration. *Nature Materials*, **6**(5), pp. 385–392.

[74] Sharma, B., Fermanian, S., Gibson, M., Unterman, S., Herzka, D.A., Cascio, B., *et al.* (2013). Human cartilage repair with a photoreactive adhesive-hydrogel composite. *Science Translational Medicine*, **5**(167), p. 167ra6.

[75] Wiltsey, C., Kubinski, P., Christiani, T., Toomer, K., Sheehan, J., Branda, A., *et al.* (2013). Characterization of injectable hydrogels based on poly(N-isopropylacrylamide)-g-chondroitin sulfate with adhesive properties for nucleus pulposus tissue engineering. *Journal of Materials Science Materials in Medicine*, **24**(4), pp. 837–847.

[76] Wu, C.J., Wilker, J.J. and Schmidt, G. (2013). Robust and adhesive hydrogels from cross-linked poly(ethylene glycol) and silicate for biomedical use. *Macromolecular Bioscience*, **13**(1), pp. 59–66.

[77] Arunbabu, D., Shahsavan, H., Zhang, W. and Zhao, B. (2013). Poly(AAc-co-MBA) hydrogel films: adhesive and mechanical properties in aqueous medium. *The Journal of Physical Chemistry B*, **117**(1), 441–449.

[78] Hermanson, G.T. (1996). Bioconjugate techniques. San Diego: Academic Press. xxv, pp. 785.

[79] Cohen, B., Pinkas, O., Foox, M. and Zilberman, M. (2013). Gelatin-alginate novel tissue adhesives and their formulation-strength effects. *Acta Biomaterialia*, **9**(11), 9004–9011.

[80] Artzi, N., Oliva, N., Puron, C., Shitreet, S., Artzi, S., bon RamosA, *et al.* (2011). *In vivo* and *in vitro* tracking of erosion in biodegradable materials using non-invasive fluorescence imaging. *Nature Materials*, **10**(9), 704–709.

# Chapter 7

# Injectable Hydrogels for Intervertebral Tissue Regeneration

T. Karukonda

*Department of Orthopaedic Surgery, University of Connecticut Health, Farmington, CT*

I. L. Moss*

*Department of Orthopaedic Surgery, University of Connecticut Health, Farmington, CT*

## 7.1 Introduction

Low back pain (LBP) is one of the most common reasons for visiting a physician in the United States. Up to 85% of adults experience back pain at some point in their lives and it is the most common cause of disability in those under the age of 45 [1]. It is estimated that the annual cost of LBP in the United States exceeds $100 billion, one-third of which is derived from direct health care expenses and the remainder of which is derived from indirect costs resulting from lost wages and productivity [1]. While there are numerous etiologies for LBP, intervertebral disc (IVD) degeneration is believed to be a major cause [2].

Degeneration of the IVD results from an imbalance in production and degradation of IVD extracellular matrix (ECM) components, resulting in structural and functional deterioration of the IVD [3]. Currently, surgical

*Author to whom correspondence should be addressed. Email: imoss@uchc.edu.

treatment for degenerative disc disease (DDD) is directed towards end-stage disease via arthrodesis or arthroplasty. Unfortunately, these treatments are invasive, carry significant risk, and have unreliable results, thereby necessitating a search for alternative treatment modalities. Biologic treatment strategies aimed at slowing or reversing the degenerative process early in its course offer a promising alternative to end-stage treatment [2].

The IVD (Figure 7.1) is composed of an inner gel-like substance known as the 'nucleus pulposus' (NP), surrounded by an organized collagenous ring known as the 'annulus fibrosus' (AF). Above and below each disc are blood vessels that penetrate the bony vertebral bodies and cartilaginous end plates (CEP) to provide the disc with nutrients. The AF is populated by elongated fibroblast-like cells, which produce primarily type I collagen organized into concentric lamellae [2]. The collagen fibrils in each layer run parallel to each other, oriented approximately 30° from the long axis of the spine, with alternating layers running in opposing directions [4]. This arrangement allows the IVD to withstand tensile and

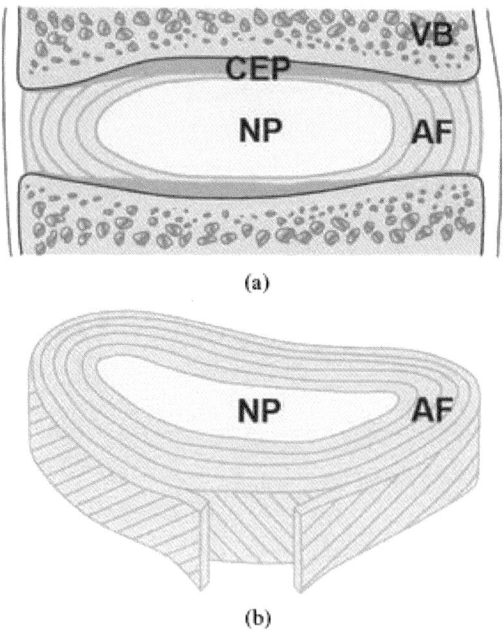


Figure 7.1 Schematic representation of an IVD. The CEP and the adjacent vertebrae (VB) are visualized in the sagittal section of the disc (A). The AF is formed by structured lamellae which encapsulate the central NP (B).

shearing loads. The NP, meanwhile, is populated by rounded, chondrocyte-like cells (cell population is predominantly notochordal at birth but shifts to chondrocytic with growth), which produce primarily type II collagen in an amorphous arrangement [2]. In addition, the ECM of the NP contains a significantly higher concentration of proteoglycan (~70%) as compared with the AF (~25%). This generates an osmotic gradient, which attracts and holds water within the tissue, conferring compressive strength to the NP [2]. While the NP contains several different proteoglycans (versican, lumican, decorin, biglycan, fibromodulin), aggrecan is the by far the most abundant, accounting for as much as 50% of the dry weight of the NP [3]. The ability of this molecule to attract and hold water is derived from the large negative charge associated with its disaccharide side chains (chondroitin sulfate (CS) and keratin sulfate). The aggrecan of the IVD also has the ability to bind to hyaluronan, with the help of link protein, to form large proteolycan aggregates, further augmenting its water-binding capacity [3].

Disk hydration is extremely important with regard to the biomechanics of the IVD. In a healthy IVD there is an even stress distribution across the end plate with both axial compression and eccentric loading [5]. The hydrated NP acts as a relatively incompressible fluid contained by the end plates and AF. With loading of the spine, force is transmitted to the NP through the compact subchondral bone of the end plate. The elevated pressure in the NP from axial compression is converted into tensile hoop stress within the organized lamellar structure of the AF. This stress is then dissipated by viscoelastic deformation of the involved tissues, including the slight deformation of the end plate and bulging of AF [5].

## 7.2 Degenerative Disc Disease

Despite the enormous clinical problem it is thought to represent, and the large body of research directed towards its understanding and treatment, there is a lack of consensus as to how DDD should be defined. There is also much debate as to which IVD changes are hallmarks of DDD and which can be considered to be consequences of 'normal' aging. It has been suggested that structural failure of the disc can serve as a basis for defining disc degeneration as it is easily detected, unambiguous, and not the inevitable result of aging [6]. The cell-mediated response to this failure, which includes elevated catabolic cytokine and enzyme levels, can be seen as a common final pathway to both aging and degeneration. In degeneration, the catabolic state is accelerated and symptomatic, but without biochemical events specific to its development [6].

The etiology of disc degeneration is multifactorial with both environmental and genetic factors playing a role in its development. Some of the environmental factors shown to contribute to DDD include heavy lifting, vibration, lack of exercise, tobacco use, and trauma [2]. Genes which have been associated with DDD include matrix metalloproteinase-3 (MMP-3), collagen I, aggrecan, vitamin D receptor, and collagen IX [2]. In addition, the molecular and cellular basis of disc degeneration is complex and involves multiple pathways and processes. These include accelerated cell senescence, decreased matrix production, increased degradative enzyme production, increased proinflammatory cytokine expression, increased apoptosis, and neural ingrowth [2].

There are a number of changes that occur within the IVD with aging and degeneration. With normal aging, the blood vessels from the vertebral endplates begin to recede, leaving the IVD with a largely avascular environment for the majority of the person's lifetime [2]. This, in turn, leads to an acidic microenvironment through the buildup of lactic acid, which results in a decreased ability of IVD cells to produce ECM [2]. Another significant change that occurs with aging and degeneration is that as the disc grows and ECM is laid down, the overall cell density decreases to a level lower than almost any other tissue in the body. This also contributes to decreased production of ECM and consequently a decreased ability of the disk to retain water [2]. Overall, this results in inferior load-bearing and may accelerate degenerative changes in the IVD.

There are at least three significant changes in the distribution and structure of collagen with degeneration of the NP [3]. First, the proportion of type I to type II collagen increases in the central NP, resulting in a more fibrous tissue. Second, the collagen fibers are subject to proteolytic cleavage by collagenases produced by disc cells. Third, the damaged collagen fibers non-enzymatically crosslink, yielding advanced glycation end products. Since many of the molecules are crosslinked, the damaged fibrils are not easily cleared from the NP, leading to accumulation of incompetent fibrous collagen. This accumulation decreases the overall swelling ability of the NP, limiting its capacity to efficiently dissipate compressive loads [7]. Other consequences of an increasingly fibrous NP and reduced swelling pressure in the IVD include decreased flow of nutrients into the tissue and increased stress placed on surviving cells within the disc further perpetuating the degenerative spiral [8].

With aging and degeneration, the aggrecan of the NP undergoes changes in both quantity and character. As cell numbers decrease and are subject to higher stresses, so the amount of aggrecan produced decreases. This phenomenon is a result of both decreasing cell numbers and a reduced rate of proteoglycan synthesis per cell [9]. Furthermore, the side chains, which are predominantly CS at birth, are replaced by keratin sulfate with age [10]. Whether this change is of benefit or not to the overall function of the NP is not yet known. The final significant change is proteolytic degradation of aggrecan and aggrecan–hyaluronan aggregates by matrix metalloproteinases (MMPs). The resulting large molecules are not as efficient at retaining water and, due to their size, are not easily cleared from the disc. They therefore accumulate and are subject to non-enzymatic crosslinking resulting in advanced glycation end products (similar to collagen, as discussed previously) [7]. The sum of these changes is further mechanical incompetence of the NP to resist compressive loads, contributing to the overall process of degeneration.

As mentioned previously, the proteolysis of collagen and proteoglycans is a significant change that occurs with DDD. The majority of matrix degradation in the NP is through the action of the MMP family of proteolytic enzymes. MMPs 1, 8, and 13 (collagenases) degrade collagen fibers, whereas MMPs 2, 3, and 9 are thought to contribute to proteoglycan degradation. The action of MMPs is inhibited by tissue inhibitors of metalloproteinases (TIMPs), which are found in the NP as well [11]. The balance between these two competing enzyme families is at the heart of the degenerating disc, with greater MMP and lesser TIMP expression seen in DDD. Other proteolytic enzyme families that contribute to NP matrix degradation to a lesser degree include ADAMTs and cathepsins [2].

Cytokines, small molecules which function as signaling molecules, have also been shown to indirectly promote the degenerative process. Interleukin-1 is an inflammatory cytokine expressed in the IVD which promotes DDD by inhibiting the production of ECM, increasing production of degradative enzymes, establishing a positive feedback loop for further cytokine production, and sensitizing IVD cells to other apoptosis triggers [2]. Another cytokine, tissue necrosis factor $\alpha$ (TNF-$\alpha$), has been shown to be associated with pain [2].

Finally, decreased cell density of the IVD with normal aging and DDD is due, in part, to increased rates of apoptosis and cell senescence [2]. Cell senescence, defined as the point at which a cell stops dividing, is a normal part of cellular aging [2]. In DDD, however, premature cell senescence is thought to occur secondary to the accumulation of irreparable DNA damage caused by oxygen free radical damage, mechanical injury, and inflammatory cytokine release [2]. These mechanisms are also thought to contribute to increased cell entry into apoptosis.

### 7.2.1 *Classification schemes for DDD*

Classification systems of the degenerating IVD are based on various methods of assessment. Two of the most commonly cited methods are the Thompson and Pfirrmann grading schemes. The five-grade system proposed by Thompson *et al.* is among the most cited and is based on gross morphology of the NP, AF, end plate, and vertebral body margin (Table 7.1) [12]. However, as this classification system was developed from examination of cadaveric specimens, no clinical correlation was established. Pfirrmann *et al.* developed a five-point scale based on MRI morphology and signal intensity

Table 7.1 Morphologic Grades of the Degenerating IVD after Thompson *et al.* 1990 [12].

| Grade | Nucleus | Annulus | End-plate | Vertebral body |
|---|---|---|---|---|
| I | Bulging gel | Discrete fibrous lamellas | Hyaline, uniformly thick | Margins rounded |
| II | White fibrous tissue peripherally | Mucinous material between lamellas | Thickness irregular | Margins pointed |
| III | Consolidated fibrous tissue | Extensive mucinous infiltration; loss of annular-nuclear demarcation | Focal defects in cartilage | Early chondrophytes or osteophytes at margins |
| IV | Horizontal cleft parallel to end-plate | Focal disruption | Fibrocartilage extending from subchondral bone; irregularity and focal sclerosis in subchondral bone | Osteophytes less than 2 mm |
| V | Clefts extend through nucleus and annulus | | Diffuse sclerosis | Osteophytes greater than 2 mm |

in a similar vein to Thompson's gross morphological system [13]. However, this classification system also has no established clinical correlation. While no single classification scheme is perfect, several issues are highlighted by these systems. First, it is clear that in order to appreciate the degree of degeneration, no single component of the IVD can be assessed in isolation. Rather, it is necessary to consider changes in the NP, AF, cartilaginous end plate, and adjacent vertebral bodies. It is also interesting that while the above-mentioned systems can categorize changes in the IVD, little if any correlation to clinical symptoms is sought. Furthermore, no clear distinction between age-related changes and pathologic degeneration is drawn.

### 7.2.2 *Pain in DDD*

While a clearer definition of IVD degeneration and an understanding of some of the biomechanical and biochemical processes involved in its pathogenesis is helpful, the source of pain in the degenerating spinal motion segment still requires clarification. Furthermore, it is often difficult to establish a correlation between the morphologic signs of degeneration and back pain [14].

The etiology of pain originating from the IVD, termed 'discogenic pain', is not well understood. It has been demonstrated that the central end plate and the periannular tissues have the most abundant nerve supply of the IVD [15]. The outer rim of the AF is thought to contain the bulk of IVD nociceptive fibers arising from the sinuvertebral nerve, a meningeal branch of the spinal nerve originating from the dorsal root ganglia [16]. Innervation of the normal disc is limited because nerve fibers do not normally penetrate deeper than the outermost layers of the AF. With the formation of annular fissures, neural ingrowth is permitted, resulting in more pervasive nociception throughout the disc [2]. In addition, normal aggrecan in the AF has been shown to inhibit neural ingrowth. Thus, with the decreased ECM seen in DDD, further promotion of neural ingrowth occurs [2]. The stretch of an abnormal AF, extravasation of foreign tissue into the peridural space, and pressure on adjacent nervous structures are thought to be potential mechanisms for pain generation. Nerve root compression or irritation can also be a mechanism for the generation of pain in DDD [17]. Finally, pain originating in arthritic facet joints, typical of late degenerate spinal motion segments, has also been implicated [18].

## 7.3 Current Therapeutic Options for DDD

### 7.3.1 *Non-operative management*

Non-operative therapy remains the treatment of choice for the vast majority of patients with LBP and is often the treatment modality initially attempted (Figure 7.2). Approximately 90% of individuals with LBP will have resolution of their symptoms within three months, with or without treatment, and most will have cessation of discomfort within six weeks [19]. While there is no standard protocol, non-operative measures include a combination of exercise, physical therapy, analgesics, anti-inflammatory medications, and epidural injections.

Physical therapy and exercise have been shown to be effective in the treatment of chronic LBP in adult patients. Hayden *et al.* published a meta-analysis on trials of exercise therapy for LBP and found it to be an effective therapy for chronic cases [20]. In addition, graded exercise therapy programs were found to be effective in patients with sub-acute pain. However, in the

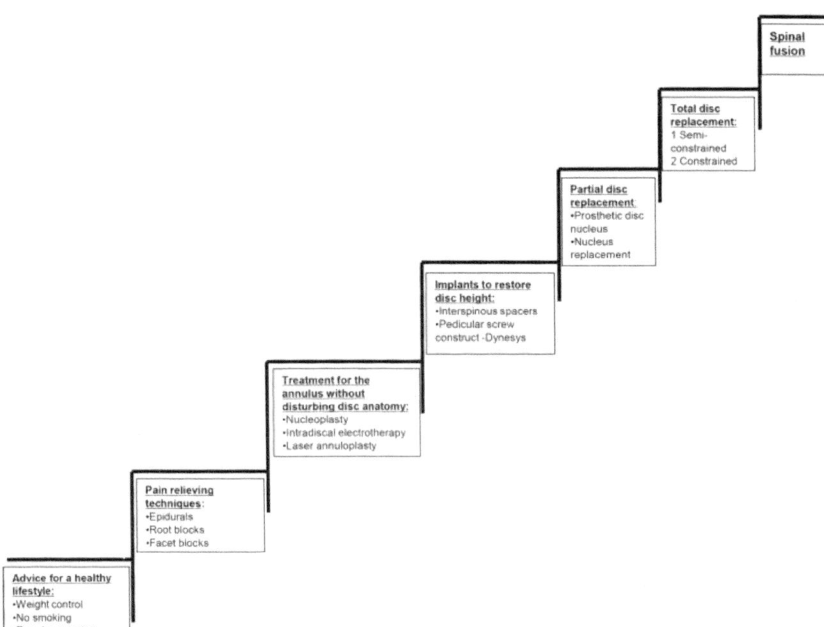


Figure 7.2 Stepladder treatment approach for management of pain from DDD. Reprinted from [63] with permission.

acute LBP population there was no evidence for the efficacy of exercise therapy. Chiropractic manipulation has been shown to have a modest effect on patients with acute LBP, however no benefits were seen in patients with chronic LBP [19].

The use of medications for the treatment of LBP is prevalent. Non-steroidal anti-inflammatory drugs (NSAIDs) have been found to be effective in the short-term relief of acute back pain [19]. In addition, the use of muscle relaxants has been shown to be efficacious in the treatment of acute back pain. Finally, the use of antidepressants in the treatment of chronic LBP has been shown to mitigate symptoms. Spinal epidural injection therapy has also been used to treat discogenic back pain, however there has been no evidence supporting its use [19].

### 7.3.2 *Surgical management*

In a proportion of patients, non-operative treatment will be unsuccessful and other avenues of therapy must be sought. The rates of hospital admission and surgery for patients with LBP have increased dramatically over recent decades and, in a select group of patients, successful outcomes have been obtained with surgical management [21].

#### 7.3.2.1 *Arthrodesis*

Currently, the most commonly performed surgical procedure for discogenic back pain is spinal fusion, with the goal of eliminating motion of the involved motion segment(s). Fusion can be achieved through a variety of surgical approaches and techniques, including anterior procedures, posterior procedures, and combined approaches. Spinal instrumentation (metal alloy implants) is often implanted for adjunct stabilization or to facilitate correction of deformity. The use of local or regional autogenous bone graft or newer bone graft alternatives provides additional surgical options for stabilization.

This form of treatment for discogenic back pain has been controversial due to a lack of consensus on its efficacy [19]. The efficacy of instrumentation is a topic of debate in the literature as well [19]. The Swedish Lumbar Spine Study Group in 2001 compared lumbar spine fusion with non-surgical management for the treatment of chronic LBP and found that fusion was more effective at relieving pain and decreasing disability than non-surgical treatment [22]. In addition, using a cost-effectiveness ratio, the group showed that lumbar fusion, while initially more expensive, was a cheaper form of treatment than non-surgical management in the long run [23]. Finally, the group compared three methods of fusion of the lumbar spine (uninstrumented posterolateral fusion, instrumented posterolateral fusion, and posterolateral with interbody fusion) and found that all the techniques reduced pain and improved function in patients with chronic LBP [24]. However, despite higher fusion rates in the circumferential groups (interbody fusion), higher rates of complications and more significant use of resources were observed, leading the investigators to conclude that there was no disadvantage in using the least demanding surgical techniques [19]. Although these studies support surgical intervention over conservative management, other studies cite conflicting findings, thus the issue remains controversial [23, 25].

In addition to the post-operative complications potentially seen with all surgical procedures (cardiac events, infection, bleeding, etc.), adjacent segment degeneration (ASD) is a concern specific to spinal fusion [19]. The elimination of motion at one spinal level can have adverse biomechanical effects on adjacent segments and has been shown to result in accelerated degeneration of the adjacent levels [26]. As a result, some patients may require subsequent surgical procedures to address symptoms arising from ASD [27, 28]. As such, the idea of motion-sparing devices has been gaining increasing popularity. Numerous surgical procedures fall into this category, including posterior soft-tissue stabilization, nucleus pulposus replacement, injectable biochemical polymer compounds, and total disk arthroplasty [19].

#### 7.3.2.2 *Total disc arthroplasty*

Total disc arthroplasty (TDA) is an approved alternative to arthrodesis for the treatment of discogenic back pain. It is designed to preserve motion in the associated spinal motion segment, thereby avoiding many of the complications described with fusion, in particular, ASD. Most designs incorporate metal alloy end plates and an ultra-high molecular weight polyethylene core [29]. There are a number of complications that have been reported with TDA leading to failure of the implant, including device migration necessitating reoperation, thereby setting the stage for further surgical complication [19]. A number of these patients continue to experience LBP despite TDA and go on to receive fusions, thereby negating the primary attempt to preserve motion in the affected region [19]. In addition, these implants do little to emulate the native IVD, leading to abnormal force distribution and subsequent complications such as vertebral body fractures.

#### 7.3.2.3 *Nucleus pulposus replacement*

NP-only replacement is another motion-preserving approach which has been developed for treating DDD. It is indicated for use in cases of symptomatic degenerating IVD with an intact AF and end plates [29]. Nucleus replacements can be divided into elastomeric and mechanical device categories. Currently many of these devices are investigational, however many of the same complications reported with TDA, such as migration or extrusion of the implant, could potentially occur as well.

#### 7.3.2.4 *Dynamic stabilization*

The final category of surgical intervention for DDD is dynamic stabilization. These devices are theorized to limit the motion of a segment to a non-painful range without overloading adjacent segments. They include dynamic interspinous devices, pedicle screw-based systems with flexible rods, and total facet arthroplasty. To date there is limited data evaluating their efficacy [30].

## 7.4 Nucleus Pulposus Regenerative Engineering

Biologic treatment strategies aimed at slowing or reversing the degenerative process early on offer a promising alternative to end-stage treatment. These can be divided into three broad categories: (1) those involving the use of cells; (2) the supplementation of matrix: and, (3) the application of therapeutic molecules [31].

### 7.4.1 *Cellular therapy*

One of the most significant features of DDD is a decrease in the number of viable cells within the IVD [2]. Therefore, in order to achieve meaningful tissue repair, new healthy cells may be introduced into the NP and/or native cells may be stimulated by bioactive molecules [2]. In addition, cells in the IVD must survive in the hostile degenerative environment and produce appropriate ECM molecules including proteoglycans and collagen type II [32]. While autologous NP cells or chondrocytes would be appropriate candidates for implantation, there is no readily available source for cell harvest.

Mesenchymal stem cells (MSCs), a primitive cell population found in many skeletal tissues, are excellent candidates for such applications. MSCs can be harvested from many autologous sources with little donor site morbidity or risk of disease transmission. The cells can be easily expanded in culture and, given their plastic nature, can be pushed toward a discogenic phenotype when exposed to an appropriate signal [33]. In addition, the primitive nature and high metabolic activity of MSCs may result in greater cell survival and a superior ability to produce adequate amounts of ECM compared to the relatively quiescent terminally differentiated native NP cell [32].

There are a number of limitations to the introduction of cells into the degenerative IVD. One issue is the uneven distribution of cells directly injected into the disc. In addition, the harsh avascular environment of the degenerative disc does little to promote the viability and biological activity of implanted cells. Finally, any ECM deposited by implanted cells would be vulnerable to the degradative milieu of the disc. Therefore, the creation of a microenvironment conducive to cell proliferation and ECM deposition and maintenance would be ideal [2, 34].

### 7.4.2 *Molecular therapy*

Another biologic process behind DDD that can be targeted is the upregulation of catabolism via enzymes and small molecules. Potential therapeutic molecules that could be introduced into the degenerative IVD include anticatabolics, mitogens, chondrogenic morphogens, and intracellular regulators [31]. These could potentially alter the overall catabolic environment of the degenerative IVD to one of anabolism.

There are a few impediments to the application of bioactive molecules in NP tissue regeneration. One limitation is the relatively short half-life of most of these molecules [2]. As such, they have a limited ability to provide a long-term stimulus which may be necessary for IVD regeneration. Another issue is the inadequate distribution of the molecule within the IVD [2]. Therefore, a delivery system capable of binding and achieving slow, even, sustained release of the active molecule(s) into the surrounding tissue would be ideal [2].

### 7.4.3 Nucleus pulposus scaffolds

The development of a scaffold to serve as a cell or drug carrier and possibly supplement the degenerating ECM may address some of the limitations associated with molecular and cellular therapy. A variety of materials are currently under investigation as scaffolds for NP tissue engineering. To be successful *in vivo*, a scaffold must provide a microenvironment conducive to cell growth as well as the migration and synthesis of ECM. In addition, a scaffold should have a structural make up that permits the transport of nutrients, metabolites, and regulatory molecules to and from the cells within it [2]. Constructs must be able to withstand the mechanical forces in the IVD in the acute phase of treatment, and will ideally remodel into the appropriate tissues in the long term [35]. For clinical application, issues concerning spinal biocompatibility and the ability to provide initial mechanical stability are also of paramount importance [32]. Many of the NP scaffolds reported on to date attempt to recapitulate the components of the IVD by combining both fibrillar molecules and glycosaminoglycans (GAG). Of particular interest are the category of emerging scaffolds known as hydrogels.

#### 7.4.3.1 *Hydrogels*

There is a growing body of research on the use of hydrogel implants for the treatment of DDD. Hydrogels are crosslinked polymer networks that have the ability to absorb and retain a great deal of water as well as maintain a distinct three-dimensional shape. Their ability to mimic *in vivo* tissue environments as well as their hydrated nature make them ideal candidates for use in IVD regeneration [36]. Many different applications of hydrogels for the treatment of DDD are currently being studied including their use as scaffolds for carrying cells, slow delivery release systems for bioactive molecules (drugs, small molecules, and growth factors), templates for the deposition of ECM, rehydration of the disk, and any combination of these. Ideally, a hydrogel scaffold would have the ability to be delivered into the degenerative IVD with minimally invasive techniques [37]. In addition, it would facilitate the restoration of disc structure and function by providing a protected microenvironment supportive of cell proliferation, ECM deposition, and the slow release of bioactive molecules within the harsh a vascular milieu of the diseased disc [38]. Hydrogels can be divided into two broad categories, those which are preformed and those which are crosslinked *in situ* (injectable).

While *in vitro* studies on cells seeded into preformed hydrogels have shown encouraging function, the concept of preformed hydrogels poses some difficulty [39, 40]. First, as cells needed to be seeded onto already cross-linked discs, adequate cellular penetration is difficult to achieve and distribution is not always even. Second, when implanting the scaffold into the spine, an incision in the AF would be necessary, requiring a more extensive procedure and putting the construct at risk of extrusion into the spinal canal.

Injectable hydrogels are ideal in that they allow minimal invasion through the AF, allow the injected material to adopt the shape of the defect, and allow the carried cells, drugs, molecules, and proteins to equally distribute throughout the tissue. In addition, an injectable hydrogel system is advantageous in that once it crosslinks *in situ*, the implant is immobile and has a lower likelihood of extravasation.

## 7.5 Current Injectable Hydrogels

### 7.5.1 *Naturally derived polymers*

Many different biomaterials are currently being investigated for use as injectable hydrogel implants in DDD. These include both naturally occurring and synthetic materials, as well as combinations of the two. Some of the more widely studied naturally occurring materials include hyaluronan, chitosan, fibrin, alginate, and collagen. In addition, there are a number of novel synthetic hydrogels which have been reported on in the literature as well as gels incorporating both naturally derived and synthetic elements, such as the use of polyethylene. Table 7.2 summarizes the various studies on naturally derived injectable hydrogels for NP regeneration.

Hyaluronic acid (HA) is a naturally occurring GAG found in the ECM of most connective tissue throughout the human body, including the IVD [41]. It is composed of repeating disaccharide units of $\beta$-1,4-linked $N$-acetyl-D-glucosamine and D-glucuronic acid. These subunits form large linear anionic polysaccharides which can attract and retain water within tissue containing HA [42]. HA was once thought to be an inert, space-occupying substance, however it has been shown that through interaction with specific binding proteins known as 'hyaladherins', HA can influence signal transduction, cell migration, and cell proliferation [42, 43]. In addition, HA is known to form massive multimolecular proteoglycan aggregates through interactions with

Table 7.2 Injectable naturally derived NP scaffolds in the literature.

| References | Authors | Journals | Material | Production | Cells tested |
|---|---|---|---|---|---|
| [50] | Mwale *et al.* | 2005 *Tissue Eng* | Chitosan x-linked with genipin | Gels at 37° | Bovine AF/NP |
| [58] | Sakai *et al.* | 2006 *Biomaterials* | Atelocollagen | Atelocallgen in DMEM, HEPES, gels at 37° | HNPSV-1 cells |
| [44] | Revell *et al.* | *J Mater Sci: Mater Med* 2007 | HYAFF and HYADD | Does not gel | *In vivo* porcine model |
| [45] | Gloria *et al.* | *J Biomater Appl* 2012 | HYAFF and HYADD | Does not gel | None |
| [46] | Li *et al.* | *Regen Med* 2014 | Fibrinogen – HA | | Bovine NP |
| [47] | Moss *et al.* | *Spine* 2011 | Thiol-modified Hyaluronan with ELP | PEGDA in DMEM gels at 37° | Human IVD cells in a rabbit model Human |
| [48] | Pereira *et al.* | *Biomaterials* 2014 | | Gels at 37° | MSCs in bovine IVDs |
| [49] | Frith *et al.* | *Biomaterials* 2014 | HA – poly(*N*–isopropylacrylamide) PEG/HA | Gels at room Temperature | Human mesenchymal progenitor cells |
| [51] | Richardson *et al.* | *Biomaterials* 2008 | | Gels at 37° | Human NP, MSC, chondrocyte |
| [52] | Dang *et al.* | *Biomaterials* 2006 | Chitosan-Glycerophosphate | Gels at 37° | Human MSC and IVD |
| [53] | Smith *et al.* | *Tissue Eng* 2014 | Hydroxybutyl chitosan | Gels at 37° | Bovine NP and MSC |
| [54] | Cheng *et al.* | *Tissue Eng* 2010 | dextran, chitosan, teleostean | Gels between 31°–34° | Rabbit NP cells |
| [56] | Naqvi and Buckley | *Tissue Eng* 2014 | Chitosan, gelatin, Glycerol phosphate | Gels at 37° | Porcine NP and MSC |
| [57] | Renani and Buckley | *Adv Biomed Res* 2012 | Chitosan, alginate, Alginate *vs.* chitosan-gelatin | Glutaraldehyde crosslinker | Human NP |

aggrecan and other ECM proteoglycans. These interactions are mechanically important for the load-bearing function of NP tissue [42]. There are a number of different injectable hydrogels which incorporate HA into their structure and have been reported on in the literature. The sole use of HA in IVD tissue engineering is not ideal for a number of reasons. Most commercially available forms of HA come as viscous liquids, which increases the likelihood that injected HA will extrude from the disc space and will be unable to provide the mechanical strength and stability required to withstand physiologic loading of the disc. In addition, the degradative environment seen in DDD could potentially lead to rapid loss of any injected HA. Therefore, current hydrogels tend to combine HA with other biomaterials in order to create a more biomechanically stable construct.

Numerous HA-containing gels have been shown to act as effective cell carriers into the IVD. Revell *et al.* injected bone marrow-derived autologous stem cells into injured porcine IVDs by using the hyaluronan-derived polymers HYAFF120 (HA combined with a molecule that initiates crosslinking on exposure to UV light) and HYADD3 (a compound with aliphatic side chains that facilitates crosslinking in water) [44]. They found viability of cells within these polymers even after crosslinking reactions had occurred. In addition, they found that this injection maintained overall histological appearance on post-mortem examination of the IVDs. Moreover, Gloria *et al.* demonstrated in a biomechanical study that these two HA derivatives emulate the viscoelastic behavior of native NP [45]. No data on IVD cell function in these constructs was reported. Other studies have demonstrated not only the biomechanical compatibility of hydrogels and carried cells with native NP tissue, but also an increase in IVD cell function. Li *et al.* showed bovine NP cell viability as well as increased NP marker gene expression in a fibrinogen–HA hydrogel *in vitro* [46]. They also demonstrated biocompatibility via the integration of the gel with native NP tissue as well as restoration of certain biomechanical properties including disc height and stiffness in a nucleotomized organ culture model. Moss *et al.* reported on a thiol-modified HA (TMHA) and elastin-like polypeptide (ELP) hydrogel crosslinked with poly(ethylene glycol) diacrylate (PEGDA) at 37°C [47]. They showed 70% viability of human IVD cells at three weeks with maintenance of IVD phenotype indicated by collagen II expression and cell morphology (Figure 7.3). They also demonstrated that the addition of ELP improved the biomechanical properties of the hydrogel. *In vivo* evaluation in a preclinical rabbit model demonstrated biocompatibility, and reduction in the severity of disc degeneration as compared to non-injured discs (Figures 7.4 and 7.5).

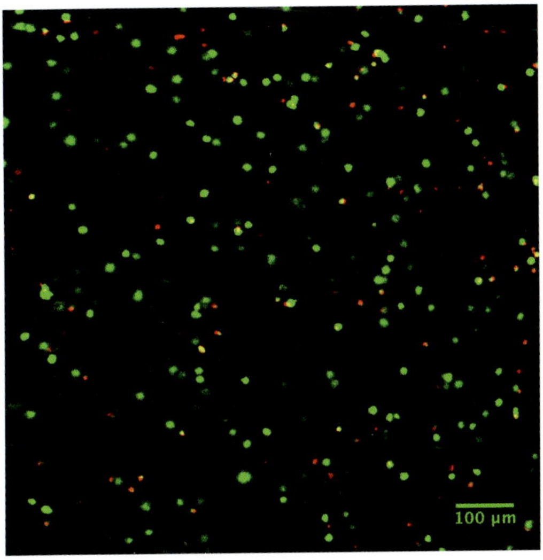


Figure 7.3 Representative confocal microscope image of live/dead staining of human IVD cells three weeks after seeding onto a TMHA/ELP4 scaffold. Live cells fluoresce green, nuclei of dead cells fluoresce red. Reprinted from [47] with permission.

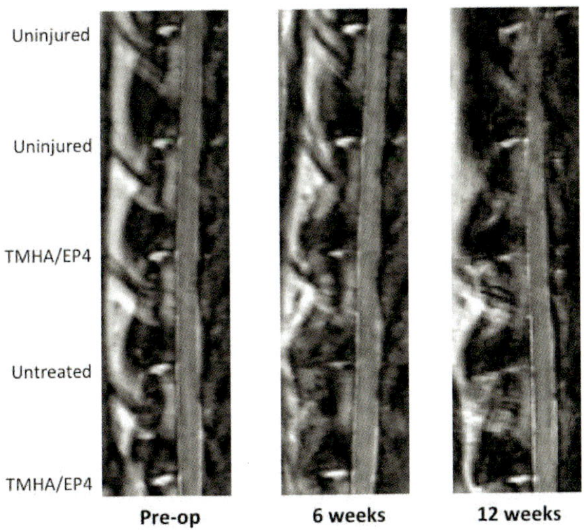


Figure 7.4 MRI images of uninjured, TMHA/EP4-injected, and injured but untreated rabbit IVDs. Uninjured and the caudal TMHA/EP4-treated IVDs demonstrate only minimal loss of area and signal intensity. The cranial TMHA/EP4-treated IVD and the untreated IVD demonstrate severe degenerative changes. Reprinted from [47] with permission.

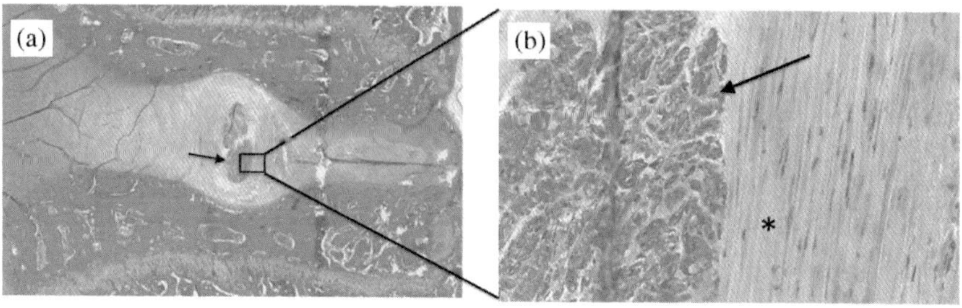


Figure 7.5 High power (a) and lower power (b) H&E-stained histological section demonstrating central placement if TMHA hydrogel (arrow) in the NP. The implanted material is surrounded by an organized fibrous tissue (*). Reprinted from [47] with permission.

Another demonstrated application of hydrogels is their ability to function not only as cell carriers, but also as slow release delivery systems for bioactive molecules such as drugs, signaling molecules, and growth factors. Pereira *et al.* reported on an HA–poly($N$-isopropylacrylamide) (HAP) hydrogel used as a delivery system for stromal cell-derived factor 1 (SDF-1) for the recruitment of human MSCs into a degenerative bovine IVD model [48]. They found increased MSC migration towards injured tissue with HAP hydrogels containing SDF-1, and found that the gel was conducive to cell proliferation and matrix deposition. Furthermore, it was shown that the gel promoted an NP-like phenotype in MSCs. This study highlighted the potential capacity of the IVD for endogenous repair when certain factors are introduced in the degenerative environment. Frith *et al.* demonstrated that the addition of pentosan polysulfate, a synthetic GAG-like molecule, to an HA-PEG hydrogel promoted viability and chondrogenesis in human mesenchymal progenitor cells [49]. While the gel was formed via enzyme-mediated crosslinking with horseradish peroxidase and hydrogen peroxide, no significant effect on cell viability was noted. They showed increased deposition of GAG and collagen II as well as NP tissue morphology after 21 days of culture. In addition, they showed the gel was biocompatible in a rat model and showed favorable biomechanics with rheologic testing.

Chitosan is another naturally occurring substance that has been widely reported on for use in injectable hydrogels for NP regeneration. Chitosan is an interesting molecule as it contains random $\beta$-1,4-linked $N$-acetylglucosamine and D-glucosamine subunits which are thought to mimic the biologic actions of hyaluronan [32]. Many hydrogel constructs containing chitosan have been shown to function as excellent vehicles for cell delivery and the promotion of ECM production. Mwale *et al.* reported on a chitosan scaffold which was crosslinked with the natural molecule genipin when heated to 37°C. Bovine NP cells entrapped within the chitosan were found to be 95% viable after 24 hours [50]. The material also exhibited an excellent biocompatibility (good integration with recipient tissue) profile. Richardson *et al.* reported 75% viability of human NP, chondrocyte, and MSC in chitosan-glycerophosphate gels after four weeks in culture as well as increased chondrogenic gene expression [51]. Dang *et al.* reported favorable viability of human IVD cells as well as increased ECM production in a hydroxybutyl chitosan gel for up to two weeks in cell culture [52].

A number of studies have also demonstrated the biomechanical suitability of chitosan-containing hydrogels. Smith *et al.* reported on a triple interpenetrating-network hydrogel comprised of dextran, chitosan, and teleostean [53]. A crosslinking agent was not needed as these materials were reported to solidify spontaneously on mixture at physiological temperatures. Bovine NP and MSCs were found to be 88% viable after 42 days with increased chondrogenic gene expression. In addition, the hydrogel demonstrated a lack of extrusion with extensive *in vitro* cyclic loading as well as integration with native NP tissue in human cadaveric discs. Cheng *et al.* reported on a chitosan, gelatin, and glycerol phosphate scaffold with gelation at 37°C [54]. They found this construct offered favorable gel strength and gelation time as compared with a chitosan glycerol phosphate gel in the absence of gelatin. They also reported viability of rabbit NP cells within the gel as well as increased GAG expression. In a separate study, Cheng *et al.* reported on the capacity of this scaffold to act as a slow delivery release system for the antioxidant ferulic acid [55]. They demonstrated decreased MMP-3 expression, increased aggrecan expression, and a decrease in apoptosis in rabbit NP cells stressed oxidatively with hydrogen peroxide.

There have been some studies that have compared chitosan with another naturally occurring polymer, alginate. Naqvi and Buckley compared porcine NP and MSC viability and gene expression in chitosan and alginate gels [56]. While both gels demonstrated favorable cell viability and ECM gene expression, they found increased cell viability and increased GAG–type II collagen gene expression in alginate gels as compared with chitosan gels. These results were supported by another study conducted by Renani *et al.*, which demonstrated increased cell viability in an alginate scaffold compared with a chitosan–gelatin scaffold [57]. Hydrogels composed of collagen have also been reported on in the literature. Sakai *et al.* reported on their experience with an atelocollagen preparation which is temperature sensitive [58]. They found that culture of cells from the HNPSV-1 human NP cell line on the scaffold led to the production of NP-like tissues.

### 7.5.2 *Synthetic polymers*

From the previous discussion it is clear that hydrogels composed of naturally derived polymers have been a popular choice for use in NP regeneration. However, there are a number of hydrogels composed of synthetic biomaterials which have also been reported on in the literature. Table 7.3 summarizes the various studies that used synthetic hydrogels for NP regeneration. It has been argued that the use of synthetic polymers is advantageous over natural polymers in that finer control can be exerted on the biomechanical properties of the construct as well as on degradation rates [36].

Kumar *et al.* reported on a synthetic photocurable hydrogel composed of poly(hydroxyl ethyl methacrylate)-co-N-(3-aminopropyl) methacrylamide grafted with polyamidoamine (p(HEMA-co-APMA)gPAA) [36]. The investigators demonstrated good cell viability of human MSCs encapsulated within the gel as well as minimal genotoxicity from the UV exposure used to solidify the gel. In addition, they were able to induce differentiation of the MSCs towards an NP phenotype demonstrated by the upregulation of gene expression of collagen II and aggrecan. Hu *et al.* reported on a silk fibroin polyurethane hydrogel [59]. They reported little cytotoxicity using rabbit bone marrow stromal cells and found that the gel was stiffer and more elastic than native NP tissue. Sivan *et al.* synthesized five sulfonate-based GAG analogues using PEGDA as a crosslinker [37]. They showed viability of bovine NP cells within the gels after 24 hours and demonstrated biomechanical stability of the gels with rheological testing. Furthermore, they reported that the constructs had viscoelastic properties similar to NP tissue. Finally, they showed that the gels were able to occupy the disc space without extrusion in a bovine explant model of degeneration.

The development of hydrogels composed of both synthetic and naturally derived polymers has also been investigated. In theory, these constructs should possess the advantages inherent to both classes of biomaterials, including the biocompatibility of naturally occurring compounds and the ability to modulate the biomechanical aspects of the synthetic compounds. Wiltsey *et al.* developed a hydrogel composed of the synthetic thermally sensitive polymer poly($N$-isopropylacrylamide) (PNIPAAm) and the natural biopolymer CS [60]. They added PEG in order to further enhance and

Table 7.3 Injectable synthetic NP scaffolds in the literature.

| References | Authors | Journal | Material | Production | Cells tested |
|---|---|---|---|---|---|
| [36] | Kumar *et al.* | *Acta Biomaterialia* 2014 | Polyhydroxyl Ethyl methacrylate-co-*N*-(3-aminopropyl) methacrylamide grafted with (p(HEMA-co-APMA)gPAA) Atelocollagen | Gels with UV exposure | Human MSCs |
| [59] | Hu *et al.* | *J Mater Sci Mater Med* 2012 | Silk fibroin polyurethane | Gels at $35^{\circ}$ | Rabbit bone marrow stromal cells |
| [37] | Sivan *et al.* | *Acta Biomaterialia* 2014 | Sulphonate based GAG analogues | PEGDA crosslinker, Gel at $37^{\circ}$ | Bovine NP cells |
| [60] | Wiltsey *et al.* | *J Mater Sci Mater Med* 2013 | Poly(*N*-isopropylacrylamide), Chondroitin sulfate | Gels above $32^{\circ}$ | Human embryonic kidney cells |

control the amount of water content and the degree of elasticity of the PNIPAAm. PNIPAAm was a solution below 32°C and a compact gel above this temperature. They hypothesized that the combination of PNIPAAm, with its optimized biomechanical profile, and CS, with its potential for enzyme degradability, anti-inflammatory activity, and nutrient absorption, would lead to a superior hydrogel design over either material in isolation. They found increased adhesive strength and decreased water loss over 14 days with the incorporation of CS into the hydrogel. In addition, they demonstrated a lack of cytotoxicity using human embryonic kidney cells *in vitro* and also demonstrated that the gel had similar compressive stiffness to native NP tissue.

## 7.6 Conclusions and Future Directions

In summary, DDD is a complex condition with high morbidity and socioeconomic cost. There are a number of conservative and surgical treatment strategies currently in use, however many are targeted towards the end stage and few attempt to intervene in the earlier stages of the disease. As such, an injectable biologic scaffold capable of halting the degenerative process at an early stage and promoting repair would be an ideal treatment modality in the future.

In order to be successful, a scaffold must provide a biologic microenvironment conducive to cell adhesion, proliferation, gene expression, and ECM production. In addition, a successful scaffold would need to provide adequate biomechanical support in the disc space without extrusion of the material and it would also need to exhibit compatibility with native IVD tissue [53].

Hydrogels are excellent candidates for use as biologic scaffolds in the NP in that they have a hydrated structure and are able to mimic *in vivo* tissue environments. There is a growing body of literature on the development of novel hydrogels for use in DDD as evidenced by the previous discussion. While many of these studies provide promising results, there are some limitations present in this body of research.

One limitation is that many of the current studies on hydrogels utilize animal cells and models such as bovine, porcine, and rabbit rather than human cells and models. While it is understandably more difficult to acquire human tissue, the NP of animals differs from that of human tissue in a number of ways including size and composition. For example, adult porcine and rabbit NP tissue contain notochordal cells, a lineage which is largely nonexistent within the adult human NP. In addition, biomechanical results on bovine, porcine, and rabbit models may not translate into a human model based on size and differences in the structure and function of the biped and quadruped spine.

Another limitation is the relative lack of studies using *in vivo* models. The IVD environment, which is complex and dynamic, plays a critical role in the development and progression of DDD. *In vitro* studies provide a relatively static environment compared to the *in vivo* environment of the native IVD and thus the properties of each gel seen *in vitro* may not necessarily correlate with how they behave in a living model.

One of the major underlying goals of current research on the effects of hydrogel implantation in the degenerative IVD is the ultimate translation of these results into an effective treatment modality in humans. With this goal in mind, future investigations should be directed toward better characterizing the effects of hydrogel injection in *in vivo* animal models of DDD. In addition, long-term studies are required in order to fully characterize the impact of this treatment modality on DDD. Due to the fact that many studies in the current literature report on the preliminary *in vitro* characterization of hydrogels, this data may be difficult to obtain at this time. However, cell viability within the gels and sustained compatibility with native IVD tissue documented over extended periods of time would be necessary in order to adopt hydrogels as a future treatment modality for DDD. Reports on the biomechanics of the hydrogel over time as well as a characterization of the IVD after long-term hydrogel implantation, ideally in a human model, should be future goals for this area of research.

With regard to the future direction of hydrogel design, the incorporation of novel biologics with hydrogels could result in optimized constructs for cell and small molecule delivery to the degenerative IVD. For example, Battig *et al.* reported on an aptamer-functionalized superporous hydrogel and its effects on the sequestration and release of the growth factor PDGF-BB [61]. They found that an aptamer, an oligonucleotide which can be selected to bind to any biologic, specific for PDGF-BB, could bind higher concentrations within the hydrogel. In addition they found that the aptamer helped to promote slower long-term release of the growth factor from the gel compared with a gel lacking the aptamer. This and other novel combinations of biologics with hydrogel technology could lead to superior outcomes. Overall, while there is a growing body of research on this subject with many promising results on novel hydrogel designs, further work must be done to establish the role of injectable hydrogels in the treatment of DDD.

## References


[1] Deyo, R.A. (2012). Back and neck pain, in D.L. Longo, *et al.* (eds.), *Harrison's Principles of Internal Medicine*. 18th edn. McGraw-Hill, New York.

[2] Kepler, C.K., Ponnapan, R.K., Tannoury, C., Risbud, M.V., and Anderson, D.G. (2013). The molecular basis of intervertebral disc degeneration. *Spine*, **13**(3), pp. 318–330.

[3] Roughley, P.J. (2004). Biology of intervertebral disc aging and degeneration: Involvement of the extracellular matrix. *Spine*, **29**(23), pp. 2691–2699.

[4] Roughley, P. *et al.* (2006). The potential of chitosan-based gels containing intervertebral disc cells for nucleus pulposus supplementation. *Biomaterials*, **27**(3), pp. 388–396.

[5] Horst, M. and Brinckmann, P. (1981). 1980 Volvo Award in biomechanics. Measurement of the distribution of axial stress on the end-plate of the vertebral body. *Spine*, **6**(3), pp. 217–232.

[6] Adams, M.A. and Roughley, P.J. (2006). What is intervertebral disc degeneration, and what causes it? *Spine*, **31**(18), pp. 2151–2161.

[7] Verzijl, N. *et al.* (2002). Crosslinking by advanced glycation end products increases the stiffness of the collagen network in human articular cartilage: a possible mechanism through which age is a risk factor for osteoarthritis. *Arthritis Rheum.*, **46**(1), pp. 114–123.

[8] Hutton, W.C. *et al.* (1999). The effect of hydrostatic pressure on intervertebral disc metabolism. *Spine*, **24**(15), pp. 1507–1515.

[9] Maeda, S. and Kokubun, S. (2000). Changes with age in proteoglycan synthesis in cells cultured *in vitro* from the inner and outer rabbit annulus fibrosus. Responses to interleukin-1 and interleukin-1 receptor antagonist protein. *Spine*, **25**(2), pp. 166–169.

[10] Sztrolovics, R., Alini, M., Roughley, P.J., and Mort, J.S. (1997). Aggrecan degradation in human intervertebral disc and articular cartilage. *Biochem. J.*, **326**(Pt 1), pp. 235–241.

[11] Goupille, P., Jayson, M.I., Valat, J.P., and Freemont, A.J. (1998). Matrix metalloproteinases: The clue to intervertebral disc degeneration? *Spine*, **23**(14), pp. 1612–1626.

[12] Thompson, J.P. *et al.* (1990). Preliminary evaluation of a scheme for grading the gross morphology of the human intervertebral disc. *Spine*, **15**(5), pp. 411–415.

[13] Pfirrmann, C.W., Metzdorf, A., Zanetti, M., Hodler, J., and Boos, N. (2001). Magnetic resonance classification of lumbar intervertebral disc degeneration. *Spine*, **26**(17), pp. 1873–1878.

[14] Keshari, K.R. *et al.* (2008). Lactic acid and proteoglycans as metabolic markers for discogenic back pain. *Spine*, **33**(3), pp. 312–317.

[15] Fagan, A., Moore, R., Vernon Roberts, B., Blumbergs, P., and Fraser, R. (2003). ISSLS prize winner: The innervation of the intervertebral disc: A quantitative analysis. *Spine*, **28**(23), pp. 2570–2576.
[16] Raj, P.P. (2008). Intervertebral disc: anatomy–physiology–pathophysiology–treatment. *Pain Pract.*, **8**(1), pp. 18–44.
[17] Chen, C. *et al.* (2004). Effects of nucleus pulposus on nerve root neural activity, mechanosensitivity, axonal morphology, and sodium channel expression. *Spine*, **29**(1), pp. 17–25.
[18] Lord, S.M., Barnsley, L., Wallis, B.J., and Bogduk, N. (1996). Chronic cervical zygapophysial joint pain after whiplash. A placebo-controlled prevalence study. *Spine*, **21**(15), pp. 1737–1744; discussion pp. 1744–1745.
[19] Madigan, L., Vaccaro, A.R., Spector, L.R., and Milam, R.A. (2009) Management of symptomatic lumbar degenerative disk disease. *JAAOS*, **17**, pp. 102–111.
[20] Hayden, J.A., van Tulder, M.W., Malmivaara, A., and Koes, B.W. (2005). Exercise therapy for treatment of non-specific low back pain. *Ann. Intern. Med.*, **142**(9), pp. 765–775.
[21] Taylor, V.M., Deyo, R.A., Cherkin, D.C., and Kreuter, W. (1994). Low back pain hospitalization. Recent United States trends and regional variations. *Spine*, **19**(11), pp. 1207–1212; discussion p. 1213.
[22] Fritzell, P., Hagg, O., Wessberg, P., and Nordwall, A. (2001). Lumbar fusion versus nonsurgical treatment for chronic low back pain: A multicenter randomized controlled trial from the Swedish lumbar spine study group. *Spine*, **26**(23), pp. 2521–2532.
[23] Fritzell, P., Hagg, O., Jonsson, D., and Nordwall, A. (2004). Cost-effectiveness of lumbar fusion and nonsurgical treatment for chronic low back pain in the Swedish Lumbar Spine Study: A multicenter, randomized, controlled trial from the Swedish Lumbar Spine Study Group. *Spine*, **29**(4), pp. 421–434; discussion p. Z3.
[24] Fritzell, P., Hagg, O., Wessberg, P., and Nordwall, A. (2002). Chronic low back pain and fusion: A comparison of three surgical techniques: A prospective multicenter randomized study from the Swedish lumbar spine study group. *Spine*, **27**(11), pp. 1131–1141.
[25] Brox, J.I. *et al.* (2007). Systematic review of back schools, brief education, and fear-avoidance training for chronic low back pain. *Spine*, **8**(6), pp. 948–958.
[26] Weinhoffer, S.L., Guyer, R.D., Herbert, M., and Griffith, S.L. (1995). Intradiscal pressure measurements above an instrumented fusion. A cadaveric study. *Spine*, **20**(5), pp. 526–531.
[27] Aota, Y., Kumano, K., and Hirabayashi, S. (1995). Postfusion instability at the adjacent segments after rigid pedicle screw fixation for degenerative lumbar spinal disorders. *J. Spinal Disord.*, **8**(6), pp. 464–473.
[28] Penta, M., Sandhu, A., and Fraser, R.D. (1995). Magnetic resonance imaging assessment of disc degeneration 10 years after anterior lumbar interbody fusion. *Spine*, **20**(6), pp. 743–747.
[29] Traynelis, V.C. (2002). Spinal arthroplasty. *Neurosurg. Focus*, **13**(2), p. E10.

[30] Don, A.S. and Carragee, E. (2008). A brief overview of evidence-informed management of chronic low back pain with surgery. *Spine*, **8**(1), pp. 258–265.
[31] Yoon, S.T. (2005). Molecular therapy of the intervertebral disc. *Spine*, **5**(6 Suppl), pp. 280S–286S.
[32] Anderson, D.G., Risbud, M.V., Shapiro, I.M., Vaccaro, A.R., and Albert, T.J. (2005). Cell-based therapy for disc repair. *Spine*, **5**(6 Suppl), pp. 297S–303S.
[33] Pittenger, M.F. *et al.* (1999). Multilineage potential of adult human mesenchymal stem cells. *Science*, **284**(5411), pp. 143–147.
[34] Kepler, C.K., Anderson, D.G., Tannoury, C., and Ponnapan, R.K. (2011). Intervertebral disk degeneration and emerging biologic treatments. *JAAOS*, **19**(9), pp. 543–553.
[35] Lotz, J.C., Staples, A., Walsh, A., and Hsieh, A.H. (2004). Mechanobiology in intervertebral disc degeneration and regeneration. *Conf. Proc. IEEE Eng. Med. Biol. Soc.*, **7**, p. 5459.
[36] Kumar, D. *et al.* (2014). Three-dimensional hypoxic culture of human mesenchymal stem cells encapsulated in a photocurable, biodegradable polymer hydrogel: A potential injectable cellular product for nucleus pulposus regeneration. *Acta Biomater.*, **10**(8), pp. 3463–3474.
[37] Sivan, S.S. *et al.* (2014). Injectable hydrogels with high fixed charge density and swelling pressure for nucleus pulposus repair: Biomimetic glycosaminoglycan analogues. *Acta Biomater.*, **10**(3), pp. 1124–1133.
[38] Frith, J.E. *et al.* (2013). An injectable hydrogel incorporating mesenchymal precursor cells and pentosan polysulphate for intervertebral disc regeneration. *Biomaterials*, **34**(37), pp. 9430–9440.
[39] Halloran, D.O. *et al.* (2008). An injectable cross-linked scaffold for nucleus pulposus regeneration. *Biomaterials*, **29**(4), pp. 438–447.
[40] Leone, G., Torricelli, P., Chiumiento, A., Facchini, A., and Barbucci, R. (2008). Amidic alginate hydrogel for nucleus pulposus replacement. *J. Biomed. Mater. Res. A*, **84**(2), pp. 391–401.
[41] Fraser, J.R., Appelgren, L.E., and Laurent, T.C. (1983). Tissue uptake of circulating hyaluronic acid. A whole body autoradiographic study. *Cell Tissue Res.*, **233**(2), pp. 285–293.
[42] Tammi, M.I., Day, A.J., and Turley, E.A. (2002). Hyaluronan and homeostasis: A balancing act. *J. Biol. Chem.*, **277**(7), pp. 4581–4584.
[43] Girish, K.S. and Kemparaju, K. (2007). The magic glue hyaluronan and its eraser hyaluronidase: A biological overview. *Life Sci.*, **80**(21), pp. 1921–1943.
[44] Revell, P.A. *et al.* (2007). Tissue engineered intervertebral disc repair in the pig using injectable polymers. *J. Mater. Sci. Mater. Med.*, **18**(2), pp. 303–308.
[45] Gloria, A., Borzacchiello, A., Causa, F., and Ambrosio, L. (2012). Rheological characterization of hyaluronic acid derivatives as injectable materials toward nucleus pulposus regeneration. *J. Biomater. Appl.*, **26**(6), pp. 745–759.
[46] Li, Z. *et al.* (2014). Biomimetic fibrin-hyaluronan hydrogels for nucleus pulposus regeneration. *Regen. Med.*, **9**(3), pp. 309–326.

[47] Moss, I.L., Gordon, L., Woodhouse, K.A., Whyne, C.M., and Yee A.J. (2011). A novel thiol-modified hyaluronan and elastin-like polypeptide composite material for tissue engineering of the nucleus pulposus of the intervertebral disc. *Spine*, **36**(13), pp. 1022–1029.

[48] Pereira, C.L. *et al.* (2014). The effect of hyaluronan-based delivery of stromal cell-derived factor-1 on the recruitment of MSCs in degenerating intervertebral discs. *Biomaterials*, **35**(28), pp. 8144–8153.

[49] Frith, J.E. *et al.* (2014). Effects of bound versus soluble pentosan polysulphate in PEG/HA-based hydrogels tailored for intervertebral disc regeneration. *Biomaterials*, **35**(4), pp. 1150–1162.

[50] Mwale, F. *et al.* (2005). Biological evaluation of chitosan salts cross-linked to genipin as a cell scaffold for disk tissue engineering. *Tissue Eng.*, **11**(1–2), pp. 130–140.

[51] Richardson, S.M., Hughes, N., Hunt, J.A., Freemont, A.J., and Hoyland, J.A. (2008). Human mesenchymal stem cell differentiation to NP-like cells in chitosan-glycerophosphate hydrogels. *Biomaterials*, **29**(1), pp. 85–93.

[52] Dang, J.M. *et al.* (2006). Temperature-responsive hydroxybutyl chitosan for the culture of mesenchymal stem cells and intervertebral disk cells. *Biomaterials*, **27**(3), pp. 406–418.

[53] Smith, L.J. *et al.* (2014). In vitro characterization of a stem-cell-seeded triple-interpenetrating network hydrogel for functional regeneration of the nucleus pulposus. *Tissue Eng. Part A*, **20**(13–14), pp. 1841–1849.

[54] Cheng, Y.H. *et al.* (2010). Thermosensitive chitosan-gelatin-glycerol phosphate hydrogels as a cell carrier for nucleus pulposus regeneration: An *in vitro* study. *Tissue Eng. Part A*, **16**(2), pp. 695–703.

[55] Cheng, Y.H., Yang, S.H., and Lin, F.H. (2011). Thermosensitive chitosan-gelatin-glycerol phosphate hydrogel as a controlled release system of ferulic acid for nucleus pulposus regeneration. *Biomaterials*, **32**(29), pp. 6953–6961.

[56] Naqvi, S.M. and Buckley, C.T. (2014). Differential response of encapsulated nucleus pulposus and bone marrow stem cells in isolation and coculture in alginate and chitosan hydrogels. *Tissue Eng. Part A*, **21**(1–2), pp. 288–299.

[57] Renani, H.B. *et al.* (2012). Determination and comparison of specifics of nucleus pulposus cells of human intervertebral disc in alginate and chitosan-gelatin scaffolds. *Adv. Biomed. Res.*, **1**(81).

[58] Sakai, D. *et al.* (2006). Atelocollagen for culture of human nucleus pulposus cells forming nucleus pulposus-like tissue *in vitro*: Influence on the proliferation and proteoglycan production of HNPSV-1 cells. *Biomaterials*, **27**(3), pp. 346–353.

[59] Hu, J. *et al.* (2012). Injectable silk fibroin/polyurethane composite hydrogel for nucleus pulposus replacement. *J. Mater. Sci. Mater. Med.*, **23**(3), pp. 711–722.

[60] Wiltsey, C. *et al.* (2013). Characterization of injectable hydrogels based on poly(*N*-isopropylacrylamide)-g-chondroitin sulfate with adhesive properties for nucleus pulposus tissue engineering. *J. Mater. Sci. Mater. Med.*, **24**(4), pp. 837–847.

[61] Battig, M.R., Huang, Y., Chen, N., and Wang, Y. (2014). Aptamer-functionalized superporous hydrogels for sequestration and release of growth factors regulated via molecular recognition. *Biomaterials*, **35**(27), pp. 8040–8048.

[62] Pattappa, G., Zhen, Li., Peroglio, M., Wismer, N., Alini, M. and Grad, S. (2012). Diversity of intervertebral disc cells: Phenotype and function. *J. Anat.*, **221**(6), pp. 480–496.

[63] Kumar, N., Shah, S., Tan, B., Juned, S., Yao, K. (2013). Discogenic axial back pain: Is there a role for nucleoplasty? *Asian Spine J.*, **7**(4), pp. 314–321.

## Chapter 8

# Injectable Hydrogels for Neural Tissue Regeneration

X. Li

*Translational Tissue Engineering Center, Whitaker Biomedical Engineering Institute, Johns Hopkins University, Baltimore MD 21231, USA*

D. Campbell-Rance

*Department of Chemical and Life Science Engineering Virginia Commonwealth University, Richmond, VA 23284, USA*

X. Liu

*Department of Chemical and Life Science Engineering Virginia Commonwealth University, Richmond, VA 23284, USA*

N. Zhang

*Department of Biomedical Engineering Virginia Commonwealth University, Richmond, VA 23284, USA*

X. Wen*

*Department of Chemical and Life Science Engineering Virginia Commonwealth University, Richmond, VA 23284, USA Institute for Advanced Materials and Nano Biomedicine (iNANO) Tongji University School of Medicine, Shanghai 200092, PR China*

*Author to whom correspondence should be addressed. Email: xwen@vcu.edu.

## 8.1 Introduction

Physical injuries and neurological disorders often lead to disruption in the architecture and proper wiring of the central nervous system (CNS). The result is regeneration failure stemming from the loss of neuronal cell bodies, axons, and supporting glial cells at the site of injury. Clinically, the CNS's inability to rewire and restore function after a physical injury are exemplified by patients suffering from traumatic brain injury (TBI) and spinal cord injury (SCI). Neurodegenerative disorders are prevalent and are characterized as hereditary or sporadic conditions which are marked by progressive nervous system dysfunctions such as loss of sensory, motor, and cognitive functions. These disorders include diseases such as Alzheimer's disease and other dementias, brain stroke, Parkinson's disease (PD), Huntington's disease, amyotrophic lateral sclerosis, and multiple sclerosis (MS) [1, 2]. Regeneration failure is due to the limited capacity of the CNS to replace neurons lost during injury or disease. The extent of this failure and the resulting functional consequences are likely to increase if there is a disruption in the blood supply to the site of injury. Disruption in blood supply usually occurs due to hemorrhage or ischemia and is likely to cause tissue damage or death, and can block messages (nerve impulses) traveling along the CNS.

Preclinical data has shown that stem cell therapy, especially of NSCs, is a promising approach for regeneration of injured/diseased tissues of the CNS. This strategy is considered versatile because NSCs are available from a number of sources and it is believed that these stem cells will provide an inexhaustible source of neurons and glia for therapies aimed at cell replacement. Sources for NSCs include embryonic stem cells (ESCs), fetal stem cells, adult stem cells, and induced pluripotent stem cells (iPSCs) [1–5]. Apart from the renowned ethical issues associated with stem cell research, the development of stem cell therapies are hindered by insufficient knowledge of disease pathology and how specific cell types in various locations of the CNS are affected. Other issues that need to be resolved before this type of therapy becomes a viable treatment option for CNS regeneration include low viability of transplanted stem cells, uncontrolled differentiation, the inability to control stem cell fate including avoidance of teratoma formation and low cell engraftment, and integration with existing neural circuitry after transplantation. Ensuring cell survival is a critical issue as only a small percentage of transplanted cells will still be viable within several days of transplantation. Usually, these cells die within the first few days of transplantation due to acute inflammation/immune response, trophic factor withdrawal, oxidative stress, excitotoxicity, hypoxia, or anoikis [4]. In an attempt to improve the survival and integration of the transplanted stem cells, researchers have developed biomaterial scaffolds which have been used to culture, transplant, and influence the differentiation of these cells. Biomaterial scaffolds having three-dimensional lattices have been engineered to support NSCs and mimic the extracellular matrix (ECM) after transplantation. Biomimetic scaffolds such as injectable hydrogels may provide a permissive microenvironment to enhance cell survival and control NSC fate, both *in vitro* and *in vivo*, by precisely delivering stem cell regulatory signals in a physiologically relevant manner [6–8]. Bioengineered strategies involving ECM-modified as well as neural-compatible hydrogels with signaling molecules and NSCs may significantly impact the future of stem cell-based treatments for CNS regeneration [9, 10].

Readers interested in a more detailed and comprehensive review of tissue engineering strategies for CNS regeneration are encourage to consult the excellent reviews available in the literature [7, 8, 11–13]. Here, the focus will primarily be on presenting a more detailed overview on the use of rationally designed hydrogels for neural tissue regeneration, with particular attention on hydrogel design criteria for engineering and manipulating stem cell fates. There will also be a discussion on the outcomes of animal studies using hydrogels as a carrier for neural stem cell transplantation. Finally, some challenges in the development of hydrogels for CNS therapy will be outlined.

## 8.2 CNS Injuries and Diseases

### 8.2.1 *Traumatic brain injury*

Instances of TBI are prevalent in modern societies, and it is a significant contributor to death and disability in the United States. The US Centers for Disease Control and Prevention estimates that over 1.5 million Americans sustain TBI annually and an estimated 5.3 million people currently live with a permanent TBI-related disability. An estimated $56.3 billion is spent on the direct and indirect cost of TBI annually, with the Department of Defense funding clinical and basic research to the tune of $500 million. Even though research on TBI has received increased public attention in recent years, there is still no clinical treatment available to regenerate tissue damaged by a TBI [14]. The reasons are complex and could be related to the fact that researchers may have focused on drugs with a very specific mechanism of action, not accounting for the heterogeneous nature of these brain injuries. The heterogeneity of TBIs may arise because of variations in the type, location, and extent of the primary injury. Additional reasons for the absence of clinical treatment for TBIs include the contributions of different pathological mechanisms, as well as delayed secondary injury mechanisms and compounding conditions, such as brain swelling, increases in intracranial pressure, and hypoxia [15].

### 8.2.2 *Spinal cord injury*

The result of acute SCI is physical disability often accompanied by psychological devastation that especially affects healthy young people in their most active years. This type of injury is especially debilitating since there are no universally accepted treatments available, and it leaves the injured person paralyzed below the level of the injury where there is disruption in nerve connection. There are approximately 450,000 individuals living with SCI in the US and an estimated 11,000 new cases are expected to be diagnosed each year. The strategic plan for therapy is to understand the course of pathological changes after SCI and identify effective therapeutics. Since restoring effective mobility is one of the ultimate goals of SCI therapy, then it follows that neuroprotection of nerves tissues must be strategic in limiting functional loss and allowing the process of regeneration to occur. Currently treatment options for SCI include administrations of high doses of methylprednisolone, surgery, hypothermia, intensive multisystem medical management, and rehabilitative care. Methylprednisolone, a neuroprotectant, is one of the main clinical interventions of SCI and its administration helps to reduce secondary injury processes. Despite its neuroprotective ability, treatment with methylprednisolone is controversial due to the many serious side effects that result, including gastric bleeding, sepsis, pneumonia, acute corticosteroid myopathy, and wound infection. Further, this treatment only boasts very minute improvements in neurological recovery [16, 17].

### 8.2.3 *Stroke*

Brain stroke occurs when a blood clot blocks an artery or if a blood vessel ruptures, leading to an area of the brain experiencing a considerable reduction in cerebral blood flow. Brain stroke is thus either ischemic or hemorrhagic, depending on the event by which it is initiated. Despite being the second leading cause of death worldwide, the majority of patients survive strokes, making this CNS disorder a major source of human disability. Brain stroke has a negative impact on American society in that it creates a socioeconomic burden of approximately $54 billion per year. Over 500,000 new cases of brain stroke are reported in the US each year and this disease accounts for approximately 6% of all health care budgets. Ischemic stroke accounts for 80% of all brain strokes, however not many effective treatment options are available. The most promising clinical treatment for ischemic stroke is thrombolysis, which is centered on the principal cause: the thrombus (clot) that blocks the cerebral artery. Therapy in which thrombolysis and neuroprotection are combined has demonstrated limited benefits in small numbers of patients. Tissue-type plasminogen activator (t-PA) is the only FDA-approved acute therapy for ischemic stroke, and has a short treatment window for administration after the onset of symptoms [18, 19]. Research efforts aimed at improving neuroprotective drugs for this type of injury have failed in clinical trials during the last two decades [20].

### 8.2.4 *Parkinson's disease*

PD is one of the most common progressive neurodegenerative diseases that affects about 2–4% of the adult population between the ages of 65 and 85. It is characterized by depletion of dopaminergic (DA) neurons in the substantia nigra as wells as a reduction in the dopamine levels in caudate and putamen. The disease is symptomatic of tremors, rigidity, dysfunction in movement and balance, bradykinesia, and loss of other cognitive abilities. Some PD patients may also suffer from anxiety, depression, autonomic disturbances, and dementia. Currently, there is no cure for PD as available therapies do not reverse the progression of the disease, only slow its progress. A number of studies have been conducted in a bid to find promising neuroprotective agents but to date no drug has been identified to have a disease-modifying effect on PD. Currently, the gold standard for symptomatic treatment of PD is L-3,4-dihdroxyphenylaline (L-DOPA), which is orally administered and is converted to dopamine in the body. However, as the degeneration of the disease progresses, this treatment loses effectiveness and adverse effects such as dyskinesias set in [21, 22].

### 8.2.5 *Multiple sclerosis*

MS is an inflammatory autoimmune disease which impacts the brain and spinal cord. The immune system attacks neurons in the CNS, causing demyelination and axon loss. Demyelination disrupts communication between the brain and the rest of the body, causing irreversible deterioration of the nerves, ultimately leading to the inability to walk. Currently, there are approximately 250,000–350,000 people in the US alone living with this debilitating disease, which accounts for an estimated annual cost that is in the multi-billion dollar range. Similar to other disorders of the CNS, MS is clinically and pathologically heterogeneous, with four different pathological patterns described by Lucchinetti *et al.* [23]: macrophage-mediated demyelination, antibody-mediated demyelination, distal oligiodendroliopathy and apoptosis, and primary oligiodndroglia degeneration. These patterns are recognized in different patients at various stages of the disease and may represent varied disease mechanisms, and consequently respond differently to MS therapies. Currently, there are no effective clinical treatments that promote remyelination, but researchers have proposed the strategy of combination therapy for MS treatment based on the heterogeneous nature of the disorder. Available treatments for MS involve the use of interferons and glatiramer acetate (GA) for relieving symptoms, reducing relapse frequency, accelerating recovery, and preserving functional competence [23]. Additional treatment for MS involves the use of immunosuppressive and immunomodulatory agents, but this has little efficacy in preventing long-term disability or restoring lost function [24].

## 8.3 Neural Stem Cells for CNS Regeneration

NSCs are cells present in the CNS of the mammalian fetus and adult, and are capable of multipotency and self-renewal. They give rise to the three cell types present in the CNS (neurons, astrocytes, and oligodendrocytes) and are responsible for the limited capacity of the developed brain to regenerate after injury/disease. NSCs can be derived from fetal stem cells, adult stem cells, ESCs, and iPSCs [25]. NSCs affect embryonic development at different stages. In the initial stage, the embryonic neural tubule is comprised of NSCs that exist as neuroepithelial stem cells. At the onset of neurogenesis, NSCs undergo symmetrical division giving rise to radial glia cells which subsequently generate neuronal lineages via asymmetrical division within the germinal ventricular zone (GVZ). Glial progenitor cells are generated after the NSCs acquire gliogenic competency, and proliferate in the subventricular zone (SVZ). The postnatal stage is marked by the transformation of radial glia to astrocytes and the disappearance of the ventricular zone. However, some portions of the SVZ are preserved into adulthood and become the site of neurogenesis in adults [5]. These sites are very important when tissue engineering approaches are used to generate the tissues of the CNS. This is achieved either through activation of the endogenous stem cells or transplantation of NSCs [26, 27]. These methods will be briefly described below, as well as a discussion of some issues involved with each method.

### 8.3.1 *Endogenous neural stem cells*

It is well established that stem cells are endogenous to the adult mammalian brain and spinal cord. This fact supports the concept of plasticity in the mature CNS [28]. As mentioned previously, adult neurogenesis are restricted to two zones: the SVZ, located along the lateral wall of the lateral ventricle [29], and the subgranular zone (SGZ), which is positioned between the dentate gyrus and the hilus of the hippocampus [30]. Newly generated neurons are supplied to the olfactory bulb and the dentate gyrus by progenitor cells in these regions. The adult spinal cord is also known as a site for NSCs which reside in the white matter parenchyma or close to the central canal, either in the ependymal or subependymal [31, 32]. Weiss *et al.* conducted *in vitro* studies using adult mammalian spinal cord to demonstrate that these cells display self-renewal and multipotent features [33]. Additionally, three different research groups conducted studies showing that non-neurogenic regions of the brain are capable of inducing neurogenesis after injury. They studied areas such as the striatum and cerebral cortex, regions susceptible to brain ischemia. These findings validate the idea that some regenerative abilities are preserved in the adult mammalian brain and spinal cord [2, 32, 34–38]. Even though these endogenous NSCs are stimulated to proliferate, migrate, and differentiate after injury, their regenerative capacities are still very low. The result is the inability of these cells to reconstruct the normal cellular/structural design of the CNS. Researchers have thus utilized agents such as neurotrophins and signaling molecules in an effort to enhance the regenerative capacity of endogenous NSCs [3, 26, 39–41].

### 8.3.2 *Neural stem cell transplantation*

NSC transplantation therapy is a feasible approach for the regeneration of damaged CNS tissues, with the results of many studies indicating that the grafted cells express the appropriate neurotransmitters and receive synaptic communication from the host neurons [27]. However, the destiny of the transplanted cells is affected by the extent and type of damage to the CNS in conjunction with the biochemical and biomechanical signals in the microenvironment surrounding the injured site. The viability and functionally of the transplanted cell is influenced by the hostile microenvironment at the site of diseased/injured tissue [4]. Direct injection of cells into the damaged site is not encouraged as this may cause further damage to the host tissue. Restoration of the normal physiological microenvironment of the transplanted cells is required to enhance their survival and desirable fate [42, 43]. The microenvironment serves as an anatomical functional compartment for NSCs and usually contains the endogenous stem cells, stromal cells, blood vessels, and a specialized ECM [44]. This specialized ECM is a three-dimensional structural lattice scaffold and is comprised of signaling molecules and moieties, bound growth factors and cytokines. The behavior of transplanted NSCs is controlled by the architecture of the structure at the lesion site, as well as mechanics, signaling moieties, and bound growth factors [45]. This knowledge can be used to address the major issue of cell transplantation by promoting the development and use of biomaterials with the specialized ECM environment that mimics their natural microenvironment [11, 12, 42].

The use of biomaterials such as hydrogels, fibrous and macroporous structures, is very attractive for this application since they are very versatile and can possess a variety of structures by simply altering their material composition, synthesis, or processing method. This versatility allows for the formation of three-dimensional environments that are suitable for NSC delivery and occupancy. In addition to this, their crosslinked hydrophilic polymeric network is beneficial for this type of application because this feature is very similar to the physical characteristics seen in soft tissues. Other important attributes of hydrogels that make them a more suitable biomaterial than other solid scaffolds include high surface tension as well as high oxygen and nutrient permeability. Interfacial tension minimizes barriers to cells and processes as they migrate into the biomaterials from the surrounding tissues.

## 8.4 Engineering Neural Stem Cells' Fate by Hydrogel Design

Compared to other organs in the body, the CNS is composed of very unique tissues because of the differences in its biomechanical properties, such as its characteristic soft physical properties and the three-dimensional structure of its ECM. The natural brain is characterized as soft matter when compared to muscle or bone. It has an elastic modulus of only 500 Pa whereas the elastic modulus of muscle or bone is on the order $10^4$ Pa and between $10^9$ and $10^{10}$ Pa, respectively [46–48]. These differences in architecture, biomechanical properties and composition of ECM significantly influence the overall cellular functions of the brain. Further supporting evidence that the cells comprising the CNS are uniquely different from those of other organs are described below. The cells of the CNS do not experience mechanical stimuli because they do not bear loads, while the cells in most other organs do. The composition of the ECM of the CNS differs from that in many other organs because components such as fibronectin and collagen are absent while there is an abundance of proteoglycans located in the intercellular spaces between neurons and glial cells [49, 50]. The majority of the CNS proteoglycans have either chondroitin sulfate or heparan sulfate side chains, along with binding growth factors and proteins such as serine protease tissue plasminogen activators (tPAs) and matrix metalloproteinases (MMPs). Additionally, other specialized ECM structures found only in the brain contain proteoglycans as well. For instance, the perineuronal net (PNN) is composed primarily of proteoglycans, as well as other components, including lecticans, hyaluronic acid, tenascin-C, and tenascin-R. This extremely specialized ECM has a number of functions; not only does it provide physical and biochemical support, but it also sustains homeostatic functions crucial to the survival of terminally differentiated neurons.

Numerous hydrogel scaffolds have been proposed and evaluated for CNS applications in an attempt to mimic the biochemical and mechanical properties of the CNS. The evaluation used naturally derived hydrogels such as collagen, hyaluronic acid (HA), agarose, alginate, chitosan, methyl cellulose, xyloglucan, Matrigel, fibrin, PolySia-based hydrogels, and peptide hydrogels, as well as synthetic hydrogels such as polyacrylamide and PEG hydrogels [8, 51]. Cell behaviors such as viability, adhesion, neurite outgrowth, proliferation, differentiation, and migration are affected by the physical and biochemical properties of these hydrogels. Important physical and biochemical properties affecting cell behavior include stiffness, morphology, structure, and biological modifications [45]. This observation should prompt research to consider the issues cited below when designing hydrogels for CNS application. Primarily, a hydrogel designed for neural tissue regeneration should promote the survival of the transplanted cell while simultaneously maintaining the intended cellular functions. Secondly, the hydrogel should be macroporous or become macroporous upon degradation to facilitate cellular infiltration of the transplanted cell and neurite outgrowth. Most importantly, the mechanical properties of the hydrogel should closely mimic that of the native CNS tissue in order to ensure that the cells within the hydrogel receive the appropriate mechanical growth cues. If hydrogels are tailored with these considerations, they should have the capacity to provide a biomimetic, three-dimensional stem cell niche at the site of damage that will allow for enhanced viability and controlled proliferation and differentiation when implanted *in vivo* [52].

### 8.4.1 *Hydrogels*

#### 8.4.1.1 *Synthetic hydrogels*

CNS regeneration studies have used both natural and synthetic scaffolds to create an artificial ECM for the transplanted neural stem cell. Synthetic hydrogels are found to have some noteworthy advantages over natural hydrogels, namely tunable mechanical properties, and facile control of scaffold structure and chemical composition. However, these synthetic hydrogels are typically bio-inert and unable to provide a microenvironment that supports cell attachment and tissue regeneration [53]. To overcome this obstacle, chemical modifications are proposed as a practical strategy to make the chemistries of hydrogels more comparable to that of the native ECM and are now being investigated.

##### 8.4.1.1.1 Polyacrylamide hydrogels

NeuroGel™, a synthetic biocompatible hydrogel of poly(N-2-(hydroxypropyl) methacryamide) (PHPMA) incorporating the RGD peptide sequence (Arg-Gly-Asp) in its structure was pioneered by Dr. Stéphane Woerly. It was developed with neural specifications to encourage structural and functional repairs to damaged CNS tissues. This hydrogel was engineered with multimodal porosity allowing transport of small and large molecules, migration of cells and blood vessels, as well as the establishment of an interface with the host tissue in order to stimulate tissue remodeling and construction within the hydrogel. Based on these features and the fact it does not require the inclusion of any trophic/growth factor, NeuroGel$^{TM}$ is deemed a desirable scaffold to bridge spinal cord lesions and promote the repair of damaged tissues [54, 55].

Healy *et al.* [56] developed another polyacrylamide hydrogel for investigating CNS regeneration, which contained poly(ethylene glycol) (PEG). This hydrogel was designed to control NSC behavior by using an architectural motif similar to that seen for laminin, vitronectin, and fibronectin: an RGD (Arg-Gly-Asp) adhesion motif. The RGD and another laminin motif, IKVAV, were grafted into the hydrogel interpenetrating polymer network to create a well-defined mechanical and chemical culturing environment for NSCs and to regulate cell behaviors such as adhesion. Results indicated that RGD controlled adult NSC self-renewal and differentiation in a dose-dependent manner; at a concentration $>5.3\,\text{pmol/cm}^2$, RGD promoted cell self-renewal and differentiation, while IKVAV only had modest effects on differentiation [56]. Additional parameters affecting stem cell behavior include the concentration of the monomers of acrylamide or the concentration of the crosslinker $N$,$N'$-methylenebis (acrylamide). Adjusting these concentrations influences the mechanical properties of the hydrogels, which can be used to observe their effects on stem cell behaviors [57, 58]. For instance, one study demonstrated that adult murine NSCs yielded mainly glial cells when cultured on a relatively stiff synthetic matrix (elastic modulus: 1,000–10,000 Pa). When cultured on a soft matrix (elastic modulus: 100–500 Pa), the stem yielded primarily neuron cells. Maximum neuron yield was obtained under conditions of neuronal differentiation on IPNs with a stiffness of 500 Pa; a stiffness similar to that of brain tissue (Figure 8.1) [58]. Langrana *et al.* created an interesting new class of polyacrylamide hydrogels featuring reversible gelation, in which DNA was used as the crosslinker [59]. The results indicated that rat spinal cord neurons extended more primary dendrites and possessed shorter axons when seeded on stiffer hydrogels (elastic modulus: 30,000 Pa). An explanation for this result is based on the involvement of focal adhesion kinase is in neuritegenesis due to this mechanical stiffness [60].

Polyacrylamide hydrogels exhibit many excellent attributes, including the ability to adjust the matrix compliance and surface coating. However, there is one major disadvantage: the hydrogel polymerization hinders its use as an injectable biomaterial. Further, human studies provided evidence that adverse effects of acrylamide and its major metabolite, glycidamide,

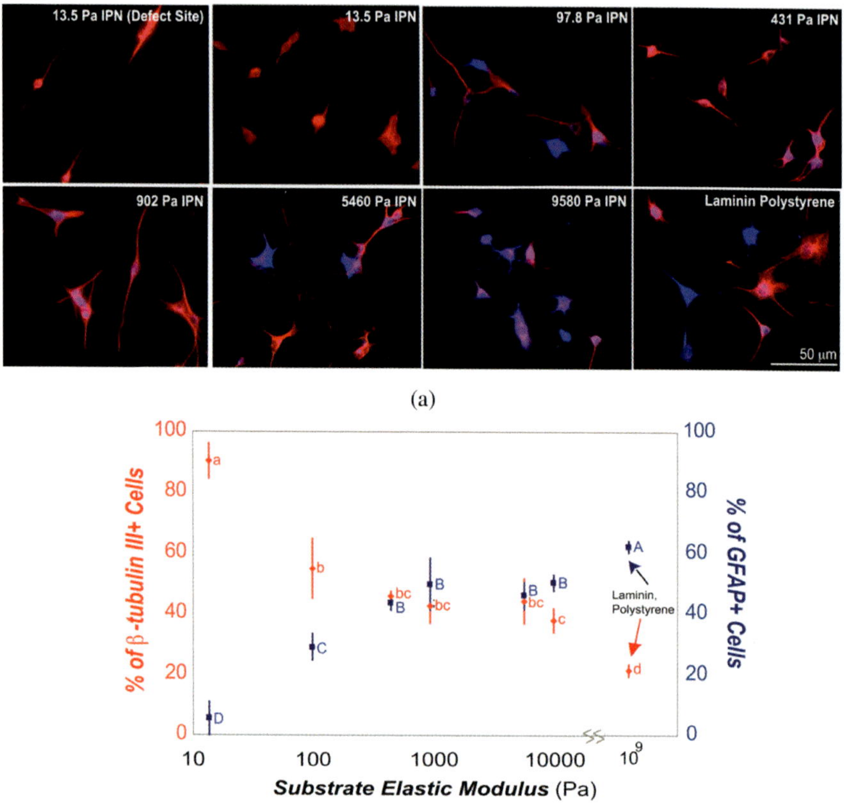


Figure 8.1 The proportion of neurons versus glia under mixed differentiation is a strong function of polyacrylamide hydrogel modulus. (a) Immunostaining images of NSCs differentiated under mixed glial and neuronal differentiation conditions (1 mM retinoic acid and 1% FBS). (b) Percentage of cells that stained positive for beta-tubulin III (red) on left axis and for GFAP (blue) on right axis. Reprinted from [58] with permission from Cell Press.

exist and cause neurotoxicity, while reproductive toxicity, genotoxicity, and carcinogenicity were seen in animal studies [69].

#### 8.4.1.1.2 Polyacrylate hydrogels

Poly(2-hydroxyethyl methacrylate) (PHEMA) is one of the super-porous, three-dimensional scaffolds designed to provide support for tissue regeneration. It is non-degradable and has certain attributes that make it a good scaffold; these include mechanical strength, elasticity, a long history of successful application in medicine, and the ability to be designed into various architectures. Even though PHEMA is tolerated to some extent *in vivo*, it is not easily conjugated with biological molecules because it lacks the desired functional groups. To overcome this situation, co-monomers containing carboxy and amino functional groups are usually incorporated in the material. Sykova *et al.* tested the cell–surface interactions of a PHEMA hydrogel scaffold which was modified with a covalently attached IKVAV adhesive peptide sequence. Amine functionality was introduced to PHEMA via NHS/EDC chemistry between co-monomers, 2-aminoethyl methacrylate (AEMA) and 2-hydroxyethyl methacrylate (HEMA) [61]. The hydrogel was further modified with other peptide sequences such as CDPGYIGSR and CQAASIKVAV in an attempt to improve neural cell adhesions and guide neurite outgrowth of primary chick dorsal root ganglia (DRG) [62]. This hydrogel was then used to fabricate nerve guide channels (NGCs) [63], with the aspiration of supporting axonal regeneration after SCI [64]. ECM components and signaling molecules were integrated within the lumen of the NGCs. The ECM components added included molecules such as collagen, fibrin, Matrigel, and methycellulose, while nerve growth factor (NGF), fibroblast growth factor 1 (FGF 1), and neurotrophin 3 (NT 3) were added to the signaling molecules. These added features were shown to influence the origin and quantity of the axons regenerated. The outcome was an improved functional recovery [65, 66]. Coils were subsequently added into these NGCs to reinforce the *in vivo* implanted hydrogel channels and prevent the NGCs collapse in the long term [67, 68].

#### 8.4.1.1.3 Poly(ethylene glycol)

PEG has many excellent qualities for its application as a polymeric hydrogel, which include high hydrophilicity, biocompatibility, non-immunogenicity, and neuroprotection [70, 71]. However, its polyether backbone causes it to be non-degradable in the body, and modifications must be made to surmount this obstacle. For example, Mahoney *et al.* synthesized hydrolysable PEG hydrogels that were modified with poly(lactic acid) (PLA) and poly(glycolic acid) (PGA) [72]. These derivatized PEG hydrogels possessed attractive properties such as neutrality, high mobility, hydration capacity, and large exclusion volume. Collectively, these properties caused the hydrogels to be highly resistant to protein adsorption. The high mobility of the hydrogels allows limited time for proteins to form positive attachments, while the surrounding water molecules prevent other molecules from contacting the polymer surface [73]. To mitigate against these limitations and enhance neural function on PEG hydrogels, biological components including collagen, poly(L-lysine) (PLL), peptides such as RGD, and signaling molecules such as FGF-2 and ciliary neurotrophic factor (CNTF), can be introduced in both two- or three-dimensional conditions [74–78]. For example, two different research groups prepared star-shaped, heparin-crosslinked PEG hydrogels to allow covalent attachment of RGD peptide as well as the non-covalent binding of signaling molecules such as FGF 2 [75, 76]. Two-dimensional *in vitro* studies were conducted on NSCs dissected from the fetus of a mouse. The findings suggested that the stem cells were attached to the surfaces of these hydrogels [75]. Another group, Sara *et al.*, developed a library of PEG hydrogels modified with PLL, which were characterized by a wide range of mechanical properties, degrees of swelling, and degradation profiles [79]. When the hydrogels were cultured with mouse neural progenitor cells (mNPCs), the cells survive for up to 17 days. Further, results indicated that the mechanical properties of the hydrogel influenced events such as migration and differentiation, with the best results correlating to hydrogels with a modulus similar to that of natural brain tissue. These properties were also influenced by PLL molecular weight, which alters the charge density of the hydrogels [79].

#### 8.4.1.2 *Biologically derived hydrogels*

Biologically derived hydrogels are prepared with substances that are found in the body, including proteins, peptides, polysaccharides, glycosaminoglycans (GAGs), and DNAs. Utilization of these natural structures enhances the hydrogels' properties which include biocompatibility, intrinsic biological activity, and degradability, by the naturally occurring enzymes. However, there are also some disadvantages to using biologically derived hydrogels, such as the difficulty of conducting a well-controlled study and immunorejection. These negative attributes result from the residual signaling molecules, and impurities and undefined constituents from animal-derived biomaterials used for creating hydrogels.

#### 8.4.1.2.1 Collagen

Collagen is the most abundant protein in the animal kingdom and constitutes about 30% of the protein found in the human body. It is a major fibrous constituent of ECM, skin, tendons, ligaments, cartilage, and bones. Its properties are diverse, somewhat analogous to the diversity that exists in the different types of collagens of which type I is the most ubiquitous. The basic unit of type I collagen consists of triple $\alpha$-helices, which self-assemble into microfibrils and then fibrils in physiological environments. Hydrogels from type I collagen are excellent candidates for use in neural tissue regeneration. Essential to its versatile and functional nature are its self-assembling tendencies, chemotactic properties which support cellular proliferation, differentiation, multiple crosslinking sites, and biodegradability. For instance, O'Connor *et al.* isolated NSCs from the rat embryonic cortex and cultured them in a pre-gel solution of type I collagen. Observation of the culture between days 1–3 showed that these cells had undergone rapid proliferation and had differentiated into both neurons and astrocytes [80]. Further, the neurons in these hydrogels developed neuronal characteristics, including neuronal polarity, neurotransmitters, ion channels/receptors, and excitability [80, 81]. In another publication, they cultured these cells in the collagen hydrogel in a rotary wall vessel bioreactor and found that the cells survived for up to ten weeks, in which time they generated neural tissue-like constructs without necrotic cores [82]. Additionally, they observed that the generated neural tissue-like constructs mimicked the architectural design and functionality of the cerebral cortex of the embryonic brain to some extent in that they produced a complex two-layered structure (Figure 8.2) [83]. NSC adhesion and survival have been improved by Iwata *et al.* who conjugated a neural stem cell adhesion molecule (NCAM) related to laminin or its derived peptides to collagen hydrogels [84, 85]. The addition of signaling molecules to the hydrogel is aimed at enhancing neural function. For example, Diester and Schmidt examined the effects of several combinations of glial cell line-derived neurotrophic factor (GDNF), neuropoietic cytokine CNTF, and NGF on neurite outgrowth. They determined that this combination of signaling molecules had a synergistic effect and neurite outgrowth was enhanced [86]. It was also shown by Shreiber *et al.* that neural function is affected by the mechanical properties of collagen hydrogels. They altered the

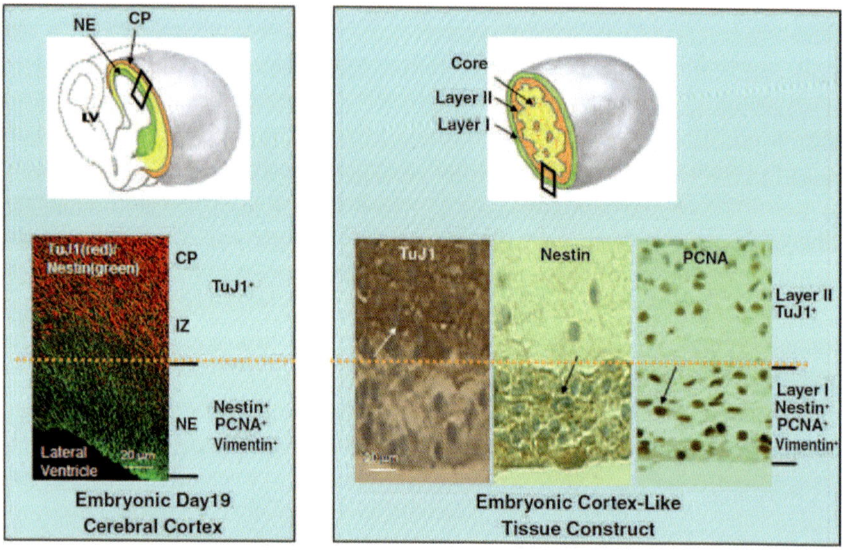


(a) (b)

Figure 8.2 Similarity in architecture between (a) E19 rat cerebral cortex and (b) RWV bioreactor-generated NPC–collagen construct of embryonic brain-like tissue. Reprinted from [83] with permission from Mary Ann Liebert, Inc. Publishers.

mechanical stiffness of collagen hydrogel using a genipin gradient and discovered that the growth of neurites was enhanced down the gradient of stiffness [87].

#### 8.4.1.2.2 Hyaluronic acid

HA is a naturally occurring, high molecular weight GAG resulting from the copolymerization of glucuronic acid and *N*-acetyl-D-glucosamine. It constitutes the ECM of mesenchymal tissues and the CNS. Being a CNS component, along with its attractive attributes including neutrality and high affinity for water, it is being investigated for its use as a hydrogel for neural tissue engineering. Crosslinkable hydrogels are possible if modifications to the carboxy functional groups on HA are controlled with groups such as polyfunctional hydrazides [88]. Hou *et al.* adopted this method and created HA hydrogels conjugated with the anti-Nogo-66 receptor(antiNgR) in order to support the attachment and survival of neural cells of chicken DRG *in vitro* [89]. *In vivo* studies were later conducted and the hydrogels were shown to hinder the formation of glial scar tissue, support formation of new blood vessels, and promote axonal growth in the damaged area of the spinal cord [90–92]. Seidlits *et al.* [93] inserted several methacrylate functionalities into HA hydrogels. The resulting hydrogels were characterized by bulk compressive moduli in the range 3–10 kPa. The hydrogels were later seeded with neural precursor cells (mNPCs) from the ventral midbrains of mice fetuses. It was shown that hydrogels having similar mechanical properties to those of the neonatal brain experienced enhanced differentiation of mNPCs into mature neurons. In contrast, when cultured in hydrogels with similar mechanical properties to the adult brain, the mNPCs differentiated into astrocytes [93]. Further, by crosslinking the hydrogels and incorporating signaling molecules such as brain-derived neurotrophic factor (BDNF), rat NSC survival in hydrogels was significantly enhanced [94].

#### 8.4.1.2.3 Agarose

Agarose is a linear polysaccharide with alternating agarobiose disaccharide units and is usually isolated from seaweed, a marine alga. Agarose is water soluble, deriving its polarity from the presence of the hydroxy groups present on the agarose backbone. Consequently, two agarose chains may interact with each other via hydrogen bonding to form a double helix. Additionally, it is a polymer with temperature-sensitive properties; at high temperature agarose is soluble in water but usually solidifies into a gel at low temperatures. The ability of agarose to form hydrogels has led to its use in tissue engineering applications. Gilbert *et al.* demonstrated that hydrogels from blends of agarose and thermoreversible methylcellulose, which solidify at physiological temperatures are ideal materials for use in neural tissue regeneration [95]. Chitosan and agarose blends are also possible and have been shown to improve neuronal attachment when an optimum concentration of chitosan (0.66–1.5 wt%) is used [96]. Luo and Shoichet have demonstrated that modification of agarose with 2-nitrobenzyl cysteine followed by conjugation with maleimide containing GRGDS was beneficial in enhancing neuronal attachment as well. This hydrogel exhibited guidance effects on three-dimensional DRG cell migration and neurite outgrowth (Figure 8.3) [97]. In further studies it was shown that these hydrogels supported the differentiation of rat NSCs to neurons and glial cells. Upon the immobilization of the platelet-derived growth factor, AA (PDGF-AA) on the GRGDS-agarose hydrogels, it was determined that the differentiation of NSCs was specifically directed towards an oligodentrocyte lineage [98]. Cullen *et al.* [99] prepared collagen-conjugated agarose hydrogels

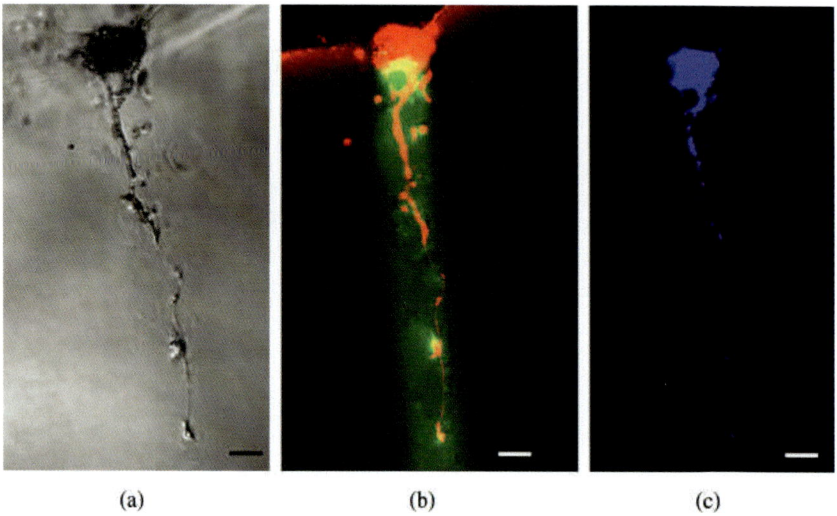

Figure 8.3 DRG cells grew within GRGDS-oligopeptide-modified agarose channels only, and not in surrounding volumes. (a) Optical microscopy, (b) confocal fluorescent microscopy, where the channel is green and the cells are red, and (c) fluorescent microscopy, where a nuclear DAPI stain (blue) confirms cell migration into the oligopeptide-modified channel. Reprinted from [97] with permission from Nature Publishing Group.

by crosslinking mixtures of collagen and molten agarose with a heterobifunctional amine reactive and photoreactive crosslinker. They discovered that various effects were possible when culturing neurons by varying the amount of collagen mixed with agarose. Dense three-dimensional neurite networks were observed at a collagen concentration of 300 ug/mL, while neurons cultured in agarose with a higher collagen concentration exhibited significantly less neurite outgrowth [99].

#### 8.4.1.2.4 Alginate

Alginate, a natural anionic polysaccharide that can also be extracted from seaweed, has two repeating monomer units, $\beta$-D-mannuronate and $\alpha$-L-guluronate joined together by $\beta$(1-4) linkage. Alginate can be crosslinked to form hydrogel in the presence of divalent ions such as $Ca^{2+}$, $Mg^{2+}$, $Sr^{2+}$, or $Ba^{2+}$, and has been shown to exhibit excellent *in vivo* compatibility with CNS tissue [100, 101]. Crosslinking in alginate hydrogels is due to the ability of these ions to bridge the G residues on different chains of the polymer. For instance, Purcell *et al.* studied the effects of alginate composition

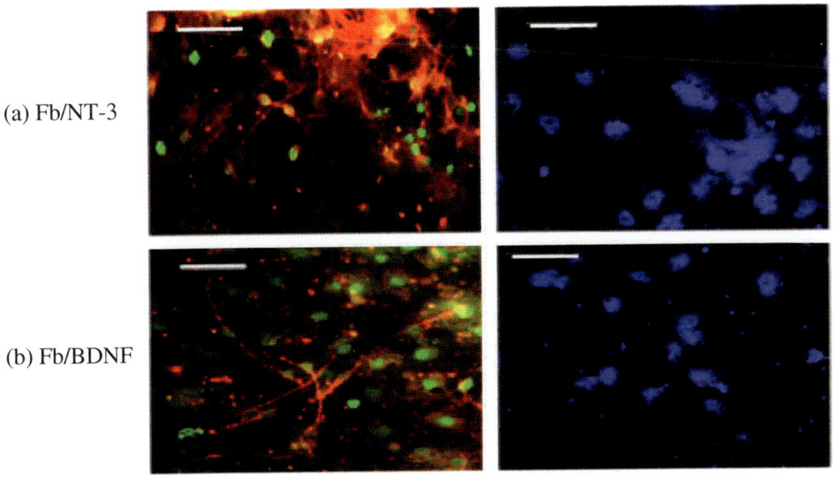


Figure 8.4 NT-3 or BDNF from genetically engineered fibroblasts encapsulated into alginate constructs differentially direct the fates of NPCs to either (a) a neuronal or (b) oligodendrocyte phenotype, respectively. Cells were stained with antibodies of anti-tubulin (red), anti-GALC (green), and anti-GFAP (blue). Reprinted from [103] with permission from the American Chemical Society.

on neurotrophic factor release, viability, and proliferation of NSCs. They demonstrated that alginate hydrogels with a higher G residue ratio proved more favorable for NSC encapsulation due to their superior mechanical stability as well as their support of neurotrophic factor release. [102]. Another group differentially directed the fates of NPCs to either oligodendrocytes or neurons by encapsulating alginate hydrogels with BDNF or NT3 genetically engineered fibroblast (Figure 8.4) [103]. Protein adsorption to alginate hydrogels is inhibited by its intrinsic net negative charge, and consequently cellular adhesion is reduced as well. For this reason, outgrowths of encapsulated DRGs are supported by adding other ECM components such as fibronectin into the alginate hydrogels [104]. The modulus of the alginate hydrogel can be tuned by varying the molecular weight of the polymer, the G residue ratio, or the concentration of the crosslinker to meet specific neural tissue engineering application needs. In some cases, creating softer alginate hydrogel is ideal because there is an inverse relationship between the hydrogel modulus of the hydrogel and the rate of proliferation of NSCs. For example, one group created soft hydrogels with an elastic modulus similar to that of the brain and reported that the expression of the neuronal marker $\beta$-tubulin III was greatly improved [105].

Notwithstanding the many excellent attributes of alginate hydrogels, their rate of degradability subsequent to *in vivo* application is a matter of concern because alginate is not naturally enzymatically degraded in mammals. Therefore, it may take months before an alginate hydrogel is completely removed from an implantation site. However, one group circumvented this issue by modifying the alginate hydrogels with alginate lyase-loaded PLGA microspheres. It was found that these hydrogels were enzymatically degraded in a controlled and tunable fashion. In return, NSCs cultured in these alginate lyase-loaded alginate hydrogels exhibited a significant increase in expansion rate [106].

#### 8.4.1.2.5 Chitosan

Chitosan is a cationic polymer obtained from alkaline deacetylation of chitin. The degree of deacetylation and the resultant positive charge significantly influence polymeric properties such as cell adhesion, degradation, and biocompatibility. Shoichet *et al.* [107] demonstrated that an increase in the degree of deacetylation resulted in a reduction of the degradation rate and an increase in the biocompatibility of chitosan hydrogels as a result of lysozymes acting on the acetylated groups. However, there are limits to the use of chitosan for CNS applications since it increases inflammatory response by activating macrophages via the receptor-mediated binding of N-acetylglucosamine. A possible consequence of this increased inflammatory response is the rapid reabsorption of the chitosan scaffold within the CNS due to the activation of microglia. Further, since chitosan is only soluble in dilute acids, the applications of chitosan under physiological conditions are further hampered. Thus, chitosan must be modified to make it water soluble. The amine groups in chitosan can be coupled with methacrylic anhydride to create water-soluble methacrylamide chitosan. Conjugation of a thiolated form of this water-soluble methacrylamide chitosan with cell adhesive peptides such as YIGSR and IKVAV resulted in enhanced neuronal adhesion and neurite outgrowth of rat superior cervical ganglion at postnatal d2 [107]. Another interesting attribute of methacrylamide chitosan is that it can be photo polymerized to form a hydrogel with tunable Young's elastic moduli ranging from <1 kPa to >30 kPa. One group tested a photo polymerized hydrogel characterized with an elastic modulus of 3.5 kPa and found that a maximum proliferation of rat NSCs was achieved. Further, NSC differentiation into all major CNS cell phenotypes was enhanced on these surfaces with varying elastic modulus: $<1$ kPa, $>7$ kPa, and 3.5 kPa, respectively [46]. Other modifications can also be made to chitosan to yield optimized properties. For example, Chenite *et al.* developed a pH-neutral chitosan solution by adding glycerophosphate salt. This chitosan solution transformed into a macroporous hydrogel upon an increase in temperature to 37°C [108]. Subsequently, this group used azidoaniline photocoupling to immobilize poly-D-lysine (PDL) onto the thermosensitive chitosan hydrogel, and cells were cultured in two- and three-dimensional structures. Two-dimensional culture showed diminished neurite outgrowth with increasing PDL, though cell survival was not affected, while in three-dimensional culture immobilized PDL improved cell survival, with an optimal concentration being 0.1%. Further increases of PDL concentration reduced both cell survival and neurite outgrowth [109].

#### 8.4.1.2.6 Fibrin

Fibrin is a naturally occurring fibrous protein responsible for the clotting of blood following injury. Fibrin is formed during the natural coagulation cascade when thrombin (the catalytic enzyme) cleaves fibrinogen to generate fibrin monomers. The produced monomers are then covalently crosslinked with factor XIII, which is activated by thrombin and $Ca^{2+}$ to form a matrix which can be subsequently degraded by molecules such as plasmin. Fibrin is a useful biomaterial because of its physical properties and its adhesive characteristics. These attributes have led to tissue regeneration applications, including for the CNS. *In vitro*, the compliance and polymerization rate of fibrin can be controlled by varying the concentration of fibrinogen and thrombin, respectively. In a publication which supports this claim, Ju *et al.* assessed the growth of rat and mouse neurons in fibrin from different sources, including human, cow, and salmon. Interestingly, they determined that the hydrogel created with salmon fibrin exhibited the greatest degree of neurite growth and the greatest resistance to degradation by cellular proteases [110]. Another group used factor XIIIa to covalently tether certain exogenous bioactive peptides to the fibrin hydrogel during crosslinking. The peptide sequences used, including RGD, IKVAV, YIGSR, HAV, and RNIAEIIKDI all supported and enhanced neurite extension of chicken embryo DRGs when used individually. However, different behavior types, including additive, inhibitory, non-interactive, and synergistic were observed when they were incorporated in combination [111]. Other molecules such as the sixth Ig-like domain of cell adhesion molecules LI can be incorporated in the fibrin matrix to promote neurite extension. This domain contains the N-terminal tranglutaminase substrate sequences, TG-L1g6. One group reported that neurite extension depended on interactions between cell surface $\alpha v\beta 3$ and RGD-sites provided by TG-L1Ig6 when attaching this molecule to fibrin hydrogels. Further, in this study, myelinated structures were almost exclusively found in these TG-L1Ig6-modified and laminin-1-containing fibrin matrices [112].

#### 8.4.1.2.7 Methylcellulose

Methylcellulose (MC) is a water-soluble non-ionic polymer derived from cellulose, which exhibits a low viscosity at 23°C but forms a soft hydrogel at 37°C. This unique gelation property of MC makes it a highly attractive biomaterial for minimally invasive tissue regeneration procedures *in vivo*. To assess MC as a scaffold for intercerebral gelation, LaPlaca *et al.* microinjected MC into the brains of rats one week after cortical impact injury. They determined that microinjection of MC did not alter the size of the injury cavity nor did it alter the pattern of the glial scar when compared to injured vehicle-injected rats [113, 114]. In another study, they attached laminin to MC using a Schiff base in an effort to enhance cell adhesion and viability [114]. More recently, they used MC scaffolds to study NSC–ECM interactions by culturing murine neurospheres into MC scaffolds functionalized with laminin-1. The tethered laminin-1 was found to significantly enhance mNSC survival, neurite outgrowth, and differentiation [115].

#### 8.4.1.2.8 Matrigel

Matrigel is a thermosensitive extract of gelatinous protein mixture derived from Engelbreth–Holm–Swarm (EHS) mouse sarcoma cells. Matrigel forms a clear three-dimensional gel at room temperature and above, but liquefies at low temperatures above 0°C. This mixture is rich in ECM proteins such as laminin, collagen, entactin, nidogen-1, and hepran sulfate proteoglycan. It also contains growth factors, such as transforming growth factor $\beta$ (TGF-$\beta$), FGF-2, insulin-like growth factor-1 (IGF-1), basic fibroblast growth factor (bFGF), and tPA. Due to its many excellent properties, this material has been investigated for many tissue engineering applications, including CNS regeneration. For instance, Liudmila *et al.* found that Matrigel has induced the sprouting of DRG neurites *in vitro* [104]. This finding was corroborated by LaPlaca *et al.* in a study where they cultured neurons from embryonic rats in Matrigel hydrogels. Like Liudmila *et al.*, they found extensive three-dimensional neurite outgrowth as well as the expression of mature neuron-specific cytoskeletal proteins. Additionally, these neurons displayed both spontaneous and induced electrophysical action potentials and exhibited functional synapse formation and network properties [116]. Despite its many excellent properties, including its ready availability, and its extensive use in *in vitro* and *in vivo* studies, it is highly probable that Matrigel will not be used in human applications because it is sourced from mouse tumor cells and its composition is relatively undetermined.

#### 8.4.1.2.9 PolySia-based hydrogels

Polysialic acid (PolySia) is an anionic polysaccharide consisting of $\alpha$-2, 8-linked sialic acid units. It is strongly expressed during the development of embryonic nerve tissue and it is transported in the CNS by the neural cell adhesion molecule (NCAM). Its principal role is to perform modulatory services in cell–cell and cell–substrate interactions in neuronal development processes [117, 118]. These properties make this material an excellent candidate for CNS regeneration applications. For instance, PolySia-based hydrogels with the hydrophobic crosslinker diepoxyoctane were prepared in order to increase the molecular mass of the polymer and decrease its hydrophilicity. Highly specific degradation was also achieved by this group by inducing phage-born endosialidases (endo-*N*-acetylneuraminidase) [119]. Another group used this material as well, but modified it with either collagen I, PLL, poly-L-ornithine-laminin, or Matrigel in order to improve the properties of these hydrogels to support growth of PC12 cells, neonatal and adult rat Schwann cells, DRG cells, and NSCs [120, 121].

#### 8.4.1.2.10 Xyloglucan

Xyloglucans are polysaccharides that are ubiquitously distributed in the cell walls of higher plants. They contain a backbone that is comprised primarily of 1,4-$\beta$-linked glucose units which are periodically substituted with 1,6-$\alpha$-linked xylose. Some may also contain galactose, fucose, and/or arabinose units in their side chains. Xyloglucan hydrogels are used extensively in biomedical applications but thermally induced gelation only occurs when 35–50% of galactose units are removed with fungal $\beta$-galactosidase.

Partial removal of galactose units reduces steric hindrance and facilitates hydrophobic interactions between groups in the side chain and the glucan backbone, resulting in thermal gelation. Two-step gelation of xyloglucans leads to three-dimensional networks which ultimately possess modulus in the range 150–480 Pka, depending on the concentration. Nisbet *et al.* investigated thermoresponsive xyloglucan hydrogel scaffolds for neural tissue engineering of the spinal cord. In this study, they functionalized the hydrogels with PDL to promote neuron adhesion and neurite outgrowth, and optimized them to mimic the mechanical properties of the native spinal cord [122]. They subsequently implanted the hydrogels with the caudate putamen of adult rat and found with increasing concentration of PDL there was an increased penetration level for astrocytes and neurites [123].

#### 8.4.1.2.11 Peptide hydrogels

Peptide hydrogels are derived from amino acids or peptide-based biomaterials. The amphiphilic nature of these peptides facilitates spontaneous self-assembly of hydrogels into a three-dimensional structure via non-covalent interactions such as hydrogen bonding, hydrophobic forces, and electrostatic interactions. Zhang *et al.* created peptide hydrogels through the spontaneous assembly of ionic self-complementary oligopeptides, which they called PuraMatrix. This was possible because of the utilized oligopeptides, which comprised alternate repeat units of positively charged groups such as lysine or arginine, and negatively charged groups such as aspartate and glutamate [124, 125]. These self-complementary oligopeptide-based hydrogels were found to support neuronal cell attachment, neurite outgrowth and differentiation, as well as synapse formation between the attached neurons, rat PC12 cells and mouse neurons [126]. In another study they incorporated bone marrow-signaling motifs into the hydrogels and observed that the survival of adult mNSCs was vastly improved [127]. In addition, peptide-based hydrogels are shown to support several important functions of human neural stem cells (hNSCs), including migration and neuronal differentiation. Silva *et al.* developed a more complex structure by self-assembly of peptide amphiphiles. The construct comprised a 16 carbon alkyl tail, several amino acid residues such as a negatively charged Glu, four Ala residues, and three Gly residues, as well as the neurite-sprouting epitope IKVAV. Self-assembly of the construct was promoted by the hydrophilic sequence, and the presence of cations in the cell culture medium facilitated hydrogen bonding as well as unfavorable contact between hydrophobic segments and water molecules. Further, as demonstrated by others, the incorporation of IKVAV into the hydrogen design limits the formation of glial scar, supports differentiation of NSCs, and promotes axon elongation. These benefits ultimately resulted in functional improvements after SCI in mice after *in vivo* transplantation (Figure 8.5) [128, 129]. Wong *et al.* also designed hydrogels based on the protein–protein interaction concept for potential applications in CNS and found that hydrogels with shear modulus (G′) in the range 9–50 Pa supported regeneration and differentiation of NSCs, and demonstrated elongated neural extension [130]. Yang *et al.* developed similar amphiphilic diblock copolypeptide hydrogels (DCHs) by combining lysine, homoarginine, or glutamate with leucine. Tunable hydrogel properties were obtained by varying the copolymer chain length or composition. DCHs are used as a hydrogel matrix for CNS regeneration because of their polyelectrolyte segments and hydrophobic helical domains that can form hydrated membranes. Another group formulated a number of DCHs with rheological properties similar to brain tissues, which were then injected into mouse forebrains. These injectable DCHs reassembled *in vivo* to form three-dimensional deposits, exhibited little or no detectable toxicity, and integrated well with brain tissue [131, 132].

### 8.4.2 *Hydrogels control neural stem cell behaviors*

Hydrogels can create an environment in which NSCs are instructed to form tissues by directing cell behaviors, such as migration, proliferation, differentiation, maintenance of phenotype, and apoptosis. This is often accomplished due to the ease of sensing and responding to the environment by both cell–matrix and cell–cell communications.

#### 8.4.2.1 *Viability*

The key indicator in determining if a material is biocompatible and therefore suitable for tissue engineering applications is cell viability when in contact with this material. When considering hydrogels for CNS regeneration, the key factor that affects neural stem cell survival is often the concentration of the hydrogel. For example, one group discovered that Pluronic F127, in a gelated (30%) form, was toxic to hNSCs and showed less viability than the 10% hydrogels. This group also tested other materials and found that

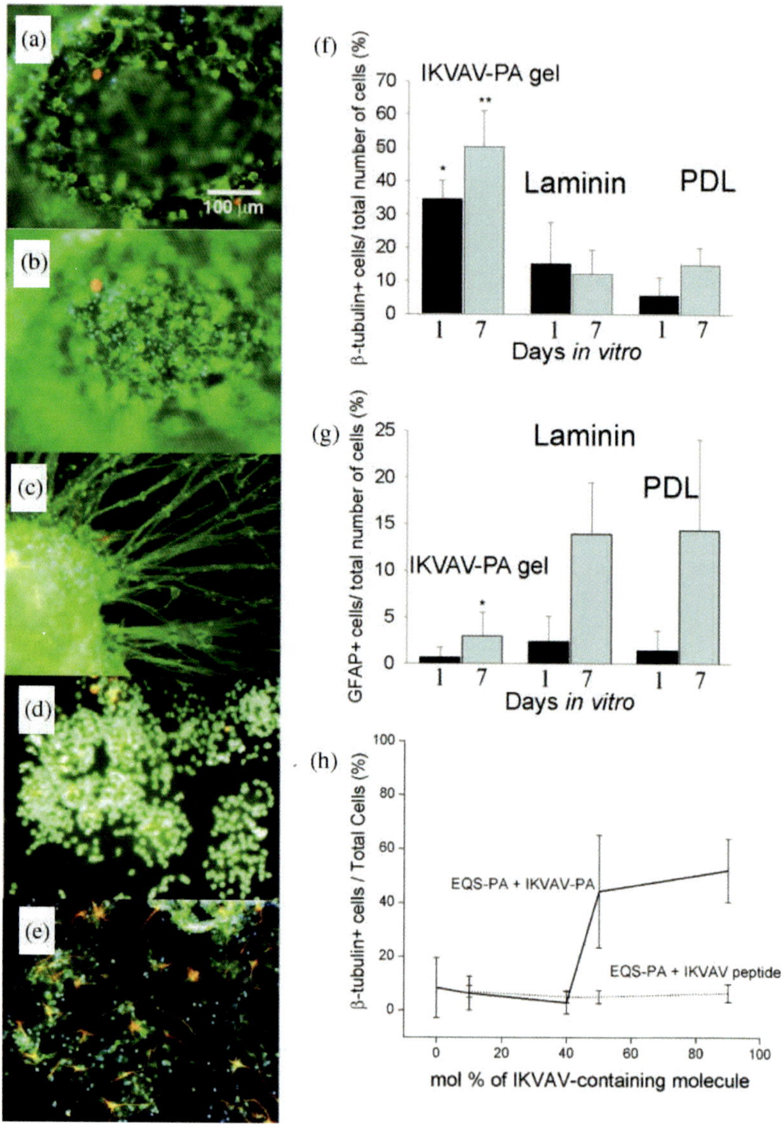


Figure 8.5 NPCs encapsulated in the hydrogels of nanofibers presenting IKVAV on their surface differentiated rapidly into neurons (green) rather than astrocytes (orange). Reprinted from [129] with permission from the American Association for the Advancement of Science.

Matrigel, though the 50% concentration showed an increase in viability, also prevented the normal capacity of hNSCs for neuronal differentiation, therefore proving toxic to neurons. Further, it was found that 1% PuraMatrix also seemed to exert a toxic effect on hNSCs, while lower concentrations, such as 0.25% and 0.5%, showed a much lower toxic effect [133].

In order to improve cell survival in/on these hydrogels, adhesive components such as PLL, laminin, or specific peptides, and bioactive factors such as BDNF, NT-3, or anti-Nogo-66 receptor can be conjugated into these hydrogels [77, 85, 89, 94, 109, 114, 134]. For example, one group immobilized PLL onto chitosan and showed improved cell survival at an optimum concentration of 0.1% [109].

#### 8.4.2.2 *Attachment*

Neural cell adhesion is very important for many cellular functions, including new tissue formation. However, the chemistry of many hydrogels often makes it difficult to support neural cell adhesion and tissue formation. This is often due to the inhibition of protein adsorption by many natural and synthetic hydrogels, including PEG and alginate. An important strategy for improving adhesion is to perform certain chemical modifications (including natural ECM-mimetic modifications) to these non-adhesive hydrogels in order to modulate cellular responses such as attachment, adhesion, and growth [62, 74, 77, 96]. One such example is the conjugation of short peptide sequences to non-adhesive gels, including RGD, GYIGSR, YIGSR, IKVAV, and CDP-GYIGSR. However, the exact neural response to these varying peptides is often different. For example, cells cultured on PHEMA hydrogels that were modified with CDPGYIGSR, a laminin-derived oligopeptides which more closely mimics the native region conformation of the active sites, exhibited greater cellular responses as compared to PHEMA hydrogels modified GYIGSR [62]. The addition of these peptides, including YIGSR, IKVAV, CDPGYIGSR, and RGD also works with biologically derived hydrogels, and has been shown to further improve neuronal adhesion [85, 107, 114].

#### 8.4.2.3 *Neurite outgrowth*

Neurite extension has also been shown to be dependent on numerous factors, including biochemical properties such as ligand type and density, as well as the biomechanical properties of the hydrogels [78, 87, 99, 104, 109, 111, 112, 122, 135–137]. For instance, one group conjugated laminin to

HA hydrogels and observed a strong, dose-dependent effect on both neurite length and outgrowth. On the other hand, fibronectin added to alginate hydrogels as well as collagen conjugated to agarose hydrogels both were shown to promote neurite outgrowth [99, 104, 135, 136]. Neurite extension can also be improved by attaching cell adhesive ligands, including RGD, IKVAV, and YIGSR. After attaching these ligands to PEG hydrogels, it was found that the concentration for each ligand needed to enhance neurite extension was different. Further, different combinations of these ligands were found to affect neurite extension to different extents [137].

On the other hand, other naturally found matrix molecules, including the family of sulfated proteoglycans, have been found to be inhibitory for axonal outgrowth [138]. These include three cell surface molecules found in CNS myelin: Nogo-A, myelin associated glycoprotein, and oligodendrocyte myelin glycoprotein. Interestingly, these three proteins bind to the same neuronal Nogo-66 receptor. When conjugated to hydrogels, this Nogo-66 receptor was able to support neural cell attachment and survival, as well as neurite outgrowth [89].

#### 8.4.2.4 *Proliferation*

Cell proliferation must be precisely controlled for tissue regeneration to occur. While enhanced proliferation is usually needed in most applications, reduced proliferation is sometimes needed, as well. An example of this is seen with astrocyte proliferation in CNS regeneration applications, as this type of proliferation at the implant/tissue interface can cause the formation of an undesirable, dense glial scar. One parameter that often affects neural stem cell proliferation is hydrogel stiffness. For example, one group found that rat NSCs that were seeded on chitosan substrates that possessed Young's elastic moduli of 3.5 kPa exhibited the maximum proliferation [46]. Another group found that NSCs within alginate hydrogels exhibited a decrease in proliferation with an increase in the modulus of the hydrogels [105]. Further, this group found that hydrogel degradation further promoted the proliferation of NSCs because of the efficient nutrient and gas exchange it provided [106]. Conjugation of certain ligands has also been shown to enhance cell proliferation. For instance, by conjugating RGD to polyacrylamide hydrogels, self-renewal and differentiation were benefited in a dose-dependent fashion [56].

#### 8.4.2.5 *Differentiation*

Many soluble cues are well known to induce NSCs to differentiate to a specific fate once incorporated within hydrogels [58, 75, 134, 139]. For instance, by incorporating NT-3 into chitosan hydrogels, an increase in NSC differentiation into neurons was observed [134, 139]. On the other hand, when FGF-2 was released from PEG hydrogels, NSCs were maintained in an undifferentiated state, even with the addition of differentiation media [75]. Similarly, Matrigel, which contains FGF-2, prevented hNSCs' normal capacity for neuronal differentiation [133]. The inclusion of other cues has also been shown to affect the lineage fate of NSCs. For example, once immobilized in an agarose matrix, PDGF-AA has been shown to control the specific differentiation of NSCs into oligodendrocytes [98]. Others have found that BDNF and NT-3, derived from genetically engineered fibroblasts, differentially direct the lineage fates of NSCs [103]. Further, it has been observed that with the incorporation of certain ligands, including the IKVAV sequence, self-assembling nanofiber hydrogels showed enhanced differentiation of neural progenitor cells [128, 129]. Another parameter that has been found to affect NSC differentiation is hydrogel stiffness [46, 58, 79]. Different hydrogel stiffnesses are favored by various cell types that are differentiated from NSCs, including neurons, astrocytes, and oligodedrocytes [46]. With regard to three-dimensional culture, optimal neural differentiation was achieved when cultured in soft hydrogels with similar elastic moduli to that of natural brain tissues [93, 105].

#### 8.4.2.6 *Migration*

Cell migration is a necessary occurrence which ultimately decides the success of tissue regeneration efforts. Attracting endogenous stem cells to injured/diseased areas has been a much researched endeavor in recent years for tissue engineering applications, including CNS regeneration. With regard to the use of hydrogels for CNS regeneration, hydrogels that allow for three-dimensional cell migration are obviously desirable. Among other things, hydrogel stiffness has also been observed to influence NSC migration [79]. Other factors that affect neural cell migration include growth factor and ligand incorporation. It has been found that with sustained, long-term, and localized delivery of hepatocyte growth factor (HGF) by incorporation within hydrogels, a therapeutic level of HGF necessary in order to allow for endogenous stem cell recruitment can be maintained at the defect site [140]. Another group found that by patterning RGD in agarose hydrogels, three-dimensional DRG cell migration and neurite outgrowth can be guided [97].

### 8.4.3 *Hydrogel design*

#### 8.4.3.1 *Hydrogel mechanical strength*

Hydrogel mechanical properties obviously play a major role in regulating various cell behaviors [141]. For this reason, it is not surprising that hydrogels with biomechanical properties that are not similar to local environmental conditions experienced by cells *in vivo* have a negative effect on cell behavior. It is well known that hydrogels that possess biomechanical properties which closely mimic those of the natural ECM provide the best environment for the differentiation of progenitor cells into the mature phenotypes inherent to that particular tissue [57]. Like many other cells and tissue types, neuron-like cells have also been shown to be affected by substrate mechanical properties. However, not all neuronal cells behave similarly, and not all hydrogels produce the same results. Though different cellular behaviors that occur by varying mechanical properties of hydrogels can be easily observed, the mechanosensing elements and pathways that affect these behaviors are still very unclear. However, many cellular components such as integrins, focal adhesions, membrane ion channels, cytoskeletal filaments, and the nucleus are known to be key molecular players in these mechanotransduction events. One of the major players is integrin, which is the transmembrane connection between the ECM and the inside of the cell. Integrins, therefore, are believed to be mechanosensitive, acting through inside-out signaling, outside-in signaling, or a combination of the two [141]. The integrins within neurons that are responsible for adhesion to the ECM are the NCAM, L1, and N-cadherin. Extracellularly, these integrins bind to specific amino acid sequences in the ECM, including RGD, which are present in all major ECM proteins, including fibronectin, collagen, laminin, and vitronectin. Intracellulary, these integrins bind to various molecules, including focal adhesion kinase, paxillin, and vinculin. The adhesive structures formed are known as focal adhesions, and link to the F-actin cytoskeleton through molecules such as tensin, talin, and $\alpha$-actinin. Then, since tension is being maintained between focal adhesions by F-actin stress fibers, the cell exerts traction forces on its underlying substrate [142]. It has been shown that focal adhesion assembly increases when external stresses are applied to integrins, which proves that these complexes are mechanosensitive. Furthermore, conformational changes of regulatory molecules within these focal adhesions are thought to mediate biochemical responses. In turn, these biochemical events activate signal transduction cascades, which regulate behaviors such as cell proliferation, differentiation, and apoptosis. One molecule that has been linked to the fibronectin-based rigidity response in fibroblasts is the receptor-like protein tyrosine phosphatase $\alpha$ (RPTP$\alpha$). A relationship between this enzyme and neurite outgrowth and rigidity sensing has also been seen, and therefore shows promise for CNS regeneration applications [143]. Additional molecules are involved in these signaling pathways, including paxillin, small GTPases, such as Rac and Rho, and focal adhesion kinase, all of which have all been shown to become activated by integrin-dependent mechanical stress. Many groups have proposed models to try to explain the molecular basis of mechanotransduction, including Ingber [144, 145], who proposed the cellular tensegrity model. A stochastic model for motility on compliant substrates was also proposed recently based on a motor-clutch system, in which molecular clutches are used by cells to physically link F-actin to the ECM [146].

#### 8.4.3.2 *Hydrogel microstructure*

Embedding cells in a three-dimensional environment introduces many issues, including a deficiency of gases and nutrients, as well as physical constraints, which may affect cell proliferation, migration, and morphogenesis. Hydrogels, however, overcome this issue, as they are defined by their interconnected porosity. This can be exaggerated by incorporating a sacrificial component that can be removed in order to create a very highly porous material. For example, Mahoney *et al.* created a network of interconnected pores within a PEG hydrogel by incorporating a fibrin network that was later enzymatically degraded. This group was then able to demonstrate a process extension from primary neural cells encapsulated within these PEG gels [147]. Another group incorporated PLGA microspheres that were loaded with alginate lyase into alginate hydrogels in order to enzymatically degrade the hydrogel in a controlled and tunable manner. This group found that rat neural progenitor cells seeded into these hydrogels showed a significant increase in the expansion rate as compared to when cultured in standard non-degrading alginate hydrogels [106].

### 8.4.3.3 *Hydrogel with cell adhesion molecules*

#### 8.4.3.3.1 Physical adhesion

Two major methods are known to assist in the physical adhesion of cells with biomaterials. These include specific interactions, which involve ligand/receptors, enzyme/substrates, and permit very precise recognition and binding, as well as non-specific interactions, which depend on charge specificity. Though specific interactions have already been touched upon, non-specific interactions have yet to be discussed. This type of bonding is often employed when cells do not have specific receptors for certain polymers, such as the cationic poly(amino acid) polylysine. It is believed that cellular adhesions with polylysine occur by means of electrostatic interactions between the anionic sites of the plasma membrane and the cationic sites of polylysine surface [148, 149]. For this reason, polylysine is often conjugated to different hydrogels to modulate NSC behaviors.

#### 8.4.3.3.2 Chemical adhesion

Adhesion molecules, which can be derived from ECM proteins such as laminin, fibronectin, and collagen, initiate receptor-mediated cell binding, thus inducing the formation of focal adhesions [150, 151]. Laminin provides one example, as it consists of three chains that provide several receptor-binding domains. The specific amino acid sequences present at the binding sites, which include RGD, IKVAV, YIGSR, RYVVLPR, and RNIAEIIKDI, have been identified as being involved in cell–receptor interactions [111]. For instance, YIGSR, which is found on the $\beta1$ laminin chain, and IKVAV, present on the C-terminal end of the $\alpha1$ laminin chain, bind 67 kDa and 110 kDa proteins on the cell membrane, respectively [152–154]. RNIAEIIKDI and RYVVLPR are also present on the laminin chain, though RNIAEIIKDI is present on the $\gamma$ laminin chain and RYVVLPR is derived from the $\beta1$ laminin chain [155]. Integrins, which have been discussed in detail above, consist of one $\alpha$ chain and one $\beta$ chain, with approximately 16 different $\alpha$ subunits and 8 different $\beta$ subunits being identified. With this in mind, various combinations of these two chains will result in different receptors that possess different affinities for certain ECM molecules. For example, $\alpha5\beta1$ and $\alpha v\beta3$ are both receptors for RGD, but $\alpha5\beta1$ recognizes the sequence on fibronectin whereas $\alpha v\beta3$ recognizes it on vitronectin [156].

Peptides, rather than whole proteins, are better suited for this application, as they are more stable, easily synthesized, and are less likely to exhibit steric hindrance after biomaterial modification [156]. Though any large-scale isolation of ECM proteins is challenging, the application of conjugating these specific peptides to hydrogels has received much attention for use in biomaterial modification for neural tissue regeneration.

#### 8.4.3.4 *Hydrogels with neurotrophic factors*

Neurotrophic factors can be incorporated into hydrogels in order to elicit the desired cellular response. Many of these neurotrophic factors, including epidermal growth factor (EGF), FGF, HGF, IGF-1, PDGF, vascular endothelial growth factor (VEGF), as well as neurotrophins, which include NGF, BDNF, NT-3, and NT 4/5, bind to tyrosine kinase receptors in order to mediate their effects. Neurotrophins also mediate their effects by binding to p75, a member of the TNF-$\alpha$ family of receptors. Other neurotrophic factors, including GDNF, neurturin, persephin, and artemin, are derived specifically from the glial cell line and exert their effects via the GDNF family receptor (GFR)-$\alpha$. Neuropoietic cytokines, which usually refers to molecules in the interleukin-6 family which share a common signal-transducing receptor subunit termed glycoprotein 130, are also plentiful, and can include IL-6, IL-11, leukemia inhibitory factor (LIF), oncostatin-M, CNTF, cardiotrophin-1, and neurotrophin-1/B-cell stimulating factor-3 [86, 157, 158]. Like most signaling molecules, all of these neurotrophic factors generally have a short life due to rapid proteolysis by peptidases *in vivo*. However, this obstacle can be overcome by providing long-term, localized delivery of these neurotrophic factors through attachment to hydrogels. In turn, this could allow for enhanced neuronal survival, proliferation, migration, differentiation, axon growth, and synaptic plasticity. Further, it has been shown that combining the use of multiple neurotrophic factors produces synergistic effects [86].

#### 8.4.3.5 *Bioprinting hydrogels*

Bioprinting is a bioengineering approach which employs computer-aided design in order to deposit, transfer, and pattern biological materials. This technique can be used to attain complex three-dimensional structures by layering and depositing cells within hydrogels, or alternately layering the two. This technique can be used specifically for CNS regeneration by alternately layering neural cells and hydrogels [159]. For instance, Yoo *et al.* [160, 161] used this method to pattern neural cells within three-dimensional collagen and fibrin hydrogels. These cells demonstrated high viability, but also showed enhanced migration, as they migrated approximately 100 $\mu$m in three days. Further, biologically active macromolecules, including growth factors and cytokines, can also be bioprinted onto hydrogels in order to elicit a desired cellular response. For example, one group printed FGF-2, CNTF, or FBS onto polyacrylamide-based hydrogels and then seeded them with primary fetal NSCs [202]. This group found that the NSCs remained undifferentiated when cultured on areas printed with FGF-2, yet expressed markers for astrocytic or smooth muscle differentiation when cultured on areas printed with CNTF or FBS, respectively. Another group photopolymerized acrylamide-based hydrogels in the presence of an acroyl-streptavidin monomer. This created planar, functionalized surfaces for binding of biotin-labeled proteins. This group used soft protein lithography, also called 'microcontact printing', to transfer the biotinylated fibronectin and laminin, as well as the laminin peptide biotin-IKVAV, onto the modified surfaces. Once seeded with rat LRM55 astroglioma cells, results indicated that the cells were only able to attach to the protein-stamped regions of the hydrogel. Further, when seeded with neurons, these cells exhibited significant neurite extension after 72 h, were able to survive for more than four weeks, and developed functionally active synapses *in vitro* [162].

### 8.4.3.6 *Hydrogels for CNS regeneration*

#### 8.4.3.6.1 Neural stem cells transplanted with hydrogels

Cell replacement therapy is a very promising tool for treating neurological disorders and injuries, such as PD, TBI, stroke, and SCI. However, there are also many issues that need to be resolved, including low viability, lack of control of cell fate, and the ability for proper cell engraftment after transplantation [163]. Though controlling these issues is extremely challenging, an appealing approach to attempt this is the use of hydrogels, which can create a specialized niche for transplanted cells. By so doing, hydrogels seem to promote neuronal differentiation, enhance the elaboration of neural processes, foster the re-formation of tissue, and promote connectivity. Hydrogels have also been shown to reduce inflammation and scarring, which also facilitates reconstitution [164].

##### 8.4.3.6.1.1 Parkinson's disease

PD is a neurodegenerative disorder that affects movement and balance, as well as thinking and other cognitive abilities [165, 166]. The debilitating symptoms arise due to a deficiency of dopamine in the striatum, resulting from the loss of DA neurons in the *substantia nigra pars compacta* and their terminals in the striatum. However, treatment by means of oral administration of L-DOPA loses effectiveness as degeneration progresses and adverse effects, including dyskinesias, begin to occur [22, 167–170]. Another treatment for PD is the use of deep brain stimulation; however, this treatment also has issues with effectiveness. This is most often due to scar encapsulation surrounding the electrode, which insulates it from nearby neurons, thus causing instability and failure in long-term stimulation [171–174]. For these reasons, one of the most promising treatment methods for PD is cell transplantation therapy [22, 175–182]. The use of stem cells, with their remarkable self-renewal ability, as well as their potential to differentiate into DA neurons, is very promising for treatment of this disease. For instance, many groups have successfully differentiated DA neurons from human embryonic [10, 183] and adult stem cells [60, 184, 185]. Though there are less restrictive conditions that are required to differentiate hESCs into DA neurons as compared to adult stem cells, hESCs are often associated with ethical issues and also have a greater chance of tumor genesis and immunorejection [19]. Though thought to be the best alternative, a number of issues must be addressed before a clinically applicable, stem cell-based therapy for PD can be developed. These issues include the survival of transplanted DA neurons, stability of DA neuronal phenotype after transplantation, synaptic integration with host circuitry, and most notably, the patient's full functional recovery. These issues, once again, are being addressed with the use of hydrogels, as they create a specialized niche for stem cell transplantation. For instance, Takahashi *et al.* [186] examined the effects of several ECM proteins, such as collagen, ornithine/laminin, and growth factor-reduced Matrigel, as well as the synthetic biomaterial PuraMatrix, on embryonic stem cell-derived neural precursor cell survival and differentiation. Results indicate that Matrigel was able to effectively support cell survival, migration, and neurite outgrowth *in vitro*. Further, this hydrogel was also shown to promote proliferation of grafted cells *in vivo*, which resulted in larger graft volume and an increase in the number of TH-positive dopaminergic neurons within the graft (Figure 8.6). This gel, however, did not induce dopaminergic differentiation directly, though it was found to reduce the amount of pan-leukocytic CD45-positive cells that invaded the graft, therefore suppressing the immune response and thus inhibiting the neuronal differentiation of grafted NPCs.

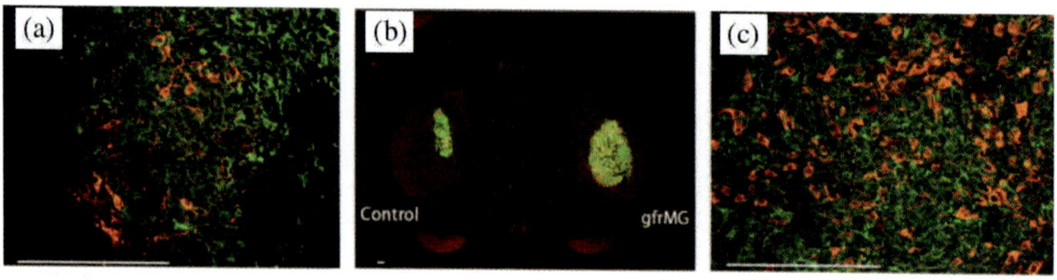


Figure 8.6 Matrigel promoted proliferation of grafted cells *in vivo*, which resulted in (b) larger graft volume and (c) an increase in the number of survival neurons within the graft than (a) the control. Grafted cells (green) survived in the striatum and some differentiated into TH-positive DA neurons (red). Reprinted from [186] with permission from Wiley-Liss, Inc.

#### 8.4.3.6.1.2 Stroke

Stroke, which is caused by a disturbance of blood supply to the brain, results in extensive cellular loss, which ultimately leads to a disintegration of the affected brain tissue and ultimately loss of brain function. By transplanting NSCs, some of the function that was lost after stroke can be recovered, though recovery is often incomplete and the lost tissue is minimally restored. However, Greenberg *et al.* [187] showed some progress by transplanting a combination of human neural cells and Matrigel, and demonstrated the potential to improve stroke outcomes. Similarly, Carmichael *et al.* [188] used hyaluronic acid-based hydrogels to transplant neural stem/progenitor cells for recovery after stroke. When compared to stem cell injection without hydrogel support, the hydrogel demonstrated enhanced survival of stem cells in this normally hostile environment and also prevented inflammatory infiltration.

#### 8.4.3.6.1.3 Traumatic brain injury

Many hydrogels, including PHEMA, PHPMA [189], HA [91, 92, 190], and xyloglucan [123], have been investigated as potential implants for TBI treatment. The main function of these hydrogels is to serve as a support structure, assist in cell infiltration and angiogenesis, and inhibit glial scarring. For instance, one group, Tate *et al.* [191] transplanted NSCs within collagen-based hydrogels that contained either laminin or fibronectin for treatment of TBI. This group found that cells delivered within these hydrogels showed enhanced distribution in the injured brain as well as improved survival compared to cells delivered in media alone. Comparing both types of hydrogels, this group found that survival was more enhanced within the laminin-based hydrogel than the fibronectin-based hydrogel. Further, it was found that after behavioral analyses were performed, mice that were treated with the laminin-based hydrogels performed significantly better at spatial learning tasks when compared to untreated mice. This supports the idea that functional recovery is directly related to donor cell survival [114, 191, 192].

#### 8.4.3.6.1.4 Spinal cord injury

Due to the multifactorial and multiphasic nature of SCI pathophysiology, treatments that are likely to be effective usually involve combinations of approaches, including neuroprotection to counteract secondary injury, provision of scaffolds to replace lost tissue, as well as methods that will enhance axonal regrowth, synaptic plasticity, inhibition of astrocytosis and remyelination [12]. In order to accomplish these goals, various hydrogels including Neurogel [193], PHEMA [194], HA [195], agarose [196], and alginate [197] have been investigated as potential implants for spinal cord repair. Further, the hydrogels are designed to exhibit bio-mechanical properties which mimic that of the native spinal cord and they are also conjugated with signaling molecules, including BDNF [194–196] and adhesive peptide as IKVAV [195] within or on these hydrogels for treatment of SCI. These implanted hydrogels should also have proper mechanical properties, similar to those of the native spinal cord [194]. For instance, Rauch *et al.* used a rat hemisection model of the SCI to demonstrate an increase in vessel density and blood–spinal cord barrier formation after endothelial cells were co-cultured with neural progenitor cells in PEG hydrogels [198, 199]. With the inclusion within fibrin hydrogels of certain growth factors including NT-3 and PDGF, another group observed enhanced survival of embryonic stem cell-derived neural progenitor cells in spinal cord lesions two weeks after hemisection injury. Other groups have also combined this heparin-binding delivery system with growth factors and found that it further promoted neural differentiation of grafted cells [200, 201]. Another group used alginate-based highly anisotropic capillary hydrogels (ACH) for this application and found that it not only integrated into the spinal cord parenchyma without a major inflammatory response, but also maintained its anisotropic structure and induced directed axon regeneration across the hydrogel. When adult neural progenitor cells were seeded into these hydrogels and implanted, the regenerative capacity of these matrices were further improved [197].

#### 8.4.3.6.2 Hydrogels for endogenous NSCs

The use of endogenous stem cells is a very appealing approach for CNS tissue regeneration. This is due to the inherent ability of these cells to avoid the immunocompatibility issues that are often seen with the use of allogenic and xenogenic cells. Further, this method is easier, safer, and more efficient than *ex vivo* approaches [140]. Though endogenous NSCs are naturally recruited after injury, this natural response is often unsatisfactory in regenerating injured/diseased tissue [36, 38]. It has been demonstrated, however, that the regeneration potential of adult endogenous NSCs can also be enhanced by administered growth factors and morphogens, including FGF-2, EGF, TGF-$\alpha$, BDNF, GDNF, BMPs, and Shh and Wnt signals [41]. For this reason, hydrogels are often investigated to deliver epigenic factors to recruit/signal endogenous stem cells for use in CNS regeneration applications.

## 8.5 Conclusions

The main goal of this review was to provide a detailed overview of various hydrogel scaffolds for neural tissue regeneration, with emphasis on hydrogel designs as a criterion for directing NSC fate. Evidence from past and current research has established that hydrogels cultured with NSCs and then transplanted at the site of injury provides an attractive strategy for treating CNS injuries/diseases. Hydrogels engineered to mimic the native CNS tissue prior to injury or disease can serve a potent microenvironment which delivers stem cell regulatory signals in a precise and physiological manner, thus directing stem cell fate *in vivo*.

Hydrogel attributes that influence its application for CNS tissue regeneration include non-immunogenicity, high biocompatibility, controlled biodegradability and mechanism of gelation. The latter is very important and could impact clinical success if not fully contemplated. Depending on the mechanism of gelation, some hydrogels are injectable and this makes them convenient for clinical use. This characteristic minimizes the invasiveness of application and strengthens the interface between tissue and hydrogel. As described above, substantial progress in the field of CNS tissue regeneration has been achieved with hydrogels that incorporated biological components such ECM proteins, growth factors, and signaling molecules. Further progress depends on our abilities to successfully integrate each component of these multifaceted systems, which includes the biomechanical properties of the hydrogels, spatial and temporal control of signaling molecules, and the packaging of major ECM motif, into discrete scalable modules for clinical use. Despite the amount of work that needs to be done, this approach is the most promising for achieving the desired goals of CNS regeneration. Furthermore, there is the option of recruiting endogenous NSCs for CNS regeneration. This strategy circumvents the immumocompatibility issues and teratoma formation frequently observed with the use of exogenous stem cells. The intrinsic regenerative capacity of endogenous stems cells are lacking and the use of hydrogels to deliver growth factors, signaling molecules or other biochemicals to manipulate endogenous stem cells to increase their regenerative capacity is very promising, despite being in the preliminary stages. Even though each approach shows promising results, optimal recovery in CNS regeneration may depend on a combination of multiple strategies.

## Acknowledgments

This work was made possible by the NIH/NINDS USA (R01 NS050243), NSF (0748129, 1055922), William H. Goodwin Endowment, and American Heart Association (10PRE4280017).

## References

[1] Lindvall, O. and Kokaia, Z. (2006). Stem cells for the treatment of neurological disorders. *Nature*, **441**, pp. 1094–1096.

[2] Goldman, S. (2005). Stem and progenitor cell-based therapy of the human central nervous system. *Nat. Biotech.*, **23**, pp. 862–871.

[3] Delcroix, G.J.R., Schiller, P.C., Benoit, J.P., and Montero-Menei, C.N. (2010). Adult cell therapy for brain neuronal damages and the role of tissue engineering. *Biomaterials*, **31**(8), pp. 2105–2120.

[4] Ronaghi, M., Erceg, S., Moreno-Manzano, V., and Stojkovic, M. (2010). Challenges of stem cell therapy for spinal cord injury: Human embryonic stem cells, endogenous neural stem cells, or induced pluripotent stem cells? *Stem Cells*, **28**, pp. 93–99.

[5] Conti, L. and Cattaneo, E. (2010). Neural stem cell systems: Physiological players or *in vitro* entities? *Nat. Rev. Neurosci.*, **11**, pp. 176–187.

[6] Potter, W., Kalil, R.E., and Kao, W.J. (2008). Biomimetic material systems for neural progenitor cell-based therapy. *Front. Biosci.*, **13**, pp 806–821.

[7] Marklein, R.A. and Burdick, J.A. (2010). Controlling stem cell fate with material design. *Adv. Mater.*, **22**, pp. 175–189.

[8] Nisbet, D.R., Crompton, K.E., Horne, M.K., Finkelstein, D.I., and Forsythe, J.S. (2008). Neural tissue engineering of the CNS using hydrogels. *J. Biomed. Mater. Res. B Appl. Biomater.*, **87B**, pp. 251–263.
[9] Teixeira, A.I., Duckworth, J.K., and Hermanson, O. (2007). Getting the right stuff: Controlling neural stem cell state and fate *in vivo* and *in vitro* with biomaterials. *Cell Res.*, **17**, pp. 56–61.
[10] Zhang, N., Yan, H., and Wen, X. (2005). Tissue-engineering approaches for axonal guidance. *Brain Res. Rev.*, **49**, pp. 48–64.
[11] Straley, K.S., Foo, C.W.P., and Heilshorn, S.C. (2009). Biomaterial design strategies for the treatment of spinal cord injuries. *J. Neurotrauma*, **27**, pp. 1–19.
[12] Nomura, H., Tator, C.H., and Shoichet, M.S. (2006). Bioengineered strategies for spinal cord repair. *J. Neurotrauma*, **23**, pp. 496–507.
[13] Norman, L.L., Stroka, K., and Aranda-Espinoza, H. (2009). Guiding axons in the central nervous system: A tissue engineering approach. *Tissue Eng. Part B Rev.*, **15**, pp. 291–305.
[14] Stein, D.G. and Wright, D.W. (2010). Progesterone in the clinical treatment of acute traumatic brain injury. *Expert Opin. Invest. Drugs*, **19**, pp. 847–857.
[15] Mueller, B.K., Mueller, R., and Schoemaker, H. (2009). Stimulating neuroregeneration as a therapeutic drug approach for traumatic brain injury. *Br. J. Pharmacol.*, **157**, pp. 675–685.
[16] Qian, T. *et al.* (2004). High-dose methylprednisolone may cause myopathy in acute spinal cord injury patients. *Spinal Cord*, **43**, pp. 199–203.
[17] Baptiste, D.C. and Fehlings, M.G. (2006). Pharmacological approaches to repair the injured spinal cord. *J. Neurotrauma*, **23**, pp. 318–334.
[18] Murray, V. *et al.* (2009). The molecular basis of thrombolysis and its clinical application in stroke. *J. Intern. Med.*, **267**, pp. 191–208.
[19] Marler, J.R. *et al.* (2000). Early stroke treatment associated with better outcome: The NINDS rt-PA stroke study. *Neurology*, **55**, pp. 1649–1655.
[20] Lapchak, P.A. (2010). A critical assessment of edaravone acute ischemic stroke efficacy trials: Is edaravone an effective neuroprotective therapy? *Expert Opin. Pharmacother.*, **11**, pp. 1753–1763.
[21] Di Stefano, A., Sozio, P., Iannitelli, A., and Cerasa, L.S. New drug delivery strategies for improved Parkinson's disease therapy. *Expert Opin. Drug Deliv.*, pp. 389–404.
[22] Olanow, C.W., Kieburtz, K., and Schapira, A.H.V. (2008). Why have we failed to achieve neuroprotection in Parkinson's disease? *Ann. Neurol.*, **64**, pp. S101–S110.
[23] Weinstock-Guttman, B. and Bakshi, R. (2004). Combination therapy for multiple sclerosis: The treatment strategy of the future? *CNS Drugs*, **18**, pp. 777–792.
[24] McDaneld, L., Fields, J., Bourdette, D., and Bhardwaj, A. (2010). Immunomodulatory therapies in neurologic critical care. *Neurocrit. Care*, **12**, pp. 132–143.
[25] Temple, S. (2001). The development of neural stem cells. *Nature*, **414**, pp. 112–117.

[26] Okano, H. and Sawamoto, K. (2008). Neural stem cells: Involvement in adult neurogenesis and CNS repair. *Phil. Trans. R. Soc. B*, **363**, pp. 2111–2122.
[27] Okano, H. (2010). Neural stem cells and strategies for the regeneration of the central nervous system. *Proc. Jpn. Acad. Ser. B Phys. Biol. Sci.*, **86**, pp. 438–450.
[28] Doetsch, F. (2003). The glial identity of neural stem cells. *Nat. Neurosci.*, **6**, pp. 1127–1134.
[29] Doetsch, F., Caillé, I., Lim, D.A., García-Verdugo, J.M., and Alvarez-Buylla, A. (1999). Subventricular zone astrocytes are neural stem cells in the adult mammalian brain. *Cell*, **97**, pp. 703–716.
[30] Seri, B., García-Verdugo, J.M., McEwen, B.S., and Alvarez-Buylla, A. (2001). Astrocytes give rise to new neurons in the adult mammalian hippocampus. *J. Neurosci.*, **21**, pp. 7153–7160.
[31] Horner, P.J. *et al.* (2000). Proliferation and differentiation of progenitor cells throughout the intact adult rat spinal cord. *J. Neurosci.*, **20**, pp. 2218–2228.
[32] Yamamoto, S., Yamamoto, N., Kitamura, T., Nakamura, K., and Nakafuku, M. (2001). Proliferation of parenchymal neural progenitors in response to injury in the adult rat spinal cord. *Exp. Neurol.*, **172**, pp. 115–127.
[33] Weiss, S. *et al.* (1996). Multipotent CNS stem cells are present in the adult mammalian spinal cord and ventricular neuroaxis. *J. Neurosci.*, **16**, pp. 7599–7609.
[34] Tonchev, A.B., Yamashima, T., Sawamoto, K., and Okano, H. (2005). Enhanced proliferation of progenitor cells in the subventricular zone and limited neuronal production in the striatum and neocortex of adult macaque monkeys after global cerebral ischemia. *J. Neurosci. Res.*, **81**, pp. 776–788.
[35] Arvidsson, A., Collin, T., Kirik, D., Kokaia, Z., and Lindvall, O. (2002). Neuronal replacement from endogenous precursors in the adult brain after stroke. *Nat. Med.*, **8**, pp. 963–970.
[36] Nakatomi, H. *et al.* (2002). Regeneration of hippocampal pyramidal neurons after ischemic brain injury by recruitment of endogenous neural progenitors. *Cell*, **110**, pp. 429–441.
[37] Kernie, S.G. and Parent, J.M. (2010). Forebrain neurogenesis after focal ischemic and traumatic brain injury. *Neurobiol. Dis.*, **37**, pp. 267–274.
[38] Ke, Y. *et al.* (2006). Early response of endogenous adult neural progenitor cells to acute spinal cord injury in mice. *Stem Cells*, **24**, pp. 1011–1019.
[39] Okano, H., Sakaguchi, M., Ohki, K., Suzuki, N., and Sawamoto, K. (2007). Regeneration of the central nervous system using endogenous repair mechanisms. *J. Neurochem.*, pp. 1459–1465.
[40] Obermair, F.J., Schroter, A., and Thallmair, M. (2008). Endogenous neural progenitor cells as therapeutic target after spinal cord injury. *Physiology*, **23**, pp. 296–304.
[41] Picard-Riera, N., Nait-Oumesmar, B., and Baron-Van Evercooren, A. (2004). Endogenous adult neural stem cells: Limits and potential to repair the injured central nervous system. *J. Neurosci. Res.*, **76**, pp. 223–231.

[42] Williams, C.A. and Lavik, E.B. (2009). Engineering the CNS stem cell microenvironment. *Regen. Med.*, **4**, pp. 865–877.
[43] Miller, F.D. and Gauthier-Fisher, A. (2009). Home at last: Neural stem cell niches defined. *Cell Stem Cell*, **4**(6), pp. 507–510.
[44] Lutolf, M.P. and Blau, H.M. (2009). Artificial stem cell niches. *Adv. Mater.*, **21**, pp. 3255–3268.
[45] Guilak, F. *et al.* (2009). Control of stem cell fate by physical interactions with the extracellular matrix. *Cell Stem Cell*, **5**, pp. 17–26.
[46] Leipzig, N.D. and Shoichet, M.S. (2009). The effect of substrate stiffness on adult neural stem cell behavior. *Biomaterials*, **30**, pp. 6867–6878.
[47] Lu, Y. *et al.* (2006). Viscoelastic properties of individual glial cells and neurons in the CNS. *Proc. Natl. Acad. Sci. USA*, **103**, pp. 17759–17764.
[48] Elkin, S.B., Azeloglu, U.E., Costa, D.K., and Morrison, B. (2007). Mechanical heterogeneity of the rat hippocampus measured by atomic force microscope indentation. *J. Neurotrauma*, **24**, pp. 812–822.
[49] Bonneh-Barkay, D. and Wiley, C.A. (2009). Brain extracellular matrix in neurodegeneration. *Brain Pathology*, **19**, pp. 573–585.
[50] Bandtlow, C.E. and Zimmermann, D.R. (2000). Proteoglycans in the developing brain: New conceptual insights for old proteins. *Physiol. Rev.*, **80**, pp. 1267–1290.
[51] Straley, K.S., Foo, C.W.P., and Heilshorn, S.C. (2010). Biomaterial design strategies for the treatment of spinal cord injuries. *J. Neurotrauma*, **27**, pp. 1–19.
[52] Lutolf, M.P., Gilbert, P.M., and Blau, H.M. (2009). Designing materials to direct stem-cell fate. *Nature*, **462**, pp. 433–441.
[53] Zhu, J. (2010). Bioactive modification of poly(ethylene glycol) hydrogels for tissue engineering. *Biomaterials*, **31**, pp. 4639–4656.
[54] Woerly, S. (2000). Restorative surgery of the central nervous system by means of tissue engineering using NeuroGel implants. *Neurosurg. Rev.*, **23**, pp. 59–77.
[55] Woerly, S., Pinet, E., de Robertis, L., van Diep, D., and Bousmina, M. (2001). Spinal cord repair with PHPMA hydrogel containing RGD peptides (NeuroGel). *Biomaterials*, **22**, pp. 1095–1111.
[56] Saha, K., Irwin, E.F., Kozhukh, J., Schaffer, D.V., and Healy, K.E. (2007). Biomimetic interfacial interpenetrating polymer networks control neural stem cell behavior. *J. Biomed. Mater. Res. A*, **81A**, pp. 240–249.
[57] Engler, A.J., Sen, S., Sweeney, H.L., and Discher, D.E. (2006). Matrix elasticity directs stem cell lineage specification. *Cell*, **126**, pp. 677–689.
[58] Saha, K. *et al.* (2008). Substrate modulus directs neural stem cell behavior. *Biophys. J.*, **95**, pp. 4426–4438.
[59] Lin, D.C., Yurke, B., and Langrana, N.A. (2004). Mechanical properties of a reversible, DNA-crosslinked polyacrylamide hydrogel. *J. Biomech. Eng.*, **126**, pp. 104–110.
[60] Jiang, F., Yurke, B., Firestein, B., and Langrana, N. (2008). Neurite outgrowth on a DNA crosslinked hydrogel with tunable stiffnesses. *Ann. Biomed. Eng.*, **36**, pp. 1565–1579.

[61] Kubinová, S. *et al.* (2010). The use of superporous Ac-CGGASIKVAVS-OH-modified PHEMA scaffolds to promote cell adhesion and the differentiation of human fetal neural precursors. *Biomaterials*, **31**, pp. 5966–5975.
[62] Yu, T.T. and Shoichet, M.S. (2005). Guided cell adhesion and outgrowth in peptide-modified channels for neural tissue engineering. *Biomaterials*, **26**, pp. 1507–1514.
[63] Dalton, P.D., Flynn, L., and Shoichet, M.S. (2002). Manufacture of poly(2-hydroxyethyl methacrylate-co-methyl methacrylate) hydrogel tubes for use as nerve guidance channels. *Biomaterials*, pp. **23**, pp. 3843–3851.
[64] Tsai, E.C., Dalton, P.D., Shoichet, M.S., and Tator, C.H. Synthetic hydrogel guidance channels facilitate regeneration of adult rat brainstem motor axons after complete spinal cord transection. *J. Neurotrauma*, **21**, pp. 789–804.
[65] Tsai, E.C., Dalton, P.D., Shoichet, M.S., and Tator, C.H. (2006). Matrix inclusion within synthetic hydrogel guidance channels improves specific supraspinal and local axonal regeneration after complete spinal cord transection. *Biomaterials*, **27**, pp. 519–533.
[66] Piotrowicz, A. and Shoichet, M.S. (2006). Nerve guidance channels as drug delivery vehicles. *Biomaterials*, **27**, pp. 2018–2027.
[67] Katayama, Y. *et al.* (2006). Coil-reinforced hydrogel tubes promote nerve regeneration equivalent to that of nerve autografts. *Biomaterials*, **27**, pp. 505–518.
[68] Nomura, H., Katayama, Y.M., Shoichet, M.S., and Tator, C.H. (2006). Complete spinal cord transection treated by implantation of a reinforced synthetic hydrogel channel results in syringomyelia and caudal migration of the rostral stump. *Neurosurgery*, **59**, pp. 183–192.
[69] Friedman, M. (2003). Chemistry, biochemistry, and safety of acrylamide. A review. *J. Agric. Food Chem.*, **51**, pp. 4504–4526.
[70] Bjugstad, K.B. *et al.* (2008). Biocompatibility of PEG-based hydrogels in primate brain. *Cell Transplant.*, **17**, pp. 409–415.
[71] Bjugstad, K.B., Lampe, K., Kern, D.S., and Mahoney, M.J. Biocompatibility of poly(ethylene glycol)-based hydrogels in the brain: An analysis of the glial response across space and time. *J. Biomed. Mater. Res. A*, **95A**, pp. 79–91.
[72] Mahoney, M.J. and Anseth, K.S. (2006). Three-dimensional growth and function of neural tissue in degradable polyethylene glycol hydrogels. *Biomaterials*, **27**, pp. 2265–2274.
[73] Whitesides, G. (1993). Poly(ethylene glycol) chemistry biotechnical and biomedical applications. *Appl. Biochem. Biotechnol.*, **41**, pp. 233–234.
[74] Mahoney, M.J. and Anseth, K.S. (2007). Contrasting effects of collagen and bFGF-2 on neural cell function in degradable synthetic PEG hydrogels. *J. Biomed. Mater. Res. A*, **81**, pp. 269–278.
[75] Freudenberg, U. *et al.* (2009). A star-PEG-heparin hydrogel platform to aid cell replacement therapies for neurodegenerative diseases. *Biomaterials*, **30**, pp. 5049–5060.

[76] Ishan, C. and Robert, J.L. (2002). Heparin-protein interactions. *Angew Chem. Int. Ed.*, **41**, pp. 390–412.

[77] Royce Hynes, S., McGregor, L.M., Ford Rauch, M., and Lavik, E.B. (2007). Photopolymerized poly(ethylene glycol)/poly(L-lysine) hydrogels for the delivery of neural progenitor cells. *J. Biomater. Sci. Polym. Ed.*, **18**(8), pp. 1017–1030.

[78] Burdick, J.A., Ward, M., Liang, E., Young, M.J., and Langer, R. (2006). Stimulation of neurite outgrowth by neurotrophins delivered from degradable hydrogels. *Biomaterials*, **27**, pp. 452–459.

[79] Hynes, S.R., Rauch, M.F., Bertram, J.P., and Lavik, E.B. (2009). A library of tunable poly(ethylene glycol)/poly(L-lysine) hydrogels to investigate the material cues that influence neural stem cell differentiation. *J. Biomed. Mater. Res. A*, **89A**, pp. 499–509.

[80] O'Connor, S.M. *et al.* (2000). Primary neural precursor cell expansion, differentiation and cytosolic Ca2+ response in three-dimensional collagen gel. *J. Neurosci. Methods*, **102**, pp. 187–195.

[81] Ma, W. *et al.* (2004). CNS stem and progenitor cell differentiation into functional neuronal circuits in three-dimensional collagen gels. *Exp. Neurol.*, **190**, pp. 276–288.

[82] Lin, H.J., O'Shaughnessy, T.J., Kelly, J., and Ma, W. (2004). Neural stem cell differentiation in a cell–collagen-bioreactor culture system. *Dev. Brain Res.*, **153**, pp. 163–173.

[83] Ma, W. *et al.* (2008). Reconstruction of functional cortical-like tissues from neural stem and progenitor cells. *Tissue Eng. Part A*, **14**, pp. 1673–1686.

[84] Miyazaki, H., Kato, K., Teramura, Y., and Iwata, H. (2008). A collagen-binding mimetic of neural cell adhesion molecule. *Bioconjugate Chem.*, **19**, pp. 1119–1123.

[85] Hiraoka, M., Kato, K., Nakaji-Hirabayashi, T., and Iwata, H. (2009). Enhanced survival of neural cells embedded in hydrogels composed of collagen and laminin-derived cell adhesive peptide. *Bioconjugate Chem.*, **20**, pp. 976–983.

[86] Deister, C. and Schmidt, C.E. (2006). Optimizing neurotrophic factor combinations for neurite outgrowth. *J. Neural Eng.*, **3**, pp. 172–179.

[87] Sundararaghavan, H.G., Monteiro, G.A., Firestein, B.L., and Shreiber, D.I. (2009). Neurite growth in 3D collagen gels with gradients of mechanical properties. *Biotechnol. Bioeng.*, **102**, pp. 632–643.

[88] Prestwich, G.D., Marecak, D.M., Marecek, J.F., Vercruysse, K.P., Ziebell, M.R. (1998). Controlled chemical modification of hyaluronic acid: Synthesis, applications, and biodegradation of hydrazide derivatives. *J. Control. Release*, **53**, pp. 93–103.

[89] Hou, S. *et al.* (2006). The enhancement of cell adherence and inducement of neurite outgrowth of dorsal root ganglia co-cultured with hyaluronic acid hydrogels modified with Nogo-66 receptor antagonist *in vitro*. *Neuroscience*, **137**, pp. 519–529.

[90] Wei, Y. *et al.* (2010). Hyaluronic acid hydrogel modified with Nogo-66 receptor antibody and poly-L-lysine to promote axon regrowth after spinal cord injury. *J. Biomed. Mater. Res. B Appl. Biomater.*, **95B**, pp. 110–117.

[91] Tian, W. *et al.* (2005). Hyaluronic acid poly-D-lysine-based three-dimensional hydrogel for traumatic brain injury. *Tissue Eng.*, **11**, pp. 513–525.
[92] Cui, F., Tian, W., Hou, S., Xu, Q., and Lee, I.S. (2006). Hyaluronic acid hydrogel immobilized with RGD peptides for brain tissue engineering. *J. Mater. Sci. Mater. Med.*, **17**, pp. 1393–1401.
[93] Seidlits, S.K. *et al.* (2010). The effects of hyaluronic acid hydrogels with tunable mechanical properties on neural progenitor cell differentiation. *Biomaterials*, **31**(10), pp. 3930–3940.
[94] Nakaji-Hirabayashi, T., Kato, K., and Iwata, H. (2009). Hyaluronic acid hydrogel loaded with genetically-engineered brain-derived neurotrophic factor as a neural cell carrier. *Biomaterials*, **30**, pp. 4581–4589.
[95] Martin, B.C., Minner, E., Wiseman, S., Klank, R.L., and Gilbert, R.J. (2008). Agarose and methylcellulose hydrogel blends for nerve regeneration applications. *J. Neural Eng.*, **5**, pp. 221–231.
[96] Cao, Z., Gilbert, R.J., and He, W. (2009). Simple agarose chitosan gel composite system for enhanced neuronal growth in three dimensions. *Biomacromolecules*, **10**, pp. 2954–2959.
[97] Luo, Y. and Shoichet, M.S. (2004). A photolabile hydrogel for guided three-dimensional cell growth and migration. *Nat. Mater.*, **3**, pp. 249–253.
[98] Aizawa, Y., Leipzig, N., Zahir, T., and Shoichet, M.S. (2008). The effect of immobilized platelet derived growth factor AA on neural stem/progenitor cell differentiation on cell-adhesive hydrogels. *Biomaterials*, **29**, pp. 4676–4683.
[99] Cullen, D., Lessing, M., and LaPlaca, M.C. (2007). Collagen-dependent neurite outgrowth and response to dynamic deformation in three-dimensional neuronal cultures. *Ann. Biomed. Eng.*, **35**, pp. 835–846.
[100] Kataoka, K. *et al.* (2004). Alginate enhances elongation of early regenerating axons in spinal cord of young rats. *Tissue Eng.*, **10**, pp. 493–504.
[101] Read, T.A. *et al.* (2001). Local endostatin treatment of gliomas administered by microencapsulated producer cells. *Nat. Biotech.*, **19**, pp. 29–34.
[102] Purcell, E.K., Singh, A., and Kipke, D.R. (2009). Alginate composition effects on a neural stem cell seeded scaffold. *Tissue Eng. Part C Methods*, **15**, pp. 541–550.
[103] Shanbhag, M. *et al.* (2010). Neural progenitor cells grown on hydrogel surfaces respond to the product of the transgene of encapsulated genetically engineered fibroblasts. *Biomacromolecules*, **11**, pp. 2936–2943.
[104] Liudmila, N.N. *et al.* (2006). Alginate hydrogel and matrigel as potential cell carriers for neurotransplantation. *J. Biomed. Mater. Res. A*, **77A**, pp. 242–252.
[105] Banerjee, A. *et al.* (2009). The influence of hydrogel modulus on the proliferation and differentiation of encapsulated neural stem cells. *Biomaterials*, **30**, pp. 4695–4699.
[106] Ashton, R.S., Banerjee, A., Punyani, S., Schaffer, D.V., and Kane, R.S. (2007). Scaffolds based on degradable alginate hydrogels and poly(lactide-co-glycolide) microspheres for stem cell culture. *Biomaterials*, **28**, pp. 5518–5525.

[107] Yu, L.M., Kazazian, K., and Shoichet, M.S. (2007). Peptide surface modification of methacrylamide chitosan for neural tissue engineering applications. *J. Biomed. Mater. Res. A*, **82A**, pp. 243–255.

[108] Chenite, A. *et al.* (2000). Novel injectable neutral solutions of chitosan form biodegradable gels *in situ*. *Biomaterials*, **21**, pp. 2155–2161.

[109] Crompton, K. *et al.* (2007). Polylysine-functionalised thermoresponsive chitosan hydrogel for neural tissue engineering. *Biomaterials*, **28**, pp. 441–449.

[110] Ju, Y.E., Janmey, P.A., McCormick, M.E., Sawyer, E.S., and Flanagan, L.A. (2007). Enhanced neurite growth from mammalian neurons in three-dimensional salmon fibrin gels. *Biomaterials*, **28**, pp. 2097–2108.

[111] Schense, J.C., Bloch, J., Aebischer, P., and Hubbell, J.A. (2000). Enzymatic incorporation of bioactive peptides into fibrin matrices enhances neurite extension. *Nat. Biotech.*, **18**, pp. 415–419.

[112] Régis, P., Fabrice, S., Jeffrey, A.H., and Heike, H. (2005). Neurite extension and in vitro myelination within three-dimensional modified fibrin matrices. *J. Neurobiol.*, **63**, pp. 1–14.

[113] Tate, M.C., Shear, D.A., Hoffman, S.W., Stein, D.G., and LaPlaca, M.C. (2001). Biocompatibility of methylcellulose-based constructs designed for intracerebral gelation following experimental traumatic brain injury. *Biomaterials*, **22**, pp. 1113–1123.

[114] Stabenfeldt, S.E., García, A.J., and LaPlaca, M.C. (2006). Thermoreversible laminin-functionalized hydrogel for neural tissue engineering. *J. Biomed. Mater. Res. A*, **77A**, pp. 718–725.

[115] Stabenfeldt, S.E., Munglani, G., García, A.J., and LaPlaca, M.C. (2010). Biomimetic microenvironment modulates neural stem cell survival, migration, and differentiation. *Tissue Eng. Part A*, **16**, pp. 3747–3758.

[116] Irons, H.R. *et al.* (2008). Three-dimensional neural constructs: A novel platform for neurophysiological investigation. *J. Neural Eng.*, **5**, pp. 333–341.

[117] Berski, S. *et al.* (2008). Synthesis and biological evaluation of a polysialic acid-based hydrogel as enzymatically degradable scaffold material for tissue engineering. *Biomacromolecules*, **9**, pp. 2353–2359.

[118] Rutishauser, U. (1998). Polysialic acid at the cell surface: Biophysics in service of cell interactions and tissue plasticity. *J. Cell Biochem.*, **70**, pp. 304–312.

[119] Stummeyer, K., Dickmanns, A., Muhlenhoff, M., Gerardy-Schahn, R., and Ficner, R. (2005). Crystal structure of the polysialic acid-degrading endosialidase of bacteriophage K1F. *Nat. Struct. Mol. Biol.*, **12**, pp. 90–96.

[120] Haile, Y. *et al.* (2007). Culturing of glial and neuronal cells on polysialic acid. *Biomaterials*, **28**, pp. 1163–1173.

[121] Haile, Y. *et al.* (2008). The effect of modified polysialic acid based hydrogels on the adhesion and viability of primary neurons and glial cells. *Biomaterials*, **29**, pp. 1880–1891.

[122] Nisbet, D.R. *et al.* (2009). Enhancing neurite outgrowth from primary neurones and neural stem cells using thermoresponsive hydrogel scaffolds

for the repair of spinal cord injury. *J. Biomed. Mater. Res. A*, **89A**, pp. 24–35.

[123] Nisbet, D.R., Rodda, A.E., Horne, M.K., Forsythe, J.S., and Finkelstein, D.I. (2010). Implantation of functionalized thermally gelling xyloglucan hydrogel within the brain: Associated neurite infiltration and inflammatory response. *Tissue Eng. Part A*, **16**, pp. 2833–2842.

[124] Zhang, S. *et al.* (1995). Self-complementary oligopeptide matrices support mammalian cell attachment. *Biomaterials*, **16**, pp. 1385–1393.

[125] Zhang, S., Holmes, T.C., Lockshin, C., and Rich, A. (1993). Spontaneous assembly of a self-complementary oligopeptide to form a stable macroscopic membrane. *Proc. Natl. Acad. Sci. USA*, **90**, pp. 3334–3338.

[126] Holmes, T.C. *et al.* (2000). Extensive neurite outgrowth and active synapse formation on self-assembling peptide scaffolds. *Proc. Natl. Acad. Sci. USA*, **97**, pp. 6728–6733.

[127] Gelain, F., Bottai, D., Vescovi, A., and Zhang, S. (2006). Designer self-assembling peptide nanofiber scaffolds for adult mouse neural stem cell 3-dimensional cultures. *PLoS ONE*, **1**, p. e119.

[128] Tysseling-Mattiace, V.M. *et al.* (2008). Self-assembling nanofibers inhibit glial scar formation and promote axon elongation after spinal cord injury. *J. Neurosci.*, **28**, pp. 3814–3823.

[129] Silva, G.A. *et al.* (2004). Selective differentiation of neural progenitor cells by high-epitope density nanofibers. *Science*, **303**, pp. 1352–1355.

[130] Wong Po Foo, C.T.S., Lee, J.S., Mulyasasmita, W., Parisi-Amon, A., and Heilshorn, S.C. (2009). Two-component protein-engineered physical hydrogels for cell encapsulation. *Proc. Natl. Acad. Sci. USA*, **106**, pp. 22067–22072.

[131] Deming, T.J. (2005). Polypeptide hydrogels via a unique assembly mechanism. *Soft Matter*, **1**, pp. 28–35.

[132] Yang, C. *et al.* (2009). Biocompatibility of amphiphilic diblock copolypeptide hydrogels in the central nervous system. *Biomaterials*, **30**, pp. 2881–2898.

[133] Thonhoff, J.R., Lou, D.I., Jordan, P.M., Zhao, X., and Wu, P. (2008). Compatibility of human fetal neural stem cells with hydrogel biomaterials in vitro. *Brain Res.*, **1187**, pp. 42–51.

[134] Li, X., Yang, Z., and Zhang, A. (2009). The effect of neurotrophin-3/chitosan carriers on the proliferation and differentiation of neural stem cells. *Biomaterials*, **30**, pp. 4978–4985.

[135] Horn, E.M. *et al.* (2007). Influence of cross-linked hyaluronic acid hydrogels on neurite outgrowth and recovery from spinal cord injury. *J. Neurosurg. Spine*, **6**, pp. 133–140.

[136] Deister, C.A., Aljabari, S., and Schmidt, C.E. (2007). Effects of collagen 1, fibronectin, laminin and hyaluronic acid concentration in multi-component gels on neurite extension. *J. Biomater Sci. Polym. Ed.*, **18**, pp. 983–997.

[137] Gunn, J.W., Turner, S.D., and Mann, B.K. (2005). Adhesive and mechanical properties of hydrogels influence neurite extension. *J. Biomed. Mater. Res. A*, **72A**, pp. 91–97.

[138] Jones, L.L., Sajed, D., and Tuszynski, M.H. (2003). Axonal regeneration through regions of chondroitin sulfate proteoglycan deposition after spinal cord injury: A balance of permissiveness and inhibition. *J. Neurosci.*, **23**, pp. 9276–9288.

[139] Yang, Z., Duan, H., Mo, L., Qiao, H., and Li, X. (2010). The effect of the dosage of NT-3/chitosan carriers on the proliferation and differentiation of neural stem cells. *Biomaterials*, **31**, pp. 4846–4854.

[140] Zhao, J., Zhang, N., Prestwich, G.D., and Wen, X. (2008). Recruitment of endogenous stem cells for tissue repair. *Macromol. Biosci.*, **8**, pp. 836–842.

[141] Discher, D.E., Janmey, P., and Wang, Y. (2005). Tissue cells feel and respond to the stiffness of their substrate. *Science*, **310**, pp. 1139–1143.

[142] Brandl, F., Sommer, F., and Goepferich, A. (2007). Rational design of hydrogels for tissue engineering: Impact of physical factors on cell behavior. *Biomaterials*, **28**, pp. 134–146.

[143] Kostic, A., Sap, J., and Sheetz, M.P. (2007). RPTPalpha is required for rigidity-dependent inhibition of extension and differentiation of hippocampal neurons. *J. Cell Sci.*, **120**, pp. 3895–3904.

[144] Ingber, D.E. (2003). Tensegrity i, Cell structure and hierarchical systems biology. *J. Cell Sci.*, **116**, pp. 1157–1173.

[145] Ingber, D.E. (2003). Tensegrity ii. How structural networks influence cellular information processing networks. *J. Cell Sci.*, **116**, pp. 1397–1408.

[146] Chan, C.E. and Odde, D.J. (2008). Traction dynamics of filopodia on compliant substrates. *Science*, **322**, pp. 1687–1691.

[147] Namba, R.M., Cole, A.A., Bjugstad, K.B., and Mahoney, M.J. (2009). Development of porous PEG hydrogels that enable efficient, uniform cell-seeding and permit early neural process extension. *Acta Biomater.*, **5**, pp. 1884–1897.

[148] Rainaldi, G., Calcabrini, A., and Santini, M.T. (1998). Positively charged polymer polylysine-induced cell adhesion molecule redistribution in K562 cells. *J. Mater. Sci. Mater. Med.*, **9**, pp. 755–760.

[149] Hartmann, W., Galla, H.J., and Sackmann, E. (1977). Direct evidence of charge-induced lipid domain structure in model membranes. *FEBS Lett.*, **78**, pp. 169–172.

[150] Meiners, S. and Mercado, M. (2003). Functional peptide sequences derived from extracellular matrix glycoproteins and their receptors. *Mol. Neurobiol.*, **27**, pp. 177–195.

[151] Lisa, A.F., Liza, M.R., Stanislava, D., Philip, H.S., and Edwin, S.M. (2006). Regulation of human neural precursor cells by laminin and integrins. *J. Neurosci. Res.*, **83**, pp. 845–856.

[152] Kleinman, H.K. *et al.* (1991). Identification of a 110-kDa nonintegrin cell surface laminin-binding protein which recognizes an a chain neurite-promoting peptide. *Arch. Biochem. Biophys.*, **290**, pp. 320–325.

[153] Liesi, P., Närvänen, A., Soos, J., Sariola, H., and Snounou, G. (1989). Identification of a neurite outgrowth-promoting domain of laminin using synthetic peptides. *FEBS Lett.*, **244**, pp. 141–148.

[154] Graf, J. *et al.* (1987). A pentapeptide from the laminin B1 chain mediates cell adhesion and binds to 67,000 laminin receptor. *Biochemistry (Mosc.)*, **26**, pp. 6896–6900.
[155] Skubitz, A.P.N., McCarthy, J.B., Zhao, Q., Yi, X., and Furcht, L.T. (1990). Definition of a sequence, RYVVLPR, within laminin peptide F-9 that mediates metastatic fibrosarcoma cell adhesion and spreading. *Cancer Res.*, **50**, pp. 7612–7622.
[156] Hersel, U., Dahmen, C., and Kessler, H. (2003). RGD modified polymers: Biomaterials for stimulated cell adhesion and beyond. *Biomaterials*, **24**, pp. 4385–4415.
[157] Boyd, J. and Gordon, T. (2003). Neurotrophic factors and their receptors in axonal regeneration and functional recovery after peripheral nerve injury. *Mol. Neurobiol.*, **27**, pp. 277–323.
[158] Abe, K. (2000). Therapeutic potential of neurotrophic factors and neural stem cells against ischemic brain injury. *J. Cereb. Blood Flow Metab.*, **20**, pp. 1393–1408.
[159] Xu, T. *et al.* (2006). Viability and electrophysiology of neural cell structures generated by the inkjet printing method. *Biomaterials*, **27**, pp. 3580–3588.
[160] Lee, Y.B. *et al.* (2010). Bio-printing of collagen and VEGF-releasing fibrin gel scaffolds for neural stem cell culture. *Exp. Neurol.*, **223**(2), pp. 645–652.
[161] Lee, W. *et al.* (2009). Three-dimensional bioprinting of rat embryonic neural cells. *Neuroreport*, **20**, pp. 798–803.
[162] Hynd, M.R., Frampton, J.P., Dowell-Mesfin, N., Turner, J.N., and Shain, W. (2007). Directed cell growth on protein-functionalized hydrogel surfaces. *J. Neurosci. Methods*, **162**, pp. 255–263.
[163] Frisch, S.M. and Francis, H. (1994). Disruption of epithelial cell–matrix interactions induces apoptosis. *J. Cell Biol.*, **124**, pp. 619–626.
[164] Park, K.I., Teng, Y.D., and Snyder, E.Y. (2002). The injured brain interacts reciprocally with neural stem cells supported by scaffolds to reconstitute lost tissue. *Nat. Biotech.*, **20**, pp. 1111–1117.
[165] Date, I. (1996). Parkinson's disease, trophic factors, and adrenal medullary chromaffin cell grafting: Basic and clinical studies. *Brain Res. Bull.*, **40**, pp. 1–19.
[166] Marion, S.A. (2001). The epidemiology of Parkinson's disease. Current issues. *Adv. Neurol.*, **86**, pp. 163–172.
[167] Swick, B.L. and Walling, H.W. (2005). Drug-induced delusions of parasitosis during treatment of Parkinson's disease. *J. Am. Acad. Dermatol.*, **53**, pp. 1086–1087.
[168] Papapetropoulos, S., Argyriou, A.A., and Ellul, J. (2005). Factors associated with drug-induced visual hallucinations in Parkinson's disease. *J. Neurol.*, **252**, pp. 1223–1228.
[169] Leopold, N.A., Polansky, M., and Hurka, M.R. (2004). Drug adherence in Parkinson's disease. *Mov. Disord.*, **19**, pp. 513–517.
[170] Kuzuhara, S. (2001). Drug-induced psychotic symptoms in Parkinson's disease. Problems, management and dilemma. *J. Neurol.*, **248** (Suppl 3), pp. 28–31.

[171] Benabid, A.L., Deuschl, G., Lang, A.E., Lyons, K.E., and Rezai, A.R. (2006). Deep brain stimulation for Parkinson's disease. *Mov. Disord.*, **21**, pp. S168–S170.
[172] Rezai, A.R. *et al.* (2006). Deep brain stimulation for Parkinson's disease: Surgical issues. *Mov. Disord.*, **21**, pp. S197–S218.
[173] Freund, H.J. (2005). Long-term effects of deep brain stimulation in Parkinson's disease. *Brain*, **128**, pp. 2222–2223.
[174] Biran, R., Martin, D.C., and Tresco, P.A. (2005). Neuronal cell loss accompanies the brain tissue response to chronically implanted silicon microelectrode arrays. *Exp. Neurol.*, **195**, pp. 115–126.
[175] Fink, J.S. (1997). Transplantation in Parkinson's disease. *Artif. Organs*, **21**, pp. 1199–1202.
[176] Borlongan, C.V. (2000). Transplantation therapy for Parkinson's disease. *Expert Opin. Investig. Drugs*, **9**, pp. 2319–2330.
[177] Freeman, T.B., Willing, A., Zigova, T., Sanberg, P.R., and Hauser, R.A. (2001). Neural transplantation in Parkinson's disease. *Adv. Neurol.*, **86**, pp. 435–445.
[178] Lindvall, O. and Hagell, P. (2001). Cell therapy and transplantation in Parkinson's disease. *Clin. Chem. Lab. Med.*, **39**, pp. 356–361.
[179] Subramanian, T. (2001). Cell transplantation for the treatment of Parkinson's disease. *Semin. Neurol.*, **21**, pp. 103–115.
[180] Borlongan, C.V. and Sanberg, P.R. (2002). Neural transplantation for treatment of Parkinson's disease. *Drug Discov. Today*, **7**, pp. 674–682.
[181] Ramachandran, A.C., Bartlett, L.E., and Mendez, I.M. (2002). A multiple target neural transplantation strategy for Parkinson's disease. *Rev. Neurosci.*, **13**, pp. 243–256.
[182] Redmond, D.E. (2002). Book review: Cellular replacement therapy for Parkinson's disease — where we are today? *Neuroscientist*, **8**, pp. 457–488.
[183] Perrier, A.L. *et al.* (2004). Derivation of midbrain dopamine neurons from human embryonic stem cells. *Proc. Natl. Acad. Sci. USA*, **101**, pp. 12543–12548.
[184] Kishi, Y. *et al.* (2005). Estrogen promotes differentiation and survival of dopaminergic neurons derived from human neural stem cells. *J. Neurosci. Res.*, **79**, pp. 279–286.
[185] Mori, H. *et al.* (2006). Effect of neurosphere size on the growth rate of human neural stem/progenitor cells. *J. Neurosci. Res.*, **84**, pp. 1682–1691.
[186] Makoto, U. *et al.* (2010). Matrigel supports survival and neuronal differentiation of grafted embryonic stem cell-derived neural precursor cells. *J. Neurosci. Res.*, **88**, pp. 542–551.
[187] Jin, K. *et al.* (2009). Transplantation of human neural precursor cells in matrigel scaffolding improves outcome from focal cerebral ischemia after delayed postischemic treatment in rats. *J. Cereb. Blood Flow Metab.*, **30**, pp. 534–544.
[188] Zhong, J. *et al.* (2010). Hydrogel matrix to support stem cell survival after brain transplantation in stroke. *Neurorehab. Neural Repair*, **24**, pp. 636–644.

[189] Lesný, P. *et al.* (2002). Polymer hydrogels usable for nervous tissue repair. *J. Chem. Neuroanat.*, **23**, pp. 243–247.
[190] Wei, Y. *et al.* (2007). Hyaluronic acid hydrogels with IKVAV peptides for tissue repair and axonal regeneration in an injured rat brain. *Biomed. Mater.*, **2**, pp. S142–S146.
[191] Tate, C.C. *et al.* (2009). Laminin and fibronectin scaffolds enhance neural stem cell transplantation into the injured brain. *J. Tissue Eng. Regen. M.*, **3**, pp. 208–217.
[192] Tate, M.C. *et al.* (2002). Fibronectin promotes survival and migration of primary neural stem cells transplanted into the traumatically injured mouse brain. *Cell Transplant.*, **11**, pp. 283–295.
[193] Novikova, L.N., Novikov, L.N., and Kellerth, J.O. (2003). Biopolymers and biodegradable smart implants for tissue regeneration after spinal cord injury. *Curr. Opin. Neurol.*, **16**, pp. 711–715.
[194] Bakshi, A. (2004). Mechanically engineered hydrogel scaffolds for axonal growth and angiogenesis after transplantation in spinal cord injury. *J. Neurosurg. Spine*, **1**, pp. 322–329.
[195] Park, J. *et al.* (2010). Nerve regeneration following spinal cord injury using matrix metalloproteinase-sensitive, hyaluronic acid-based biomimetic hydrogel scaffold containing brain-derived neurotrophic factor. *J. Biomed. Mater. Res. A*, **93A**, pp. 1091–1099.
[196] Jain, A., Kim, Y.T., McKeon, R.J., and Bellamkonda, R.V. (2006). *In situ* gelling hydrogels for conformal repair of spinal cord defects, and local delivery of BDNF after spinal cord injury. *Biomaterials*, **27**, pp. 497–504.
[197] Prang, P. *et al.* (2006). The promotion of oriented axonal regrowth in the injured spinal cord by alginate-based anisotropic capillary hydrogels. *Biomaterials*, **27**, pp. 3560–3569.
[198] Ford, M.C. (2005). A macroporous hydrogel for the coculture of neural progenitor and endothelial cells to form functional vascular networks *in vivo*. *Proc. Natl. Acad. Sci. USA*, **103**, pp. 2512–2517.
[199] Rauch, M.F. *et al.* (2009). Engineering angiogenesis following spinal cord injury: A coculture of neural progenitor and endothelial cells in a degradable polymer implant leads to an increase in vessel density and formation of the blood–spinal cord barrier. *Eur. J. Neurosci.*, **29**, pp. 132–145.
[200] Johnson, P.J., Tatara, A., Shiu, A., and Sakiyama-Elbert, S.E. (2010). Controlled release of neurotrophin-3 and platelet-derived growth factor from fibrin scaffolds containing neural progenitor cells enhances survival and differentiation into neurons in a subacute model of SCI. *Cell Transplant.*, **19**, pp. 89–101.
[201] Willerth, S.M., Faxel, T.E., Gottlieb, D.I., and Sakiyama-Elbert, S.E. The effects of soluble growth factors on embryonic stem cell differentiation inside of fibrin scaffolds. *Stem Cells*, **25**(9), pp. 2235–2244.
[202] Ilkhanizadeh S., Teixeira A.I., Hermanson O. (2007). Inkjet printing of macromolecules on hydrogels to steer neural stem cell differentiation. *Biomaterials*, **28**(27), pp. 3936–3943.

# Chapter 9

# Injectable Hydrogels for Articular Cartilage Regeneration

B. Choi

*Section of Biomaterials Division of Advanced Prosthodontics*
*Weintraub Center for Resconstructive Biotechnology*
*University of California*
*Los Angeles*

M. Lee*

*Section of Biomaterials Division of Advanced Prosthodontics*
*Weintraub Center for Resconstructive Biotechnology*
*University of California*
*Los Angeles*

## 9.1 Introduction

Articular cartilage is a connective tissue consisting of less than 2% cells, 20–35% extracellular matrix (ECM), and 65–80% water [1]. Mature articular cartilage has a very poor self-healing capacity after injury, due to its avascular and aneural environment [2]. Therefore, an increasing number of adults suffer from joint disease such as osteoarthritis, which is a leading cause of pain and disability among adults in the US [3]. Currently available clinical cartilage repair procedures are microfracture for marrow stimulation, osteochondral autograft transfer (OAT), fresh osteochondral allograft, autologous chondrocyte implantation (ACI), and particulated chondral allograft [4]. However, these techniques have limitations, such as cell leakage and the

*Author to whom correspondence should be addressed. Email: leemin@ucla.edu.

development of fibrocartilaginous tissue with poor mechanical viscoelastic properties. Thus, numerous approaches have been investigated in order to enhance cartilage regeneration, including various three-dimensional systems that can localize cells (chondrocytes or mesenchymal stem cells (MSCs)) and cytokines at the site of the defect [5]. Specifically, injectable and *in situ* gelling systems have been widely investigated and are preferred for cartilage tissue engineering due to their high water content (similar to ECM), capability of mimicking tissue environments by encapsulating cells and bioactive molecules, and/or controlling physicochemical properties. In addition, their administration is patient friendly and minimally invasive, and they boast the property of easy formation in any desired shape at the site of injury [6, 7].

## 9.2 Crosslinking Methods to Form Hydrogels

Hydrogels can be classified into chemical or physical gels according to their crosslinking methods. When hydrogels are prepared by covalent crosslinking they form chemical gels, while physical intermolecular association induces reversible physical gels. Various crosslinking approaches for preparing hydrogels are used to manage specific properties, such as gelation time, mechanical modulus, and the biocompatibility of the ensuing hydrogels, which are important for tissue engineering applications.

### 9.2.1 *Chemical hydrogels*

Redox-initiated polymerization, photopolymerization, classical organic reactions between functional groups, and enzymatic reactions are typical chemical crosslinking methods. Cells can be encapsulated in certain hydrogels, such as poly(ethylene glycol) (PEG), oligo(PEG fumarate) (OPF), poly(lactide-co-ethylene oxide-co-fumarate) (PLEOF), chitosan derivatives, or carboxybetaine hydrogels through redox-initiated polymerization using ammonium persulfate (APS)/N,N,N′,N′-tetramethylethylenediamine (TEMED) or APS/ascorbic acid [8–13]. These redox initiated chemical hydrogels were found to be feasible for tissue engineering and drug delivery applications. Recently, an injectable APS/TEMED redox-polymerized methylcellulose hydrogel showed potential as a filler material [14]. In these systems, an increase in initiator concentration led not only to a decrease in the gelling time but also reduced cell viability.

Photopolymerization of vinyl groups bearing polymers via ultraviolet (UV) or visible blue light (VBL) irradiation with various photoinitiators is one of the most common and effective encapsulation methods for cells or bioactive molecules in biomedical applications. The hydrogel can be formed under mild conditions of physiological pH and temperature. Various photo initiators, such as 2-hydroxy-1-[4-(hydroxyethoxy)phenyl]-2-methyl 1-propanone (Irgacure 2959) [15–17], lithium acyl phosphinate (LAP) [18, 19], methyl benzoylformate (MBF) [20], or 2,2-dimethoxy-2-phenylacetophenone (DMPA) [21] have been used to produce UV-sensitive hydrogels. UV-initiated free radical polymerization systems showed lower cytotoxicity than redox-initiated polymerization. However, UV-irradiation can damage cells, proteins, or tissues; to overcome this issue, visible light-inducible hydrogel systems have been developed. Exposure to visible light is less thermogenic, but causes less cell damage. In addition, visible light penetrating through human skin provided a greater depth of cure than UV [22, 23]. Visible light can penetrate the epidermal and dermal layers of the human skin and reach the hypodermal layer, while most UV light cannot penetrate the dermal layer. Riboflavin (RF; vitamin B2) [24, 25], eosin-Y [26, 27], and ruthenium (Ru (II))/sodium persulfate (SPS) [28] have been used as a visible light initiators. Hu *et al.*, reported VBL-induced hydrogel systems using photocrosslinkable chitosan (methacrylated glycol chitosan; MeGC) and VBL initiators of camphoquinone (CQ), fluorescein (FR), and RF [24]. This study demonstrated that the RF-initiated MeGC hydrogel system provided the most favorable cell viability and mechanical properties among the groups tested. Park *et al.* [25] and Choi *et al.* [100, 101] further modified the RF-crosslinkable hydrogel systems by adding cartilaginous ECM compositions such as hyaluronic acid (HA) and type II collagen (Col II) into MeGC. The incorporation of HA or Col II into MeGC hydrogel greatly improved the chondrogenesis of encapsulated cells, suggesting the potential of the hydrogel system as biomaterials to promote cartilage regeneration.

Crosslinking via chemical reactions between functional groups present in the water-soluble monomers or macromers can also produce hydrogels. Classical organic reactions between functional groups such as Michael addition, click reaction, Schiff base formation, epoxide coupling, genipin coupling, and disulfide exchange reaction have been used to prepare hydrogels. Michael addition can induce hydrogel formation by the addition of nucleophiles (amine or thiol group) to $\alpha,\beta$-unsaturated carbonyl compounds or $\alpha,\beta$-unsaturated sulfones in water. Some examples of compounds that can be crosslinked via Michael addition reaction are functionalized polymers, such as 4-arm PEG-acrylate with 4-arm PEG-thiol [29], 4-arm PEG maleimide, 4-arm PEG-vinylsulfone (VS), or 4-arm PEG-acrylate with enzyme-cleavable dithiol-crosslinker [29–31], methacrylated poly(vinyl alcohol) (PVA) with thiol end-capped PVA [32], vinyl-modified poly $N$-isopropylacrylamide (NIPAAm) with thiol-modified pNIPAAm [33], VS-functionalized saccharide-peptide polymer with cysteine (Cys)-functionalized saccharide-peptide polymer [34], and methacrylate- and maleimide-functionalized HA (MeMAHA) with Cys-functionalized MMP degradable peptide [17]. Copper [Cu(I)]-catalyzed azide-alkyne cycloaddition is one of the most popular click chemistry reactions for preparing hydrogels. For instance, PVA-based microgels were prepared from azide-bearing PVA and alkyne-bearing PVA [35], enzymatically sensitive hydrogels were synthesized from 4-arm or 8-arm PEG-alkyne and protease-sensitive bis-azido peptide [36], micropatterned hydrogel surface was fabricated using 4-arm PEG-alkyne and azide-functionalized plasma polymer film [37, 38], thermosensitive hydrogel was synthesized using azide-modified P(NIPAAm-co-HEMA) and alkyne-modified $\beta$-cyclodextrin [39], and cartilage ECM mimetic hydrogel was synthesized from azide-functionalized HA and chondroitin sulfate (CS), with alkyne-functionalized gelatin via Cu(I)-catalyzed click chemistry [40]. However, the problem of Cu(I) cytotoxicity must be resolved before these click chemistry-induced hydrogels can be used for biomedical applications. Thus, Cu(I)-free click reactions have been developed to create tissue engineering scaffolds [41, 42]. The Diels–Alder reaction, highly selective [4 + 2] cycloaddition between a diene and a dienophile without a catalyst, is also known as a click-type reaction. Hydrogels prepared by Diels–Alder reaction using PEG-tetrazine/dinorbornene cell degradable peptide [43], furan-modified HA (HA-furan)/maleimide-modified HA [44, 46], and HA-furan/dimaleimido-PEG [45, 47] have been investigated for tissue engineering applications.

Because of their high specificity for substrates, enzymatic crosslinkages result in few side reactions. As a result, enzymatically crosslinkable hydrogels have been the subject of increasing interest. Horseradish peroxidase (HRP), a single-chain $\beta$-type hemoprotein, catalyzes the coupling of phenols to aniline derivatives in the presence of hydrogen peroxide ($H_2O_2$) [48]. Hydroxyphenyl propionic acid (HPA)-functionalized 8-arm PEG has been crosslinked with HRP and $H_2O_2$ to form hydrogels [49]. An HRP/$H_2O_2$-catalyzed reaction has been adopted to design hydrogel systems based on HA [47, 50], dextran [51–53], and alginate [54]. Transglutaminase (TG) is a catalyst that crosslinks proteins depending on calcium *in vivo* [55]. TG-catalyzed crosslinking between the carboxamide and primary amines has been used to prepare hydrogels, from gelatin or gelatin/HA/CS for a 3D culture system [56–58]. These enzymatic crosslinks provide fast gelation with less cytotoxicity.

### 9.2.2 *Physical hydrogels*

Hydrogels formed by physical association are called physical, reversible, and stimuli-responsive hydrogels. Heat, ions, inclusion complex, stereocomplex, and/or complimentary binding can induce hydrogel formation by forming physical junctions via molecular entanglement, crystalline order, or intermolecular interactions. These hydrogel systems should be biocompatible with the host as well as the incorporated bioactive agents.

Polymers containing well-balanced hydrophilic and hydrophobic blocks can physically crosslink as temperature increases by forming associations of hydrophobic domains. PEG has been used as a hydrophilic block in most thermogelling systems. Thermogelling polymer aqueous solution is a free-flowing sol at a low temperature and becomes a semisolid gel as the temperature increases. Thermogelling systems, which show a sol-gel transition at around physiological temperature, have been widely developed for injectable biomedical applications. At low temperature, the thermosensitive polymer aqueous solution is easy to mix with cells, drugs, and/or bioactive molecules, followed by an injection of the mixture to the target site to form a hydrogel as the solution warms. The target site can be a subcutaneous layer for a protein drug delivery, a tumor tissue for an anticancer drug delivery, or a damaged/defected site for tissue repair. Thermogelling systems have been developed based on polyacrylates, polyesters, polyethers, polypeptides, and natural poylmers [7, 59–63]. Among these systems, polypeptide-based thermogelling systems with enzymatic degradability have several advantages when compared with polyester-based hydrogels. For instance, polypeptide thermogels usually maintain neutral pH during degradation, while the pH of polyester thermogels decreases due to their acidic degradation products. A decrease in pH can be a problem for biomedical applications, since it can decrease cell viability or protein drug stability. In addition, enzyme-sensitive degradation of the polypeptide-based thermosensitive polymers provides the polymer with hydrolytic stability in aqueous solution as a ready-to-use formulation [64]. The degradation rate of these polypeptide-based hydrogels can be tailored by enzyme specificity to the peptide sequences. Finally, polypeptides have unique secondary structures including $\alpha$-helix, $\beta$-sheet, triple helix, and random coil, allowing various nano-assemblies followed by sol-gel transition. These nano-assemblies can give unique nanostructures in the hydrogels. Therefore, polypeptide-based hydrogels can provide biomimetic ECM assemblies with unique nanostructures, which can affect proliferation and/or differentiation of encapsulated cells [65].

Crosslinking by addition of ions creates reversible hydrogels. Alginate hydrogel formation using calcium ions can be carried out at mild conditions of room temperature and physiological pH. Alginate gels have been used for encapsulating cells and protein drugs [66]. Salts in media also trigger the folding of MAX8 (VKVKVKVKVDPLPTKVEVKVKV-$NH_2$) and HLT2 (VLTKVKTKVDPLPTKVEVKVLV-$NH_2$) peptides into a $\beta$-hairpin conformation that induces hydrogel formation [67].

Inclusion complexes between cyclodextrins (CDs) and guest molecules induce physical associations to form physical hydrogels. There are three subtypes of $\alpha$-, $\beta$-, $\gamma$-CDs consisting of 6, 7, and 8 glucopyranose units, where the internal diameters of the resulting cavities are 5.7 Å, 7.8 Å, and 9.5 Å, respectively [68]. Different-sized guest molecules can be selectively inserted into the inner cavity of CD inclusion complexes with the proper diameter. Sequential inclusion complexations were used to form an *in situ* hydrogel [69]. First, Pluronics® were immobilized on a cellulose nanocrystal via interaction between the PPG block of Pluronic® and $\beta$-CD on the nanocrystal surface, and then uncovered PEG blocks of Pluronic® were inserted into cavities of $\alpha$-CD followed by *in situ* hydrogel formation.

Stereocomplexation between enantiomers by stereoselective van der Waals interaction have been used in the design of *in situ* hydrogel systems. Mixing two polymer aqueous solutions containing poly(L-lactide) (PLLA) block and poly(D-lactide) (PDLA) blocks induces formation of a stereocomplexed hydrogel. Stereocomplexes of PLLA-PEG-PLLA, PDLA-b-cationic poly(carbonate)-PDLA (PDLA-CPC-PDLA), and PDLA-PEG-PDLA triblock copolymers provided a thermosensitive antimicrobial hydrogel [70]. The 8-armed star block copolymers of PEG-(NHCO)-$(PDLA)_8$ and PEG-(NHCO)-$(PLLA)_8$ were mixed to form stereocomplexed spontaneous hydrogel [71].

Complementary binding such as antigen/antibody interactions, ligand/receptor interactions, and base-pairing interactions between oligonucleotides can also form *in situ* hydrogels. Acrylamides attached either an antigen (rabbit immunoglobulin G; rabbit IgG) or its specific antibody (goat anti-rabbit IgG) formed semi-interpenetrating network hydrogel [72]. Gyrase subunit B (GyrB) was dimerized by the addition of the aminocoumarin antibiotic coumermycin, resulting in hydrogel formation [73]. Multi-arm star-shaped PEG functionalized with thymine and adenine self-assembled via base pairing of thymine and adenine were used to form a hydrogel and showed potential for targeted growth factor delivery and cell encapsulation [74].

### 9.2.3 *Combining chemical and physical crosslinking (dual gelling)*

Physical hydrogels generally have much weaker mechanical strength when compared to the chemical hydrogels, which limits their use in tissue engineering. A combination of chemical and physical crosslinking has been used to overcome this limitation. A thiolated CS (CS-TGA)/PEG-diacrylate (PEGDA)/$\beta$-glycerophosphate disodium salt ($\beta$-GP) mixture formed hydrogel by Michael addition, disulfide bond forming, and temperature increase to 37°C [75]. The PNIPAAm-co-glycidyl methacrylate (GMA)/polyamidoamine (PAMAM) mixed solution was formed hydrogel by a physical and chemical dual gelation [76]. An injectable, degradable, and dual-gelling macromer of thermoresponsive PNIPAAm-based macromere with pendant phosphate groups functionalized with chemically crosslinkable methacrylate groups was suggested as a promising biomaterial for defect-filling, cellular encapsulation, and bone tissue engineering [77].

## 9.3 The Structure of Articular Cartilage

As described in Section 9.1, articular cartilage is a hyaline cartilage composed of water (65–80%) and ECM (20–35%) with a few chondrocytes at a density in the range 1.4–3 $\times$ $10^7$ cells/g [1]. Specifically, the ECM of articular cartilage is composed of 60% fibrous Col II, which provides resistance to shear and tensile forces, 30% proteoglycans, which attract water to resist compression, and about 10% other proteins. The compressive modulus of human articular cartilage is 4–19 MPa [78–80]. Articular cartilage, a load-bearing tissue, can experience forces up to six times the individual's bodyweight and compressive stresses near 10 MPa [81]. Typically, articular cartilage can be divided into four different zones: superficial zone (tangential zone), transitional zone (middle zone), radial zone (deep zone), and calcified cartilage zone (Figure 9.1) [82]. In the superficial zone, collagen

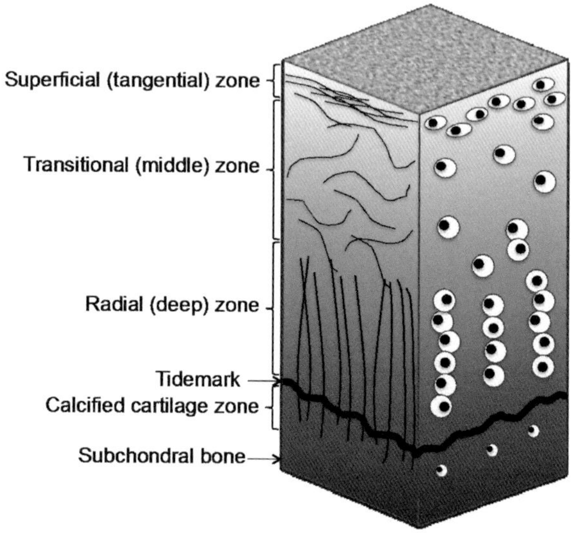


Figure 9.1 Zones of articular cartilage.

fibers are densely packed and oriented horizontally. In the transitional and radial zones, there is a high concentration of aggrecan, the major proteoglycan in cartilage, and the collagen fibers are oriented randomly, which is essential to absorb shock. In the calcified cartilage zone, the collagen fibers are oriented perpendicular to the bone surface. This orientation of collagen fibers is important in the resistance against compression forces. The chondrocytes in different zones differ in size, shape, orientation, and metabolic activity [83]. The superficial zone is composed of thin collagen fibers aligned parallel to the joint surface, with spindle-shaped chondrocytes. More spherical chondrocytes with larger and disordered collagen fibers are in the middle zone. In the deep zone, radial columns of chondrocytes align along collagen fibers, which are vertical to the surface of joint. Cells are smaller in the thin calcified cartilage zone, which is just above the subchondral bone.

### 9.3.1 *Classification of articular cartilage defects and evaluation of repaired cartilage defects*

Articular cartilage defects can be classified according to their depth. One of the relevant classifications was developed by the International Cartilage Repair Society (ICRS). The ICRS classified articular cartilage defects

Table 9.1 ICRS cartilage defect grading.

| Grade | ICRS classification |
|---|---|
| 0 | Normal cartilage |
| 1 | Nearly normal Superficial lesions, soft indentation and/or superficial fissures and cracks |
| 2 | Abnormal Lesions extending to less than half of the cartilage depth |
| 3 | Severely abnormal Lesion extending half or more of the cartilage depth, but not into the subchondral bone |
| 4 | Severely abnormal Osteochondral lesion |

Reproduced with permission from ICRS.

into five grades from 0 (normal healthy cartilage) to 4 (severely abnormal) (Table 9.1) [84, 85].

Repaired cartilage defects can be evaluated through macroscopic and histological scoring systems to measure the success of the repair. The most common macroscopic method of evaluation of repaired cartilage is the ICRS macroscopic scoring system, which was designed to evaluate the macroscopic consequence of cartilage repair to simplify and focus the scoring system on clinical needs (Table 9.2) [84, 86, 87]. Many histological scoring systems, such as the Mankin *et al.* [88] and O'Driscoll *et al.* [89] systems, have been used for the histological assessment of cartilage repair. However, these scoring systems have low intra- and inter-reader reliability. Recently, the ICRS has developed a new histology scoring system, ICRS II, for assessing cartilage repair tissue with enhanced reliability [90]. The ICRS II scoring system comprises 14 criteria related to chondrocyte phenotype, tissue structure, and other features, such as vascularization or mineralization (Table 9.3).

## 9.4 Application of Hydrogels for Cartilage Tissue Regeneration

Injectable and chemically crosslinkable polysaccharide-derived hydrogels have been increasingly studied for cartilage tissue engineering [91]. Chitosan is an analog of glycosaminoglycan (GAG), which has the advantages of biocompatibility, biodegradability, no pathological inflammatory response, no induced infection, intrinsic antibacterial ability, and low immunological reactions [92]. In addition, primary amine groups make various

Table 9.2 ICRS cartilage repair assessment: macroscopic evaluation score.

| Criteria | | Points |
|---|---|---|
| Degree of defect repair | * In level with surrounding cartilage | 4 |
| Protocol A | * 75% repair of defect depth | 3 |
| | * 50% repair of defect depth | 2 |
| | * 25% repair of defect depth | 1 |
| | * 0% repair of defect depth | 0 |
| Protocol B | * 100% survival of initially grafted surface | 4 |
| | * 75% survival of initially grafted surface | 3 |
| | * 50% survival of initially grafted surface | 2 |
| | * 25% survival of initially grafted surface | 1 |
| | * 0% (plugs are lost or broken) | 0 |
| Integration to border zone | * Complete integration with surrounding cartilage | 4 |
| | * Demarcating border <1 mm | 3 |
| | * 3/4 of graft integrated, 1/4 with a notable border > 1 mm | 2 |
| | * 1/2 of graft integrated with surrounding cartilage, 1/2 with a notable border >1 mm | 1 |
| | * From no contact to 1/4 of graft integrated with surrounding cartilage | 0 |
| Macroscopic appearance | * Intact smooth surface | 4 |
| | * Fibrillated surface | 3 |
| | * Small, scattered fissures or cracks | 2 |
| | * Several, small or few but large fissures | 1 |
| | * Total degeneration of grafted area | 0 |
| Overall repair assessment | Grade I: normal | 12 P |
| | Grade II: near normal | 11–8 P |
| | Grade III: abnormal | 7–4 P |
| | Grade IV: severely abnormal | 3–1 P |

*Notes*: • Protocol A: ACI; periosteal or perichondrial transplantation; subchondral drilling; microfracturing; carbon fiber implants; others.
*Source*: • Protocol B: Mossaicplasty; OAT; osteochondral allografts; others.
Reproduced with permission from ICRS.

chemical modifications possible. The VBL crosslinkable chitosan hydrogel system has been developed using MeGC and RF, a VBL initiator [24]. Stable chitosan hydrogels could be formed with only 40 s irradiation of visible light with great cell viabilities. Incorporation of HA, the most abundant polysaccharide in cartilage and synovial fluid, into a VBL-crosslinkable MeGC hydrogel enhanced deposition of cartilaginous ECM (Figure 9.2) [25]. Photocrosslinkable methacrylated HA has been used to encapsulate cells for chondrogenesis and cartilage regeneration. A recent study involving a methacrylated HA hydrogel functionalized

Table 9.3 ICRS cartilage repair assessment: ICRS II-histological evaluation score.

| Histological parameter | Score (%) |
|---|---|
| 1. Tissue morphology (viewed under polarized light microscopy) | 0: Full-thickness collagen fibers<br>100: Normal cartilage birefringence |
| 2. Matrix staining (metachromasia) | 0: No staining<br>100: Full metachromasia |
| 3. Cell morphology | 0: No round/oval cells<br>100: Mostly round/oval cells |
| 4. Chondrocyte clustering (four or more grouped cells) | 0: Present<br>100: Absent |
| 5. Surface architecture | 0: Delamination or major irregularity<br>100: Smooth surface |
| 6. Basal integration | 0: No integration<br>100: Complete integration |
| 7. Formation of a tidemark | 0: No calcification front<br>100: Tidemark |
| 8. Subchondral bone abnormalities/marrow fibrosis | 0: Abnormal<br>100: Normal marrow |
| 9. Inflammation | 0: Present<br>100: Absent |
| 10. Abnormal calcification/ossification | 0: Present<br>100: Absent |
| 11. Vascularization (within the repaired tissue) | 0: Present<br>100: Absent |
| 12. Surface/superficial assessment | 0: Total loss or complete disruption<br>100: Resemble intact articular cartilage |
| 13. Mid/deep zone assessment | 0: Fibrous tissue<br>100: Normal hyaline cartilage |
| 14. Overall assessment | 0: Bad (fibrous tissue)<br>100: Good (hyaline cartilage) |

*Notes*: Each criterion should be evaluated based on the visual analog scale [90] and graded from 0 to 100%.
*Source*: The best and the worst features are described for each parameter.
Reproduced with permission from ICRS.

with $N$-cadherin mimetic peptides established its significance in regulating the onset of chondrogenesis and cartilage matrix deposition by the encapsulated MSCs [93]. Photocrosslinkable hydrogels using PEG derivatives have also been used for cartilage repair. PEG dimethacrylate and methacrylated CS showed improved cartilage repair after six months compared to an empty, untreated defect [94]. The physicochemical properties of photodegradable, PEG-based chemical hydrogels could be dynamically controlled by light [12]. Specifically, the crosslinking density of the hydrogel was decreased via photodegradation, which facilitated spreading or migration of the embedded cells. Encapsulated MSCs also showed enhanced

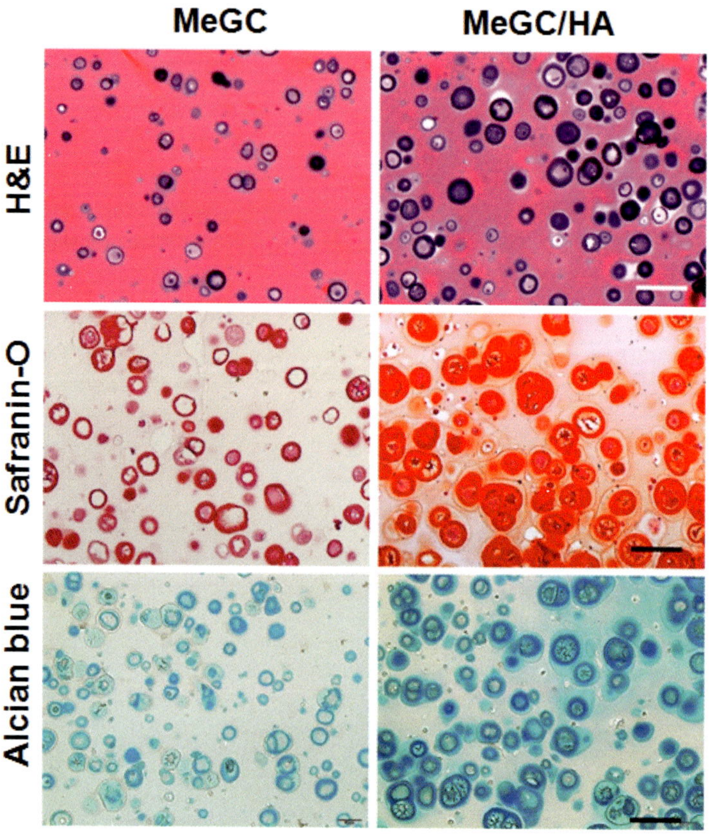


Figure 9.2 Histological analysis of three-dimensional cultured chondrocytes in MeGC and MeGC/HA (HA, 350 kDa) hydrogel after 21 days in culture. Scale bar is 100 $\mu$m. Reprinted from [25] with permission from Elsevier.

chondrogenesis when a cell adhesive sequence, Arg-Gly-Asp (RGD), was photocleaved from the hydrogels. It has been shown that the mesh size of the photocrosslinked hydrogels will not only control diffusion of nutrients and signaling molecules from the surrounding environment to the embedded cells, but also control diffusion of tissue-specific molecules secreted by the cells [79]. Multifunctionalized $\alpha$-CD nanobeads regulated the stem cell differentiation fate of human adipose derived stem cells (hADSCs) encapsulated in PEG hydrogels [95]. Ultimately, the alcohol, hydrophobic methyl group, and phosphate-substituted $\alpha$-CD nanobeads stimulated chondrogenic, adipogenic, and osteogenic differentiation, respectively.

A reverse thermogelling (L/DL)-PA-Pluronic®-(L/DL)-PA aqueous solution with L/DL ratio of 60:40 (PII) was reported as a promising matrix for chondrocyte three-dimensional culture [65]. (L/DL)-PA-Pluronic®-(L/DL)-PA hydrogel showed a higher population of nanofibrous structures as the L-Ala content increased (PI > PII > PIII) and chondrocytes cultured in PII hydrogel expressed higher chondrogenic markers of Col II and sulfated GAG (sGAG) compared to chondrocytes cultured in the commercially available Matrigel™ (Figure 9.3). Additionally, the initial modulus of (L/DL)-PA-Pluronic®-(L/DL)-PA hydrogel, which was controlled by the polymer concentration, showed a marked influence on biocompatibility, proliferation, and chondrogenesis (Figure 9.4) [96]. Specifically, the storage modulus of hydrogel was controlled into 350–380 Pa, 2,100–2,300 Pa, and 5,300–5,700 Pa according to the initial polymer concentration of 7.0 wt%, 10.0 wt%, and 15 wt%, respectively. Cartilaginous ECM production was

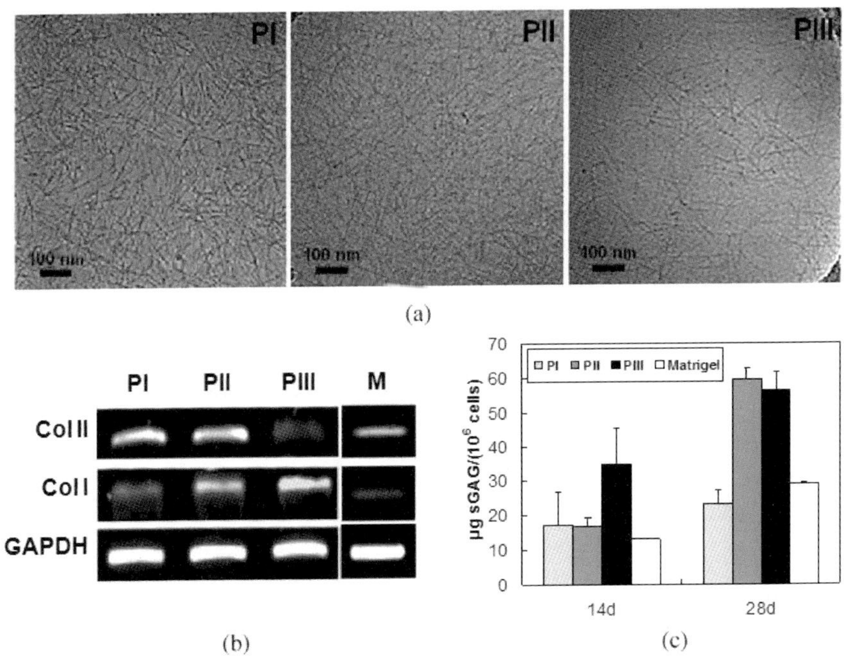


Figure 9.3 (a) Nanofibrous interior morphology of (L/DL)-PA-Pluronic® -(L/DL)-PA thermogel. Expression of chondrogenic markers of type II collagen (b) and sulfated glycosaminoglycan (sGAG) (c). PI, L/DL = 100/0; PII, L/DL = 60/40; PIII, L/DL = 0/100; M, Matrigel™ (control). Reprinted from [65] with permission from Elsevier.

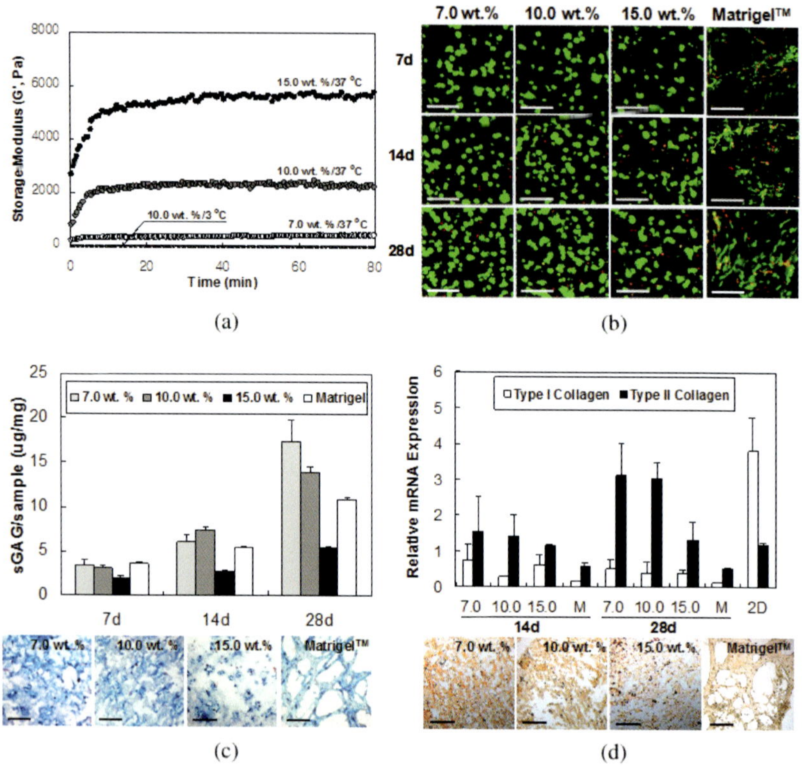


Figure 9.4 (a) Storage modulus of (L/DL)-PA-Pluronic®-(L/DL)-PA (L/DL = 60/40) aqueous solutions as a function of initial polymer concentration at 37°C. (b) Cell viability in the hydrogels analyzed using a live/dead staining. Green: live cells, red: dead cells. Expression of chondrogenic markers of sGAG (c) and type II collagen (d). M, Matrigel™ (control). The scale bar is 200 $\mu$m. Reprinted from [96] with permission from the Royal Society of Chemistry.

enhanced in the hydrogels with 7.0 wt% and 10.0 wt% compared to the stiff hydrogel of 15.0 wt%.

Incorporation of transforming growth factor-$\beta$1 (TGF-$\beta$1) into hydrogels has been shown to increase the proliferation and cartilaginous ECM production of encapsulated chondrocytes in OPF hydrogels [97] and chondrogenesis of MSCs in self-assembling peptide (AcN–(KLDL)$_3$–CNH$_2$ or (KLDL)$_3$) hydrogels [98]. Thiolated TGF-$\beta$1 was successfully incorporated into chemically crosslinked PEG hydrogels and this system showed maintenance of bioactivity of tethered TGF-$\beta$1 and enhanced chondrogenesis of human MSCs [19].

## 9.5 Conclusions

Many injectable chemically and/or physically crosslinked hydrogels produced under mild conditions have been developed and studied for cartilage tissue engineering. Injectable hydrogel systems are highly hydrated, minimally invasive, easy to form in any desired shapes, and able to easily encapsulate cells and/or bioactive molecules. Injectable and *in situ* forming hydrogels have been suggested as promising biomaterials for cartilage repair. However, despite over 20 years of cartilage tissue engineering applications, stable repair of articular cartilage defects still remains a challenge [99]. Most hydrogels tend to be brittle, and exhibit order of magnitude lower mechanical strength than that of cartilage. Therefore, increased understanding of and considerations about various hydrogel systems will advance the development of the design of durable and bioactive *in situ* gelling hydrogels for cartilage tissue engineering.

## References

[1] Lopez, O., Amrami, K.K., Manduca, A., and Ehman, R.L. (2008). Characterization of the dynamic shear properties of hyaline cartilage using high-frequency dynamic MR elastography, *Magn. Res. Med.*, **59**, pp. 356–364.

[2] Bhosale, A.M., and Richardson, J.B. (2008). Articular cartilage: Structure, injuries and review of management. *Br. Med. Bull.*, **87**, pp. 77–95.

[3] Hootman, J.M., Helmick, C.G., and Brady, T.J. (2012). A public health approach to addressing arthritis in older adults: The most common cause of disability. *Am. J. Pub. Health*, **102**, pp. 426–433.

[4] Gomoll, A.H. and Minas, T. (2014). The quality of healing: Articular cartilage, *Wound Wound Rep. Regen.*, **22** (Suppl 1), pp. 30–38.

[5] Lyon, R. and Liu, X.C. (2014). Future treatment strategies for cartilage repair. *Clinics Sports Med.*, **33**, pp. 335–352.

[6] Hoffman, A.S. (2002). Hydrogels for biomedical applications. *Adv. Drug Del. Rev.*, **54**, pp. 3–12.

[7] Ko, D.Y., Shinde, U.P., Yeon, B., and Jeong, B. (2013). Recent progress of in situ formed gels for biomedical applications. *Prog. Polym. Sci.*, **38**, pp. 672–701.

[8] Shin, H., Temenoff, J.S., and Mikos, A.G. (2003). *In vitro* cytotoxicity of unsaturated oligo[poly(ethylene glycol) fumarate] macromers and their cross-linked hydrogels. *Biomacromolecules*, **4**, pp. 552–560.

[9] Temenoff, J.S. *et al.* (2004). *In vitro* osteogenic differentiation of marrow stromal cells encapsulated in biodegradable hydrogels. *J. Biomed. Mater. Res. Part A*, **70**, pp. 235–244.

[10] Hong, Y. *et al.* (2007). Covalently crosslinked chitosan hydrogel: Properties of *in vitro* degradation and chondrocyte encapsulation. *Acta Biomater.*, **3**, pp. 23–31.
[11] He, X., Ma, J., and Jabbari, E. (2008). Effect of grafting RGD and BMP-2 protein-derived peptides to a hydrogel substrate on osteogenic differentiation of marrow stromal cells. *Langmuir*, **24**, pp. 12508–12516.
[12] Kloxin, A.M., Kasko, A.M., Salinas, C.N., and Anseth, K.S. (2009). Photodegradable hydrogels for dynamic tuning of physical and chemical properties. *Science*, **324**, pp. 59–63.
[13] Chien, H.W., Tsai, W.B., and Jiang, S.Y. (2012). Direct cell encapsulation in biodegradable and functionalizable carboxybetaine hydrogels. *Biomaterials*, **33**, pp. 5706–5712.
[14] Gold, G.T. *et al.* (2014). Injectable redox-polymerized methylcellulose hydrogels as potential soft tissue filler materials. *J. Biomed. Mater. Res. Part A.*
[15] Chen, Y.C. *et al.* (2012). Functional human vascular network generated in photocrosslinkable gelatin methacrylate, hydrogels. *Adv. Funct. Mater.*, **22**, pp. 2027–2039.
[16] Geng, X.H., Mo, X.M., Fan, L.P., Yin, A.L., and Fang, J. (2012). Hierarchically designed injectable hydrogel from oxidized dextran, amino gelatin and 4-arm poly(ethylene glycol)-acrylate for tissue engineering application. *J. Mater. Chem.*, **22**, pp. 25130–25139.
[17] Khetan, S. *et al.* (2013). Degradation-mediated cellular traction directs stem cell fate in covalently crosslinked three-dimensional hydrogels. *Nature Mater.*, **12**, pp. 458–465.
[18] Fairbanks, B.D., Singh, S.P., Bowman, C.N., and Anseth, K.S. (2011). Photodegradable, photoadaptable hydrogels via radical-mediated disulfide fragmentation reaction. *Macromolecules*, **44**, pp. 2444–2450.
[19] McCall, J.D., Luoma, J.E., and Anseth, K.S. (2012). Covalently tethered transforming growth factor beta in PEG hydrogels promotes chondrogenic differentiation of encapsulated human mesenchymal stem cells. *Drug Del. Transl. Res.*, **2**, pp. 305–312.
[20] Silva-Correia, J. *et al.* (2013). Biocompatibility evaluation of ionic- and photo-crosslinked methacrylated gellan gum hydrogels: *In vitro* and *in vivo* study. *Adv. Health. Mater.*, **2**, pp. 568–575.
[21] Barrow, M. and Zhang, H.F. (2013). Aligned porous stimuli-responsive hydrogels via directional freezing and frozen UV initiated polymerization. *Soft Matter*, **9**, pp. 2723–2729.
[22] Elisseeff, J. *et al.* (1999). Transdermal photopolymerization for minimally invasive implantation. *Proc. Nat. Acad. Sci. USA*, **96**, pp. 3104–3107.
[23] Sowa, P., Rutkowska-Talipska, J., Rutkowski, K., Kosztyla-Hojna, B., and Rutkowski, R. (2013). Optical radiation in modern medicine. *Post. Derm. Alerg.*, **30**, pp. 246–251.
[24] Hu, J. *et al.* (2012). Visible light crosslinkable chitosan hydrogels for tissue engineering. *Acta Biomater.*, **8**, pp. 1730–1738.

[25] Park, H., Choi, B., Hu, J., and Lee, M. (2013). Injectable chitosan hyaluronic acid hydrogels for cartilage tissue engineering. *Acta Biomater.*, **9**, pp. 4779–4786.
[26] Purcell, B.P., Elser, J.A., Mu, A., Margulies, K.B., and Burdick, J.A. (2012). Synergistic effects of SDF-1alpha chemokine and hyaluronic acid release from degradable hydrogels on directing bone marrow derived cell homing to the myocardium. *Biomaterials*, **33**, pp. 7849–7857.
[27] Shih, H. and Lin, C.C. (2013). Visible-light-mediated thiol-ene hydrogelation using eosin-Y as the only photoinitiator. *Macromol. Rapid Commun.*, **34**, pp. 269–273.
[28] Lim, K.S., Alves, M.H., Poole-Warren, L.A., and Martens, P.J. (2013). Covalent incorporation of non-chemically modified gelatin into degradable PVA-tyramine hydrogels. *Biomaterials*, **34**, pp. 7097–7105.
[29] Li, Y. *et al.* (2012). Nanostructured PEG-based hydrogels with tunable physical properties for gene delivery to human mesenchymal stem cells. *Biomaterials*, **33**, pp. 6533–6541.
[30] Phelps, E.A. *et al.* (2012). Maleimide cross-linked bioactive PEG hydrogel exhibits improved reaction kinetics and cross-linking for cell encapsulation and *in situ* delivery. *Adv. Mater.*, **24**, pp. 64–70.
[31] Wang, H. *et al.* (2013). Multifunctional biohybrid hydrogels for cell culture and controlled drug release. *Chem. Commun.* (*Camb.*), **49**, pp. 7448–7450.
[32] Tortora, M., Cavalieri, F., Chiessi, E., and Paradossi, G. (2007) Michael-type addition reactions for the *in situ* formation of poly(vinyl alcohol)-based hydrogels. *Biomacromolecules*, **8**, pp. 209–214.
[33] Wang, Z.C. *et al.* (2010). *In situ* formation of thermosensitive PNIPAAm-based hydrogels by Michael-type addition reaction. *ACS Appl. Mater. Intcr.*, **2**, pp. 1009–1018.
[34] Chawla, K., Yu, T.B., Liao, S.W., and Guan, Z. (2011). Biodegradable and biocompatible synthetic saccharide-peptide hydrogels for three-dimensional stem cell culture. *Biomacromolecules*, **12**, pp. 560–567.
[35] Kupal, S.G., Cerroni, B., Ghugare, S.V., Chiessi, E., and Paradossi, G. (2012) Biointerface properties of core-shell poly(vinyl alcohol)-hyaluronic Acid Microgels Based on Chemoselective Chemistry. *Biomacromolecules*, **13**, pp. 3592–3601.
[36] van Dijk, M., van Nostrum, C.F., Hennink, W.E., Rijkers, D.T., and Liskamp, R.M. (2010). Synthesis and characterization of enzymatically biodegradable PEG and peptide-based hydrogels prepared by click chemistry. *Biomacromolecules*, **11**, pp. 1608–1614.
[37] Adzima, B.J. *et al.* (2011). Spatial and temporal control of the alkyne-azide cycloaddition by photoinitiated Cu(II) reduction. *Nature Chem.*, **3**, pp. 256–259.
[38] Chen, R.T. *et al.* (2012). Photoinitiated alkyne-azide click and radical cross-linking reactions for the patterning of PEG hydrogels. *Biomacromolecules*, **13**, pp. 889–995.

[39] Xu, X.D. *et al.* (2009). Modular synthesis of thermosensitive P(NIPAAm-co-HEMA)/beta-CD based hydrogels via click chemistry. *Macromol. Rapid Commun.*, **30**, pp. 157–164.

[40] Hu, X., Li, D., Zhou, F., and Gao, C. (2011). Biological hydrogel synthesized from hyaluronic acid, gelatin and chondroitin sulfate by click chemistry. *Acta Biomater.*, **7**, pp. 1618–1626.

[41] Deforest, C.A., Sims, E.A., and Anseth, K.S. (2010). Peptide-functionalized click hydrogels with independently tunable mechanics and chemical functionality for 3D cell culture. *Chem. Mater.*, **22**, pp. 4783–4790.

[42] Fan, Y., Deng, C., Cheng, R., Meng, F., and Zhong, Z. (2013). *In Situ* forming hydrogels via catalyst-free and bioorthogonal "Tetrazole-Alkene" photo-click chemistry. *Biomacromolecules*, **14**, pp. 2814–2821.

[43] Alge, D.L., Azagarsamy, M.A., Donohue, D.F., and Anseth, K.S. (2013). Synthetically tractable click hydrogels for three-dimensional cell culture formed using Tetrazine–Norbornene chemistry. *Biomacromolecules*, **14**, pp. 949–953.

[44] Wei, H.-L. *et al.* (2010). Facile preparation of poly(*N*-isopropylacrylamide)-based hydrogels via aqueous Diels–Alder click reaction. *Polymer*, **51**, pp. 1694–1702.

[45] Nimmo, C.M., Owen, S.C., and Shoichet, M.S. (2011). Diels-Alder Click cross-linked hyaluronic acid hydrogels for tissue engineering. *Biomacromolecules*, **12**, pp. 824–830.

[46] Tan, H., Rubin, J.P., and Marra, K.G. (2011). Direct synthesis of biodegradable polysaccharide derivative hydrogels through aqueous Diels–Alder chemistry. *Macromol. Rapid Commun.*, **32**, pp. 905–911.

[47] Yu, F. *et al.* (2014). An injectable hyaluronic acid/PEG hydrogel for cartilage tissue engineering formed by integrating enzymatic crosslinking and Diels–Alder "click chemistry". *Polymer Chem.*, **5**, p. 1082.

[48] Kobayashi, S., Uyama, H., and Kimura, S. (2001). Enzymatic polymerization. *Chemical Rev.*, **101**, pp. 3793–3818.

[49] Menzies, D.J. *et al.* (2013). Tailorable cell culture platforms from enzymatically cross-linked multifunctional poly(ethylene glycol)-based hydrogels. *Biomacromolecules*, **14**, pp. 413–423.

[50] Kurisawa, M., Chung, J.E., Yang, Y.Y., Gao, S.J., and Uyama, H. (2005). Injectable biodegradable hydrogels composed of hyaluronic acid-tyramine conjugates for drug delivery and tissue engineering. *Chem. Commun. (Camb.)*, pp. 4312–4314.

[51] Jin, R., Hiemstra, C., Zhong, Z., and Feijen, J. (2007). Enzyme-mediated fast *in situ* formation of hydrogels from dextran-tyramine conjugates. *Biomaterials*, **28**, pp. 2791–2800.

[52] Jin, R. *et al.* (2010). Enzymatically crosslinked dextran-tyramine hydrogels as injectable scaffolds for cartilage tissue engineering. *Tissue Eng. Part A*, **16**, pp. 2429–2440.

[53] Moreira Teixeira, L.S. *et al.* (2012). The effect of platelet lysate supplementation of a dextran-based hydrogel on cartilage formation. *Biomaterials*, **33**, pp. 3651–3661.

[54] Devolder, R., Antoniadou, E., and Kong, H. (2013). Enzymatically cross-linked injectable alginate-g-pyrrole hydrogels for neovascularization. *J. Control. Release*, **172**, pp. 30–37.

[55] Sanborn, T.J., Messersmith, P.B., and Barron, A.E. (2002). *In situ* crosslinking of a biomimetic peptide-PEG hydrogel via thermally triggered activation of factor XIII. *Biomaterials*, **23**, pp. 2703–2710.

[56] Chen, P.Y. *et al.* (2014). Fabrication of large perfusable macroporous cell-laden hydrogel scaffolds using microbial transglutaminase, *Acta Biomater.*, **10**, pp. 912–920.

[57] De Colli, M. *et al.* (2012). A biomimetic porous hydrogel of gelatin and glycosaminoglycans cross-linked with transglutaminase and its application in the culture of hepatocytes. *Biomed. Mater.*, **7**, p. 055005.

[58] Fang, J.Y., Tan, S.J., Yang, Z., Tayag, C., and Han, B. (2014). Tumor bioengineering using a transglutaminase crosslinked hydrogel. *PLoS One*, **9**, p. e105616.

[59] Vermonden, T.M., N.A., Van, M.J., and Hennink, W.E. (2006). Rheological studies of thermosensitive triblock copolymer hydrogels. *Langmuir*, **22**, pp. 10180–10184.

[60] Liu, C.D., Zhang, Z.X., Liu, K.L., Ni, X.P., and Li, J. (2013). Biodegradable thermogelling poly(ester urethane)s consisting of poly(1,4-butylene adipate), poly(ethylene glycol), and poly(propylene glycol). *Soft Matter*, **9**, pp. 787–794.

[61] Yu, L., Zhang, H., and Ding, J. (2006). A subtle end-group effect on macroscopic physical gelation of triblock copolymer aqueous solutions. *Angew Chem. Int. Ed. Engl.*, **45**, pp. 2232–2235.

[62] Megeed, Z. *et al.* (2004). *In vitro* and *in vivo* evaluation of recombinant silk-elastinlike hydrogels for cancer gene therapy. *J. Control. Release*, **94**, pp. 433–445.

[63] Park, M.H., Joo, M.K., Choi, B.G., and Jeong, B. (2012). Biodegradable thermogels. *Acc. Chem. Res.*, **45**, pp. 424–433.

[64] Moon, H.J., Choi, B.G., Park, M.H., Joo, M.K., and Jeong, B. (2011). Enzymatically degradable thermogelling poly(alanine-co-leucine)–poloxamer–poly(alanine-co-leucine). *Biomacromolecules*, **12**, pp. 1234–1242.

[65] Choi, B.G. *et al.* (2010). *In situ* thermal gelling polypeptide for chondrocytes 3D culture. *Biomaterials*, **31**, pp. 9266–9272.

[66] Kim, W.S. *et al.* (2012). Adipose tissue engineering using injectable, oxidized alginate hydrogels. *Tissue Eng. Part A*, **18**, pp. 737–743.

[67] Sinthuvanich, C., Haines-Butterick, L.A., Nagy, K.J., and Schneider, J.P. (2012). Iterative design of peptide-based hydrogels and the effect of network electrostatics on primary chondrocyte behavior. *Biomaterials*, **33**, pp. 7478–7488.

[68] Appel, E.A., del Barrio, J., Loh, X.J., and Scherman, O.A. (2012). Supramolecular polymeric hydrogels. *Chem. Soc. Rev.*, **41**, pp. 6195–6214.

[69] Lin, N. and Dufresne, A. (2013). Supramolecular hydrogels from *in situ* host-guest inclusion between chemically modified cellulose nanocrystals and cyclodextrin. *Biomacromolecules*, **14**, pp. 871–880.

[70] Li, Y. *et al.* (2013). Broad-spectrum antimicrobial and biofilm-disrupting hydrogels: Stereocomplex-driven supramolecular assemblies. *Angew Chem. Int. Ed. Engl.*, **52**, pp. 674–678.

[71] Buwalda, S.J., Calucci, L., Forte, C., Dijkstra, P.J., and Feijen, J. (2012). Stereocomplexed 8-armed poly(ethylene glycol)–poly(lactide) star block copolymer hydrogels: Gelation mechanism, mechanical properties and degradation behavior. *Polymer*, **53**, pp. 2809–2817.

[72] Miyata, T., Asami, N., and Uragami, T (1999). A reversibly antigen-responsive hydrogel. *Nature*, **399**, pp. 766–769.

[73] Ehrbar, M., Schoenmakers, R., Christen, E.H., Fussenegger, M., and Weber, W. (2008). Drug-sensing hydrogels for the inducible release of biopharmaceuticals. *Nature Mater.*, **7**, pp. 800–804.

[74] Tan, H.P., Xiao, C., Sun, J.C., Xiong, D.S., and Hu, X.H. (2012). Biological self-assembly of injectable hydrogel as cell scaffold via specific nucleobase pairing. *Chem. Commun.*, **48**, pp. 10289–10291.

[75] Chen, C., Wang, L., Deng, L., Hu, R., and Dong, A. (2013). Performance optimization of injectable chitosan hydrogel by combining physical and chemical triple crosslinking structure. *J. Biomed. Mater. Res. Part A*, **101**, pp. 684–693.

[76] Ekenseair, A.K. *et al.* (2012). Synthesis and characterization of thermally and chemically gelling injectable hydrogels for tissue engineering. *Biomacromolecules*, **13**, pp. 1908–1915.

[77] Watson, B.M., Kasper, F.K., Engel, P.S., and Mikos, A.G. Synthesis and characterization of injectable, biodegradable, phosphate-containing, chemically cross-linkable, thermoresponsive macromers for bone tissue engineering. *Biomacromolecules*, **15**, pp. 1788–1796.

[78] Armstrong, C.G. and Mow, V.C. (1982). Variations in the intrinsic mechanical properties of human articular cartilage with age, degeneration, and water content. *J. Bone Joint Surg. Am.*, **64**, pp. 88–94.

[79] Bryant, S.J. and Anseth, K.S. (2002). Hydrogel properties influence ECM production by chondrocytes photoencapsulated in poly(ethylene glycol) hydrogels. *J. Biomed. Mater. Res.*, **59**, pp. 63–72.

[80] Shepherd, D.E. and Seedhom, B.B. (1999). The 'instantaneous' compressive modulus of human articular cartilage in joints of the lower limb. *Rheumatology*, **38**, pp. 124–132.

[81] Brand, R.A. (2005). Joint contact stress: A reasonable surrogate for biological processes? *Iowa Orthop. J.*, **25**, pp. 82–94.

[82] James, C.B. and Uhl, T.L. (2001). A review of articular cartilage pathology and the use of glucosamine sulfate. *J. Athl. Train.*, **36**, pp. 413–419.

[83] Buckwalter, J.A., Mankin, H.J., and Grodzinsky, A.J. (2005). Articular cartilage and osteoarthritis. *Instr. Course Lect.*, **54**, pp. 465–480.

[84] Brittberg, M. and Peterson, L. (1998). Introduction of an articular cartilage classification. *ICRS Newsletter*, **1**, pp. 5–8.

[85] Brittberg, M. and Winalski, C.S. (2003). Evaluation of cartilage injuries and repair. *J. Bone Joint Surg. Am. Vol.*, **85-A** (Suppl 2), pp. 58–69.

[86] Peterson, L. *et al.* (2000). Two- to 9-year outcome after autologous chondrocyte transplantation of the knee. *Clinic. Orthop. Rel. Res.*, **374**, pp. 212–234.

[87] van den Borne, M.P. *et al.* (2007). International Cartilage Repair Society (ICRS) and Oswestry macroscopic cartilage evaluation scores validated for use in Autologous Chondrocyte Implantation (ACI) and microfracture. *Osteoarth. Cart.*, **15**, pp. 1397–1402.

[88] Mankin, H.J., Dorfman, H., Lippiello, L., and Zarins, A. (1971). Biochemical and metabolic abnormalities in articular cartilage from osteo-arthritic human hips. II. Correlation of morphology with biochemical and metabolic data. *J. Bone Joint Surg. Am.*, **53**, 523–537.

[89] O'Driscoll, S.W. *et al.* (2001). Validation of a simple histological-histochemical cartilage scoring system. *Tissue Eng.*, **7**, pp. 313–320.

[90] Mainil-Varlet, P. *et al.* (2010). A new histology scoring system for the assessment of the quality of human cartilage repair: ICRS II. *Am. J. Sports Med.*, **38**, pp. 880–890.

[91] Zhao, W., Jin, X., Cong, Y., Liu, Y., and Fu, J. (2013). Degradable natural polymer hydrogels for articular cartilage tissue engineering. *J. Chem. Tech. Biotech.*, **88**, pp. 327–339.

[92] Kumar, M.N., Muzzarelli, R.A., Muzzarelli, C., Sashiwa, H., and Domb, A.J. (2004). Chitosan chemistry and pharmaceutical perspectives. *Chem. Rev.*, **104**, pp. 6017–6084.

[93] Bian, L., Guvendiren, M., Mauck, R.L., and Burdick, J.A. (2013). Hydrogels that mimic developmentally relevant matrix and N-cadherin interactions enhance MSC chondrogenesis. *Proc. Nat. Acad. Sci. USA*, **110**, pp. 10117–10122.

[94] Wang, D.A. *et al.* (2007). Multifunctional chondroitin sulphate for cartilage tissue-biomaterial integration. *Nature Mater.*, **6**, 385–392.

[95] Singh, A., Zhan, J., Ye, Z., and Elisseeff, J.H. (2013). Modular multifunctional poly(ethylene glycol) hydrogels for stem cell differentiation. *Adv. Funct. Mater.*, **23**, pp. 575–582.

[96] Choi, B.G. *et al.* (2011). Thermal gelling polyalanine–poloxamine–polyalanine aqueous solution for chondrocytes 3D culture: Initial concentration effect. *Soft Matter*, **7**, pp. 456–462.

[97] Park, H., Temenoff, J.S., Holland, T.A., Tabata, Y., and Mikos, A.G. (2005). Delivery of TGF-beta1 and chondrocytes via injectable, biodegradable hydrogels for cartilage tissue engineering applications. *Biomaterials*, **26**, pp. 7095–7103.

[98] Kopesky, P.W. *et al.* (2014). Sustained delivery of bioactive TGF-beta1 from self-assembling peptide hydrogels induces chondrogenesis of encapsulated bone marrow stromal cells. *J. Biomed. Mater. Res. Part A*, **102**, pp. 1275–1285.

[99] Huey, D.J., Hu, J.C., and Athanasiou, K.A. (2012). Unlike bone, cartilage regeneration remains elusive. *Science*, **338**, pp. 917–921.

[100] Choi, B., Kim, S., Lin, B. *et al.* (2015). Visible-light-initiated hydrogels preserving cartilage extracellular signaling for inducing chondrogenesis of mesenchymal stem cells. *Acta Biomater.*, **12,** pp. 30–41.
[101] Choi, B., Kim, S., Lin, B. *et al.* (2014). Cartilaginous extracellular matrix-modified chitosan hydrogels for cartilage tissue engineering. *ACS Appl. Mater. Interfaces,* **6**, pp. 20110–20121.

Chapter 10

# Injectable Hydrogels for Cardiac Tissue Regeneration Post-Myocardial Infarction

G. N. Grover

*Department of Bioengineering*
*Sanford Consortium for Regenerative Medicine*
*University of California, San Diego*

K. L. Christman*

*Department of Bioengineering*
*Sanford Consortium for Regenerative Medicine*
*University of California, San Diego*

## 10.1 Introduction

Myocardial infarction (MI) is the interruption of blood flow to heart tissue as a result of an occlusion in one of the coronary arteries. The ensuing ischemia results in cell death downstream of the occlusion site, and since the heart has limited regenerative capacity, the initial inflammatory response is resolved with a collagenous scar. Wall thinning, hypertrophy of remaining viable myocardium, and dilation of the LV chamber, collectively termed 'negative LV remodeling' (Figure 10.1), often occur and lead to heart failure (HF) [1–4]. Despite the prevalence of MI-associated HF [5], there are no therapies that directly address the negative LV remodeling process that leads to HF, and inadequate treatment options exist for HF patients. Therapies for acute MI include angioplasty and stents to open the

*Author to whom correspondence should be addressed. Email: christman@eng.ucsd.edu.

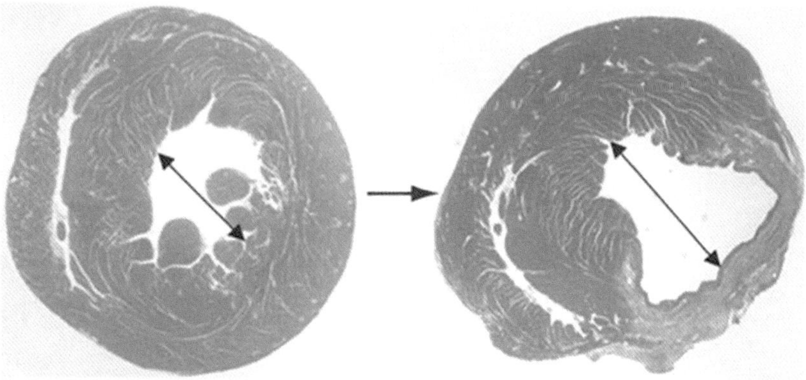

Figure 10.1 A horizontal cross-section of a heart pre-MI (left) and post-MI (right) stained with hematoxylin and eosin. Adapted from [9] with permission from Macmillan Publishers Ltd.

blocked coronary artery and restore blood flow, as well as pharmacologics that reduce the load on the heart. However, none of these treatments are capable of fully preventing the downward spiral of negative LV remodeling that leads to HF. For end-stage HF, heart transplants are effective; however, this approach is limited by the supply of donor hearts [6]. In addition, LV assist devices (LVADs), mechanical pumps that take some of the workload off of the failing heart, have been employed. These devices have historically been used only as a bridge to transplant since they suffer from various complications and require the use of an extracorporeal battery pack [7, 8].

Clinical trials examining potential new therapies for treating MI and HF have largely been focused on cell therapies. Cells have either been injected intravenously with the goal of having the cells home to the damaged region, or injected via catheter into the ischemic tissue and border zone of the infarct in the hope of improving cardiac function [10, 11]. The catheter-based procedures used to deliver cells to the heart are considered minimally invasive treatment approaches, and do not require general anesthesia. Major limitations of cell therapy are poor retention and survival of viable cells in the tissue post-injection, with recent studies in clinic trials producing mixed results [11–14]. Given that the majority of cells die within 24–48 hours, the small improvements in cardiac function have largely been attributed to paracrine effects [15]. Tissue engineering-based strategies using biomaterial scaffolds to prevent anoikis have therefore more recently been examined in preclinical studies. These include the use of injectable biomaterials and surgically placed epicardial patches to improve cell survival with the goal of enhancing this paracrine effect or truly achieving engrafted cells that contribute to contraction. Acellular biomaterials for encouraging endogenous cell infiltration and cardiac repair or delivering therapeutics such as growth factors have also been explored [16, 17].

A broad survey of the field of injectable hydrogels for cardiac tissue regeneration is provided in this chapter. Following an initial review of the animal models used to study these materials and potential delivery routes, the different naturally derived and synthetic materials that have been explored for treating MI and HF will be discussed.

## 10.2 Animal Models of Myocardial Infarction

Animal models of MI are separated into small animal models [18] and large animal models [19]. Small animal models include mice, rats, and rabbits, while large animal models include pigs, sheep, and dogs. Small animal models, especially rats and mice, are widely used as costs are much lower, allowing the researcher to increase the number of animals used to improve the statistical power. However, there are also limitations to these small animal models. Rodents have a much smaller heart, which has a higher heart rate and oxygen consumption. There can also be differences in the regulatory proteins within the heart and how these proteins respond to drugs. For example, drugs found to effectively target certain ion channels in mice can have little to no effect in humans [20, 21]. As a result, the use of large animal models of MI and HF, especially pig models, which more closely match human cardiovascular physiology, is essential prior to clinical translation. For thorough discussions on these differences and types of animal models, the reader is referred to the following reviews [18, 19].

The studies presented below cover the use of injectable hydrogels alone as well as with therapeutic molecules and cells delivered to infarcts in small and large animal models of MI. Two general types of procedures are used to induce MIs in animals: total occlusion and occlusion–reperfusion. Total occlusion MI models involve the permanent ligation of a coronary artery, whereas occlusion–reperfusion models consist of temporarily occluding the coronary artery, followed by reperfusion. The latter model is more representative of a patient who receives an angioplasty, while the former is more representative of a patient who misses the reperfusion window in the clinic and is more severe. These models can be further divided into chronic MI/HF models, where the materials are delivered after a scar has fully formed, or acute MI models where the materials are delivered immediately or 1–2 weeks post-MI before the negative LV remodeling process is complete. While immediate delivery is easier in terms of study design since it does not require follow-up surgery, it should be noted that this is not clinically relevant as this therapy window could not be achieved in patients.

The effects of hydrogel injection in animal models of MI can be characterized in a variety of ways. However, there is no standard set of characterization criteria for quantifying the effects of injectable materials in the heart [16]. Since the heart dilates as it undergoes negative LV remodeling and proceeds to HF, LV geometry is assessed with measurements such as LV dimensions, areas, and volumes at systole and diastole. LV dimension (end-systolic dimension (ESD) or end-diastolic dimension (EDD)) is the internal diameter, while area (end-systolic area (ESA) or end-diastolic dimension (EDA)) is typically the short axis area, with both measurements usually taken at the level of the papillary muscles. Volumes (end-systolic volume (ESV) or end-diastolic volume (EDV)) are calculated using mathematical assumptions of the shape of the ventricle and based on two-dimensional images. Cardiac function is typically reported as either fractional shortening (FS), which is calculated as ((EDD − ESD) ÷ EDD), fractional area change (FAC), which is calculated as ((EDA − ESA) ÷ EDA), or ejection fraction (EF), which is calculated as ((EDV − ESV) ÷ EDV). Stroke volume (SV = EDV − ESV) is also sometimes reported as a measure of cardiac function. These measures of LV geometry and global cardiac function can be obtained with magnetic resonance imaging (MRI) or echocardiography (ultrasound). While not typically performed, these imaging modalities are also capable of assessing regional cardiac function by evaluating local strains or wall movement. Hemodynamic measurements, which report LV pressures, have also been used for assessing global cardiac function. Additional measurements to assess remodeling of the ventricle include wall thickness and infarct size. These are often determined in large animals with imaging procedures due to the size of the heart, while for small animals they are often determined by histological analysis of tissue sections of the heart. Cellular retention (in the case of cellular delivery), endogenous cellular infiltration, and neovascularization are often assessed by histology and immunohistochemistry.

## 10.3 Delivery of Injectable Materials to the Heart

Injectable hydrogels can be grouped based upon the mode of gelation, which can occur via self-assembly or aggregation of small molecules or macromolecules. Hydrogel networks can also be formed by polymerization of monomers and crosslinkers, typically by radical polymerization. Finally, prepolymers can be condensed via a variety of chemical reactions to synthesize the hydrogel network [22]. All of these approaches can render a material injectable, or capable of being injected through a syringe and needle and gelling rapidly within in tissue; however, this does not necessarily allow a material to be injected into the heart via catheter.

There are three approaches for delivering materials to the heart (Figure 10.2) [13]. Direct injection requires surgical access and involves

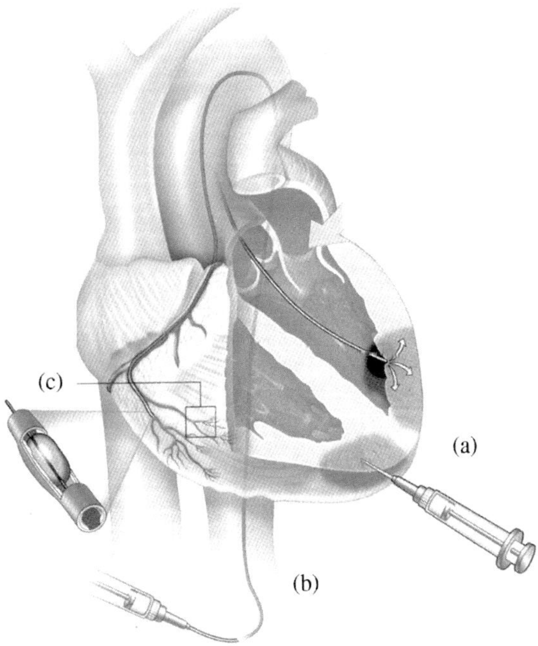


Figure 10.2 Methods for injection of materials/cells into the heart. (a) Direct epicardial injection via syringe requires surgical access, but allows for precise control of the location of the injections. (b) Transendocardial injection with a catheter allows for needle-based injection into the infarct and/or border zone through the endocardium. (c) Intracoronary injection with a catheter involves inflating a transluminal coronary angioplasty balloon to temporarily occlude the vessel and material/cells are infused into the infarct through the damaged, leaky vessels. Reprinted from [13] with permission from Elsevier.

direct injection of the material through the epicardium into the damaged area (Figure 10.2a). This method is used in the majority of the animal studies presented in this chapter given that mainly rodent models are employed, which are too small for catheter access. Direct injection allows for precise control over the amount and location of the injection, but requires an invasive and potentially risky surgery. There are two approaches for delivering materials to the heart with a catheter: transendocardial injection (Figure 10.2b) and intracoronary infusion (Figure 10.2c). Transendocardial delivery involves guiding the catheter from the femoral artery access point into the LV chamber where the material is injected multiple times via a retractable needle directly into the infarct and border zone areas [12, 23]. With intracoronary infusion, the catheter is positioned in the blood vessel upstream of the infarct and an angioplasty balloon is used to temporarily occlude the vessel while the materials and/or cargo are infused through the vessels into the damaged tissue [24, 25]. This latter approach requires access to a re-opened vessel and is only applicable in the acute MI setting, which has leaky vasculature. These procedures were originally developed to deliver cells to the heart.

In order for the delivery of injectable materials to the heart to become a minimally invasive procedure in the clinic, they must be able to be delivered via these cardiac catheter methods, and must also be hemocompatible since both approaches result in some material being released into the systemic circulation, particularly with intracoronary infusion. This also requires that the injectable hydrogel be premixed and capable of injection through a 27 G needle, which is the typical catheter size. The material must also stay in liquid form and not clog the catheter while being held at 37°C during the procedure, which in the case of transendocardial delivery can take up to an hour since between 15 and 30 injections are routinely performed. With intracoronary infusion the material must be capable of infusing through leaky vessels into the peri-infarct and then gelling. The majority of the materials presented in this chapter gel rapidly upon mixing, thus limiting the potential for minimally invasive delivery to the heart. In fact, only three materials have been delivered via a catheter to the heart in infarct models: alginate [26] by intracoronary infusion, and decellularized myocardial matrix [27] and an ureido-pyrimidinone terminated triblock copolymer of poly(urea-block-PEG-block-urea) [28] by transendocardial injection. Additionally, oxime-crosslinked hydrogels have demonstrated the potential to be injected multiple times through a catheter at 37°C over the course of hours *in vitro*, suggesting the potential for *in vivo* catheter delivery [29].

## 10.4 Naturally Derived Injectable Materials

### 10.4.1 *Protein-based injectable materials*

Protein-based materials are widely used in tissue engineering as they are enzymatically degradable, making them biodegradable. Enzymes or temperature triggers gelation of these materials, and they form nanofibrous networks similar to the native extracellular matrix (ECM). While versatile and widely used, a drawback of protein-based materials is the limited range of mechanical properties, rapid degradation, batch-to-batch variability, and protein heterogeneity. In addition, many of the protein-based materials are from xenogeneic sources and thus have additional immune response concerns, although this has not hindered clinical translation in other applications.

#### 10.4.1.1 *Fibrin*

Fibrin is a major component of blood clots and a natural scaffold for wound healing. Mixing fibrinogen with thrombin results in rapid formation of fibrin hydrogels with tunable properties including gelation time, biodegradability, mechanical strength, and angiogenic activity [30]. Christman *et al.* were the first to inject fibrin into the myocardium [31]. The left coronary artery of rats was occluded for 17 min and then reperfused. One week post-MI, fibrin glue (50 $\mu$L) was injected into the infarct, and the animals were analyzed five weeks post-treatment. Fibrin glue was found to be capable of preserving wall thickness and FS [31]. Follow-up studies by Christman *et al.* demonstrated that injection of fibrin glue decreased infarct size and increased arteriole density within the infarct (Figure 10.3) [32]. This was the first series of papers to demonstrate that injecting a biomaterial alone was capable of maintaining cardiac function post-MI. Neovascularization with fibrin alone in the infarct was confirmed by Huang *et al.* [33]. They performed a comparative study of fibrin glue, collagen, and Matrigel where infarcts were formed in rats via 30-minute occlusion followed by reperfusion of the left coronary artery, and then injection of the materials (50 $\mu$L) one week post-MI. The most interesting result of this study was the fact

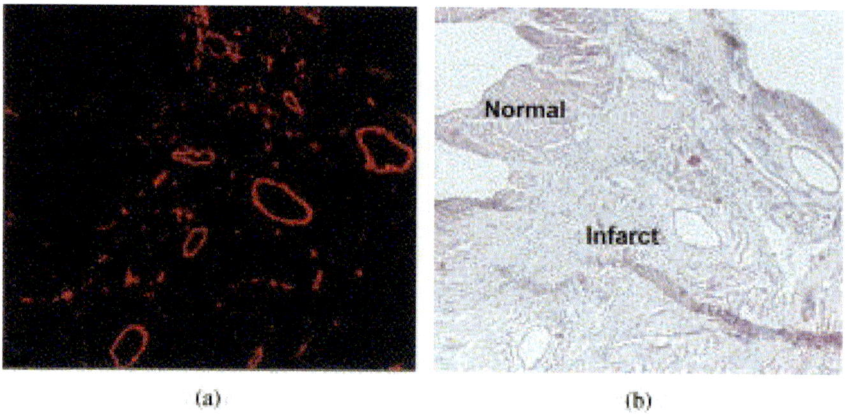


Figure 10.3 Arteriole formation in fibrin gel injected within the infarct. (a) Anti-smooth muscle actin-labeled arterioles. (b) Corresponding hematoxylin and eosin stained section. 100 × magnification. Reprinted from [32] with permission from Elsevier.

that all three materials resulted in the formation of new vessels within the infarct (Figure 10.4) [33]. The combination of these studies was pivotal in demonstrating that biomaterials alone could promote neovascularization of ischemic tissue in the heart.

Fibrin has also been used as a carrier for growth factors and plasmids to infarcted myocardium [16, 17]. Fibrin was used to deliver transforming growth factor-beta (TGF-$\beta$), a heparin binding protein. The TGF-$\beta$ was bound to heparin-decorated poly(lactic-co-glycolic acid) microspheres, and the protein-bound microspheres were co-delivered with bone marrow mononuclear cells (BMMNCs) and fibrin [34]. TGF-$\beta$ is rapidly activated within the damaged myocardium, is thought to be a key regulator in the inflammatory response in the heart, and promotes ECM deposition in the infarct by upregulating collagen and fibronectin synthesis [35]. Co-delivery of TGF-$\beta$ with bone marrow stem cells promoted cardiac-like differentiation of the encapsulated cells [34]. Christman *et al.* delivered plasmid encoding pleiotrophin, which is an angiogenic cytokine released in response to ischemic injury in the heart, in fibrin glue, demonstrating increased neovascularization compared to the plasmid alone [36].

Co-injection of cells with rapidly gelling materials such as fibrin can increase cellular retention and viability in ischemic myocardium. Christman *et al.* demonstrated that the fibrin hydrogels were capable of effectively delivering skeletal myoblasts with improved survival versus no fibrin carrier; however, cardiac function and LV geometry were preserved for both fibrin

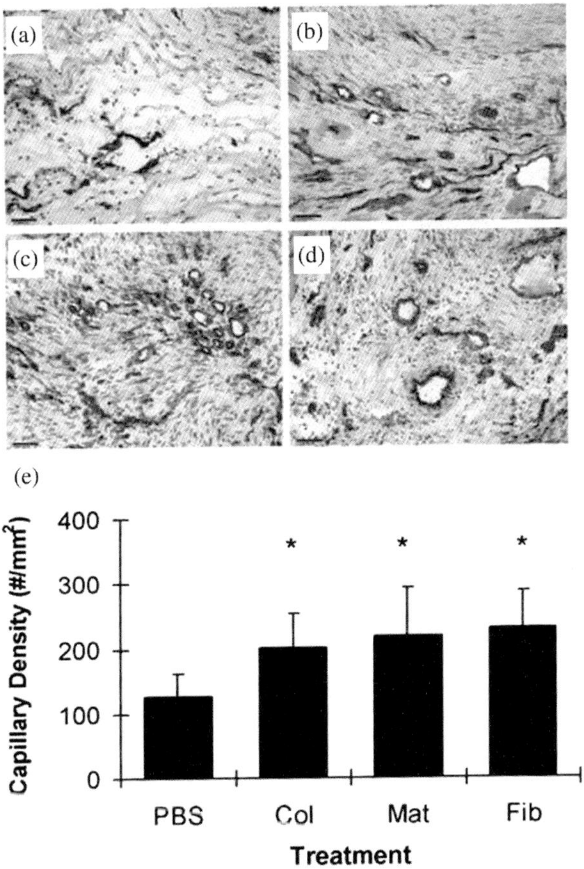


Figure 10.4 Neovascularization post-material injection. Vessel staining in infarct area for (a) saline, (b) Matrigel, (c) fibrin, and (d) collagen. (e) Average capillary density. PBS: phosphate buffered saline; Col: collagen; Mat: Matrigel; Fib: fibrin. $*p < 0.05$ compared to saline group. Scale bar is 50 $\mu$m. Reprinted from [33] with permission.

and fibrin–skeletal myoblasts [32]. Ryu *et al.* demonstrated similar results when delivering BMMNCs with and without fibrin [37], while Chekanov *et al.* found that injection of endothelial cells plus fibrin resulted in improved LV function and neovascularization compared to saline [38].

#### 10.4.1.2 *Collagen*

Collagen is the most abundant protein in the human body, and has been extensively studied for biomedical applications. It contains integrin-binding domains that allow cellular adhesion, and collagen-based materials can be enzymatically degraded *in vitro* and *in vivo*. Collagen is solubilized in acidic solution and when brought to physiological pH, it self-assembles into a nanofibrous protein hydrogel [39–41]. Dai *et al.* injected bovine dermal collagen (100 $\mu$L, 65 mg/mL) into the infarct of rats one week after ligation of the proximal left coronary artery [42]. Six weeks post-injection, the collagen-injected animals displayed increased wall thickness, higher stroke volume (SV), and greater EF compared to controls. Interestingly, in this study no angiogenesis was observed in the collagen group [42]. This is most likely due to the high concentration of collagen that was injected, which would have formed a dense protein network, thus limiting cellular infiltration into the injected material. This is supported by the comparative study mentioned previously in which Huang *et al.* compared fibrin glue, collagen, and Matrigel. Collagen (50 $\mu$L, 1 mg/mL) was injected one week post-MI and histological analysis indicated a significant increase in myofibroblast density for the collagen group. All three groups had increased neovascularization compared to saline injections, demonstrating that collagen injections into the infarct are capable of promoting growth of new vessels if at the appropriate density (Figure 10.4) [33]. Collagen matrices have also been shown to improve cellular retention and survival upon injection into the infarct [33, 43–45].

#### 10.4.1.3 *Matrigel*

Matrigel is a complex ECM mixture commonly used for cell culture, and is capable of forming hydrogels for *in vitro* and *in vivo* applications. It is isolated from a mouse sarcoma cell line and is composed of collagen IV, laminin, heparin sulfate proteoglycans, entactin, and growth factors [46, 47]. Huang *et al.* compared single injections of Matrigel with fibrin and collagen. Growth factor-reduced Matrigel was mixed with cell culture media and injected into the infarct of rats one week post-MI. It showed increased new vessel formation and trended toward increased myofibroblast infiltration versus saline (Figure 10.4) [33]. Ou *et al.* [48] obtained similar results with multiple injections of Matrigel ($5 \times 50$ $\mu$L) into the infarct and border zone in rats immediately after permanent ligation of the left coronary artery. Immediate treatment after injury resulted in an increase in EF, thicker LV wall, increased arteriole density and recruitment of c-kit+ and CD34+ stem cells, demonstrating that Matrigel supported stem cell recruitment to the infarct [48]. Matrigel has also been used as a vehicle to deliver cells to the heart. Growth factor-reduced Matrigel (50 μL) with green fluorescent protein-positive mouse embryonic stem cells (mESCs) was injected immediately after injury in a mouse model of MI. Interestingly, this study demonstrated improvement in heart function after two weeks with the Matrigel–cells and cells only groups, but Matrigel alone produced similar results to the saline group [49]. Laflamme *et al.* used a prosurvival cocktail (PSC) to deliver cardiomyocytes derived from hESCs in a rat model of MI. The PSC was a complex mixture of drugs, peptides, and proteins mixed with Matrigel, and was designed to improve the survival of the cardiomyocytes after injection into the infarct. Interestingly, four weeks post-injection the cells and PSC slowed progression to HF and improved global function compared to animals receiving PSC only and non-cardiac cells [50]. While these studies have generated positive results, a major concern over the use of Matrigel is tumor formation in the injection sites since it is derived from a mouse sarcoma cell line [51, 52]. This issue is likely to limit the translation of this material into the clinic.

#### 10.4.1.4 *Decellularized ECM hydrogels*

One of the goals for designing tissue engineering scaffolds is to recapitulate the environment of the native ECM, which contains a complex composition of proteins, proteoglycans, and polysaccharides that is different for each tissue. Since this complexity is currently not achievable with standard protein-based or synthetic materials, tissue engineers have recently employed decellularization techniques to isolate the intact ECM [53]. Functional constructs following reseeding of decellularized organs has now been demonstrated for the heart [54], lung [55, 56], and kidney [57]. Decellularized ECM can also be rendered into injectable hydrogels through partial enzymatic digestion. These liquid forms of decellularized ECM still retain their complex biomolecule composition after digestion [58] and self-assemble into nanofibrous networks similar to native ECM when brought to physiological conditions or injected *in vivo* [59, 60].

A cardiac-specific injectable hydrogel has recently been developed [60] and studied in both rat and porcine models of MI [27, 61] Singelyn *et al.* decellularized porcine heart tissue and then enzymatically digested it into a liquid form (Figure 10.5). This material self-assembled in nanofibrous gels at 37°C, and was shown to maintain a complex protein and proteoglycan composition similar to native heart tissue. Injection of the myocardial matrix hydrogel (6 mg/mL, 75 μL) facilitated the recruitment of c-kit+

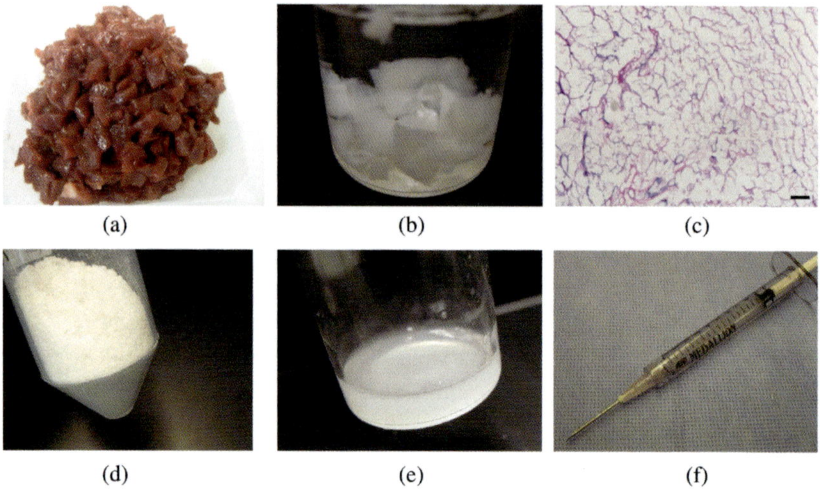


Figure 10.5 (a) Sliced porcine ventricular myocardium. (b) Decellularized cardiac ECM. (c) Hematoxylin and eosin staining shows cellular removal. (d) Decellularized ECM after lyophilization and milling into a fine powder, (e) rendered into a liquid by partial digestion with pepsin, and (f) ready for injection in a syringe. Reprinted from [27] with permission from Elsevier.

stem cells, increased areas of cardiomyocytes surviving in the infarct, and preserved cardiac function when injected two weeks post-MI in rats [27]. Seif-Naraghi *et al.* then translated this material to a porcine model of MI in which the material was delivered to the infarct two weeks post-MI via multiple transendocardial catheter injections, similar to a procedure that would be performed on humans in the clinic. Post-treatment, pigs that received the myocardial matrix hydrogel showed increased cardiac muscle and reduced infarct fibrosis (Figure 10.6). This resulted in improvements in both global and regional cardiac function at three months post-injection. Additional biocompatibility and hemocompatibility studies and hemocompatibility studies were performed, paving the way for clinical translation of this material [61].

Seif-Naraghi *et al.* also investigated a method of obtaining an autologous injectable hydrogel from pericardial tissue [62]. This is an interesting approach clinically since the pericardium can be resected during cardiac surgery without detriment to the patient since it is not sutured closed after accessing the heart [63, 64]. Application of this material in the heart would eliminate concerns about a negative immune response since it is autologous. Pericardial tissue was decellularized and rendered injectable similar

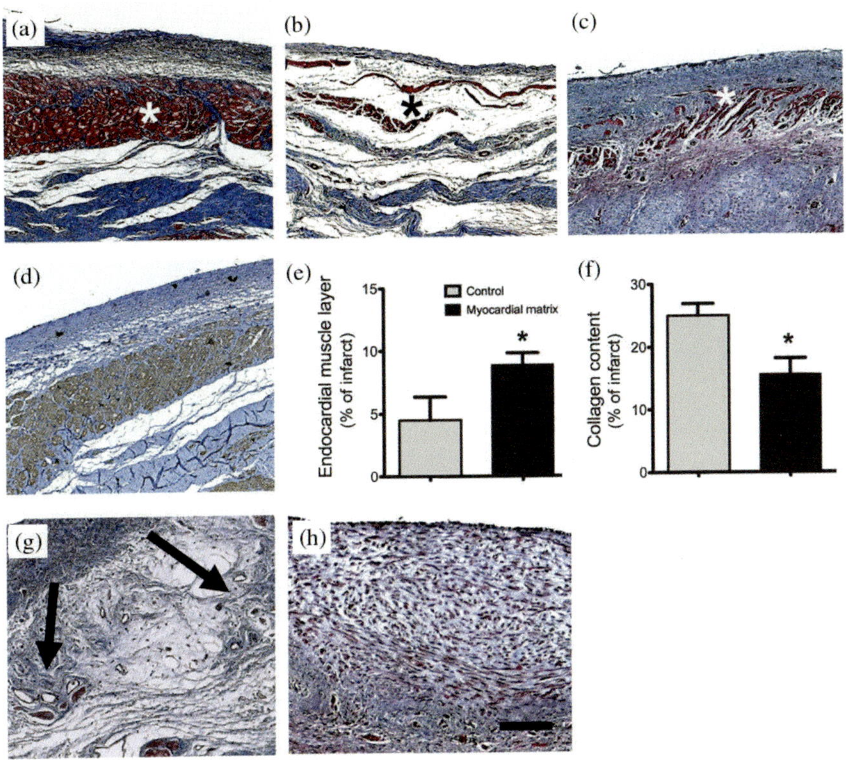


Figure 10.6 Myocardial matrix in a porcine MI model. Masson's trichrome staining of infarcted pigs heart after treatment with (a) myocardial matrix hydrogel, * indicates thick band of muscle at the endocardium; (b) non-injected control, * indicates loose fibrillar layer beneath endothelium; or, (c) saline control, * indicates moderately thickened endocardium with small amount of muscle (red). (d) Cardiac troponin-T stained section adjacent to (e) that indicates the presence of cardiac muscle. (f) The area of the layer of endocardial muscle as a proportion of the infarct. (g) Average percent collagen content in the infarcts. (h) Myocardial matrix injected hearts contain foci of neovascularization (arrows), but (i) saline and no injection contained no similar areas of neovascularization. Scale bar is 200 $\mu$m. Reprinted from [61] with permission from AAAS.

to the myocardial matrix hydrogel. Injection of the material into healthy rat hearts demonstrated a significant increase in arterioles and the presence of c-kit+ cells in the injection site after two weeks [62]. This material was also shown to enhance the retention and delivery of basic fibroblast growth factor (bFGF), suggesting ECM hydrogels may provide a natural platform for growth factor delivery to the heart since they retain sulfated glycosaminoglycans [65].

Decellularized porcine small intestinal submucosa (SIS) in particulate and digested form has also been injected into animal models of MI. Particulate SIS was emulsified and injected into rat hearts immediately post-MI [66]. Histological analysis and echocardiography showed that SIS increased neovascularization and preserved cardiac function. These results were attributed to enhanced recruitment of c-kit cells, myofibroblasts, and macrophages to the infarct as a result of the SIS particles [66]. Okada *et al.* tested two different forms of an SIS matrix, SIS-B and SIS-C, in a mouse immediately after MI. SIS-B was stiffer and contained more endogenous bFGF than SIS-C, which resulted in small infarcts and increased capillary density [67]. These studies demonstrate that SIS is capable of promoting neovascularization in the infarct after injection immediately post-MI in rats; however, a patch form of SIS has been shown to form collagenous-like tissue in the heart, potentially because of non-cardiac-specific cues [68].

### 10.4.2 *Polysaccharide-based injectable materials*

Polysaccharide-based materials are widely used in biomedical and tissue engineering applications. Polysaccharides are isolated from plant and animal sources and are typically highly branched chains of sugars. Processing techniques for polysaccharides involve isolation of biomolecules with different molecular weights and branching, resulting in inherent heterogeneity of the isolated material. These characteristics affect the properties and biocompatibility of these materials. An advantage of these types of materials is that they can be easily modified with a wide variety of functional groups. However, similar to protein-based materials, there are concerns with batch-to-batch variability and the overall homogeneity [69].

#### 10.4.2.1 *Alginate*

Alginate is a naturally occurring polysaccharide obtained from seaweed, which undergoes a phase transition from liquid to gel in the presence of calcium cations. Alginate hydrogels have been extensively studied *in vivo* and used to deliver drugs and cells [26, 70, 71]. Calcium gluconate and alginate (molecular weight of 20–50 kDa) was injected into rat hearts one week and two months after total occlusion of the coronary artery, representing acute and chronic MI models, respectively. Eight weeks post-injection the early injection group had limited expansion of the LV, while the late injection group had improved wall thickness [72]. Yu *et al.* examined alginate in an occlusion–reperfusion rat MI model, injecting the material five weeks post-MI. Vessel formation and degradation of the hydrogel was observed over time, but the positive effects of the alginate hydrogel were seen eventually to decrease [73]. Leor *et al.* examined catheter delivery of alginate by intracoronary infusion in a porcine model. Four days post-MI, three different volumes of calcium gluconate/alginate solution were injected: 1 mL, 2 mL, and 4 mL. The material gelled in the infarct because of the leaky, damaged vasculature and the high concentration of calcium in the acute MI setting (Figure 10.7). Two months post-treatment, the 2 mL injection group had a reversal in LV diastolic and systolic dilation, but no changes in FS were

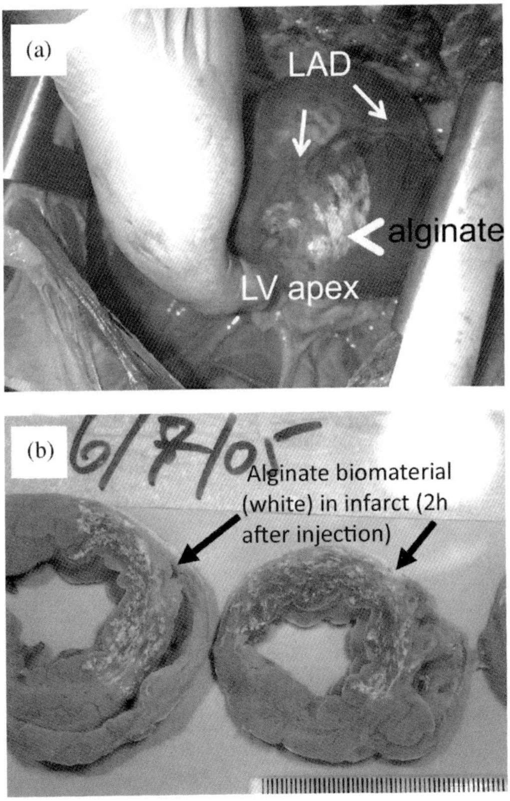


Figure 10.7 Intracoronary infusion of alginate in a porcine acute MI model. (a) Infarcted porcine heart two hours after intracoronary injection of calcium-crosslinked alginate solution. LAD: Left anterior descending coronary artery. (b) Heart sections showing deposition of alginate within the infarct (white areas). Reprinted from [26] with permission from Elsevier.

observed. This study demonstrated the feasibility of catheter delivery for an injectable hydrogel in an MI for the first time [26]. The positive results from the above rodent and porcine studies have resulted in alginate becoming the first injectable acellular biomaterial for treating MI and its use is in clinical trials in humans. The pilot study demonstrated the preliminary efficacy and safety of the material after implantation in 27 patients. The product, now labeled IK-5001, is currently in a phase II clinical trial [74].

Alginate is also an attractive material for injectable hydrogels because it can be easily functionalized. For substantial cell adhesion, it is necessary to conjugate bioactive molecules to alginate since unmodified alginate does not contain cellular adhesion sites. A popular method for facilitating cell adhesion to non-cell adhesive materials is to functionalize the materials with integrin binding peptides such as RGD and YIGSR, which are peptide sequences derived from native proteins in the ECM [71]. Yu *et al.* injected alginate and integrin-binding peptide functionalized alginate in a rat reperfusion model of MI five weeks post-MI. It was thought that the RGD-modified material would facilitate endogenous cellular infiltration into the hydrogel over unmodified alginate. Five weeks after treatment both functionalized and non-functionalized alginate displayed improvements in FS, increased wall thickness, and reduced LV dimensions compared to bovine serum albumin (BSA) injections. Interestingly, the modified alginate showed increased neovascularization within the infarct compared to the unmodified alginate [75]. Tsur-Gang *et al.* performed a similar study but observed conflicting results, with the peptide-modified alginate outperforming unmodified alginate. The difference in the studies is hypothesized to be due to the difference in the viscosity of the alginate from the extent of peptide modifications, which altered the distribution of the material and reduced coverage of the infarct [76]. Alginate has also been modified to deliver various heparin-binding growth factors to the infarct to promote neovascularization within the infarct [77, 78].

#### 10.4.2.2 *Chitosan*

Chitosan is a biocompatible, biodegradable linear polysaccharide. It is synthesized by treating chitin isolated from the shells of crustaceans with sodium hydroxide. Chitosan forms thermoresponsive hydrogels upon mixing with glycerol phosphate and glyoxal. Thermoresponsive chitosan hydrogels have been investigated as drug delivery platforms in many tissues [79, 80].

Thermoresponsive chitosan was injected one week post-MI in a rat model. Four weeks post-injection the chitosan gel was partially degraded within the infarct. The chitosan treatment group showed significant improvements in wall thickness, infarct size, cardiac geometry, and vessel density within the infarct [81]. Chitosan was also used to deliver bFGF post-MI. Four weeks after injection the growth factor–chitosan group showed improved cardiac geometry, decreased infarct size, and decreased fibrosis compared to bFGF alone [82]. Chitosan was also used to deliver undifferentiated ESCs in a rodent MI model. The chitosan group increased cell retention and cell survival compared to PBS alone. Chitosan and ESCs improved cardiac geometry compared to acellular chitosan [81, 83].

#### 10.4.2.3 *Hyaluronic acid*

Hyaluronic acid (HA) is a branched polysaccharide involved in angiogenesis, cell migration, and scar reduction [84]. HA is isolated from animal or bacteria sources and is easily modified with functional groups, resulting in covalently crosslinked injectable hydrogels. HA-based materials have been extensively investigated for biomedical applications [85–87]. Ifkovits *et al.* functionalized HA with methacrylates (MeHA). *In vivo* injection of MeHA with ammonium persulate and tetramethylethylenediamine resulted in a radical crosslinked hydrogel. MeHA modified with different amounts of methacrylates was injected 30 minutes after ligation of the left anterior coronary artery and second diagonal coronary artery in an ovine MI model. Methacrylate functionalization affected MeHA distribution within the infarct. Higher modified MeHA significantly reduced infarct expansion and trended toward less LV dilation [88]. Tous *et al.* functionalized HA with hydroxyethyl methacrylate (HEMA-HA) and tested the radical crosslinked hydrogel in the same ovine model as the previous study. Two different percentages of HEMA-functionalized HA were examined. The higher functionalized HEMA-HA had a slower degradation rate *in vivo* and showed a higher vessel concentration eight weeks post-injection. The higher HEMA-HA also showed a visible increase in the inflammation response compared to MeHA, which was believed to be the result of the extended release of HA degradation products [85]. Injectable HA has also been used to encapsulate and deliver therapeutics to the heart. Purcell *et al.* injected HA post-MI with stromal cell-derived factor-1a (SDF-1a). This study demonstrated that co-injection of HA and SDF-1a increased endogenous bone marrow cell (BMC) infiltration into the infarct [89].

### 10.4.3 *Mixed hydrogels*

The above categories of naturally-derived materials have also been mixed together to synthesize injectable hydrogels with new properties for the heart. By blending naturally derived materials it is possible to synthesize a mixed hydrogel that retains the properties of both components [90]. These properties include mode of degradation, stiffness, cellular adhesion sites, and biocompatibility. For example, Mukherjee *et al.* injected a fibrin–alginate hydrogel one week post-MI in a swine model [91]. A double barrel syringe was used to deliver the material with one barrel containing fibrinogen, fibronectin, factor XIII, plasminogen, and gelatin-grafted alginate dissolved in aprotinin, and the other barrel containing thrombin and calcium chloride. The fibrin–alginate group was found to have increased wall thickness one week post-injection. Reduction in infarct expansion was observed three and four weeks after treatment. It was also determined that there was a significant reduction of soluble collagen in the mixed hydrogel group. It was believed that this was the result of decreased proteolytic degradation of collagen due to lower levels of matrix metalloproteinase-2 (MMP-2) [91].

## 10.5 Synthetically-Derived Injectable Materials

In this chapter, 'synthetic materials' refer to materials that are synthesized from small molecule precursors, resulting in materials with predetermined structure and functionality. The advantage of synthetic materials is that the structure can be tuned and controlled more than natural materials, in which the structure is predetermined. These types of materials are also considered homogenous, have limited batch-to-batch variability, and the synthesis can be easily scaled up. Synthetic materials are also generally viewed to have increased stability compared to naturally derived materials. When designing these types of materials it is important to consider biocompatibility, biodegradability, and bioaccumulation. However, it is still unclear how the materials' properties correlate to *in vivo* responses in the heart.

### 10.5.1 *Self-assembling peptides*

Synthetic peptides with a specific sequence of amino acids can self-assemble into nanofibrous materials. The ability of these designer peptides to self-assemble into well-defined nanostructured materials has resulted in the use of self-assembling peptides (SAPs) in biomedical nanotechnology, cell culture, and molecular electronics applications [92, 93]. If SAPs are properly designed, they also have shear-thinning properties, which could make them amenable to catheter delivery [94]. Davis *et al.* [95] demonstrated that SAPs of RAD16-II were capable of self-assembling into a nanofibrous material upon injection into healthy myocardium and promoted cellular infiltration into the material. The SAPs were also capable of delivering neonatal cardiomyocytes. Co-injection of SAPs with cardiomyocytes improved the recruitment of endogenous cells similar to what had been seen for naturally derived injectable materials [95]. SAPs are also capable of delivering proteins to the infarct. Davis *et al.* tethered insulin growth factor-1 (IGF-1) to SAPs. Injection of growth factor-functionalized SAPs with neonatal cardiomyocytes improved systolic function 21 days post-treatment compared to untethered IGF-1 and SAPs, demonstrating the utility of these materials to deliver growth factors with cells to prevent decline in cardiac function post-MI [96]. Lin *et al.* verified that SAPs alone were capable of preserving function post-MI. They also used SAPs to deliver BMMNCs in miniature pigs immediately after MI. Co-injections of the cells and nanofibers increased capillary density (Figure 10.8) and improved systolic and diastolic function compared to cells alone and nanofibers alone [97]. Lin *et al.* used SAPs with vascular endothelial growth factor (VEGF) in rat and pig models where the coronary arteries were permanently occluded. The SAPs were able to prevent leakage of VEGF into the circulatory system compared to VEGF alone, thus preventing vascular leakage in other organs in rats. The SAPs supported engraftment of myofibroblasts and SAPs with VEGF supported an angiogenic environment 28 days post-injection in rats. Seven days post-treatment, the rats were injected intravenously with BMCs. The SAPs with VEGF supported recruitment of the BMCs to the infarct. Finally, in pigs the SAPs alone did not prevent decline in cardiac function while SAPs with VEGF improved cardiac performance and arteriogenesis post-MI [98].

SAPs can also be used as a vehicle to deliver multiple growth factors. Angiogenesis is regulated by multiple growth factors. The ability to deliver more than one growth factor to ischemic tissue has been shown to enhance neovascularization. Using hydrogel scaffolds to deliver growth factor can result in extended and tunable release as well as protect these fragile proteins from degradation [99, 100]. Co-delivery of platelet-derived growth factor and basic fibroblast growth factor in a rat MI model increased neovascularization and reduced infarct size at four and eight weeks after injection [101]. Guo *et al.* covalently attached IGF-1 and RGD to SAPs for

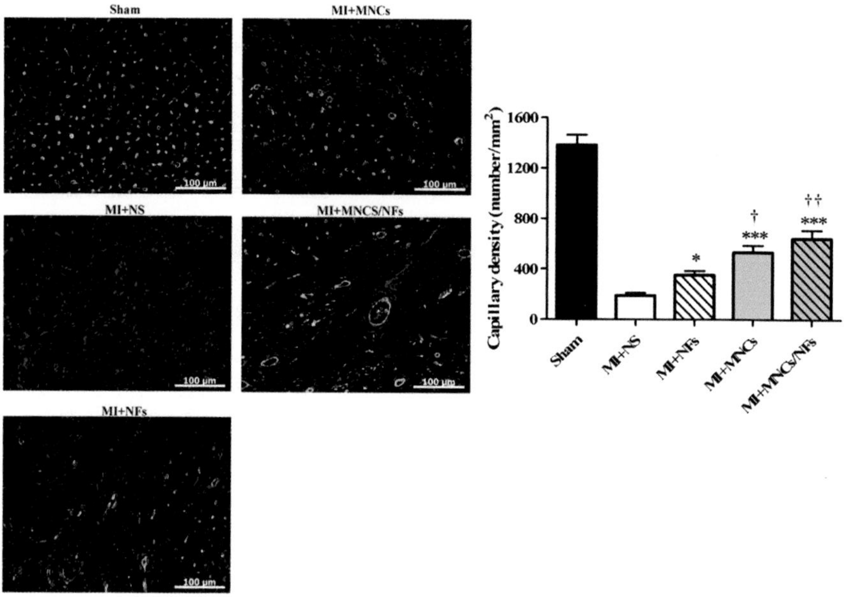


(a) (b)

Figure 10.8 Injection of SAPs with BMCs in a porcine MI model. (a) Immunostaining of isolectin (green) overlapped with cardiomyocytes stained with cardiac tropomyosin (red) at the border zone; nuclei are stained blue. (b) Quantification of the capillary density in the peri-infarct. $^{*}p < 0.05$, $^{***}p < 0.001$ versus MI+NFs; $^{\dagger}p$, 0.05, $^{\dagger\dagger}p < 0.01$ versus MI+NFs. Reprinted from [97] with permission.

delivering BMCs to the infarct. RGD-SAPs increased cell survival, and cells encapsulated within RGD-SAPs had positive staining for cardiac troponin T and connexin-43 [102]. SAPs have also been shown to be efficient scaffolds for delivering therapeutic cargos and cells; however, these materials alone have not been found to improve heart function post-MI.

### 10.5.2 *Poly(ethylene glycol)-containing materials*

Poly(ethylene glycol) (PEG) is a widely studied synthetic polymer in tissue and biomedical engineering. PEG is a non-degradable, water-soluble polymer with terminal alcohols that can be easily modified with different functional groups; this modification is essential to allow PEG to form hydrogels and be incorporated into other biomaterials. The type of functional group dictates the method of PEG incorporation into hydrogels, as well as the mode of degradation. PEG-based hydrogels are bioinert and resist protein adsorption and cell adhesion [103]. For example, PEG can be functionalized with vinyl sulfones (PEG-VS), which selectively react with thiols in the presence of other nucleophiles at mild pH [104] PEG-VS was mixed with dithiothreitol (DTT), resulting in non-degradable injectable PEG hydrogels. This hydrogel system was injected two minutes post-MI in a rat model. Wall thickness was significantly increased as well as EDD for the PEG-VS compared to saline four weeks after injection; however, 13 weeks post-treatment the performance of the hydrogel group was similar to the saline group [105]. Similar results were seen by Rane *et al.* using a non-degradable PEG hydrogel formed by amidation. The PEG hydrogel was injected into rats one week post-MI. Wall thickness was increased with the PEG group compared to saline seven weeks after treatment; however, the PEG group showed a decline in function similar to the saline group [106]. These results demonstrate that injection of an inert, non-degradable material can increase infarct wall thickness, but this does not relate to improvements in cardiac function, thereby suggesting that bioactivity of injectable polymers is likely an important factor for improving cardiac function.

PEG hydrogels have also been used to deliver proteins to the infarct and can be rendered enzymatically degradable when the PEG is crosslinked with a peptide sequence that is susceptible to cleavage by enzymes released from cells. Salimath *et al.* crosslinked star maleimide-PEG with a thiolated enzymatically degradable peptide to deliver hepatocyte growth factor (HGF) and VEGF immediately after MI in rats. Three weeks after injection of the dual growth factor-loaded enzymatically degradable hydrogel there was a significant decrease in fibrosis and increase in neovascularization compared to single growth factor delivery (Figure 10.9). There was also an improvement in heart function [107]. An enzymatically degradable PEG hydrogel was used as a carrier for thymosin $\beta$-4 as well as endothelial and smooth muscle-like cells derived from hESCs. This injectable system decreased infarct size and increased vessel density. EF, ED volume, and ES volume were also improved compared to controls [108].

The presence of terminal alcohol groups on linear and star PEG polymers allows for synthesis of stimuli responsive block copolymers upon functionalization of the termini with other types of polymers. A triblock copolymer of methyl-PEG-block-polycaprolactone-block-methyl-PEG (mPEG-PCL-mPEG) was designed to form hydrogels upon injection *in vivo* with $\alpha$-cyclodextrin. mPEG-PCL-mPEG with erythropoietin was injected into rats immediately after inducing MI. The protein-PEG system showed improvement in heart function and geometry compared to PEG

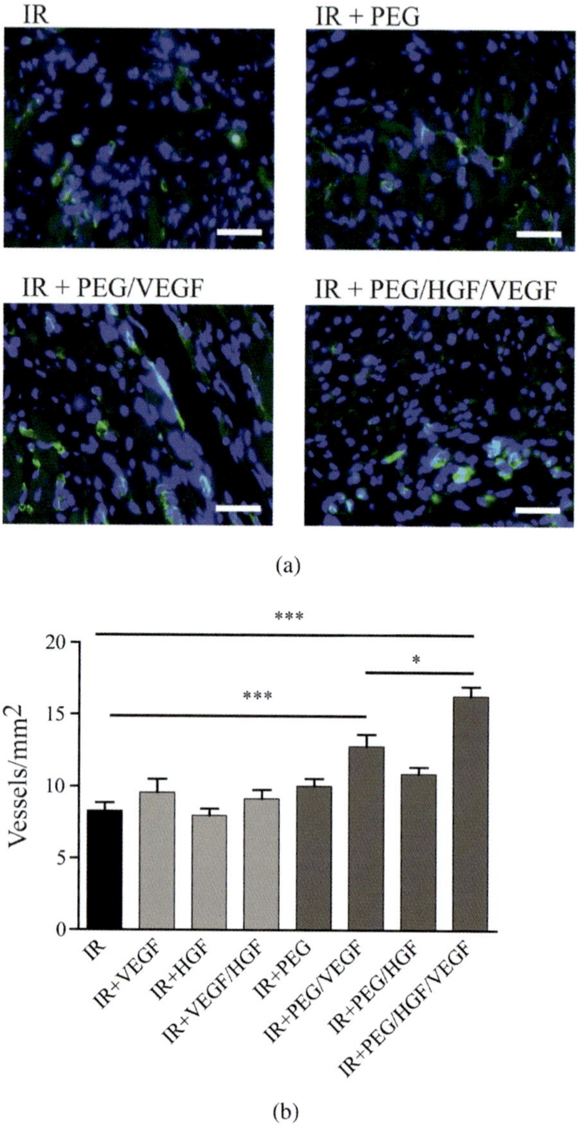


Figure 10.9 Delivery of growth factors in PEG in ischemia-reperfusion (IR) model. (a) Representative images of neovascularization within tissue sections stained with isolectin for endothelial cells (green) and a nuclear stain (blue); scale bar is 30 $\mu$m. (b) Average vessels per mm$^2$ in tissue sections. $^{***}p < 0.001$, $^{*}p < 0.05$. Reprinted from [107] with permission.

and protein alone [109]. This polymer system without erythropoietin was injected in rabbit hearts one week post-MI. Infarct size, LV end-diastolic diameter, LV end-systolic diameter, and EF were significantly improved, but no increase in neovascularization was observed [110]. A biodegradable, thermoresponsive hydrogel of oligo(PEG-fumarate) (OPF) was used to deliver naive mESCs post-MI. OPF hydrogels alone and OPF-mESCs showed significant reduction in collagen deposition and infarct size as well as improvement in LV function [111]. Wu *et al.* investigated a thermosensitive triblock polymer, poly(D-valerolactone)-block-PEG-block-poly(D-valerolactone) (PVL-PEG-PVL), which assembled into a hydrogel at 37°C in a rat one week post-MI [112]. The triblock hydrogel system conjugated to VEGF showed a decrease in scar area and thickness, an improvement in heart function, and an increase in vessel density after 35 days [112]. Bastings *et al.* have synthesized a pH-responsive triblock copolymer, polyurea-block-PEG-block-urea terminated with ureido-pyrimidinone units (urea-PEG-urea). This triblock copolymer quickly forms nanofibrous gels at pH $<8.5$. It was demonstrated that this material was capable of injection through a one-meter catheter and capable of delivering growth factors HGF and IGF-1 post-MI in pigs. Transendocardial delivery of growth factors with the triblock copolymer reduced infarct size (Figure 10.10) [28].

### 10.5.3 *N-Isopropylacrylamide-containing materials*

Polymerization of *N*-isopropylacrylamide results in polymers that have temperature inducible phase transitions. Copolymerization of *N*-isopropylacrylamide with other monomers and conjugation of poly(*N*-isopropylacrylamide) to other materials transfers the thermoresponsive trait to the new material. As a result poly(*N*-isopropylacrylamide) (PNIPAAm)-containing materials have been extensively studied *in vitro* and *in vivo* for biomedical applications, including cell sheets and thermoresponsive drug-conjugates [90, 113–115]. Poly(N-isopropylacrylamide-co-acrylic acid) was crosslinked with a matrix metalloproteinase degradable peptide and RGD peptide as a carrier for BMMNCs into mice post-MI. While the polymer and cell group showed improvement in EF, the polymer alone showed the best increase in EF post-MI [116]. Fujimoto *et al.* [117] synthesized a biodegradable thermoresponsive copolymer composed of poly(N-isopropylacrylamide-co-acrylic acid-co-hydroxyethyl methacrylate-poly(trimethylene carbonate)) (PNIPAAm-co-AA-co-HEMATC). The random copolymer formed a hydrogel at body temperature and was completely

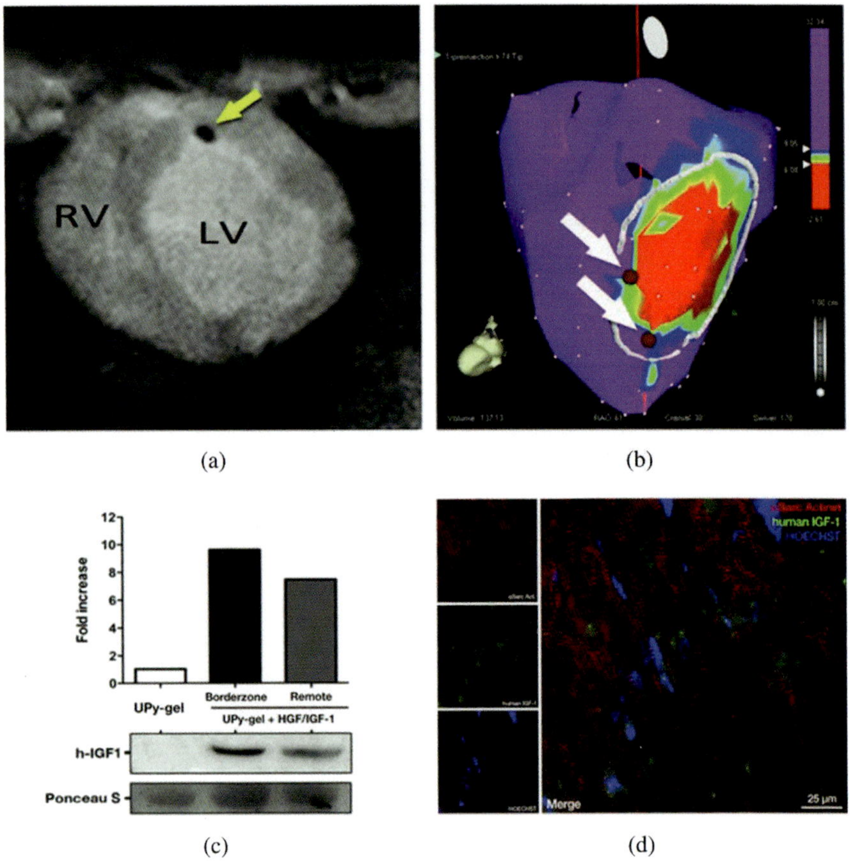


Figure 10.10 Transendocardial injection of hydrogen-bonding triblock copolymer loaded with HGF and IGF-1. (a) Short axis cardiac MRI showing the gel (yellow arrow). (b) Electromechanical mapping of the left ventricle showing the injection locations in the border zone of the infarct (white arrows). (c) Western blot analysis showing an increase in IGF-1 in the border zone after injection of growth factor-loaded hydrogel. (d) Immunofluorescent staining of the heart post-injection showing uptake of IGF-1 (green) by cardiomyocytes, $\alpha$-sarcomeric actin (red), and nuclei (blue). Reprinted from [28] with permission.

degraded after five months. The copolymer was injected into rats two weeks post-MI. Increased wall thickness and higher capillary density were significantly observed in the hydrogel group compared to saline eight weeks post-injection. Cellular infiltration was also increased and a layer of smooth muscle was observed next to the remaining material [117].

## 10.6 Natural and Synthetic Polymer Hybrid Materials

Hydrogels composed of mixtures of natural and synthetic polymer blends are finding increased use in cardiac tissue engineering applications. Hybrid hydrogels result in materials that combine the bioactivity of the natural polymer with other properties of the synthetic component. These hybrid hydrogels represent a new class of materials for cardiac tissue engineering with interesting properties such as increased stiffness, decreased rate of enzymatic degradation, and tunable pore size [90, 114, 118, 119]. Alginate has been blended with an electroactive polymer, polypyrrole. The polypyrrole–alginate blend showed an increase in arterioles compared to alginate and saline five weeks post-injection in a rat MI model [120]. Fibrin functionalized with PEG to produce PEGylated fibrin hydrogels has been used to deliver growth factors and cells in murine models of MI [121]. Acrylated HA was crosslinked with thiolated PEG (HA-PEG) and injected into rats two weeks post-MI. Four weeks post-treatment the HA-PEG hybrid had led to a significant decrease in infarct size, increase in EF, and neovascularization. Histological analysis also showed an increase in wall thickness, even though the HA-PEG had been completely degraded [122]. A recent study used a similar hydrogel system, thiolated HA and PEG-diacrylate, in rats and demonstrated a negative inflammatory response to the hydrogel [123].

## 10.7 Outlook and Future Directions

The field of injectable hydrogels for cardiac tissue regeneration is relatively new. The above studies demonstrate that this treatment approach has the potential for treating MI and HF. A wide variety of materials with a broad range of properties and additives have been studied in different animals (Table 10.1). Furthermore, these studies have used different MI models and time points of assessment. The studies have produced mixed results, but there are some interesting trends among these different approaches with injectable materials (Table 10.2). For example, many of the materials without any additives are capable of promoting neovascularization within the infarct including fibrin, collagen, Matrigel, decellularized cardiac matrix, SIS, alginate, chitosan, HEMA-HA, SAPs, PVL-PEG-PVL, PNIPAAm-co-AA-co-HEMATC, fibrin PEG, and HA-PEG. Interestingly, purely synthetic PVL-PEG-PVL and PNIPAAm-co-AA-co-HEMATC

Table 10.1 Survey of materials injected into the heart.

| **Injectable material** | **Additives** | **Animal model** |
|---|---|---|
| Fibrin | alone, SkMs, BMCs, ACSs, BMMNCs, bFGF, HGF, TGF$\beta$-1, pleiotrophin plasmid | mouse, rat, dog, sheep |
| Collagen | alone, BMCs | rat, pig |
| Matrigel | alone, hESCs derived CMs, hESCs + survival cocktail, ESCs | mouse, rat |
| Gelatin | alone, bFGF | rat, dog |
| Collagen type I + Matrigel | alone, neonatal CMs | rat |
| Decelled cardiac ECM | alone | rat, pig |
| Decelled pericardial matrix | alone | rat |
| SIS | alone | mouse, rat |
| Alginate | alone, integrin binding peptides, VEGF, PDGF-BB, IGF-1, HGF, hMSCs, polypyrrole | rat, pig, human |
| Chitosan | alone, bFGF, ESCs | rat |
| HA | alone | rat, sheep |
| Fibrin-alginate | alone | pig |
| SAPs | alone, SkM + PDGF-BB, BMMNCs, IGF-1, PDGF-BB, bFGF | mouse, rat, pig |
| Calcium hydroxyapitite | alone | sheep |
| PEG and PEG copolymers | alone, BMSCs, hESC + thymosin $\beta$4, thymosin $\beta$4, VEGF, erythropoietin | rat, rabbit, pig |
| PNIPAAm systems | alone, BMMNCs, BMSCs, bFGF | rat, rabbit |
| Fibrin-PEG | alone | mice |
| HA-PEG | alone | rat |

induce neovascularization while the other PEG- and PNIPAAm-based materials do not. This highlights the ability of a wide variety of different materials to induce neovascularization within the infarcted myocardium and some of these materials have also seen preservation/improvement in global cardiac function and/or LV geometry. However, neovascularization does not necessarily lead to preservation or improvement in global function as can be seen with fibrin, collagen, Matrigel, alginate, HEMA-HA, and SAPs. With alginate and SAPs, an improvement was seen in LV geometry but not global function. Certain materials are capable of inducing migration of progenitor-like stem cells into the injected material. These materials can be protein

Table 10.2 Outcome of selected injectable materials in the heart post-MI.

| Injected material | Additives | Animal | MI | Post-MI | Post-Injection | Dec. infarct Size | LV area or vol | EF, FAC, FS | Reg Fxn | NV | Trp | CD34 ckit | Tissue regn | Delivered by catheter | Refs. |
|---|---|---|---|---|---|---|---|---|---|---|---|---|---|---|---|
| Fibrin | alone | rat | OR | 1 week | 5 weeks | ✓ | — | ✓ | — | ✓ | — | — | — | — | 31–33 |
| | plasmid | rat | OR | 1 week | 5 weeks | — | — | — | — | ✓ | — | — | — | — | 36 |
| | SkM | rat | OR | 1 week | 5 weeks | ✓ | — | — | — | ✓ | — | — | — | — | 31 |
| | BMMNCs | rat | CI | 3 weeks | 8 weeks | — | — | — | — | ✓ | — | — | ✓ | — | 37 |
| | ECs | sheep | AC | 4 weeks | 8 weeks | — | — | ✓ | — | ✓ | — | — | — | — | 38 |
| Collagen | alone | rat | PL | 1 week | 6 weeks | — | — | ✓ | — | — | — | — | — | — | 42 |
| | alone | rat | OR | 1 week | 5 weeks | — | — | — | — | ✓ | — | — | — | — | 33 |
| Matrigel | alone | rat | OR | 1 week | 5 weeks | — | — | — | — | ✓ | — | — | — | — | 33 |
| | alone | rat | PL | immed | 4 weeks | — | — | ✓ | — | ✓ | — | ✓ | — | — | 48 |
| | mESCs | mouse | PL | immed | 2 weeks | — | — | ✓ | — | — | — | — | — | — | 49 |
| | CMs + PSC | rat | OR | 4 days | 4 weeks | — | ✓ | ✓ | — | — | — | — | ✓ | — | 50 |
| DCM | alone | rat | OR | 2 weeks | 4 weeks | — | — | ✓ | — | — | ✓ | ✓ | — | — | 27 |
| | alone | pig | CE | 2 weeks | 3 months | — | — | ✓ | ✓ | ✓ | ✓ | — | ✓ | ✓ | 61 |
| SIS | powder | rat | OR | multiple | multiple | — | — | ✓ | — | ✓ | — | ✓ | — | — | 66 |
| | SIS-B gel | mouse | PL | immed | 6 weeks | ✓ | — | ✓ | — | ✓ | — | — | — | — | 67 |
| | SIS-C gel | mouse | PL | immed | 6 weeks | — | — | — | — | — | — | — | — | — | 67 |
| Alginate | alone | pig | OR | 4 days | 2 months | — | ✓ | — | — | ✓ | — | — | — | ✓ | 26 |
| | pep. Mod. | rat | OR | 5 weeks | 5 weeks | — | — | ✓ | — | ✓ | — | — | — | — | 75 |
| | pyrrole | rat | OR | 1 week | 5 weeks | — | — | — | — | ✓ | — | — | — | — | 120 |
| Chitosan | alone | rat | PL | 1 week | 4 weeks | ✓ | ✓ | ✓ | — | ✓ | — | — | — | — | 81 |
| | mESCs | rat | PL | 1 week | 4 weeks | ✓ | ✓ | ✓ | — | ✓ | ✓ | — | — | — | 81 |
| | bFGF | rat | PL | 1 week | 4 weeks | ✓ | — | ✓ | — | ✓ | — | — | — | — | 82 |
| HA | methacrylated | sheep | PL | immed | 8 weeks | ✓ | — | — | — | — | — | — | — | — | 88 |
| | HEMA | sheep | PL | immed | 8 weeks | — | — | — | — | ✓ | — | — | — | — | 85 |
| Fibrin-Alginate | alone | pig | PL | 1 week | 3 weeks | — | — | — | — | — | — | — | — | — | 91 |
| SAPs | alone | mini-pig | PL | immed | 4 weeks | — | ✓ | — | — | ✓ | — | — | — | — | 97 |
| | IGF-1 + neoCMs | rat | PL | immed | 3 weeks | — | ✓ | ✓ | — | — | — | — | — | — | 96 |
| | BMMNCs | mini-pig | PL | immed | 4 weeks | ✓ | ✓ | — | — | ✓ | — | — | — | — | 97 |
| | VEGF | rat | PL | immed | 2 weels | ✓ | ✓ | ✓ | — | ✓ | ✓ | — | — | — | 98 |
| | VEGF | pig | PL | immed | 4 weeks | ✓ | — | ✓ | — | ✓ | — | — | — | — | 98 |
| | PDGF + bFGF | rat | PL | immed | 8 weeks | ✓ | — | — | — | ✓ | — | — | — | — | 101 |
| | RGD + rMSCs | rat | PL | immed | 4 weeks | — | — | ✓ | — | — | ✓ | — | — | — | 102 |

(*Continued*)

Table 10.2 (*Continued*)

| Injected material | Additives | Animal | MI | Post-MI | Post-Injection | Dec. infarct Size | LV area or vol | EF, FAC, FS | Reg Fxn | NV | Trp | CD34 ckit | Tissue regn | Delivered by catheter | Refs. |
|---|---|---|---|---|---|---|---|---|---|---|---|---|---|---|---|
| PEG | alone | rat | PL | immed | 13 weeks | — | ✓ | — | — | — | — | — | — | — | 105 |
| | alone | rat | PL | 11 days | 7.5 weeks | — | — | — | — | — | — | — | — | — | 106 |
| | RGD+VPM+GFs | rat | OR | immed | 3 weeks | ✓ | ✓ | ✓ | — | ✓ | — | ✓ | — | — | 107 |
| | RGD+MMP+T$\beta$4 | rat | PL | immed | 6 weeks | ✓ | ✓ | ✓ | — | ✓ | — | — | — | — | 108 |
| PEG-PCL-PEG | alone | rabbits | PL | immed | 4 weeks | ✓ | ✓ | ✓ | — | — | — | — | — | — | 110 |
| | rhEPO | rat | PL | immed | 4 weeks | ✓ | ✓ | ✓ | — | ✓ | — | ✓ | — | — | 109 |
| OPF | alone | rat | PL | 1 week | 4 weeks | ✓ | ✓ | ✓ | — | — | — | — | — | — | 111 |
| | mESCs | rat | PL | 1 week | 4 weeks | ✓ | ✓ | ✓ | — | ✓ | ✓ | — | — | — | 111 |
| PVL-PEG-PVL | VEGF | rat | PL | 1 week | 5 weeks | ✓ | ✓ | ✓ | — | ✓ | — | — | — | — | 112 |
| urea-PEG-urea | HGF+VEGF | pig | OR | 4 weeks | 4 weeks | — | — | — | — | — | — | — | — | ✓ | 28 |
| PNIPAAm | MMP+RGD | mouse | PL | immed | 6 weeks | — | — | ✓ | — | — | — | — | — | — | 116 |
| | BMMNCs | mouse | PL | immed | 6 weeks | — | — | ✓ | — | — | — | — | — | — | 116 |
| pNIP copoly | alone | rat | PL | 2 weeks | 8 weeks | — | — | ✓ | — | ✓ | — | — | ✓ | — | 117 |
| fibrin-PEG | HGF+BMSCs | mouse | PL | immed | 4 weeks | — | — | ✓ | — | ✓ | — | — | — | — | 121 |
| HA-PEG | alone | rat | PL | 2 weeks | 4 weeks | ✓ | — | ✓ | — | ✓ | — | — | — | — | 122 |

DCM = decellularized cardiac matrix; pNIP copoly = PNIPAAm-co-AA-co-HEMATC; MI = MI-model of injury; OR = occlusion/reperfusion; PL = permanent ligation; CI = cyroinjury; CE = coil embolism; Post-MI = the time between injury and injection of material; immed = immediately; mult = multiple time points; post-injection = the time lapse between treatment and assessment; LV area or vol = preservation or improvement in LV dimension, area, or volume; EF, FAC, FS = preservation or improvement in EF, FAC, or FS; Reg Fxn = improvement in regional function; NV = neovascularization within the infarct; Trp = the presence of troponin positive cells within the infarct. CD34 ckit = the presence of CD34+ or c-kit+ cells within the infarct. Tissue Regn = tissue regeneration.

based like Matrigel, decellularized cardiac matrix, and SIS. Synthetic-based materials that induced progenitor cell infiltration contained enzymatically degradable linkages, integrin-binding peptides, and growth factors, or used self-assembly of a triblock copolymer to deliver erythropoietin. Decellularized cardiac matrix, chitosan with mESCs, SAPs with VEGF, SAPs with RGD and rMSCs, and OPF with mESCs have demonstrated the presence of troponin-positive cells in the infarct post-injection. Of note is that the decellularized cardiac matrix alone produced this effect while all of the other systems required addition of growth factors or stem cells.

While all of the above factors are important parameters to assess the ability of injectable materials to treat MI, the materials should promote the generation of new and functional cardiac tissue. Tissue regeneration within the infarct post-MI has been demonstrated with only a few injectable hydrogels. Co-injection of fibrin with BMMNCs resulted in new tissue formation within the infarct, but this tissue was not cardiac muscle. PNIPAAm-co-AA-co-HEMATC produced a muscle-like area adjacent to the polymer material that stained positive for alpha-smooth muscle actin. Matrigel co-injected with cardiomyocytes derived from hESCs plus a PSC resulted in human myocardial tissue within a rat heart. Finally, the decellularized cardiac matrix, which is also deliverable via catheter, generated a thick band of cardiac muscle at the endocardium, which improved regional as well as global cardiac function without the use of additional proteins or stem cells.

Many challenges must be met before safe application of the majority of these materials in human patients can be achieved. There are a variety of different methods to induce MI-like injuries in animal models. The different models also report a broad range of successful outcomes for injected materials, and the differences in these models make cross comparison of different studies difficult and unreliable. It is known that the heart possesses stem cells [124–126], but the exact function of these cells in healthy and damaged hearts is still unclear. The mechanism of endogenous cellular recruitment by injectable materials is also poorly understood. It is clear that simple or singular component materials, largely composed of only one type of material, are not capable of regenerating cardiac tissue. The field is shifting toward materials with increased complexity and tunable properties, such as mechanical strength and rate of degradation. An improved understanding of how and why material properties and composition result in tissue regeneration is essential for designing future injectable materials. The dose and timing of delivery of injectable materials also needs to be investigated, with an eye towards what is feasible in the clinic. This is also true of cellular delivery with injectable hydrogels. A suitable human cell type needs to be identified since the lack of human cell sources remains a major hurdle for cell-based treatment approaches. While small animal models of the disease are useful, there needs to be an increased focus on clinical translatability into large animals and humans. Ideally injectable materials will be developed that can be delivered by minimally invasive approaches, such as via catheter.

There is a pressing need to develop new clinical therapies to prevent HF post-MI, but there is a lack of understanding in the field of cardiac injectable materials as to what material properties are ideal to prevent it. Properties that are generally considered important are stiffness, elasticity, biodegradability, and biocompatibility. To promote tissue regeneration, the material also needs to recruit endogenous cells and promote the generation of functional tissue. The reason that materials promote these responses in the heart is still unclear, as is the correlation between the inherent properties of the material and the body's response to it. A large number of materials highlighted in this chapter have been tested *in vivo* for other applications, while some of the materials were specifically designed for injection into the heart post-MI. Of all of the materials described, only three have demonstrated the ability to be deliverable via catheter, thus facilitating clinical translation as a minimally invasive treatment approach. Of these three materials, only one acellular material, the decellularized cardiac ECM hydrogel, has been able to regenerate cardiac muscle *in vivo* and demonstrate delivery via catheter (Table 10.2). This highlights the need within the field to begin to focus on approaches that are translatable and have regenerative properties *in vivo*.

## Acknowledgments

The authors thank Dr. Sonya Seif-Naraghi for helpful comments toward preparing this manuscript. Funding was provided in part by the National Institutes of Health (R01HL113468).

## References

[1] Yang, F. *et al.* (2002). Myocardial infarction and cardiac remodelling in mice. *Exp. Phys.*, **87**(5), pp. 547–555.

[2] Mann, D.L. (1999). Mechanisms and models in heart failure: A combinatorial approach. *Circulation*, **100**(9), pp. 999–1008.

[3] Sutton, M.G. and Sharpe, N. Left ventricular remodeling after myocardial infarction: Pathophysiology and therapy. *Circulation*, **101**(25), pp. 2981–2988.

[4] Goldthwaite, C.A. (2006). Mending a broken heart: Stem cells and cardiac repair. *Regen. Med.*

[5] Roger, V.L. *et al.* (2012). Heart disease and stroke statistics – 2012 update: A report from the American Heart Association. *Circulation*, **125**(1), pp. e2–e220.

[6] Klotz, S., Loeher, A., Drees, G., and Scheld, H.H. (2006) [Surgical therapy of end-stage heart failure. State of the art 2006]. *Herz.*, **31**(5), pp. 445–454.

[7] Burkhoff, D., Klotz, S., and Mancini, D.M. (2006). LVAD-induced reverse remodeling: Basic and clinical implications for myocardial recovery. *J. Card. Fail.*, **12**(3), pp. 227–239.

[8] Rizzieri, A.G., Verheijde, J.L., Rady, M.Y., and McGregor, J.L. (2008). Ethical challenges with the left ventricular assist device as a destination therapy. *Philos. Ethics Humanit. Med.*, **3** pp. 20–28.

[9] Laflamme, M.A. and Murry, C.E. Regenerating the heart. *Nature Biotech.*, **23**(7), pp. 845–856.

[10] Schulman, I.H. and Hare, J.M. (2012). Key developments in stem cell therapy in cardiology. *Regen. Med.*, **7**(Suppl 6), pp. 17–24.

[11] Sherman, W., Martens, T.P., Viles-Gonzalez, J.F., and Siminiak, T. (2006). Catheter-based delivery of cells to the heart. *Nature Clinic. Pract. Cardiovas. Med.*, **3**(Suppl 1), pp. S57–S64.

[12] Silva, G.V. *et al.* (2004). Catheter-based transendocardial delivery of autologous bone-marrow-derived mononuclear cells in patients listed for heart transplantation. *Texas Heart Inst. J.*, **31**(3), pp. 214–219.

[13] Stamm, C., Nasseri, B., Choi, Y.-H., and Hetzer, R. (2009). Cell therapy for heart disease: Great expectations, as yet unmet. *Heart Lung Circ.*, **18**(4), pp. 245–256.

[14] Povsic, T.J. and O'Connor, C.M. (2010). Cell therapy for heart failure: The need for a new therapeutic strategy. *Expert Rev. Cardiovas. Therapy*, **8**(8), pp. 1107–1126.

[15] Gnecchi, M. *et al.* (2006). Evidence supporting paracrine hypothesis for Akt-modified mesenchymal stem cell-mediated cardiac protection and functional improvement. *Faseb J.*, **20**(6), pp. 661–669.

[16] Rane, A.A. and Christman, K.L. (2011). Biomaterials for the treatment of myocardial infarction a 5-year update. *J. Am. Coll. Cardiol.*, **58**(25), pp. 2615–2629.

[17] Christman, K.L. and Lee, R.J. (2006). Biomaterials for the treatment of myocardial infarction. *J. Am. Coll. Cardiol.*, **48**(5), pp. 907–913.

[18] Patten, R.D. and Hall-Porter, M.R. (2009). Small animal models of heart failure development of novel therapies, past and present. *Circ. Heart Fail.*, **2**(2), pp. 138–144.

[19] Dixon, J.A. and Spinale, F.G. (2009). Large animal models of heart failure a critical link in the translation of basic science to clinical practice. *Circ. Heart Fail.*, **2**(3), pp. 262–271.

[20] Fedorov, V.V. *et al.* (2011). Effects of KATP channel openers diazoxide and pinacidil in coronary-perfused atria and ventricles from failing and non-failing human hearts. *J. Mol. Cell. Cardiol.*, **51**(2), pp. 215–225.
[21] Glukhov, A.V., Flagg, T.P., Fedorov, V.V., Efimov, I.R., and Nichols, C.G. (2010). Differential K(ATP) channel pharmacology in intact mouse heart. *J. Mol. Cell. Cardiol.*, **48**(1), pp. 152–160.
[22] Li, Y., Rodrigues, J., and Tomas, H. (2012). Injectable and biodegradable hydrogels: Gelation, biodegradation and biomedical applications. *Chem. Soc. Rev.*, **41**(6), pp. 2193–2221.
[23] Kornowski, R. *et al.* (2000). Electromagnetic guidance for catheter-based transendocardial injection: A platform for intramyocardial angiogenesis therapy. Results in normal and ischemic porcine models. *J. Am. Coll. Cardiol.*, **35**(4), pp. 1031–1039.
[24] Qian, H. *et al.* (2007). Intracoronary delivery of autologous bone marrow mononuclear cells radiolabeled by 18F-fluoro-deoxy-glucose: Tissue distribution and impact on post-infarct swine hearts. *J. Cell. Biochem.*, **102**(1), pp. 64–74.
[25] Tatsumi, T. *et al.* (2007). Intracoronary transplantation of non-expanded peripheral blood-derived mononuclear cells promotes improvement of cardiac function in patients with acute myocardial infarction. *Circulation*, **71**(8), pp. 1199–1207.
[26] Leor, J. *et al.* (2009). Intracoronary injection of *in situ* forming alginate hydrogel reverses left ventricular remodeling after myocardial infarction in swine. *J. Am. Coll. Cardiol.*, **54**(11), pp. 1014–1023.
[27] Singelyn, J.M. *et al.* (2012). Catheter-deliverable hydrogel derived from decellularized ventricular extracellular matrix increases endogenous cardiomyocytes and preserves cardiac function post-myocardial infarction. *J. Am. Coll. Cardiol.*, **59**(8), pp. 751–763.
[28] Bastings, M.M. *et al.* (2013). A fast pH-switchable and self-healing supramolecular hydrogel carrier for guided, local catheter injection in the infarcted myocardium. *Adv. Health. Mater.*, **3**(1), pp. 70–78.
[29] Grover, G.N., Braden, R.L., and Christman, K.L. (2013). Oxime cross-linked injectable hydrogels for catheter delivery. *Adv. Mater.*, **25**(21), pp. 2937–2942.
[30] Janmey, P.A., Winer, J.P., and Weisel, J.W. (2009). Fibrin gels and their clinical and bioengineering applications. *J. R. Soc. Interface*, **6**(30), pp. 1–10.
[31] Christman, K.L., Fok, H.H., Sievers, R.E., Fang, Q., and Lee, R.J. (2004). Fibrin glue alone and skeletal myoblasts in a fibrin scaffold preserve cardiac function after myocardial infarction. *Tissue Eng.*, **10**(3–4), pp. 403–409.
[32] Christman, K.L. *et al.* (2004). Injectable fibrin scaffold improves cell transplant survival, reduces infarct expansion, and induces neovasculature formation in ischemic myocardium. *J. Am. Coll. Cardiol.*, **44**(3), pp. 654–660.
[33] Huang, N.F., Yu, J., Sievers, R., Li, S., and Lee, R.J. (2005). Injectable biopolymers enhance angiogenesis after myocardial infarction. *Tissue Eng.*, **11**(11–12), pp. 1860–1866.

[34] Yang, H.S. *et al.* (2012). *In situ* cardiomyogenic differentiation of implanted bone marrow mononuclear cells by local delivery of transforming growth factor-1. *Cell Transplant.*, **21**(1), pp. 299–312.
[35] Bujak, M. and Frangogiannis, N.G. (2007). The role of TGF-beta signaling in myocardial infarction and cardiac remodeling. *Cardiovas. Res.*, **74**(2), pp. 184–195.
[36] Christman, K.L. *et al.* (2005). Enhanced neovasculature formation in ischemic myocardium following delivery of pleiotrophin plasmid in a biopolymer. *Biomaterials*, **26**(10), pp. 1139–1144.
[37] Ryu, J.H. *et al.* (2005). Implantation of bone marrow mononuclear cells using injectable fibrin matrix enhances neovascularization in infarcted myocardium. *Biomaterials*, **26**(3), pp. 319–326.
[38] Chekanov, V. *et al.* (2003). Transplantation of autologous endothelial cells induces angiogenesis. *PACE*, **26**(1), pp. 496–499.
[39] Parenteau-Bareil, R., Gauvin, R., and Berthod, F. (2010). Collagen-based biomaterials for tissue engineering applications. *Materials*, **3**(3), pp. 1863–1887.
[40] Achilli, M., Lagueux, J., and Mantovani, D. (2010). On the effects of UV-C and pH on the mechanical behavior, molecular conformation and cell viability of collagen-based scaffold for vascular tissue engineering. *Macromol. Biosci.*, **10**(3), pp. 307–316.
[41] Achilli, M. and Mantovani, D. (2010). Tailoring mechanical properties of collagen-based scaffolds for vascular tissue engineering: The effects of pH, temperature and ionic strength on gelation. *Polymer.*, **2**(4), pp. 664–680.
[42] Dai, W.D., Wold, L.E., Dow, J.S., and Kloner, R.A. (2005). Thickening of the infarcted wall by collagen injection improves left ventricular function in rats. *J. Am. Coll. Cardiol.*, **46**(4), pp. 714–719.
[43] Kutschka, I. *et al.* (2006). Collagen matrices enhance survival of transplanted cardiomyoblasts and contribute to functional improvement of ischemic rat hearts. *Circulation*, **114**, pp. I167–I173.
[44] Kutschka, I. *et al.* (2007). *In vivo* optical bioluminescence imaging of collagen-supported cardiac cell grafts. *J. Heart Lung Transpl.*, **26**(3), pp. 273–280.
[45] Zhang, Y. *et al.* (2008). Collagen-based matrices improve the delivery of transplanted circulating progenitor cells development and demonstration by *ex vivo* radionuclide cell labeling and *in vivo* tracking with positron-emission tomography. *Circ. Cardiovasc. Imag.*, **1**(3), pp. 197–204.
[46] Hughes, C.S., Postovit, L.M., and Lajoie, G.A. (2010). Matrigel: A complex protein mixture required for optimal growth of cell culture. *Proteomics*, **10**(9), pp. 1886–1890.
[47] Kleinman, H.K. and Martin, G.R. (2005). Matrigel: Basement membrane matrix with biological activity. *Semin. Cancer Biol.*, **15**(5), pp. 378–386.
[48] Ou, L.L. *et al.* (2011). Intracardiac injection of matrigel induces stem cell recruitment and improves cardiac functions in a rat myocardial infarction model. *J. Cell. Mol. Med.*, **15**(6), pp. 1310–1318.

[49] Kofidis, T. *et al.* (2005). Novel injectable bioartificial tissue facilitates targeted, less invasive, large-scale tissue restoration on the beating heart after myocardial injury. *Circulation*, **112**(9), pp. I173–I177.
[50] Laflamme, M.A. *et al.* (2007). Cardiomyocytes derived from human embryonic stem cells in pro-survival factors enhance function of infarcted rat hearts. *Nature Biotech.*, **25**(9), pp. 1015–1024.
[51] Fridman, R. *et al.* (1991). Enhanced tumor-growth of both primary and established human and murine tumor-cells in athymic mice after coinjection with matrigel. *J. Natl. Cancer Inst.*, **83**(11), pp. 769–774.
[52] Prokhorova, T.A. *et al.* (2009). Teratoma formation by human embryonic stem cells is site dependent and enhanced by the presence of matrigel. *Stem Cells Dev.*, **18**(1), pp. 47–54.
[53] Gilbert, T.W., Sellaro, T.L., and Badylak, S.F. (2006). Decellularization of tissues and organs. *Biomaterials*, **27**(19), pp. 3675–3683.
[54] Ott, H.C. *et al.* (2008). Perfusion-decellularized matrix: Using nature's platform to engineer a bioartificial heart. *Nat. Med.*, **14**(2), pp. 213–221.
[55] Ott, H.C. *et al.* (2010). Regeneration and orthotopic transplantation of a bioartificial lung. *Nat. Med.*, **16**(8), pp. 927–933.
[56] Petersen, T.H. *et al.* (2010). Tissue-engineered lungs for *in vivo* implantation. *Science*, **329**(5991), pp. 538–541.
[57] Song, J.J. *et al.* (2013). Regeneration and experimental orthotopic transplantation of a bioengineered kidney. *Nat. Med.*, **19**(5), pp. 646–651.
[58] DeQuach, J.A. *et al.* (2010). Simple and high yielding method for preparing tissue specific extracellular matrix coatings for cell culture. *PLoS One*, **5**(9), p. e13039.
[59] Freytes, D.O., Martin, J., Velankar, S.S., Lee, A.S., and Badylak, S.F. (2008). Preparation and rheological characterization of a gel form of the porcine urinary bladder matrix. *Biomaterials*, **29**(11), pp. 1630–1637.
[60] Singelyn, J.M. *et al.* (2009). Naturally derived myocardial matrix as an injectable scaffold for cardiac tissue engineering. *Biomaterials*, **30**(29), pp. 5409–5416.
[61] Seif-Naraghi, S.B. *et al.* (2013). Safety and efficacy of an injectable extracellular matrix hydrogel for treating myocardial infarction. *Sci. Transl. Med.*, **5**(173).
[62] Seif-Naraghi, S.B., Salvatore, M.A., Schup-Magoffin, P.J., Hu, D.P., and Christman, K.L. (2010). Design and characterization of an injectable pericardial matrix gel: A potentially autologous scaffold for cardiac tissue engineering. *Tissue Eng. Part A*, **16**(6), pp. 2017–2027.
[63] Duran, C.M., Gometza, B., Kumar, N., Gallo, R., and Martin-Duran, R. (1995). Aortic valve replacement with freehand autologous pericardium. *J. Thorac. Cardiovasc. Surg.*, **110**(2), pp. 511–516.
[64] Fuster, V., Walsh, R.A., and Harrington, R.A. (2011). *Hurst's the Heart.* 13th ed. McGraw-Hill Medical, New York.
[65] Seif-Naraghi, S.B., Horn, D., Schup-Magoffin, P.J., and Christman, K.L. (2012). Injectable extracellular matrix derived hydrogel provides a platform

for enhanced retention and delivery of a heparin-binding growth factor. *Acta Biomater.*, **8**(10), pp. 3695–3703.
[66] Zhao, Z.Q. *et al.* (2010). Improvement in cardiac function with small intestine extracellular matrix is associated with recruitment of C-kit cells, myofibroblasts, and macrophages after myocardial infarction. *J. Am. Coll. Cardiol.*, **55**(12), pp. 1250–1261.
[67] Okada, M. *et al.* (2010). Differential efficacy of gels derived from small intestinal submucosa as an injectable biomaterial for myocardial infarct repair. *Biomaterials*, **31**(30), pp. 7678–7683.
[68] Ravichandran, R., Venugopal, J.R., Sundarrajan, S., Mukherjee, S., and Ramakrishna, S. (2012). Minimally invasive cell-seeded biomaterial systems for injectable/epicardial implantation in ischemic heart disease. *Int. J. Nanomed.*, **7**, pp. 5969–5994.
[69] Delair, T. (2012). *In situ* forming polysaccharide-based 3D-hydrogels for cell delivery in regenerative medicine. *Carbohyd. Polym.*, **87**(2), pp. 1013–1019.
[70] Rowley, J.A., Madlambayan, G., and Mooney, D.J. (1999). Alginate hydrogels as synthetic extracellular matrix materials. *Biomaterials*, **20**(1), pp. 45–53.
[71] Augst, A.D., Kong, H.J., and Mooney, D.J. (2006). Alginate hydrogels as biomaterials. *Macromol. Biosci.*, **6**(8), 623–633.
[72] Landa, N. *et al.* (2008). Effect of injectable alginate implant on cardiac remodeling and function after recent and old infarcts in rat. *Circulation*, **117**(11), pp. 1388–1396.
[73] Yu, J.S. *et al.* (2009). Restoration of left ventricular geometry and improvement of left ventricular function in a rodent model of chronic ischemic cardiomyopathy. *J. Thorac. Cardiovas. Surg.*, **137**(1), pp. 180–187.
[74] Inc. IH. IK-5001 for the Prevention of Remodeling of the Ventricle and Congestive Heart Failure After Acute Myocardial Infarction (PRESERVATION) (2012–2014). *Study NCT01226563.* Available at: http://ClinicalTrials.gov/ct2/show/NCT01226563. Accessed August 8, 2013.
[75] Yu, J. *et al.* (2009). The effect of injected RGD modified alginate on angiogenesis and left ventricular function in a chronic rat infarct model. *Biomaterials*, **30**(5), pp. 751–756.
[76] Tsur-Gang, O. *et al.* (2009). The effects of peptide-based modification of alginate on left ventricular remodeling and function after myocardial infarction. *Biomaterials*, **30**(2), pp. 189–195.
[77] Ruvinov, E., Leor, J., and Cohen, S. (2011). The promotion of myocardial repair by the sequential delivery of IGF-1 and HGF from an injectable alginate biomaterial in a model of acute myocardial infarction. *Biomaterials*, **32**(2), pp. 565–578.
[78] Hao, X. *et al.* (2007). Angiogenic effects of sequential release of VEGF-A165 and PDGF-BB with alginate hydrogels after myocardial infarction. *Cardiovas. Res.*, **75**(1), pp. 178–185.
[79] Kim, I.Y. *et al.* (2008). Chitosan and its derivatives for tissue engineering applications. *Biotechnol. Adv.*, **26**(1), pp. 1–21.

[80] Bhattarai, N., Gunn, J., and Zhang, M.Q. (2001). Chitosan-based hydrogels for controlled, localized drug delivery. *Adv. Drug Del. Rev.*, **62**(1), pp. 83–99.
[81] Lu, W.N. *et al.* (2009). Functional improvement of infarcted heart by co-injection of embryonic stem cells with temperature-responsive chitosan hydrogel. *Tissue Eng. Part A*, **15**(6), pp. 1437–1447.
[82] Wang, H.B. *et al.* (2010). Improved myocardial performance in infarcted rat heart by co-injection of basic fibroblast growth factor with temperature-responsive Chitosan hydrogel. *J. Heart Lung Transpl.*, **29**(8), pp. 881–887.
[83] Lu, S.H. *et al.* (2010). Both the transplantation of somatic cell nuclear transfer- and fertilization-derived mouse embryonic stem cells with temperature-responsive chitosan hydrogel improve myocardial performance in infarcted rat hearts. *Tissue Eng. Part A*, **16**(4), pp. 1303–1315.
[84] Toole, B.P. (2004). Hyaluronan: From extracellular glue to pericellular cue. *Nat. Rev. Cancer*, **4**(7), pp. 528–539.
[85] Tous, E. *et al.* (2011). Influence of injectable hyaluronic acid hydrogel degradation behavior on infarction-induced ventricular remodeling. *Biomacromolecules*, **12**(11), pp. 4127–4135.
[86] Burdick, J.A. and Prestwich, G.D. (2011). Hyaluronic acid hydrogels for biomedical applications. *Adv. Mater.*, **23**(12), pp. H41–H56.
[87] Xu, X., Jha, A.K., Harrington, D.A., Farach-Carson, M.C., and Jia, X.Q. (2012). Hyaluronic acid-based hydrogels: From a natural polysaccharide to complex networks. *Soft Matter*, **8**(12), pp. 3280–3294.
[88] Ifkovits, J.L. *et al.* (2010). Injectable hydrogel properties influence infarct expansion and extent of postinfarction left ventricular remodeling in an ovine model. *Proc. Natl. Acad. Sci. USA*, **107**(25), pp. 11507–11512.
[89] Purcell, B.P., Elser, J.A., Mu, A.B., Margulies, K.B., and Burdick, J.A. (2012). Synergistic effects of SDF-1 alpha chemokine and hyaluronic acid release from degradable hydrogels on directing bone marrow derived cell homing to the myocardium. *Biomaterials*, **33**(31), pp. 7849–7857.
[90] Hoffman, A.S. (2002). Hydrogels for biomedical applications. *Adv. Drug Del. Rev.*, **54**(1), pp. 3–12.
[91] Mukherjee, R. *et al.* (2008). Targeted myocardial microinjections of a biocomposite material reduces infarct expansion in pigs. *Ann. Thorac. Surg.*, **86**(4), pp. 1268–1277.
[92] Gazit, E. (2007). Self-assembled peptide nanostructures: The design of molecular building blocks and their technological utilization. *Chem. Soc. Rev.*, **36**(8), pp. 1263–1269.
[93] Matson, J.B. and Stupp, S.I. (2012). Self-assembling peptide scaffolds for regenerative medicine. *Chem. Commun.*, **48**(1), pp. 26–33.
[94] Maude, S., Ingham, E., and Aggeli, A. (2013). Biomimetic self-assembling peptides as scaffolds for soft tissue engineering. *Nanomed. UK*, **8**(5), pp. 823–847.
[95] Davis, M.E. *et al.* (2005). Injectable self-assembling peptide nanofibers create intramyocardial microenvironments for endothelial cells. *Circulation*, **111**(4), pp. 442–450.

[96] Davis, M.E. *et al.* (2006). Local myocardial insulin-like growth factor 1 (IGF-1) delivery with biotinylated peptide nanofibers improves cell therapy for myocardial infarction. *Proc. Natl. Acad. Sci. USA*, **103**(21), pp. 8155–8160.
[97] Lin, Y.D. *et al.* (2010). Intramyocardial peptide nanofiber injection improves postinfarction ventricular remodeling and efficacy of bone marrow cell therapy in pigs. *Circulation*, **122**(11), pp. S132–S141.
[98] Lin, Y.D. *et al.* (2012). Instructive nanofiber scaffolds with VEGF create a microenvironment for arteriogenesis and cardiac repair. *Sci. Transl. Med.*, **4**(146).
[99] Lee, K., Silva, E.A., and Mooney, D.J. (2011). Growth factor delivery-based tissue engineering: General approaches and a review of recent developments. *J. R. Soc. Interface*, **8**(55), pp. 153–170.
[100] Saif, J. *et al.* (2010). Combination of injectable multiple growth factor-releasing scaffolds and cell therapy as an advanced modality to enhance tissue neovascularization. *Arterioscl. Throm. Vas.*, **30**(10), pp. 1897–1904.
[101] Kim, J.H. *et al.* (2011). The enhancement of mature vessel formation and cardiac function in infarcted hearts using dual growth factor delivery with self-assembling peptides. *Biomaterials*, **32**(26), pp. 6080–6088.
[102] Guo, H.D., Cui, G.H., Wang, H.J., and Tan, Y.Z. (2010). Transplantation of marrow-derived cardiac stem cells carried in designer self-assembling peptide nanofibers improves cardiac function after myocardial infarction. *Biochem. Bioph. Res. Co.*, **399**(1), pp. 42–48.
[103] Zhu, J.M. (2010). Bioactive modification of poly(ethylene glycol) hydrogels for tissue engineering. *Biomaterials*, **31**(17), pp. 4639–4656.
[104] Hermanson, G.T. (1996). *Bioconjugate Techniques*. Academic Press, San Diego.
[105] Dobner, S., Bezuidenhout, D., Govender, P., Zilla, P., and Davies, N. (2009). A synthetic non-degradable polyethylene glycol hydrogel retards adverse post-infarct left ventricular remodeling. *J. Card. Fail.*, **15**(7), pp. 629–636.
[106] Rane, A.A. *et al.* (2011). Increased infarct wall thickness by a bio-inert material is insufficient to prevent negative left ventricular remodeling after myocardial infarction. *PLoS ONE*, **6**(6), p. e21571.
[107] Salimath, A.S. *et al.* (2012). Dual delivery of hepatocyte and vascular endothelial growth factors via a protease-degradable hydrogel improves cardiac function in rats. *PLoS ONE*, **7**(11), p. e50980.
[108] Kraehenbuehl, T.P. *et al.* (2011). Human embryonic stem cell-derived microvascular grafts for cardiac tissue preservation after myocardial infarction. *Biomaterials*, **32**(4), pp. 1102–1109.
[109] Wang, T. *et al.* (2009). The inhibition of postinfarct ventricle remodeling without polycythaemia following local sustained intramyocardial delivery of erythropoietin within a supramolecular hydrogel. *Biomaterials*, **30**(25), pp. 4161–4167.
[110] Jiang, X.-J. *et al.* (2009). Injection of a novel synthetic hydrogel preserves left ventricle function after myocardial infarction. *J. Biomed. Mater. Res. Part A*, **90A**(2), pp. 472–477.

[111] Wang, H.B. *et al.* (2012). Injectable biodegradable hydrogels for embryonic stem cell transplantation: Improved cardiac remodelling and function of myocardial infarction. *J. Cell. Mol. Med.*, **16**(6), pp. 1310–1320.
[112] Wu, J. *et al.* (2011). Infarct stabilization and cardiac repair with a VEGF-conjugated, injectable hydrogel. *Biomaterials*, **32**(2), pp. 579–586.
[113] Jhang, J., Pelton, R., and Deng, Y.L. (1995). Temperature-dependent contact angles of water on poly($N$-isopropylacrylamide) gels. *Langmuir*, **11**(6), pp. 2301–2302.
[114] Hoffman, A.S. (2004). Applications of "Smart Polymers" as Biomaterials, in: B.D. Ratner, A.S. Hoffman, F.J. Schoen, and J.E. Lemons (eds.), *Biomaterials Science*. Elsevier Academic Press, San Diego.
[115] Hoffman, A.S. and Stayton, P.S. (2007). Conjugates of stimuli-responsive polymers and proteins. *Prog. Polym. Sci.*, **32**(8–9), pp. 922–932.
[116] Wall, S.T., Yeh, C.-C., Tu, R.Y.K., Mann, M.J., and Healy, K.E. (2010). Biomimetic matrices for myocardial stabilization and stem cell transplantation. *J. Biomed. Mater. Res. Part A*, **95A**(4), pp. 1055–1066.
[117] Fujimoto, K.L. *et al.* (2009). Synthesis, characterization and therapeutic efficacy of a biodegradable, thermoresponsive hydrogel designed for application in chronic infarcted myocardium. *Biomaterials*, **30**(26), pp. 4357–4368.
[118] Jia, X.Q. and Kiick, K.L. (2009). Hybrid multicomponent hydrogels for tissue engineering. *Macromol. Biosci.*, **9**(2), pp. 140–156.
[119] Vandermeulen, G.W.M. and Klok, H.A. (2004). Peptide/protein hybrid materials: Enhanced control of structure and improved performance through conjugation of biological and synthetic polymers. *Macromol. Biosci.*, **4**(4), pp. 383–398.
[120] Mihardja, S.S., Sievers, R.E., and Lee, R.J. (2008). The effect of polypyrrole on arteriogenesis in an acute rat infarct model. *Biomaterials*, **29**(31), pp. 4205–4210.
[121] Zhang, G., Hu, Q., Braunlin, E.A., Suggs, L.J., and Zhang, J. (2008). Enhancing efficacy of stem cell transplantation to the heart with a PEGylated fibrin biomatrix. *Tissue Eng. Part A*, **14**(6), pp. 1025–1036.
[122] Yoon, S.J. *et al.* (2009). Regeneration of ischemic heart using hyaluronic acid-based injectable hydrogel. *J. Biomed. Mater. Res. B*, **91B**(1), pp. 163–171.
[123] Young, J.L. *et al.* (2013). *In vivo* response to dynamic hyaluronic acid hydrogels. *Acta Biomater.*, **9**(7), pp. 7151–7157.
[124] Beltrami, A.P. *et al.* (2003). Adult cardiac stem cells are multipotent and support myocardial regeneration. *Cell*, **114**(6), pp. 763–776.
[125] Hierlihy, A.M., Seale, P., Lobe, C.G., Rudnicki, M.A., and Megeney, L.A. (2002). The post-natal heart contains a myocardial stem cell population. *Febs. Lett.*, **530**(1–3), pp. 239–243.
[126] Penn, M.S. (2007). *Stem cells and myocardial regeneration*. Humana Press, Totowa, N.J.

# Index